The Most Successful Way of JEE Preparation

DPP

Daily Practice Problems

JEE Main & Advanced

Trigonometry & Its Applications

with Properties of Triangles

MATHEMATICS Vol-3

The Most Successful Way of JEE Preparation

DPP

Daily Practice Problems

JEE Main & Advanced

Trigonometry & Its Applications

with Properties of Triangles

MATHEMATICS Vol-3

Amit M Agarwal

arihant

ARIHANT PRAKASHAN (SERIES), MEERUT

arihant
ARIHANT PRAKASHAN (SERIES), MEERUT

Administrative & Production Offices

Corporate Office 'Ramchhaya' 4577/15, Agarwal Road, Darya Ganj, New Delhi -110002
Tele: 011- 47630600, 43518550; Fax: 011- 23280316

Head Office Kalindi, TP Nagar, Meerut (UP) - 250002
Tel: 0121-2401479, 2512970, 4004199; Fax: 0121-2401648

Sales & Support Offices

Agra, Ahmedabad, Bengaluru, Bareilly, Chennai, Delhi, Guwahati, Hyderabad, Jaipur, Jhansi, Kolkata, Lucknow, Meerut, Nagpur & Pune

ISBN : 978-93-13193-51-7

Price : ₹180

Published by Arihant Publications (I) Ltd.

Production Team

Publishing Managers	Mahendra Singh Rawat & Keshav Mohan	*Inner Designer*	Mazher Chaudhary
Project Head	Sorabh Chaudhary	*Page Layouting*	Sunil Sharma, Hemant
Cover Designer	Shanu Mahad	*Proof Reading*	Himanshu

For further information about the products from Arihant, log on to www.arihantbooks.com or email to info@arihantbooks.com

/arihantpub /@arihantpub Arihant Publications /arihantpub

ABOUT DPP

Daily Practice Problems

Indian Institutes of Technology better known as IITs have always been the dream destination of each science group students who want to pursue their career as Engineers. These are the institutes which have got very high value not even in India but in the whole world. To get the admission in such institutes one need to face entrance Examinations; JEE Main & JEE Advanced, out of which JEE Main is considered as the preliminary exam & JEE Advanced is the final exam to get into the IITs. Approximately per year 12 lakh students enroll themselves for JEE Main and only the top 2.25 lakh in JEE Main merit can attempt for JEE Advanced, the fact clearly express the degree of toughness of JEE Main & JEE Advanced.

To have success in such an exam one needs to be very planned & systematic in its preparation. A very popular wise saying 'Rome was not built in a day' suggests that to achieve a big target like IITs we need to work continuously. For continuous work, one need to have proper planning that how much work he/she has to do on regular basis, i.e., on daily basis. Each student preparing for JEE has many books but unfortunately none of them gives the students a plan for daily studies, which is the most effective success key for JEE.

Keeping the daily practice needs of the students in mind we have come up with DPP Series (Daily Practice Problems) having approximately 26 books. The primary focus of this series is on achieving success through practice & proper planning. For that we have divided the whole syllabus of JEE Main & JEE Advanced; Physics, Chemistry & Mathematics, into different volumes (around 8 volumes of Mathematics & 9 volumes each of Physics & Chemistry), each volume has Daily Practice Problems Sheets having specific questions on various Topics of the individual chapter of the volume ensuring the complete Practice & Assessment of the topic.

DPP

Daily Practice Problems

In this way we have tried to give you an effective daily planning for JEE keeping in mind the daily hours that a student can devote to his practice, you are just required to practice one sheet of each subject on daily basis. This habit of daily practice will keep you away from exam stress and thoughts like; how much I have prepared/am I preparing properly or not/ do I need to study more on daily basis or what I am doing is ok, etc., which regularly tease your mind and can give you panic at any stage of your preparation.

Some special features that we have tried to incorporate in our Daily Practice Problems (DPP) series are discussed below

1. Micro Level coverage of all the topics of a Chapter/ Unit given in JEE Main & Advanced Syllabus.
2. Each DPP has various types of questions; Subjective Questions, Single/ Multiple Correct Options, Statements type, Comprehension based, Integer Value Answer type, Matching, etc., making you practice & ready for all the question formats of JEE.
3. New & innovative questions in each DPP challenging & improving your problem solving skills.
4. Along with Topical Coverage, Revisal DPPs for JEE Main & JEE Advanced with each Chapter, covering multidisciplinary questions proving comprehensive practice.
5. JEE Main & JEE Advanced Archive (collection of last 13 years' exams questions) with each chapter.
6. Complete Solutions for each DPP.

We hope that DPP series will prove to be very beneficial for the students preparing for JEE, by giving them daily practice in a systematic & planned manner. The suggestions/criticism from the students, teachers & parents about DPP series will be welcomed whole heartedly and we promise to update this series accordingly from time to time.

DPP
Daily Practice Problems

DAILY TASKS . . .

1 TRIGONOMETRIC RATIOS AND IDENTITIES

2 HEIGHT AND DISTANCE

DPP
Daily Practice Problems

3 TRIGONOMETRIC EQUATIONS

4 SOLUTIONS OF TRIANGLES

1. Trigonometric Ratios and Identities

DPP

DAILY PRACTICE PROBLEMS

DPP-1 Measurement of Angles

Subjective Questions

Working Space

Direction *(Q. Nos. 1-5) These questions are subjective in nature, need to be solved completely on notebook.*

1 Express the following angles in degree measure:

i. $\left(\frac{5\pi}{3}\right)^C$ ii. $\left(\frac{-5\pi}{24}\right)^C$

iii. $(-4)^C$

2 Express the following angles in radian measure:

i. 105° ii. −47°30′

iii. 42°57′16′′

3 The difference between two acute angles of a right angle triangle is $\frac{3\pi}{10}$ rad. Find the angles in degree.

4 Find the length of an arc of a circle of radius 6 cm subtending an angle of 15° at the centre.

5 Find the degree measure of angle subtended at the centre of a circle of diameter 100 cm by an arc of length 22 cm.

Only One Option Correct Type

Direction *(Q. Nos. 6-10) This section contains 5 multiple choice questions. Each question has four choices (a), (b), (c) and (d), out of which ONLY ONE is correct.*

6 If the minute hand of a clock is 14 cm long, then the distance moved by tip of the hand in 30 min, is

a. 44 cm b. 46 cm
c. 48 cm d. None of these

7 The angle between the hour hand and minute hand in circular measure at 4 pm in clock, is

a. $\frac{2\pi}{3}$ b. $\frac{4\pi}{3}$
c. $\frac{\pi}{2}$ d. None of these

8 If the angles of a triangle are in the ratio 3 : 4 : 5, then the greatest angle in radian is

a. $\frac{3\pi}{5}$ b. $\frac{\pi}{12}$
c. $\frac{5\pi}{12}$ d. None of these

Working Space

9 A horse is tied to post by a rope. If the horse moves along circular path always keeping the tight and describes 88 m, when it has traced out 72° at centre, then the length of rope is

a. 50 m
b. 60 m
c. 70 m
d. 80 m

10 The angle between the minute hand and hour hand of a clock, when the time is 7 : 30 pm is

a. 60°
b. 45°
c. 30°
d. None of these

One or More than One Options Correct Type

Direction (Q. Nos. 11-18) *This section contains 8 multiple choice questions. Each question has four choices (a), (b), (c) and (d), out of which ONE or MORE THAN ONE are correct.*

11 The diameter of a circle is 120 cm, if length of the chord of the arc is 60 cm, then the length of an arc of circle is

a. 15π cm
b. 20π cm
c. 30π cm
d. None of these

12 If OQ makes 4 revolutions in 1s, then the angular velocity in radians per second is

a. π
b. 2π
c. 4π
d. 8π

13 If a train is moving on the circular path of 1500 m radius at the rate of 66 km/h, then it has turned an angle in 10 s is

a. $\left(\frac{11}{9}\right)^C$
b. $\left(\frac{11}{90}\right)^C$
c. $\left(\frac{13}{90}\right)^C$
d. None of these

14 The distance from the eye at which a coin of 2.2 cm diameter should be held so as to conceal the full moon with angular diameter 30′, is

a. 1.217cm
b. 1.317cm
c. 2.217cm
d. None of these

15 The wheel of a railway carriage is 40 cm in diameter and makes 7 revolutions in a second, then speed of train is

a. 980 cm/s
b. 880 cm/s
c. 780 cm/s
d. 680 cm/s

16 If the arcs of the same length in two circles subtend angles 65° and 110° at the centre, then the ratio of radii of circles is/are

a. 22 : 13
b. 11 : 13
c. $x : y$ (when $x = 44$, $y = 26$)
d. 22 : 15

17 The perimeter of a sector of a circle is equal to the length of the arc of a semi-circle having the same radius, then the angle of sector is/are

a. 114°32′44′′
b. 65°27′16″
c. $(\pi - 2)^C$
d. $(\pi - 3)^C$

Working Space

18 Assuming that a person of normal sight can read print at such a distance that the letters subtend an angle of 5′ at his eye, then the height of letters that he can read a distance of 12 m is/are

a. 1.7 cm
b. 2.7 cm
c. $\frac{4\pi}{9}$ cm
d. $\frac{5\pi}{9}$ cm

One Integer Value Correct Type

Direction *(Q. Nos. 19 and 20) This section contains 2 questions. When worked out will result in an integer from 0 to 9 (both inclusive).*

19 The angles of a triangle are in AP and the number of degrees in the least is to the number of radians in the greatest is $60:\pi$. If the difference between least and greatest angles is $\left(\frac{\lambda\pi}{6}\right)^C$, then the value of λ is

20 For each natural number K, let C_K denotes the circle with radius K cm and centre at origin. On the circle C_K, a particle moves K cm in the counter-clockwise direction. After completing its motion on C_K, the particle moves on C_{K+1} in the radial direction. The motion of the particle continues in this manner. The particle starts at (1, 0). If the particle crosses the positive direction of the X-axis for the first time on the circle C_n, then n is equal to

Comprehension Type

Direction *(Q. Nos. 21 and 22) This section contains a paragraph, which describing theory, experiments, data, etc. Two questions related to the paragraph have been given. Each question has only one correct answer among the four given options (a), (b), (c) and (d).*

Passage

$$\pi^C = 180°$$

$$\Rightarrow \quad 1^C = \left(\frac{180}{\pi}\right)^\circ$$

$$= 57°17'44.8''$$

$$= 57° \text{ (approx.)}$$

On the basis of above information, answer the following questions :

21 The value of $\cos 1^C$ is

a. $< \frac{1}{2}$
b. $\frac{1}{2}$
c. $> \frac{1}{2}$
d. 0

22 The number of sides of two regular polygons are as 5 : 4 and the difference between their angles is $\frac{\pi}{40}$, then the number of sides in polygons respectively are

a. 20, 16
b. 10, 8
c. 25, 20
d. None of these

DPP-2 Trigonometric Functions using Identities

Subjective Questions

Working Space

***Direction** (Q. Nos. 1-4) These questions are subjective in nature, need to be solved completely on notebook.*

1 Prove that $(\operatorname{cosec}\theta - \sin\theta)(\sec\theta - \cos\theta)(\tan\theta + \cot\theta) = 1$.

2 Prove that $\dfrac{\sin^3\theta + \cos^3\theta}{\sin\theta + \cos\theta} + \dfrac{\sin^3\theta - \cos^3\theta}{\sin\theta - \cos\theta} = 2$.

3 If $x = a\sec\theta\cos\phi$, $y = b\sec\theta\sin\phi$ and $z = c\tan\theta$, then show that

$$\frac{x^2}{a^2} + \frac{y^2}{b^2} - \frac{z^2}{c^2} = 1.$$

4 If $\cos^2\alpha - \sin^2\alpha = \tan^2\beta$, then show that $\tan^2\alpha = \cos^2\beta - \sin^2\beta$.

Only One Option Correct Type

***Direction** (Q. Nos. 5-15) This section contains 11 multiple choice questions. Each question has four choices (a), (b), (c) and (d), out of which ONLY ONE is correct.*

5 If $\sin^6\theta + \cos^6\theta - 1 = \lambda\sin^2\theta\cos^2\theta$, then the value of λ is

a. 1
b. 2
c. 3
d. None of these

6 The value of $(1 + \tan\alpha\tan\beta)^2 + (\tan\alpha - \tan\beta)^2$ is

a. $\sec^2\alpha \cdot \sec^2\beta$
b. $\sec\alpha \cdot \sec\beta$
c. $\sec^2\alpha \cdot \sec\beta$
d. $\sec\alpha \cdot \sec^2\beta$

7 The value of $\dfrac{2\sin\theta\cos\theta - \cos\theta}{1 - \sin\theta + \sin^2\theta - \cos^2\theta}$ is

a. $\tan\theta$
b. $\cot\theta$
c. $\sin\theta$
d. None of these

8 If $a\cos\theta - b\sin\theta = c$, then the value of $a\sin\theta + b\cos\theta$ is

a. $\sqrt{a^2 + b^2 - c^2}$
b. $-\sqrt{a^2 + b^2 - c^2}$
c. Both (a) and (b)
d. None of these

9 The value of $3(\sin x - \cos x)^4 + 6(\sin x + \cos x)^2 + 4(\sin^6 x + \cos^6 x)$ is **[IIT JEE 1995]**

a. 7
b. 9
c. 13
d. None of these

Working Space

10 If $\cos\theta + \sin\theta = \sqrt{2}\cos\theta$ and $\cos\theta - \sin\theta = k\sin\theta$, then k is equal to

a 1 b. 2
c. $\sqrt{2}$ d. $\sqrt{3}$

11 $\sec^2\theta = \dfrac{4xy}{(x+y)^2}$ is possible only when $x : y$ is **[IIT JEE 1996]**

a. 1 : 1 b. 1 : 2
c. 2 : 1 d. 2 : 3

12 If $\sin\theta + \operatorname{cosec}\theta = 2$, then the value of $\sin^{20}\theta + \operatorname{cosec}^{20}\theta$ is

a. 1 b. 2
c. 3 d. 4

13 If $\sec\theta + \tan\theta = 1.5$ and $\sec\theta = a$, $\tan\theta = b$, then the value of (a, b) is

a. $\left(\dfrac{1}{12}, \dfrac{5}{12}\right)$ b. $\left(\dfrac{13}{12}, \dfrac{1}{12}\right)$
c. $\left(\dfrac{13}{12}, \dfrac{5}{12}\right)$ d. None of these

14 Let $F_k(x) = \dfrac{1}{k}(\sin^k x + \cos^k x)$, where $x \in R$ and $k \geq 1$, then the value of $F_4(x) - F_6(x)$ is **[IIT JEE 2014]**

a. $\dfrac{1}{4}$ b. $\dfrac{1}{6}$
c. $\dfrac{1}{8}$ d. $\dfrac{1}{12}$

15 If $a\cos^3\theta + 3a\cos\theta\sin^2\theta = m$ and $a\sin^3\theta + 3a\cos^2\theta\sin\theta = n$, then the value of $(m+n)^{2/3} + (m-n)^{2/3}$ is

a. $2a^{2/3}$ b. $2a^{1/3}$
c. $2a^2$ d. $a^{2/3}$

One or More than One Options Correct Type

***Direction** (Q. Nos. 16-18) This section contains 3 multiple choice questions. Each question has four choices (a), (b), (c) and (d), out of which ONE or MORE THAN ONE are correct.*

16 If $10\sin^4\alpha + 15\cos^4\alpha = 6$, then $27\operatorname{cosec}^6\alpha + 8\sec^6\alpha$ is not equal to

a. 200 b. 250
c. 300 d. 350

17 If $\dfrac{\sin^4 x}{2} + \dfrac{\cos^4 x}{3} = \dfrac{1}{5}$, then **[IIT JEE 2009]**

a. $\tan^2 x = \dfrac{2}{3}$ b. $\dfrac{\sin^8 x}{8} + \dfrac{\cos^8 x}{27} = \dfrac{1}{125}$
c. $\tan^2 x = \dfrac{1}{8}$ d. $\dfrac{\sin^8 x}{8} + \dfrac{\cos^8 x}{27} = \dfrac{2}{125}$

18 If $\cot\theta + \tan\theta = x$ and $\sec\theta - \cos\theta = y$, then

a. $\sin\theta \cdot \cos\theta = \dfrac{1}{x}$
b. $\sin\theta \cdot \tan\theta = y$
c. $(x^2y)^{2/3} - (xy^2)^{2/3} = 1$
d. $(x^2y)^{1/3} + (xy^2)^{1/3} = 1$

Working Space

Comprehension Type

Direction (Q. Nos. 19 and 20) This section contains a paragraph, which describing theory, experiments, data, etc. Two questions related to the paragraph have been given. Each question has only one correct answer among the four given options (a), (b), (c) and (d).

Passage

If *a*, *b*, *c* are non-zero real numbers and are in GP, then

$$\frac{b}{a} = \frac{c}{b} \Rightarrow b^2 = ac$$

On the basis of above information, answer the following questions :

19 If $\sin A$, $\cos A$, $\tan A$ are in GP, then $\cot^6 A - \cot^2 A$ is equal to

a. 1 b. –1
c. 0 d. 2

20 If $\tan A$, $2\tan A + 2$, $3\tan A + 3$ are in GP, then $\tan A$ is equal to

a. 4 b. – 4
c. – 2 d. None of these

One Integer Value Correct Type

Direction (Q. Nos. 21 and 22) This section contains 2 questions. When worked out will result in an integer from 0 to 9 (both inclusive).

21 If $\tan\theta + \sin\theta = m$ and $\tan\theta - \sin\theta = n$ and $(m^2 - n^2)^2 = 4\lambda mn$, then λ is equal to

22 If $\sin A + \sin^2 A + \sin^3 A = 1$, then the value of $\cos^6 A - 4\cos^4 A + 8\cos^2 A$ is

DPP-3 Sign of Trigonometric Ratios and Allied Angles

Subjective Questions

Working Space

Direction *(Q. Nos. 1-3) These questions are subjective in nature, need to be solved completely on notebook.*

1 Prove that

$$\left(1+\cos\frac{\pi}{4}\right)\left(1+\cos\frac{3\pi}{4}\right)\left(1+\cos\frac{5\pi}{4}\right)\left(1+\cos\frac{7\pi}{4}\right)=\frac{1}{4}.$$

2 If $x = y\cos\frac{2\pi}{3} = z\cos\frac{4\pi}{3}$, then prove that $xy + yz + zx = 0$.

3 Show that $\sec^2\theta + \operatorname{cosec}^2\theta \geq 4$.

Only One Option Correct Type

Direction *(Q. Nos. 4-13) This section contains 10 multiple choice questions. Each question has four choices (a), (b), (c) and (d), out of which ONLY ONE is correct.*

4 Which statement is correct, when *a, b, m* and *n* are non-zero real numbers?

a. $4\sin^2\theta = 5$

b. $(a^2+b^2)\cos\theta = 2ab$

c. $(m^2+n^2)\operatorname{cosec}\theta = m^2-n^2$

d. $\sin\theta = 2.375$

5 The value of $e^{\log_{10}\tan 1^\circ + \log_{10}\tan 2^\circ + \log_{10}\tan 3^\circ + \ldots + \log_{10}\tan 89^\circ}$ is

a. 0 b. e

c. $\frac{1}{e}$ d. None of these

6 The set of all possible values of α in $[-\pi, \pi]$ such that $\sqrt{\frac{1-\sin\alpha}{1+\sin\alpha}} = \sec\alpha - \tan\alpha$, is

a. $\left[-\frac{\pi}{2}, \frac{\pi}{2}\right]$ b. $\left(-\frac{\pi}{2}, \frac{\pi}{2}\right)$

c. $\left(\frac{\pi}{2}, \pi\right)$ d. None of these

7 If $4n\alpha = \pi$, then the value of $\cot\alpha \cdot \cot 2\alpha \cdot \cot 3\alpha \cdot \ldots \cdot \cot(2n-1)\alpha$ is

a. 0 b. 1

c. 2 d. None of these

Working Space

8 The number of real solutions of the equation $\cos^7 x + \sin^4 x = 1$ in interval $[-\pi, \pi]$, is

a. 0 b. 1
c. 2 d. 3

9 The value of $\cos^2 5° + \cos^2 10° + \cos^2 15° + \ldots + \cos^2 85° + \cos^2 90°$ is

a. 7 b. 8
c. 9 d. None of these

10 If $x \in R - \{0\}$, then $\cos\theta$ is

a. $> x + \frac{1}{x}$ b. $= x + \frac{1}{x}$
c. $\neq x + \frac{1}{x}$ d. None of these

11 The value of

$$3\left[\sin^4\left(\frac{3\pi}{2} - \alpha\right) + \sin^4(3\pi + \alpha)\right] - 2\left[\sin^6\left(\frac{\pi}{2} + \alpha\right) + \sin^6(5\pi - \alpha)\right] \text{ is}$$

a. 0 b. 1
c. −1 d. 2

12 If $\frac{\sin^3\theta - \cos^3\theta}{\sin\theta - \cos\theta} - \frac{\cos\theta}{\sqrt{1+\cot^2\theta}} - 2\tan\theta\cot\theta = -1, \theta \in [0, 2\pi]$, then θ belongs to

a. $[0, \pi] - \left\{\frac{\pi}{4}, \frac{\pi}{2}\right\}$
b. $(0, \pi)$
c. $(0, \pi) - \left\{\frac{\pi}{4}, \frac{\pi}{2}\right\}$
d. None of the above

13 The number of solutions for which $e^{\sin x} - e^{-\sin x} = 4$, is

a. 0 b. 1
c. 2 d. 3

One or More than One Options Correct Type

Direction *(Q. Nos. 14-17) This section contains 4 multiple choice questions. Each question has four choices (a), (b), (c) and (d), out of which ONE or MORE THAN ONE are correct.*

14 If $0 \leq x \leq \pi$ and $81^{\sin^2 x} + 81^{\cos^2 x} = 30$, then x is equal to

a. $\frac{\pi}{6}$ b. $\frac{\pi}{3}$
c. $\frac{5\pi}{6}$ d. $\frac{2\pi}{3}$

15 If $\sin(\alpha + \beta) = 1$ and $\sin(\alpha - \beta) = \frac{1}{2}$, where $\alpha, \beta \in \left[0, \frac{\pi}{2}\right]$, then

a. $\tan(\alpha + 2\beta) = -\sqrt{3}$
b. $\tan(2\alpha + \beta) = -\frac{1}{\sqrt{3}}$
c. $\tan(\alpha + \beta) = 1$
d. None of the above

Working Space

16 If for all real values of x,

$$\frac{4x^2 + 1}{64x^2 - 96x \sin\alpha + 5} < \frac{1}{32}, \text{ then } \alpha \text{ lies in the interval}$$

a. $\left(0, \frac{\pi}{3}\right)$ b. $\left(\frac{\pi}{3}, \frac{2\pi}{3}\right)$
c. $\left(\frac{2\pi}{3}, \pi\right)$ d. $\left(\frac{4\pi}{3}, \frac{5\pi}{3}\right)$

17 If $\theta_i > 0$ for $1 \le \theta \le n$ and $\theta_1 + \theta_2 + \theta_3 + \ldots + \theta_n = \pi$, then the greatest value of sum $\sin\theta_1 + \sin\theta_2 + \sin\theta_3 + \ldots + \sin\theta_n$ is equal to

a. n b. π
c. $n \sin\left(\frac{\pi}{n}\right)$ d. None of these

Matching List Type

Direction *(Q. No. 18) Choices for the correct combination of elements from Column I and Column II are given as options (a), (b), (c) and (d), out of which one is correct.*

18. Match the following columns and select the correct option from the codes given below.

	Column I		Column II
i.	$\tan\theta(\cot\theta\cos\theta + \sin\theta)$	p.	$\cot\theta$
ii.	$\frac{\tan\theta \operatorname{cosec}^2\theta}{1+\tan^2\theta}$	q.	$\tan\theta$
iii.	$\frac{\operatorname{cosec}^2\theta - 1}{\operatorname{cosec}\theta\cos^2\theta}$	r.	$\sec\theta$
iv.	$\frac{\cot\theta\sec^2\theta}{1+\cot^2\theta}$	s.	$\operatorname{cosec}\theta$
		t.	$\frac{\operatorname{cosec}\theta}{\cot\theta}$

Codes

	i	ii	iii	iv
a.	r	p	s	q
b.	p	r	s	q
c.	s	p	s	q
d.	r	p	q	s

One Integer Value Correct Type

Direction *(Q. Nos. 19 and 20) This section contains 2 questions. When worked out will result in an integer from 0 to 9 (both inclusive).*

19 If $\sec A \tan B + \tan A \sec B = 2$, then the value of $(\sec A \sec B + \tan A \tan B)^2$ is

20 If $0 < A < \frac{\pi}{2}$ and $\sin A + \cos A + \tan A + \cot A + \sec A + \operatorname{cosec} A = 7$, $\sin A$ and $\cos A$ are roots of equation $4x^2 - 3x + \alpha = 0$, then the value of $\left|\frac{25\alpha}{4}\right|$ is

DPP-4 Sign of T-Ratios and T-Ratios of Standard Angles

Subjective Questions

Working Space

Direction (Q. Nos. 1-3) *These questions are subjective in nature, need to be solved completely on notebook.*

1 If $\cot \theta = \frac{12}{5}$, θ lies in III quadrant, then find the values of other trigonometry ratios.

2 If $\sin \theta = \frac{12}{13}, \frac{\pi}{2} < \theta < \pi$, then prove that $\sec \theta + \tan \theta = -5$.

3 If $\sin \theta = \frac{3}{5}$, $\tan \phi = \frac{1}{2}$ and $\frac{\pi}{2} < \theta < \pi < \phi < \frac{3\pi}{2}$, then find the value of $8 \tan \theta - \sqrt{5} \sec \phi$.

Only One Option Correct Type

Direction (Q. Nos. 4-13) *This section contains 10 multiple choice questions. Each question has four choices (a), (b), (c) and (d), out of which ONLY ONE is correct.*

4 The value of

$\sin^2 \frac{\pi}{18} + \sin^2 \frac{\pi}{9} + \sin^2 \frac{7\pi}{18} + \sin^2 \frac{4\pi}{9}$ is

a. 1
b. 2
c. 3
d. None of these

5 If $\sin \theta = \frac{2t}{1+t^2}$ and θ lies in II quadrant, then $\cos \theta$ is equal to

a. $\frac{1-t^2}{1+t^2}$
b. $\frac{t^2-1}{1+t^2}$
c. $-\frac{|1-t^2|}{1+t^2}$
d. $\frac{1+t^2}{|1-t^2|}$

6 If $\tan \theta = x - \frac{1}{4x}$, then $\sec \theta - \tan \theta$ is equal to

a. $-2x, \frac{1}{2x}$
b. $-\frac{1}{2x}, 2x$
c. $2x$
d. $2x, \frac{1}{2x}$

7 If $\tan \theta + \sec \theta = \sqrt{3}$, $0 < \theta < \pi$, then θ is equal to

a. $\frac{5\pi}{6}$
b. $\frac{2\pi}{3}$
c. $\frac{\pi}{6}$
d. $\frac{\pi}{3}$

Working Space

8 If $\frac{3\pi}{4} < \alpha < \pi$, then $\sqrt{2 \cot \alpha + \frac{1}{\sin^2 \alpha}}$ is equal to

a. $1 - \cot \alpha$ b. $1 + \cot \alpha$
c. $-1 + \cot \alpha$ d. $-1 - \cot \alpha$

9 If $x = \sin \theta \,|\sin \theta|$ and $y = \cos \theta \,|\cos \theta|$, where $\frac{99\pi}{2} < \theta < 50\pi$, then

a. $x - y = 1$ b. $x + y = -1$
c. $x + y = 1$ d. $y - x = 1$

10 If $x \sin 45° \cos^2 60° = \frac{\tan^2 60° \operatorname{cosec} 30°}{\sec 45° \cot^2 30°}$, then x is equal to

a. 9 b. 10
c. 11 d. None of these

11 If $0 < A, B < \frac{\pi}{2}$ and $\frac{\sin A}{\sin B} = \sqrt{2}$ and $\frac{\tan A}{\tan B} = \sqrt{3}$, then

a. $A = 45°$ and $B = 30°$
b. $A = 30°$ and $B = 45°$
c. $A = 60°$ and $B = 45°$
d. None of the above

12 If $\cot(\alpha + \beta) = 0$, then the value of $\sin(\alpha + 2\beta)$ is

a. $\sin \alpha$ b. only $\cos \beta$
c. $\sin \alpha$ or $\cos \beta$ d. None of these

13 If θ lies in the II quadrant, then the value of $\sqrt{\frac{1 - \sin \theta}{1 + \sin \theta}} + \sqrt{\frac{1 + \sin \theta}{1 - \sin \theta}}$ is

a. $2 \sec \theta$ b. $-2 \sec \theta$
c. $2 \operatorname{cosec} \theta$ d. None of these

One or More than One Options Correct Type

***Direction** (Q. Nos. 14-16) This section contains 3 multiple choice questions. Each question has four choices (a), (b), (c) and (d), out of which ONE or MORE THAN ONE are correct.*

14 The value of $\sqrt{\frac{1 + \cos \theta}{1 - \cos \theta}}$ is

a. $\operatorname{cosec} \theta + \cot \theta$, $0 < \theta < \pi$
b. $-(\operatorname{cosec} \theta + \cot \theta)$, $\pi < \theta < 2\pi$
c. always $(\operatorname{cosec} \theta - \cot \theta)$
d. None of the above

15 If $\sec \alpha = -\frac{5}{4}$ and $\frac{\pi}{2} < \alpha < \pi$, then $\frac{7 - 5 \cot \alpha}{9 - 4\sqrt{\sec^2 \alpha - 1}}$ is equal to

a. $\frac{41}{18}$ b. $\frac{31}{18}$
c. $\frac{41}{36}$ d. None of these

16 If θ lies in I quadrant and $x = p \cos \theta + q \sin \theta$, $y = p \sin \theta - q \cos \theta$ such that $x^2 + 4xy + y^2 = ap^2 + bq^2$, where a and b are constants. Then,

a. $a = -1$ and $b = 3$ b. $\theta = 45°$
c. $a = 3$ and $b = -1$ d. $\theta = 60°$

Working Space

One Integer Value Correct Type

Direction *(Q. Nos. 17-19) This section contains 3 questions. When worked out will result in an integer from 0 to 9 (both inclusive).*

17 If $2k - 4 = 0$ and $\sin x + \operatorname{cosec} x = k$, $0 < x \le \frac{\pi}{2}$, $\forall\, n \in N$, then the value of $\sin^n x + \operatorname{cosec}^n x$ is

18 If $0 \le x \le \frac{\pi}{2}$ and $\cos x + \cos 2x + \cos 3x = 3$, then the value of $\sin x + \sin 2x + \sin 3x$ is

19 If number of all possible values of θ, where $0 < \theta < \pi$ for which the system of equations $(y + z)\cos 3\theta = xyz \sin 3\theta$, $x \sin 3\theta = \frac{2\cos 3\theta}{y} + \frac{2\sin 3\theta}{z}$ and $xyz \sin 3\theta = (y + 2z)\cos 3\theta + y \sin 3\theta$ have a solution (x_0, y_0, z_0) with $y_0 z_0 \ne 0$, is **[IIT JEE 2010]**

True/False Statements

Direction *(Q. No. 20) This section contains 1 question. This question has statement which is TRUE or FALSE.*

20 If $A = \sin^8 \theta + \cos^{14} \theta$, then $0 \le A \le 1$, $\forall\, \theta$. **[IIT JEE 1980]**

DPP-5 T-Ratios using Addition and Subtraction Formulae

Fill in the Blanks

Working Space

Direction *(Q. Nos. 1-3) This section contains 3 sentences. Complete the given sentences filling the blanks with suitable words, informations, data, etc.*

1 If $\cos(A+B) = \alpha \cos A \cos B + \beta \sin A \sin B$, then $(\alpha, \beta) =$ ....

2 If α lies in II quadrant, β lies in III quadrant and $\tan(\alpha+\beta) > 0$, then $(\alpha+\beta)$ lies in .... quadrants.

3 If $\tan A = \frac{1}{2}$, $\tan B = \frac{1}{3}$ and $\cos 2A = \sin \lambda B$, then $\lambda =$ ....

Subjective Questions

Direction *(Q. Nos. 4-6) These questions are subjective in nature, need to be solved completely on notebook.*

4 Prove that $\cos\left(\frac{3\pi}{4} - A\right)\cos\left(\frac{3\pi}{4} - B\right) - \sin\left(\frac{3\pi}{4} - A\right)\sin\left(\frac{3\pi}{4} - B\right) = -\sin(A+B)$.

5 If A and B are positive acute angles and $\cos A = \frac{1}{7}$, $\cos B = \frac{13}{14}$, then show that $A - B = \frac{\pi}{3}$.

6 If $3 \tan A \tan B = 1$, then prove that $\frac{\cos(A-B)}{\cos(A+B)} = 2$.

Only One Option Correct Type

Direction *(Q. Nos. 7-18) This section contains 12 multiple choice questions. Each question has four choices (a), (b), (c) and (d), out of which ONLY ONE is correct.*

7 If $\tan\alpha = \frac{m}{m+1}$ and $\tan\beta = \frac{1}{2m+1}$, then $\alpha + \beta$ is equal to **[IIT JEE 1978]**

a. $\frac{\pi}{3}$ b. $\frac{\pi}{4}$ c. $\frac{\pi}{6}$ d. None of these

8 If $\cot\alpha = \frac{1}{2}$, $\sec\beta = \frac{-5}{3}$, where $\pi < \alpha < \frac{3\pi}{2}$ and $\frac{\pi}{2} < \beta < \pi$, then the value of $\tan(\alpha+\beta)$ and quadrant in which $(\alpha+\beta)$ terminates is

a. $\tan(\alpha+\beta) = \frac{2}{11}$, $(\alpha+\beta) \in$ I quadrant
b. $\tan(\alpha+\beta) = \frac{3}{11}$, $(\alpha+\beta) \in$ II quadrant
c. $\tan(\alpha+\beta) = 1$, $(\alpha+\beta) \in$ II quadrant
d. None of these

Working Space

9 If $\cos(\alpha+\beta)=\frac{4}{5}$, $\sin(\alpha-\beta)=\frac{5}{13}$ and $\alpha, \beta \in \left(0, \frac{\pi}{4}\right)$, then the value of $\tan 2\alpha$ is **[IIT JEE 1979]**

a. $\frac{56}{33}$
b. $\frac{65}{33}$
c. $\frac{55}{34}$
d. None of these

10 If $\tan 70° = \tan 20° + \lambda \tan 50°$, then the value of λ is

a. 1
b. 2
c. 3
d. None of these

11 If $3\sin(2A+B)=\sin B$, then the value of $2\tan A + \tan(A+B)$ is equal to

a. 0
b. 1
c. 2
d. 3

12 If $\tan\beta = \frac{n\sin\alpha\cos\alpha}{1-n\sin^2\alpha}$ and $\tan(\alpha-\beta)=k\tan\alpha$, then k is equal to

a. $n+1$
b. $1-n$
c. $n-1$
d. $1-\frac{1}{n}$

13 If $2\tan\beta + \cot\beta = \tan\alpha$ and $\cot\beta = n\tan(\alpha-\beta)$, then n is equal to

a. 1
b. 2
c. 3
d. 4

14 If $A+B+C=180°$ and $\tan A+\tan B+\tan C=\mu\tan A\tan B\tan C$, $\mu\in R$, then μ is

a. -1
b. 1
c. 2
d. None of these

15 If $\frac{\sin(x+y)}{\sin(x-y)}=k\left(\frac{\tan x+\tan y}{\tan x-\tan y}\right)$, then k is equal to

a. 0
b. 1
c. 2
d. 3

16 If $\cos 2\alpha\cos 2\beta+\sin^2(\alpha-\beta)-\sin^2(\alpha+\beta)=\cos\lambda(\alpha+\beta)$, then λ is equal to

a. 2
b. $\frac{1}{2}$
c. 3
d. $\frac{3}{4}$

17 If $\alpha+\beta=\frac{\pi}{2}$ and $\beta+\gamma=\alpha$, then $\tan\alpha$ is equal to **[IIT JEE 2001]**

a. $2(\tan\beta+\tan\gamma)$
b. $\tan\beta+\tan\gamma$
c. $\tan\beta+2\tan\gamma$
d. $2\tan\beta+\tan\gamma$

18 Given both θ and ϕ are acute angles such that $\sin\theta=\frac{1}{2}$ and $\cos\phi=\frac{1}{3}$, then $\theta+\phi$ belongs to **[IIT JEE 2004]**

a. $\left(\frac{\pi}{2}, \frac{2\pi}{3}\right)$
b. $\left[\frac{\pi}{2}, \frac{2\pi}{3}\right]$
c. $\left(\frac{\pi}{4}, \frac{\pi}{2}\right)$
d. None of these

One or More than One Options Correct Type

Working Space

Direction (Q. Nos. 19-21) *This section contains 3 multiple choice questions. Each question has four choices (a), (b), (c) and (d), out of which ONE or MORE THAN ONE are correct.*

19 If $\tan\alpha$ and $\tan\beta$ are the roots of the equation $x^2 + px + q = 0 (p \neq 0)$, then

a. $\sin^2(\alpha+\beta) + p\sin(\alpha+\beta)\cos(\alpha+\beta) + q\cos^2(\alpha+\beta) = q$

b. $\tan(\alpha+\beta) = \dfrac{p}{q-1}$

c. $\cos(\alpha+\beta) = 1-q$

d. $\sin(\alpha+\beta) = -p$

20 If $\dfrac{\tan A}{1} = \dfrac{\tan B}{2} = \dfrac{\tan C}{3} (\neq 0)$ and $A + B + C = \pi$, then

a. $\tan A = \pm 1$, $\tan B = \pm 2$, $\tan C = \pm 3$

b. minimum value of $\tan A + \tan B + \tan C$ is -6

c. maximum value of $\tan A + \tan B + \tan C$ is 6

d. $\tan A + \tan B + \tan C = 0$, $\forall A, B, C \in R$

21 If in a ΔABC, $\tan A + \tan B + \tan C = 6$ and $\tan A \tan B = 2$, then the values of $\tan A$, $\tan B$, $\tan C$ are

a. 1, 2, 3
b. 2, 1, 3
c. 5, 8, 9
d. None of these

One Integer Value Correct Type

Direction (Q. Nos. 22-24) *This section contains 3 questions. When worked out will result in an integer from 0 to 9 (both inclusive).*

22 If $\tan(\pi\cos\theta) = \cot(\pi\sin\theta)$ and $\cos\left(\theta - \dfrac{\pi}{4}\right) = \pm\dfrac{1}{\lambda\sqrt{\mu}}$, then the value of $\lambda + \mu$ is

23 If $\tan\theta_1$, $\tan\theta_2$, $\tan\theta_3$, $\tan\theta_4$ are roots of the equation $x^4 - x^3\sin 2\beta + x^2\cos 2\beta - x\cos\beta - \sin\beta = 0$ and $\tan(\theta_1 + \theta_2 + \theta_3 + \theta_4) = \cot(n\beta)$, then n is equal to

24 If $\sqrt{3}\sec 290° + \operatorname{cosec} 250° = \sqrt{3}\lambda$, then the value of $9\lambda^4 + 3\lambda^2 - 270$ is

DPP-6 Trigonometric Functions using Greatest and Least Values of T-Ratios

Subjective Questions

Working Space

***Direction** (Q. Nos. 1-4) These questions are subjective in nature, need to be solved completely on notebook.*

1 Prove that the minimum value of $3\cos x + 4\sin x + 5$ is 0.

2 If $\sin\theta_1 + \sin\theta_2 + \sin\theta_3 = 3$, then find the value of $\cos\theta_1 + \cos\theta_2 + \cos\theta_3$.

3 If $\sin x + \sin^2 x = 1$, then find the value of $\cos^8 x + 2\cos^6 x + \cos^4 x$.

4 If $x = r\sin\theta\cos\phi$, $y = r\sin\theta\sin\phi$ and $z = r\cos\theta$, then prove that $x^2 + y^2 + z^2$ is independent of θ and ϕ.

Only One Option Correct Type

***Direction** (Q. Nos. 5-14) This section contains 10 multiple choice questions. Each question has four choices (a), (b), (c) and (d), out of which ONLY ONE is correct.*

5 The least value of $2\sin^2\theta + 3\cos^2\theta$ is

a. 1 b. 2 c. 3 d. 5

6 If $\cos x + \sec x = -2$, then for a positive integer n, $\cos^n x + \sec^n x$ is

a. always 2 b. -2, if n is odd c. 2, if n is even d. Both (b) and (c)

7 α, β, γ are real numbers satisfying $\alpha + \beta + \gamma = \pi$. The minimum value of given expression $\sin\alpha + \sin\beta + \sin\gamma$ is **[IIT JEE 1995]**

a. 0 b. -3 c. positive d. negative

8 If A, B, C are acute angles of a triangle such that $A + B + C = 180°$ and $\cot A \cot B \cot C = \lambda$, then

a. $\lambda \le \frac{1}{3\sqrt{3}}$ b. $\lambda \ge \frac{1}{3\sqrt{3}}$ c. $\lambda = \frac{1}{\sqrt{3}}$ d. $\lambda < \frac{1}{3}$

9 If $A = \sin^2\theta + \cos^4\theta$, then for all real values of θ is **[IIT JEE 1980, AIEEE 2011]**

a. $1 \le A \le 2$ b. $\frac{3}{4} \le A \le 1$ c. $\frac{13}{16} \le A \le 1$ d. $\frac{3}{4} \le A \le \frac{13}{16}$

10 The minimum value of $\sec^2\theta + \operatorname{cosec}^2\theta - 4$ is

a. 2 b. 4 c. 6 d. None of these

11 Which of the following is correct?

a. $\sin 1° > \sin 1$ b. $\sin 1° < \sin 1$ c. $\sin 1° = \sin 1$ d. $\sin 1° = \frac{\pi}{180}\sin 1$

12 If $3\sin\theta + 5\cos\theta = 5$, then the value of $5\sin\theta - 3\cos\theta$ is

a. 3 b. -3 c. Both (a) and (b) d. 8

13 If $\sin^2 x = 1 - \sin x$, then the value of $\cos^{12} x + 3\cos^{10} x + 3\cos^8 x + \cos^6 x$ is

a. -1 b. 1 c. 0 d. None of these

14 If $P = \cos(\cos x) + \sin(\cos x)$, then the least and greatest value of P are

a. -1 and 1 b. 0 and 2 c. $-\sqrt{2}$ and $\sqrt{2}$ d. 0 and $\sqrt{2}$

Working Space

One or More than One Options Correct Type

Direction (Q. Nos. 15-17) *This section contains 3 multiple choice questions. Each question has four choices (a), (b), (c) and (d), out of which ONE or MORE THAN ONE are correct.*

15 If $x = \text{cosec}^2\theta$, $y = \sec^2\theta$ and $z = \dfrac{1}{1 - \sin^2\theta\cos^2\theta}$, then

a. $xyz = xz + y$ b. $xyz = xy + z$

c. $xyz = x + y + z$ d. $xyz = yz + x$

16 Let $\theta \in \left(0, \dfrac{\pi}{4}\right)$ and $t_1 = (\tan\theta)^{\tan\theta}$, $t_2 = (\tan\theta)^{\cot\theta}$, $t_3 = (\cot\theta)^{\tan\theta}$ and $t_4 = (\cot\theta)^{\cot\theta}$, then **[IIT JEE 2006]**

a. $t_1 > t_2 > t_3 > t_4$ b. $t_4 > t_3 > t_1 > t_2$ c. $t_3 > t_1 > t_2 > t_4$ d. $t_2 > t_3 > t_1 > t_4$

17 The ratio of greatest value of $2 - \cos x + \sin^2 x$ to its least value is

a. 13 : 4 b. 1 : 4 c. 26 : 8 d. None of these

Comprehension Type

Direction (Q. Nos. 18-21) *This section contains a paragraph, which describing theory, experiments, data, etc. Four questions related to the paragraph have been given. Each question has only one correct answer among the four given options (a), (b), (c) and (d).*

Passage

If $P_n = \sin^n\theta + \cos^n\theta$, where $n \in W$ (whole number) and $\theta \in R$ (real number).

On the basis of above information, answer the following questions :

18 If $P_n - P_{n-2} = kP_{n-4}$, then k is equal to

a. 1 b. $-\sin^2\theta\cos^2\theta$ c. $\sin^2\theta$ d. $\cos^2\theta$

19 If $P_1 = m$, then the value of $4(1 - P_6)$ is

a. $3(m-1)^2$ b. $3(m^2-1)^2$ c. $3(m+1)^2$ d. $3(m^2+1)^2$

20 If $P_{n-2} - P_n = \sin^2\theta\cos^2\theta\, P_\lambda$, then the value of λ is

a. $(n-1)$ b. $(n-2)$ c. $(n-3)$ d. $(n-4)$

21 The value of $2P_6 - 3P_4 + 10$ is

a. 0 b. 6 c. 9 d. 15

Fill in the Blanks

Direction (Q. Nos. 22-24) *This section contains 3 sentences. Complete the given sentences filling the blanks with suitable words, informations, data, etc.*

22 ...... $\le \sin\theta \le$ ......

23 $\cos\theta$ and $\tan\theta$ are negative in ...... quadrant.

24 Maximum and minimum value of $a\sin\theta + b\cos\theta$ are ...... and ...... respectively.

DPP-7 Addition and Subtraction Formulae using Sign of T-Ratios and Allied Angles

Subjective Questions

Working Space

Direction *(Q. Nos. 1-3) These questions are subjective in nature, need to be solved completely on notebook.*

1 If $\tan A - \tan B = x$ and $\cot B - \cot A = y$, then prove that $\cot(A - B) = \frac{1}{x} + \frac{1}{y}$

2 If $m \tan(\theta - 30^\circ) = n \tan(\theta + 120^\circ)$, then prove that $\frac{m+n}{m-n} = 2\cos 2\theta$.

3 Prove that $\tan 40^\circ + \tan 20^\circ + \sqrt{3}\tan 20^\circ \tan 40^\circ = \sqrt{3}$.

Only One Option Correct Type

Direction *(Q. Nos. 4-15) This section contains 12 multiple choice questions. Each question has four choices (a), (b), (c) and (d), out of which ONLY ONE is correct.*

4 If $\tan\theta = \frac{\sin\alpha - \cos\alpha}{\sin\alpha + \cos\alpha}$, then $\sin\alpha + \cos\alpha$ and $\sin\alpha - \cos\alpha$ must be equal to

a. $\sqrt{2}\cos\theta, \sqrt{2}\sin\theta$
b. $\sqrt{2}\sin\theta, \sqrt{2}\cos\theta$
c. $\sqrt{2}\sin\theta, \sqrt{2}\sin\theta$
d. $\sqrt{2}\cos\theta, \sqrt{2}\cos\theta$

5 If $\cos(A - B) = \frac{3}{5}$ and $\tan A \tan B = 2$, then

a. $\cos A \cos B = \frac{1}{5}$
b. $\cos A \cos B = -\frac{1}{5}$
c. $\sin A \sin B = -\frac{1}{5}$
d. $\sin A \sin B = \frac{1}{5}$

6 If $A + B + C = 180^\circ$, then $\sec A(\cos B \cos C - \sin B \sin C)$ is equal to

a. 0
b. −1
c. 1
d. None of these

7 If $\sin(3\pi\cos\theta) = \cos(3\pi\sin\theta)$, then $\sin 2\theta$ is equal to

a. $\pm\frac{3}{4}$
b. $+\frac{4}{3}$
c. $\pm\frac{1}{3}$
d. None of these

Working Space

8 If $\tan 69^\circ + \tan 66^\circ - \tan 69^\circ \tan 66^\circ = 2k$, then k is equal to

a. -1 b. $\frac{1}{2}$
c. $-\frac{1}{2}$ d. None of these

9 The value of $\operatorname{cosec} 10^\circ - \sqrt{3} \sec 10^\circ$ is equal to

a. $\frac{1}{2}$ b. 2
c. 4 d. 8

10 If $\tan \frac{\pi}{9}$, x and $\tan \frac{5\pi}{18}$ are in AP and $\tan \frac{\pi}{9}$, y, $\tan \frac{7\pi}{18}$ are also in AP, then

a. $2x = y$ b. $x > y$
c. $x = y$ d. None of these

11 If $ABCD$ is a convex quadrilateral such that $4 \sec A + 5 = 0$, then the quadratic equation whose roots are $\tan A$ and $\operatorname{cosec} A$ is

a. $12x^2 - 11x - 15 = 0$ b. $12x^2 + 11x - 15 = 0$
c. $12x^2 - 29x + 15 = 0$ d. None of these

12 The number of values of x in the interval $[0, 5\pi]$ satisfying the equation $3\sin^2 x - 7\sin x + 2 = 0$, is

a. 0 b. 5
c. 6 d. 10

13 If $\alpha \in \left(0, \frac{\pi}{2}\right)$, then $\sqrt{x^2 + 1} + \frac{\tan^2 \alpha}{\sqrt{x^2 + 1}}$ is always greater than or equal to

a. $2 \tan \alpha$ b. 1
c. 2 d. $\sec^2 \alpha$

14 Let A and B denote the statements:

$$A : \cos \alpha + \cos \beta + \cos \gamma = 0$$
$$B : \sin \alpha + \sin \beta + \sin \gamma = 0$$

If $\cos (\beta - \gamma) + \cos (\gamma - \alpha) + \cos (\alpha - \beta) = -\frac{3}{2}$, then **[AIEEE 2009]**

a. A is true and B is false b. A is false and B is true
c. Both A and B are true d. Both A and B are false

15 In a ΔPQR, if $3 \sin P + 4 \cos Q = 6$ and $4 \sin Q + 3 \cos P = 1$, then the $\angle R$ is equal to **[AIEEE 2012]**

a. $\frac{5\pi}{6}$ b. $\frac{\pi}{6}$
c. $\frac{\pi}{4}$ d. None of these

One or More than One Options Correct Type

Direction *(Q. Nos. 16-19) This section contains 4 multiple choice questions. Each question has four choices (a), (b), (c) and (d), out of which ONE or MORE THAN ONE are correct.*

16 The value of $\frac{\cot 76^\circ \cot 16^\circ + 3}{\cot 76^\circ + \cot 16^\circ}$ is

a. $\tan 16^\circ$ b. $\cot 76^\circ$
c. $\tan 46^\circ$ d. $\cot 44^\circ$

Working Space

17 If in a ΔABC, $\tan A + \tan B + \tan C = 6$ and $\tan A \tan B = 2$, then $\sin^2 A : \sin^2 B : \sin^2 C$ is

a. 8 : 9 : 5
b. 8 : 5 : 9
c. 5 : 8 : 9
d. 5 : 8 : 5

18 If $\sec\theta + \tan\theta = 1$, then one root of the equation $a(b - c)x^2 + b(c - a)x + c(a - b) = 0$ is

a. $\tan\theta$
b. $\sec\theta$
c. $\cos\theta$
d. $\sin\theta$

19 If $k_1 = \dfrac{\sin\theta}{\cos 3\theta} + \dfrac{\sin 3\theta}{\cos 9\theta} + \dfrac{\sin 9\theta}{\cos 27\theta}$ and $k_2 = \tan 27\theta - \tan\theta$, then

a. $k_2 = 2k_1$
b. $k_2 = k_1 + 4$
c. $k_1 = k_2$
d. None of these

One Integer Value Correct Type

Direction *(Q. No. 20) This section contains 1 question. When worked out will result in an integer from 0 to 9 (both inclusive).*

20 In an acute angled ΔABC, the least value of $\sec A + \sec B + \sec C$ is

Matching List Type

Direction *(Q. No. 21) Choices for the correct combination of elements from Column I and Column II are given as options (a), (b), (c) and (d), out of which one is correct.*

21 Match the following columns and select the correct option from the codes given below.

	Column I		Column II
i.	The maximum and minimum values of $3\sin 2\theta + 4\cos 2\theta + 3$ for all real values of θ are λ and μ respectively, then	p.	$\lambda + \mu = 2$
ii.	If maximum and minimum values of $5\cos\theta + 3\cos\left(\theta + \dfrac{\pi}{3}\right) + 3$ for all real values of θ are λ and μ respectively, then	q.	$\lambda - \mu = 6$
iii.	If maximum and minimum values of $1 + \sin\left(\dfrac{\pi}{4} + \theta\right) + 2\cos\left(\dfrac{\pi}{4} - \theta\right)$ for all real values of θ are λ and μ respectively, then	r.	$\lambda + \mu = 6$
		s.	$\lambda - \mu = 10$
		t.	$\lambda - \mu = 14$

Codes

	i	ii	iii
a.	r,s	r,t	p,q
b.	p,q	r,s	t
c.	s,t	p,r	q,r
d.	t	p,s	r,s

DPP-8 Transformation Formulae-I

Subjective Questions

Working Space

Direction *(Q. Nos. 1-4) These questions are subjective in nature, need to be solved completely on notebook.*

1 Prove that $4\sin\theta\sin\left(\frac{\pi}{3}+\theta\right)\sin\left(\frac{2\pi}{3}+\theta\right)=\sin 3\theta$.

2 Show that
$\sin(B-C)\cos(A-D)+\sin(C-A)\cos(B-D)+\sin(A-B)\cos(C-D)=0$.

3 Prove that $\dfrac{1+\cos 2x+\cos 4x+\cos 6x}{\cos 3x\cos 2x\cos x}=4$.

4 Prove that $\sin\alpha+\sin\left(\alpha+\frac{2\pi}{3}\right)+\sin\left(\alpha+\frac{4\pi}{3}\right)=0$.

Only One Option Correct Type

Direction *(Q. Nos. 5-14) This section contains 10 multiple choice questions. Each question has four choices (a), (b), (c) and (d), out of which ONLY ONE is correct.*

5 If $\cos 20°\cos 40°\cos 60°\cos 80°=p$ and $\sin 10°\sin 30°\sin 50°\sin 70°=q$, then

a. $p=2q$ b. $\frac{p}{q}=1$

c. $q=2p$ d. None of these

6 If $\tan\frac{2\pi}{18}\tan\frac{4\pi}{18}\tan\frac{8\pi}{18}=\sqrt{k}$, then k is equal to

a. 1 b. 2

c. 3 d. None of these

7 If $\dfrac{\sin(A-C)+2\sin A+\sin(A+C)}{\sin(B-C)+2\sin B+\sin(B+C)}=\dfrac{k\sin A}{\sin B}$, then k is equal to

a. 0 b. 1

c. 2 d. None of these

8 If $\sin 2\theta+\sin 2\phi=\frac{1}{2}$ and $\cos 2\theta+\cos 2\phi=\frac{3}{2}$, then $\cos^2(\theta-\phi)$ is equal to

a. $\frac{3}{8}$ b. $\frac{5}{8}$ c. $\frac{3}{4}$ d. $\frac{5}{4}$

9 The value of $\sin 78°-\sin 66°-\sin 42°+\sin 6°$ is

a. $\frac{1}{2}$ b. $-\frac{1}{2}$ c. -1 d. None of these

10 If $\sin x+\sin y=\sqrt{3}(\cos y-\cos x)$, then $\sin 3x+\sin 3y$ is equal to

a. $2\sin 3x$ b. 0 c. 1 d. None of these

Working Space

11 If $\sin(y+z-x)$, $\sin(z+x-y)$ and $\sin(x+y-z)$ are in AP, then $\tan x$, $\tan y$, $\tan z$ are in

a. AP b. GP c. HP d. None of these

12 If $-\frac{\pi}{2}<\theta<\frac{\pi}{2}$ and $\theta \neq \pm\frac{\pi}{4}$, then the value of $\cot\left(\frac{\pi}{4}+\theta\right)\cot\left(\frac{\pi}{4}-\theta\right)$ is equal to

a. 0 b. 1 c. −1 d. 2

13 The maximum value of $\left|\sqrt{\sin^2 x+2a^2}-\sqrt{2a^2-1-\cos^2 x}\right|$, where $a, x \in R$ is

a. $\sqrt{3}$ b. $\sqrt{2}$ c. 1 d. $\sqrt{5}$

14 If $\sin\theta+\cos\theta=a$ and $\sin^3\theta+\cos^3\theta=b$, then we get $\lambda a^3+\mu b+\nu c=0$ and λ, μ, ν are independent of θ, then the value of $\lambda^3+\mu^3+\nu^3$ is

a. −6 b. −18 c. −36 d. −98

One or More than One Options Correct Type

Direction *(Q. Nos. 15-17) This section contains 3 multiple choice questions. Each question has four choices (a), (b), (c) and (d), out of which ONE or MORE THAN ONE are correct.*

15 The value of $\left(\frac{\cos A+\cos B}{\sin A-\sin B}\right)^n+\left(\frac{\sin A+\sin B}{\cos A-\cos B}\right)^n$ is/are

a. $2\cot^n\left(\frac{A-B}{2}\right)$, if n is even b. $2\cot^n\left(\frac{A-B}{2}\right)$, if n is odd

c. 0, when n is odd d. None of these

16 If $\frac{x}{y}=\frac{\cos A}{\cos B}$, $A \neq B$, then

a. $\tan\left(\frac{A+B}{2}\right)=\frac{x\tan A+y\tan B}{x+y}$ b. $\tan\left(\frac{A-B}{2}\right)=\frac{x\tan A-y\tan B}{x+y}$

c. $\frac{\sin(A+B)}{\sin(A-B)}=\frac{y\sin A+x\sin B}{y\sin A-x\sin B}$ d. $x\cos A+y\cos B=0$

17 The minimum value of $2^{\sin\theta}+2^{-\cos\theta}$ is λ at $\theta=\phi$, then

a. $\theta=\frac{7\pi}{4}, \lambda=2^{1-\frac{1}{\sqrt{2}}}$ b. $\theta=\frac{15\pi}{4}, \lambda=2^{1-\frac{1}{\sqrt{2}}}$

c. $\theta=\frac{7\pi}{4}, \lambda=2^{1+\frac{1}{\sqrt{2}}}$ d. None of these

One Integer Value Correct Type

Direction *(Q. Nos. 18-20) This section contains 3 questions. When worked out will result in an integer from 0 to 9 (both inclusive).*

18 If $a\sin\theta=b\sin\left(\theta+\frac{2\pi}{3}\right)=c\sin\left(\theta+\frac{4\pi}{3}\right)$, then the value of $ab+bc+ca$ is

19 If $\frac{\tan(\theta+\alpha)}{a}=\frac{\tan(\theta+\beta)}{b}=\frac{\tan(\theta+\gamma)}{c}$, then the value of $\left(\frac{a+b}{a-b}\right)\sin^2(\alpha-\beta)+\left(\frac{b+c}{b-c}\right)\sin^2(\beta-\gamma)+\left(\frac{c+a}{c-a}\right)\sin^2(\gamma-\alpha)$ is

20 The expression $\frac{\cos 6x+6\cos 4x+15\cos 2x+10}{\cos 5x+5\cos 3x+10\cos x}=\lambda\cos x$, then the value of λ is

DPP-9 Transformation Formulae-II

Subjective Questions

Working Space

Direction (Q. Nos. 1-4) *These questions are subjective in nature, need to be solved completely on notebook.*

1 If $\cos x + \cos y + \cos z = 0$ and $\sin x + \sin y + \sin z = 0$, then prove that $\cot\left(\frac{x+y}{2}\right) = \cot z$.

2 If $\tan A = 2\tan B + \cot B$, then prove that $2\tan(A - B) = \cot B$.

3 Prove that $\frac{\tan 70° - \tan 20°}{\tan 50°} = 2$.

4 Find the value of $\tan 5x \tan 3x \tan 2x$.

Only One Option Correct Type

Direction (Q. Nos. 5-15) *This section contains 11 multiple choice questions. Each question has four choices (a), (b), (c) and (d), out of which ONLY ONE is correct.*

5 If $\cos(A + B)\sin(C - D) = \cos(A - B)\sin(C + D)$, then $\tan A \tan B \tan C$ is equal to

a. $\tan D$
b. $-\tan D$
c. $\cot D$
d. None of the above

6 If A, B, C are in AP, then $\frac{\sin A - \sin C}{\cos C - \cos A}$ is equal to

a. $\tan B$
b. $\cot B$
c. $\tan 2B$
d. None of the above

7 If $A + B = C$, then $\tan A \tan B \tan C$ is equal to

a. $\tan C - \tan A - \tan B$
b. $\tan B - \tan A - \tan C$
c. $\tan A - \tan B - \tan C$
d. None of the above

8 If $\sin 2\theta = \lambda \sin 2\phi$, then $\frac{\tan(\theta + \phi)}{\tan(\theta - \phi)}$ is equal to

a. $\frac{\lambda - 1}{\lambda + 1}$
b. $\lambda + 1$
c. $\frac{\lambda + 1}{\lambda - 1}$
d. None of these

Working Space

9 If $\sin A + \sin B = a$ and $\cos A + \cos B = b$, then $\cos(A + B)$ is equal to

a. $\dfrac{a^2 + b^2}{a^2 - b^2}$

b. $\dfrac{2ab}{a^2 + b^2}$

c. $\dfrac{b^2 - a^2}{a^2 + b^2}$

d. $\dfrac{a^2 - b^2}{a^2 + b^2}$

10 If $A + B = \dfrac{\pi}{4}$, then $(\tan A + 1)(\tan B + 1)$ is equal to

a. 1

b. 2

c. $\sqrt{3}$

d. -1

11 The value of

$$\sin y \cos\left(\frac{\pi}{2} - x\right) - \cos\left(\frac{\pi}{2} - y\right) \cos x + \cos x \sin\left(\frac{\pi}{2} - y\right) + \cos y \cos\left(\frac{\pi}{2} - x\right)$$

is equal to zero. If

a. $x = 0$

b. $y = 0$

c. $x - \dfrac{\pi}{4} = y$

d. $x - \dfrac{3\pi}{4} = y$

12 If $\tan A = \dfrac{1}{2}$ and $\tan B = \dfrac{1}{3}$, then the value of $(A + B)$ is

a. $\dfrac{\pi}{6}$

b. π

c. 0

d. $\dfrac{\pi}{4}$

13 If $A + B + C = \pi$, then the value of $\sin A \sin(B - C) + \sin B \sin(C - A) + \sin C \sin(A - B)$ is

a. 1

b. 2

c. 0

d. None of these

14 If $\dfrac{\cos 18° - \sin 18°}{\sin 27°} = k$, then k is equal to

a. 1

b. $\sqrt{2}$

c. $\sqrt{3}$

d. None of these

15 If $\sin\theta = n \sin(\theta + 2\alpha)$, then $\dfrac{\tan(\theta + \alpha)}{\tan\alpha}$ is equal to

a. $\dfrac{1+n}{1-n}$

b. $\dfrac{1-n}{1+n}$

c. $\dfrac{n-1}{n+1}$

d. None of these

One or More than One Options Correct Type

Direction *(Q. Nos. 16-18) This section contains 3 multiple choice questions. Each question has four choices (a), (b), (c) and (d), out of which ONE or MORE THAN ONE are correct.*

16 If $\cos\alpha = \dfrac{3}{5}$ and $\cos\beta = \dfrac{5}{13}$, then

a. $\cos(\alpha + \beta) = -\dfrac{33}{65}$

b. $\sin(\alpha + \beta) = \dfrac{56}{65}$

c. $\cos(\alpha - \beta) = \dfrac{63}{65}$

d. None of these

Working Space

17 If $\sin\alpha = \sin\beta$ and $\cos\alpha = \cos\beta$, then

a. $\sin\left(\frac{\alpha-\beta}{2}\right)=0$

b. $\sin\left(\frac{\alpha+\beta}{2}\right)=0$

c. $\cos\left(\frac{\alpha-\beta}{2}\right)=1$

d. $\cos\left(\frac{\alpha+\beta}{2}\right)=0$

18 If $a\sin^2 x + b\cos^2 x = c$, $b\sin^2 y + a\cos^2 y = d$ and $a\tan x = b\tan y$, then

a. $\tan^2 x = \frac{c-b}{a-c}$

b. $(d-a)=\tan^2 y\cdot(b-d)$

c. $\frac{a^2}{b^2}=\frac{(a-d)(c-a)}{(b-c)(d-b)}$

d. None of the above

One Integer Value Correct Type

Direction *(Q. Nos. 19-21) This section contains 3 questions. When worked out will result in an integer from 0 to 9 (both inclusive).*

19 If $y=(1+\tan A)(1-\tan B)$ and $A-B=\frac{\pi}{4}$, then the value of $(y+1)^{y-1}$ is

20 If $\cos\alpha+\cos\beta+\cos\gamma+\cos(\alpha+\beta+\gamma)=\lambda\cos\left(\frac{\alpha+\beta}{2}\right)\cos\left(\frac{\beta+\gamma}{2}\right)\cos\left(\frac{\gamma+\alpha}{2}\right)$, then the value of λ is

21 In quadrilateral $ABCD$, if $\sin\left(\frac{A+B}{2}\right)\cos\left(\frac{A-B}{2}\right)+\sin\left(\frac{C+D}{2}\right)\cos\left(\frac{C-D}{2}\right)=2$, then $\Sigma\cos\frac{A}{2}\cos\frac{B}{2}$ is

DPP-10 Trigonometric Ratios using Half-Angle Formulae

Fill in the Blanks

Working Space

***Direction** (Q. Nos. 1-3) This section contains 3 sentences. Complete the given sentences filling the blanks with suitable words, informations, data, etc.*

1 If $\frac{1-\tan^m \theta}{1+\tan^2 \theta} = \cos 2\theta$, then $m = \ldots.$.

2 If $\frac{1+\lambda \cos 2A}{2} = \sin^2 A$, then $\lambda = \ldots$.

3 If $\tan \frac{x}{2} = \frac{\lambda}{\mu}$, then the value of $\lambda \sin x + \mu \cos x$ is .... .

True/False Statements

***Direction** (Q. Nos. 4-6) This section contains 3 questions. Each question has statement which is TRUE or FALSE.*

4 $\tan \theta \sin\left(\frac{\pi}{2}+\theta\right) \cos\left(\frac{\pi}{2}-\theta\right) - k \cos^2 \theta = 1$ is possible only, when $k = -1$.

5 The value of $\cos \frac{\pi}{7} \cos \frac{5\pi}{7} \cos \frac{3\pi}{7}$ is $-\frac{1}{8}$.

6 If $\frac{\pi}{9} < \theta < \frac{\pi}{2}$, then the value of $\sqrt{1-\sin 2\theta} = \cos \theta - \sin \theta$.

Subjective Questions

***Direction** (Q. Nos. 7-10) These questions are subjective in nature, need to be solved completely on notebook.*

7 If $\pi < \theta < \frac{3\pi}{2}$, then find the value of $\sqrt{\frac{1-\cos 2\theta}{1+\cos 2\theta}}$.

8 If $\tan x = -\frac{4}{3}$, x lies in II quadrant, then find the value of $\sin \frac{x}{2}$.

9 Prove that $\sin^4 \frac{\pi}{8} + \sin^4 \frac{3\pi}{8} + \sin^4 \frac{5\pi}{8} + \sin^4 \frac{7\pi}{8} = \frac{3}{2}$.

10 If $A = 2\sin^2 \theta - \cos 2\theta$ and $A \in [\alpha, \beta]$, then find the values of α and β.

Only One Option Correct Type

Working Space

Direction *(Q. Nos. 11-21) This section contains 11 multiple choice questions. Each question has four choices (a), (b), (c) and (d), out of which ONLY ONE is correct.*

11 The value of $\sqrt{3}$ cosec 20° − sec 20° is

a. 1 b. 2 c. 3 d. 4

12 The value of $\left(\cot\frac{x}{2}-\tan\frac{x}{2}\right)^2(1-2\tan x\cot 2x)$ is

a. 1 b. 2 c. 3 d. 4

13 If $\cos 7^\circ\cos 14^\circ\cos 28^\circ\cos 56^\circ=\frac{\sin 68^\circ}{k\cos 83^\circ}$, then k is equal to

a. 4 b. 8
c. 16 d. None of these

14 If $A=125^\circ$, then $2\cos\frac{A}{2}$ is equal to

a. $-\sqrt{1+\sin A}-\sqrt{1-\sin A}$ b. $-\sqrt{1+\sin A}+\sqrt{1-\sin A}$
c. $\sqrt{1+\sin A}-\sqrt{1-\sin A}$ d. $\sqrt{1+\sin A}+\sqrt{1-\sin A}$

15 If $\sec 2\theta=p+\tan 2\theta$, then the value of $\sin^2\theta$ is equal to

a. $\frac{1}{2}\left(\frac{p-1}{p+1}\right)^2$ b. $\left(\frac{p-1}{p+1}\right)^2$
c. $\frac{1}{2}\left(\frac{p+1}{p-1}\right)^2$ d. None of these

16 If $\sin x+\cos x=\frac{1}{5}$, then $\tan 2x$ is equal to **[AIEEE 2006]**

a. $\frac{25}{17}$ b. $\frac{7}{25}$ c. $\frac{25}{7}$ d. $\frac{24}{7}$

17 Let $0<x<\frac{\pi}{4}$, then $\sec 2x-\tan 2x$ is equal to **[IIT JEE 1994]**

a. $\tan\left(x-\frac{\pi}{4}\right)$ b. $\tan\left(\frac{\pi}{4}-x\right)$ c. $\tan\left(x+\frac{\pi}{4}\right)$ d. $\tan^2\left(x+\frac{\pi}{4}\right)$

18 If $\cos(\theta-\alpha)$, $\cos\theta$, $\cos(\theta+\alpha)$ are in HP, then $\cos\theta\cdot\sec\frac{\alpha}{2}$ is equal to **[IIT JEE 1997]**

a. $\pm\sqrt{2}$ b. $\pm\sqrt{3}$
c. $\pm\frac{1}{\sqrt{2}}$ d. None of these

19 If $F(x)=\cos^2 x\sec^2 x$, then

a. $F(x)<1$ b. $F(x)=1$
c. $1<F(x)<2$ d. $F(x)\geq 2$

20 If $\sin\alpha=\frac{4}{5}$ and $450^\circ<\alpha<540^\circ$, then $\sin\frac{\alpha}{2}$ is equal to

a. $-\frac{2}{\sqrt{5}}$ b. $\frac{1}{5\sqrt{2}}$ c. $\frac{1}{3\sqrt{5}}$ d. $\frac{1}{5\sqrt{3}}$

21 If $0<\alpha<\frac{\pi}{6}$ and $\sin\alpha+\cos\alpha=\frac{\sqrt{7}}{2}$, $\tan\frac{\alpha}{2}=\frac{\sqrt{k}-2}{3}$, then the value of k is

a. 5 b. 7 c. 11 d. None of these

One or More than One Options Correct Type

Working Space

Direction (Q. Nos. 22-24) *This section contains 3 multiple choice questions. Each question has four choices (a), (b), (c) and (d), out of which ONE or MORE THAN ONE are correct.*

22 If $A + B = \frac{\pi}{3}$ and $\cos A + \cos B = 1$, then

a. $\cos(A - B) = \frac{1}{3}$

b. $(\cos A - \cos B) = \sqrt{\frac{2}{3}}$

c. $\cos(A - B) = -\frac{1}{3}$

d. $(\cos A - \cos B) = \frac{1}{2\sqrt{3}}$

23 If $\tan x = \frac{2b}{a - c}$ $(a \neq c)$,

$y = a\cos^2 x + 2b \sin x \cos x + c \sin^2 x$ and

$z = a \sin^2 x - 2b \sin x \cos x + c \cos^2 x$, then

a. $y = z$

b. $y + z = a + c$

c. $y - z = a - c$

d. $y - z = (a - c)^2 + 4b^2$

24 In a ΔABC, $A > B$. If the measures of $\angle A$ and $\angle B$ satisfy the equation $3 \sin x - 4 \sin^3 x - k = 0$, $0 < k < 1$, then

a. $A + B = \frac{\pi}{3}$

b. $A + B = \frac{\pi}{6}$

c. $\angle C = \frac{2\pi}{3}$

d. $\angle C = \frac{5\pi}{6}$

Comprehension Type

Direction (Q. Nos. 25-28) *This section contains a paragraph, which describing theory, experiments, data, etc. Four questions related to the paragraph have been given. Each question has only one correct answer among the four given options (a), (b), (c) and (d).*

Passage

If $\cos \alpha \cdot \cos 2\alpha \cdot \cos 2^2\alpha \cdot \cos 2^3\alpha \ldots \cos 2^{n-1}\alpha = \frac{\sin 2^n \alpha}{2^n \sin \alpha}$. If $\alpha \neq n\pi$, where $n \in I$

On the basis of above information, answer the following questions :

25 The value of $\cos \frac{2\pi}{7} \cos \frac{4\pi}{7} \cos \frac{6\pi}{7}$ is

a. $-\frac{1}{2}$ b. $\frac{1}{2}$ c. $\frac{1}{4}$ d. $\frac{1}{8}$

26 If $\alpha = \frac{\pi}{13}$, then the value of $\prod_{r=1}^{6} \cos r\alpha$ is

a. $\frac{1}{64}$ b. $-\frac{1}{64}$ c. $\frac{1}{32}$ d. $-\frac{1}{8}$

27 The value of $\sin \frac{\pi}{14} \sin \frac{3\pi}{14} \sin \frac{5\pi}{14} \sin \frac{7\pi}{14} \sin \frac{9\pi}{14} \sin \frac{11\pi}{14} \sin \frac{13\pi}{14}$ is **[IIT JEE 1991]**

a. 1

b. $\frac{1}{8}$

c. $\frac{1}{32}$

d. $\frac{1}{64}$

28 The value of $128\sqrt{3} \sin \frac{\pi}{48} \cos \frac{\pi}{48} \cos \frac{\pi}{24} \cos \frac{\pi}{12} \cos \frac{\pi}{6}$ is

a. 8

b. 10

c. 6

d. None of these

DPP-11 Trigonometric Ratios of Multiple Angles

Subjective Questions

Direction (Q. Nos. 1-3) These questions are subjective in nature, need to be solved completely on notebook.

Working Space

1 If $(\cos A + \cos B)^2 + (\sin A - \sin B)^2 = k^2 \cos^2\left(\dfrac{A+B}{2}\right)$, then find the value of k.

2 If $\tan(45^\circ + A) + \tan(45^\circ - A) = \mu \sec 2A$, then prove that $\mu = 2$.

3 If $\sin 2\theta = k$, then prove that $\dfrac{\tan^3\theta}{1+\tan^2\theta} + \dfrac{\cot^3\theta}{1+\cot^2\theta} = \dfrac{2-k^2}{k}$.

Fill in the Blanks

Direction (Q. Nos. 4-6) This section contains 3 sentences. Complete the given sentences filling the blanks with suitable words, information, data, etc.

4 If $(\sin A + \cos A)^2 = 1 - m \sin nA$, then $n - m = \ldots\ldots$

5 If $\tan 3A = \dfrac{3\tan A + k\tan^3 A}{1 - 3\tan^2 A}$, then $k = \ldots\ldots$

6 If $\cot 75^\circ = 2 - \sqrt{\lambda}$, then $\lambda = \ldots\ldots$

True/False Statements

Direction (Q. Nos. 7-9) This section contains 3 questions. Each question has statement which is TRUE or FALSE.

7 $\sin 22\dfrac{1^\circ}{2} = \dfrac{1}{2}\sqrt{2-\sqrt{2}}$

8 $\tan A + 2\cot 2A = \cot A$

9 $\tan A \pm \tan B = \dfrac{\sin(A \pm B)}{\cos A \cos B}$, when $A = n\pi + \dfrac{\pi}{2}$

Only One Option Correct Type

Direction (Q. Nos. 10-19) This section contains 10 multiple choice questions. Each question has four choices (a), (b), (c) and (d), out of which ONLY ONE is correct.

10 If $n \in N$, then the value of $(1+\sec 2\theta)(1+\sec 4\theta)(1+\sec 8\theta)\ldots(1+\sec 2^n\theta)$ is

a. $\tan 2^n\theta \cot\theta$ b. $\cot 2^n\theta \tan\theta$

c. $\tan 2^n\theta \tan\theta$ d. $\tan 2n\theta \cot\theta$

Working Space

11 $\left(1+\cos\frac{\pi}{8}\right)\left(1+\cos\frac{3\pi}{8}\right)\left(1+\cos\frac{7\pi}{8}\right)\left(1+\cos\frac{5\pi}{8}\right)$ is equal to **[IIT JEE 1984]**

a. $\frac{1}{2}$ b. $\cos\frac{\pi}{8}$
c. $\frac{1}{8}$ d. $\frac{1+\sqrt{2}}{2\sqrt{2}}$

12 If $\tan(\theta+50°)\tan\theta\tan(\theta-50°)=\tan(\theta+100°)$, then the smallest positive value of θ is

a. 45° b. 30°
c. 60° d. None of these

13 If $\tan A+2\tan 2A+4\tan 4A+8\cot 8A=k\cot A$, then the value of k is **[IIT JEE 1988]**

a. 2 b. 3
c. 1 d. None of these

14 If $\cos^3\theta+\cos^3(120°+\theta)+\cos^3(240°+\theta)=p\cos q\theta$, then the value of $8p-q$ is

a. 2 b. 3
c. 4 d. None of these

15 If $\tan\theta+\tan\left(\frac{\pi}{3}+\theta\right)-\tan\left(\frac{\pi}{3}-\theta\right)=k\tan 3\theta$, then k is equal to

a. 2 b. 3
c. 6 d. None of these

16 If $m^2\cos\frac{2\pi}{15}\cos\frac{4\pi}{15}\cos\frac{8\pi}{15}\cos\frac{14\pi}{15}=n^2$, then $\frac{m^2-n^2}{n^2}$ is equal to

a. 10 b. 15
c. 20 d. None of these

17 If $\cos\theta=\frac{1}{2}\left(a+\frac{1}{a}\right)$ and $\cos 3\theta=\lambda\left(a^3+\frac{1}{a^3}\right)$, then λ is equal to

a. $\frac{1}{4}$ b. $\frac{1}{2}$
c. 1 d. None of these

18 If $(2^n+1)\theta=\pi$, then $2^n\cos\theta\cos 2\theta\cos 2^2\theta\ldots\cos 2^{n-1}\theta$ is equal to

a. 1 b. −1
c. $\frac{1}{2}$ d. None of these

19 The value of $\frac{\cos 5\theta+\cos 4\theta}{-1+2\cos 3\theta}$ is equal to

a. $\cos 2\theta+\cos\theta$ b. $\cos 2\theta$
c. $2\cos 2\theta-\cos\theta$ d. None of these

One or More than One Options Correct Type

Direction *(Q. Nos. 20-22) This section contains 3 multiple choice questions. Each question has four choices (a), (b), (c) and (d), out of which ONE or MORE THAN ONE are correct.*

20 If $\frac{\sec 8\theta-1}{\sec 4\theta-1}=\frac{\tan p\theta}{\tan q\theta}$, then

a. $p:q=4:1$ b. $p:q=2:3$
c. $2p=8q$ d. None of these

Working Space

21 If $\frac{\tan 3\theta}{\tan \theta} = k$, then the value of $\frac{\sin 3\theta}{\sin \theta}$ is

a. $\frac{2k}{k-1}$ b. $k < \frac{1}{3}$

c. $k > 3$ d. None of these

22 For $0 < \theta < \frac{\pi}{2}$, the solution(s) of

$$\sum_{m=1}^{6} \operatorname{cosec}\left(\theta + \frac{(m-1)\pi}{4}\right) \operatorname{cosec}\left(\theta + \frac{m\pi}{4}\right) = 4\sqrt{2} \text{ is/are}$$

[IIT JEE 2009]

a. $\frac{\pi}{4}$ b. $\frac{\pi}{6}$

c. $\frac{\pi}{12}$ d. $\frac{5\pi}{12}$

One Integer Value Correct Type

Direction (Q. Nos. 23 and 24) *This section contains 2 questions. When worked out will result in an integer from 0 to 9 (both inclusive).*

23 The maximum value of the expression **[IIT JEE 2010]**

$$F(\theta) = \frac{1}{\sin^2 \theta + 3 \sin \theta \cos \theta + 5 \cos^2 \theta} \text{ is}$$

24 If $\frac{\cos \theta_1}{\cos \theta_2} + \frac{\sin \theta_1}{\sin \theta_2} = \frac{\cos \theta_0}{\cos \theta_2} + \frac{\sin \theta_0}{\sin \theta_2} = 1$, where θ_1 and θ_0 do not differ by an odd multiple of π, then the value of $\left|\frac{\cos \theta_1 \cos \theta_0}{\cos^2 \theta_2} + \frac{\sin \theta_1 \sin \theta_0}{\sin^2 \theta_2}\right|$ is

DPP-12 T-Ratios of Sub-Multiple Angles

True/False Statements

Direction (Q. Nos. 1-3) *This sectio contains 3 questions. Each question has statement which is TRUE or FALSE.*

1 $\tan(60° - A) \tan A \tan(60° + A) = \tan 3A$.

2 $\sin\theta = \frac{-1-\sqrt{5}}{4}$

Working Space

3 If $\cos 36° = \frac{1}{4}(\sqrt{5}+1)$, then the value of $\sin 54° = \frac{\sqrt{5}+1}{4}$.

Subjective Questions

Direction (Q. Nos. 4-6) *These questions are subjective in nature, need to be solved completely on notebook.*

4 Prove that $\sin 18° = \frac{\sqrt{5}-1}{4}$.

5 Prove that $\frac{1}{\tan 3A - \tan A} - \frac{1}{\cot 3A - \cot A} = \cot 2A$.

6 Prove that $(\sin 55° - \sin 19°) + (\sin 53° - \sin 17°) = \cos 1°$.

Only One Option Correct Type

Direction (Q. Nos. 7-15) *This section contains 9 multiple choice questions. Each question has four choices (a), (b), (c) and (d), out of which ONLY ONE is correct.*

7 $\sin 12° \sin 48° \sin 54°$ is equal to **[IIT JEE 1982]**

a. $\frac{1}{8}$ b. $-\frac{1}{8}$

c. $\frac{1}{4}$ d. None of these

8 If $\sin^2 24° - \sin^2 6° = k \sin 18°$, then the value of k is

a. 1 b. $\frac{1}{2}$

c. $\frac{1}{3}$ d. 2

9 If $\cos A + \cos B + \cos C = 0$ and $\cos 3A + \cos 3B + \cos 3C = \lambda \cos A \cos B \cos C$, then λ is

a. 8 b. 10

c. 12 d. None of these

Working Space

10 The value of tan 6° tan 42° tan 66° tan 78° is

a. 0 b. 1
c. 2 d. None of these

11 If $\sin^3 A + \sin^3(120° + A) + \sin^3(240° + A) = -\frac{p}{q}\sin 3A$, then $p : q$ is

a. 2 : 3 b. 2 : 5
c. 3 : 4 d. 3 : 5

12 tan 9° − tan 27° − tan 63° + tan 81° is equal to

a. 1 b. 2
c. 3 d. 4

13 Which of the number(s) is/are rational? **[IIT JEE 1998]**

a. sin 15° b. cos 15°
c. sin 15° cos 15° d. sin 15° cos 75°

14 The value of $\cot\alpha + \cot(60° + \alpha) - \cot(60° - \alpha)$ is

a. $\sin 3\alpha$ b. $\tan 3\alpha$
c. $3\cot 3\alpha$ d. None of these

15 The numerical value of

$(\tan 20° + \tan 40° + \sqrt{3}\tan 20° \tan 40°)^2$ is

a. $\sqrt{3}$ b. 3
c. −3 d. None of these

One or More than One Options Correct Type

Direction *(Q. Nos. 16-19) This section contains 4 multiple choice questions. Each question has four choices (a), (b), (c) and (d), out of which ONE or MORE THAN ONE are correct.*

16 If $\cot 7\frac{1}{2}° = \sqrt{n_1} + \sqrt{n_2} + \sqrt{n_3} + \sqrt{n_4}$, then

a. $n_1 + n_4 = 8$ b. $2n_4 - n_2 - n_3 = 5$
c. $n_1 + n_2 + n_3 = 9$ d. $n_1 + n_2 + n_3 + n_4 = 15$

17 The value of 4 sin 27° is/are

a. $\sqrt{8 - 2\sqrt{10 - 2\sqrt{5}}}$ b. $\sqrt{8 + 2\sqrt{10 - 2\sqrt{5}}}$
c. $\sqrt{5 + \sqrt{5}} - \sqrt{3 - \sqrt{5}}$ d. None of these

18 The value of $\frac{\cos 9° + \sin 9°}{\cos 9° - \sin 9°}$ is/are

a. tan 54° b. cot 36°
c. − tan 126° d. None of these

19 If $x = \sin 130° \cos 80°$, $y = \sin 80° \cos 130°$, $z = 1 + xy$, then which of the following is/are not true?

a. $x > 0, y > 0, z > 0$ b. $x > 0, y < 0, 0 < z < 1$
c. $x > 0, y < 0, z > 1$ d. $x < 0, y < 0, 0 < z < 1$

Working Space

Matching List Type

Direction *(Q. No. 20) Choices for the correct combination of elements from Column I and Column II are given as options (a), (b), (c) and (d), out of which one is correct.*

20 Match the conditions/expressions in Column I with statement in Column II.

$\dfrac{\sin 3\alpha}{\cos 2\alpha}$ is **[IIT JEE 1992]**

	Column I		Column II
i.	Positive	p.	$\left(\frac{13\pi}{48}, \frac{14\pi}{48}\right)$
ii.	Negative	q.	$\left(\frac{14\pi}{48}, \frac{18\pi}{48}\right)$
		r.	$\left(\frac{18\pi}{48}, \frac{23\pi}{48}\right)$
		s.	$\left(0, \frac{\pi}{2}\right)$

Codes

	i	ii
a.	r	p
b.	p	q
c.	r	s
d.	s	r

One Integer Value Correct Type

Direction *(Q. Nos. 21 and 22) This section contains 2 questions. When worked out will result in an integer from 0 to 9 (both inclusive).*

21 If $\sin \frac{\pi}{18} \sin \frac{5\pi}{18} \sin \frac{7\pi}{18} = \frac{1}{\lambda}$, then the value of λ is **[IIT JEE 1993]**

22 Suppose $\sin^3 x \sin 3x = \sum_{m=0}^{n} C_m \cos nx$ is an identity in x, where $C_0, C_1, \ldots, C_n$ are constants and $C_n \neq 0$. Then, the value of n is **[IIT JEE 1983]**

DPP-13 Sum of Three or More Angles

Subjective Questions

Working Space

Direction (Q. Nos. 1-3) *These questions are subjective in nature, need to be solved completely on notebook.*

1 If $A + B + C = 180°$, then prove that
$\sin 2A + \sin 2B + \sin 2C = 4 \sin A \sin B \sin C$.

2 If $A + B + C = 180°$, then prove that $\tan^2 \frac{\theta}{2} = \tan \frac{B}{2} \tan \frac{C}{2}$,
when $\cos \theta (\sin B + \sin C) = \sin A$.

3 If A, B, C are angles of a ΔABC, then prove that
$\sin 2A + \sin 2B - \sin 2C = 4 \cos A \cos B \sin C$.

Only One Option Correct Type

Direction (Q. Nos. 4-14) *This section contains 11 multiple choice questions. Each question has four choices (a), (b), (c) and (d), out of which ONLY ONE is correct.*

4 If in a ΔABC, $\frac{\sin A + \sin B + \sin C}{\sin A + \sin B - \sin C} = 2\lambda \cot \frac{A}{2} \cot \frac{B}{2}$, then the value of λ is

a. -1 b. $\frac{1}{2}$
c. 1 d. -2

5 If $A + B + C = 180°$, then the value of $\frac{\cos A}{\sin B \sin C} + \frac{\cos B}{\sin C \sin A} + \frac{\cos C}{\sin A \sin B}$ is

a. 1 b. 2
c. 3 d. 4

6 In ΔABC, $\frac{1 - \cos A + \cos B + \cos C}{1 - \cos C + \cos A + \cos B}$ is equal to

a. $\tan \frac{A}{2} \cot \frac{C}{2}$ b. $\tan \frac{A}{2} \cot \frac{B}{2}$
c. $\tan \frac{A}{2} \tan \frac{C}{2}$ d. $\tan \frac{B}{2} \cot \frac{C}{2}$

7 In a ΔABC, if $\tan\left(\frac{B + C - A}{4}\right) \tan\left(\frac{C + A - B}{4}\right) \tan\left(\frac{A + B - C}{4}\right) = 1$, then the value of $\cos A + \cos B + \cos C$ is

a. 1 b. -1
c. 2 d. None of these

8 If $A + B + C = 180°$, then $\sin \frac{A}{2} \sin \frac{B}{2} \sin \frac{C}{2}$ is

a. $\geq \frac{1}{8}$ b. $\geq \frac{1}{4}$ c. $\leq \frac{1}{8}$ d. $\leq \frac{1}{4}$

Working Space

9 If in a ΔABC, $\cot \frac{A}{2} + \cot \frac{B}{2} + \cot \frac{C}{2} = \lambda \cot \frac{A}{2} \cot \frac{B}{2} \cot \frac{C}{2}$, then the value of λ is **[IIT JEE 2000]**

a. 2
b. −2
c. 1
d. None of these

10 If $\alpha = \frac{2\pi}{7}$, then $\tan \alpha \tan 2\alpha + \tan 2\alpha \tan 4\alpha + \tan 4\alpha \tan \alpha$ is equal to

a. 3
b. 5
c. 7
d. − 7

11 In ΔABC, if $\tan A + \tan B + \tan C = 3\sqrt{3}$, then ΔABC is a/an

a. obtuse angled triangle
b. right angled triangle
c. equilateral triangle
d. None of these

12 The value of $\cos \frac{2\pi}{7} + \cos \frac{4\pi}{7} + \cos \frac{6\pi}{7}$ is equal to

a. 1
b. −1
c. $\frac{1}{2}$
d. $-\frac{1}{2}$

13 If $A + B + C = \frac{3\pi}{2}$, then $\cos 2A + \cos 2B + \cos 2C$ is equal to

a. $1 - 4 \cos A \cos B \cos C$
b. $4 \sin A \sin B \sin C$
c. $1 + 2 \cos A \cos B \cos C$
d. $1 - 4 \sin A \sin B \sin C$

14 If $\alpha + \beta + \gamma = 2\pi$, then **[IIT JEE 1979]**

a. $\tan \frac{\alpha}{2} + \tan \frac{\beta}{2} + \tan \frac{\gamma}{2} = \tan \frac{\alpha}{2} \tan \frac{\beta}{2} \tan \frac{\gamma}{2}$

b. $\tan \frac{\alpha}{2} \tan \frac{\beta}{2} + \tan \frac{\beta}{2} \tan \frac{\gamma}{2} + \tan \frac{\gamma}{2} \tan \frac{\alpha}{2} = 1$

c. $\tan \frac{\alpha}{2} + \tan \frac{\beta}{2} + \tan \frac{\gamma}{2} = -\tan \frac{\alpha}{2} \tan \frac{\beta}{2} \tan \frac{\gamma}{2}$

d. None of the above

One or More than One Options Correct Type

Direction *(Q. Nos. 15-17) This section contains 3 multiple choice questions. Each question has four choices (a), (b), (c) and (d), out of which ONE or MORE THAN ONE are correct.*

15 If in a ΔABC, $\cos 3A + \cos 3B + \cos 3C = 1$, then one angle must not be exactly equal to

a. $\frac{\pi}{3}$
b. $\frac{2\pi}{3}$
c. π
d. $\frac{4\pi}{3}$

16 ABC is a triangle such that

$\sin (2A + B) = \sin (C - A) = -\sin (B + 2C) = \frac{1}{2}$. If A, B, C are in AP, then **[IIT JEE 1990]**

a. $\angle A = 45°$
b. $\angle B = 60°$
c. $\angle C = 75°$
d. None of these

17 In ΔABC, the value of $\cos A + \cos B + \cos C$ is

a. > 1
b. $\leq \frac{3}{2}$
c. $\geq \frac{2}{3}$
d. None of these

Comprehension Type

Working Space

Direction *(Q. Nos. 18-21) This section contains a paragraph, which describing theory, experiments, data, etc. Four questions related to the paragraph have been given. Each question has only one correct answer among the four given options (a), (b), (c) and (d).*

Passage

Let $n \in N$ and $\alpha, \beta \in R$, it is true that

$$\sin\alpha + \sin(\alpha+\beta) + \sin(\alpha+2\beta) + \ldots + \sin[\alpha+(n-1)\beta] = \frac{\sin\left(\frac{n\beta}{2}\right)}{\sin\left(\frac{\beta}{2}\right)} \sin\left[\frac{\alpha+\alpha+(n-1)\beta}{2}\right]$$

and

$$\cos\alpha + \cos(\alpha+\beta) + \cos(\alpha+2\beta) + \ldots + \cos[\alpha+(n-1)\beta] = \frac{\sin\left(\frac{n\beta}{2}\right)}{\sin\left(\frac{\beta}{2}\right)} \cos\left[\frac{\alpha+\alpha+(n-1)\beta}{2}\right]$$

On the basis of above information, answer the following questions :

18 The value of $\sin\frac{\pi}{7} + \sin\frac{2\pi}{7} + \sin\frac{3\pi}{7}$ is

a. $\cot\frac{\pi}{14}$

b. $\frac{1}{2}\cot\frac{\pi}{14}$

c. $\tan\frac{\pi}{14}$

d. $\frac{1}{2}\tan\frac{\pi}{14}$

19 $\sin\frac{\pi}{n} + \sin\frac{3\pi}{n} + \sin\frac{5\pi}{n} + \ldots$ upto n terms is equal to

a. 1

b. 0

c. $\frac{n}{2}$

d. None of these

20 The value of $\cos\frac{\pi}{11} + \cos\frac{3\pi}{11} + \cos\frac{5\pi}{11} + \cos\frac{7\pi}{11} + \cos\frac{9\pi}{11}$ is

a. 0

b. 1

c. $\frac{1}{2}$

d. None of these

21 $\sum_{r=0}^{n} \sin^2\frac{r\pi}{n}$ is equal to

a. $\frac{n+1}{2}$

b. $\frac{n-1}{2}$

c. $\frac{n}{2}$

d. None of these

SET - 1

Revisal Problems for JEE Main

Only One Option Correct Type

Working Space

Direction *(Q. Nos. 1-7) This section contains 7 multiple choice questions. Each question has four choices (a), (b), (c) and (d), out of which ONLY ONE option is correct.*

1 For what and only what value of $\alpha \in (0, \pi)$ is the inequality $\sin \alpha \cos^3 \alpha > \sin^3 \alpha \cos \alpha$ valid?

a. $\alpha \in \left(0, \frac{\pi}{4}\right)$
b. $\alpha \in \left(0, \frac{\pi}{2}\right)$
c. $\alpha \in \left(\frac{\pi}{4}, \frac{\pi}{2}\right)$
d. None of these

2 The maximum value of $\sin\left(x + \frac{\pi}{6}\right) + \cos\left(x + \frac{\pi}{6}\right)$ in the interval $\left(0, \frac{\pi}{2}\right)$ is attained at

a. $\frac{\pi}{12}$
b. $\frac{\pi}{6}$
c. $\frac{\pi}{3}$
d. $\frac{\pi}{2}$

3 If n is a fixed positive integer such that $\sin \frac{\pi}{2n} + \cos \frac{\pi}{2n} = \frac{\sqrt{n}}{2}$, then

a. $n = 4$
b. $n = 5$
c. $n = 6$
d. None of these

4 The ratio of greatest value of $2 - \cos x + \sin^2 x$ to its least value, is

a. 13 : 4
b. 1 : 4
c. 9 : 4
d. None of these

5 If $\sin x$ and $\cos x$ are roots of $x^2 - px + q = 0$, then the value of $\sin^3 x + \cos^3 x$ is

a. $p^3 - 3pq$
b. $p^2 - 3pq$
c. $q^2 - 3pq$
d. None of these

6 If $\tan x \tan y = a$ and $x + y = \frac{\pi}{6}$, then $\tan x$ and $\tan y$ satisfy the equation

a. $x^2 - \sqrt{3}(1 - a)x + a = 0$
b. $\sqrt{3}x^2 - (1 - a)x + a\sqrt{3} = 0$
c. $x^2 - \sqrt{3}(1 + a)x - a = 0$
d. None of the above

7 If $\cot(\theta - \alpha)$, $3\cot \theta$, $\cot(\theta + \alpha)$ are in AP and θ is not an integral multiple of $\frac{\pi}{2}$, then $\frac{304\sin^2\theta}{3\sin^2\alpha}$ is equal to

a. 150
b. 152
c. 154
d. None of these

SET - 2

Revisal Problems for JEE Advanced

Only One Option Correct Type

Working Space

Direction *(Q. Nos. 1-7) This section contains 7 multiple choice questions. Each question has four choices (a), (b), (c) and (d), out of which ONLY ONE option is correct.*

1 If $a\cos^2 3\alpha + b\cos^4 \alpha = 16\cos^6 \alpha + 9\cos^2 \alpha$ is an identity, then (a, b) is

a. (1, 24)
b. (2, 12)
c. (24, 1)
d. None of these

2 If $0 < \theta < \pi$, then $\sqrt{2+\sqrt{2+\sqrt{2+\cdots+\sqrt{2(1+\cos\theta)}}}}(\ldots n \text{ times}) = p$ and $q = \lim_{n\to\infty} p$, then $\dfrac{3+2q}{(2+3q)^2}$ is

a. $\frac{5}{64}$
b. $\frac{7}{64}$
c. $\frac{9}{64}$
d. None of these

3 If $\cot\frac{\alpha}{2}, \cot\frac{\beta}{2}$ are roots of the equation $15x^2 - 26x + 8 = 0$, then $\cos(\alpha+\beta)$ is

a. $\frac{-627}{725}$
b. $\frac{627}{725}$
c. -1
d. None of these

4 If $\tan A, \tan B$ and $\tan C$ are the roots of equation $x^3 - \alpha x^2 - \beta = 0$ and $(1+\tan^2 A)(1+\tan^2 B)(1+\tan^2 C) = k - 1$, then k is

a. $(\alpha-\beta)^2$
b. $2+(\alpha-\beta)^2$
c. $1+(\alpha-\beta)^2$
d. None of the above

5 If $\cos\theta = \dfrac{a\cos\phi + b}{a + b\cos\phi}$ and $\tan\frac{\theta}{2} = \sqrt{\frac{m}{n}}\tan\frac{\phi}{2}$, then $a : b$ is

a. $m^2+n^2 : n^2-m^2$
b. $m^2+n^2 : m^2-n^2$
c. $m^2-n^2 : m^2+n^2$
d. None of the above

6 If $\lim_{x\to 3}\dfrac{3x^2-9F(x)}{x-3} = \lambda$, where $F'(3)=1, F(3)=3$ and $\operatorname{cosec} 10° - \sqrt{3}\sec 10° = \mu$, then $\lambda - \mu$ is

a. 5
b. 10
c. 7
d. None of these

7 If A, B, C are acute positive angles and $\dfrac{(\sin A+\sin B)(\sin B+\sin C)(\sin C+\sin A)}{\sin A\sin B\sin C} \ge p$, then the value of p is

a. 3
b. 5
c. 8
d. None of these

Working Space

One or More than One Options Correct Type

Direction *(Q. Nos. 8-16) This section contains 9 multiple choice questions. Each question has four choices (a), (b), (c) and (d), out of which ONE or MORE THAN ONE are correct.*

8 Let $F_n(\theta) = \tan\frac{\theta}{2}(1+\sec\theta)(1+\sec 2\theta)(1+\sec 4\theta)\cdots(1+\sec 2^n\theta)$, then

a. $F_3\left(\frac{\pi}{32}\right) = 1$ b. $F_4\left(\frac{\pi}{64}\right) = 1$ c. $F_5\left(\frac{\pi}{128}\right) = 1$ d. $F_6\left(\frac{\pi}{256}\right) = 1$

9 If $\cos^2\theta + \alpha, \sin^2\theta + \alpha$ are roots of the equation $x^2 + b(2x+1) = 0$ and $\cos^2\theta + \beta, \sin^2\theta + \beta$ are roots of the equation $x^2 + 4x + 2 = 0$, then b is equal to

a. 1 b. −1 c. 2 d. −2

10 If $\cos x = \tan y$, $\cos y = \tan z$ and $\cos z = \tan x$, then $\sin x$ is equal to

a. $2\sin 18°$ b. $\frac{\sqrt{5}-1}{2}$
c. $2\cos 72°$ d. None of these

11 Let n be an odd integer. If $\sin n\theta = \sum_{r=0}^{n} b_r \sin^r\theta$, $\forall \theta \in R$, then

a. $b_0 = 0$ b. $b_1 = n$
c. $2b_1 - b_0 = 2n$ d. None of these

12 If $a = \sin\frac{\pi}{18}\cdot\sin\frac{5\pi}{18}\cdot\sin\frac{7\pi}{18}$ and x is the solution of equations $y = 2[x] + 2$ and $y = 3[x-2]$, where [·] denotes the greatest integer function, then a is/are

a. $[x]$ b. $\frac{1}{[x]}$ c. $2[x]$ d. $\frac{1}{8}$

13 If $\sin^4\alpha\cos^2\alpha = \sum_{k=0}^{3} C_k \cos 2k\alpha$, then

a. $C_1 + C_3 = 0$ b. $C_0 + C_2 = 0$
c. $C_0 + C_1 + C_2 + C_3 = 0$ d. None of these

14 If the mapping $F(x) = ax + b$, $a < 0$ maps $[-1, 1]$ onto $[0, 2]$ and $A = \cos^2\theta + \sin^4\theta$, $\forall\theta$, then which of the options is/are wrong?

a. $F\left(\frac{1}{4}\right) \le A \le F(0)$ b. $F(0) \le A \le F(-2)$
c. $F\left(\frac{1}{3}\right) \le A \le F(0)$ d. $F(-1) < A \le F(-2)$

15 If $3\sin\beta = \sin(2\alpha + \beta)$, then

a. $[\cot\alpha + \cot(\alpha+\beta)][\cot\beta - 3\cot(2\alpha+\beta)] = 6$
b. $\sin\beta = \cos(\alpha+\beta)\sin\alpha$
c. $2\sin\beta = \sin(\alpha+\beta)\cos\alpha$
d. $\tan(\alpha+\beta) = 2\tan\alpha$

16 If $x = \sec\theta - \tan\theta$ and $y = \text{cosec}\,\theta + \cot\theta$, then

a. $x = \frac{y+1}{y-1}$ b. $x = \frac{y-1}{y+1}$
c. $y = \frac{1+x}{1-x}$ d. $xy + x - y + 1 = 0$

One Integer Value Correct Type

Direction *(Q. Nos. 17-26) This section contains 10 questions. When worked out will result in an integer from 0 to 9 (both inclusive).*

17 If $A + B + C = \pi$ and $\frac{\sin 2A + \sin 2B + \sin 2C}{\sin A + \sin B + \sin C} = \lambda\sin\frac{A}{2}\cdot\sin\frac{B}{2}\cdot\sin\frac{C}{2}$, then the value of λ is

Working Space

18 If $\tan\frac{3\pi}{11} + 4\sin\left(\frac{2\pi}{11}\right) = \lambda$, then the value of $[\lambda^2]$ is, (where [·] denotes the greatest integer function)

19 If $4\cos 36° + \cot 7\frac{1}{2}° = \sqrt{n_1} + \sqrt{n_2} + \sqrt{n_4} + \sqrt{n_5} + \sqrt{n_6}$, then the value of $\frac{\sum_{i=1}^{6} n_i}{7}$ is

20 If $\tan^2\left(\frac{\pi}{16}\right) + \tan^2\left(\frac{2\pi}{16}\right) + \tan^2\left(\frac{3\pi}{16}\right) + \cdots + \tan^2\left(\frac{7\pi}{16}\right) = \lambda$ and $x^y + y^x = \lambda$, then the value of $\left(\frac{x+y}{5}\right)$ is

21 If $\cos\alpha$ is the slope of bisector of the angle in first quadrant between X and Y-axes and $\tan\beta$ is an integral solution of the equation $4x^2 - 16x + 15 < 0$ and the value of $\sin(\alpha+\beta) : \sin(\alpha-\beta)$ is $-\lambda$, then λ is

22 The set of values of $\lambda \in R$ such that $\tan^2\theta + \sec\theta = \lambda$ holds for some θ is $[a, \infty)$, then the value of a is

23 If $A + B + C = \pi$ and minimum value of $\frac{\cos\left(\frac{A-B}{2}\right)}{\cos\left(\frac{A+B}{2}\right)} + \frac{\cos\left(\frac{B-C}{2}\right)}{\cos\left(\frac{B+C}{2}\right)} + \frac{\cos\left(\frac{C-A}{2}\right)}{\cos\left(\frac{C+A}{2}\right)}$ is $2k$, then the value of k is

24 If α is the common positive root of equations $x^2 - ax + 12 = 0$, $x^2 - bx + 15 = 0$ and $x^2 - (a+b)x + 36 = 0$ and $\cos x + \cos 2x + \cos 3x = \alpha$, then the value of $\sin x + \sin 2x + \sin 3x$ is

25 If $\tan\theta = \sqrt{\frac{3}{2}}$ and the sum of the infinite series $1 + 2(1-\cos\theta) + 3(1-\cos\theta)^2 + 4(1-\cos\theta)^3 + \ldots \infty$ is $\frac{\lambda}{2}$, then the value of λ is

26 If $\alpha, \beta, \gamma, \delta$ are the smallest positive angles in ascending order of magnitude which have their sines equal to the positive quantity k and the value of $4\sin\frac{\alpha}{2} + 3\sin\frac{\beta}{2} + 2\sin\frac{\gamma}{2} + \sin\frac{\delta}{2} = p\sqrt{1+k}$, then p is

Matching List Type

***Direction** (Q. No. 27) Choices for the correct combination of elements from Column I and Column II are given as options (a), (b), (c) and (d), out of which one is correct.*

27 Match the following columns and select the correct option from the codes given below.

	Column I		Column II
i.	$\sin A \sin B \sin C \le$	p.	$1 + \cos C$
ii.	$\sin^2 A + \sin^2 B + \sin^2 C \le$	q.	$\frac{3\sqrt{3}}{8}$
iii.	$\sin A \sin B \sin C$ is always	r.	$\frac{9}{4}$
iv.	$\sin^2 A + \sin^2 B \le$	s.	positive
		t.	negative

Codes

	i	ii	iii	iv
a.	q	r	s	p
b.	p	q	r	t
c.	q	r	t	p

Working Space

Statement Type

***Direction** (Q. Nos. 28 and 29) This section is based on Statement I and Statement II. Select the correct answer from the codes given below.*

a. Both Statement I and Statement II are correct and Statement II is the correct explanation of Statement I
b. Both Statement I and Statement II are correct but Statement II is not the correct explanation of Statement I
c. Statement I is correct but Statement II is incorrect
d. Statement II is correct but Statement I is incorrect

28 **Statement I** The maximum value of $\prod_{i=1}^{n} \cos\alpha_i$ under the restriction $0 \le \alpha_1, \alpha_2, \alpha_3, \cdots, \alpha_n \le \frac{\pi}{2}$ is $\frac{1}{2^{\pi/2}}$.

Statement II $\prod_{i=1}^{n} \cot\alpha_i = 1$.

29 **Statement I** $\cos^3\alpha + \cos^3\left(\alpha + \frac{2\pi}{3}\right) + \cos^3\left(\alpha + \frac{4\pi}{3}\right) = 3\cos\alpha\cos\left(\alpha + \frac{2\pi}{3}\right)\cos\left(\alpha + \frac{4\pi}{3}\right)$

Statement II If $a + b + c = 0 \Leftrightarrow a^3 + b^3 + c^3 = 3abc$.

Comprehension Type

***Direction** (Q. Nos. 30-33) This section contains a paragraph, which describing theory, experiments, data, etc. Four questions related to the paragraph have been given. Each question has only one correct answer among the four given options (a), (b), (c) and (d).*

Passage

If $\cos\frac{\pi}{7}, \cos\frac{3\pi}{7}, \cos\frac{5\pi}{7}$ are roots of the equation $8x^3 - 4x^2 - 4x + 1 = 0$.

On the basis of above information, answer the following questions :

30 The value of $\sin\frac{\pi}{14}\sin\frac{3\pi}{14}\sin\frac{5\pi}{14}$ is

a. $\frac{1}{4}$
b. $\frac{1}{8}$
c. $\frac{\sqrt{7}}{4}$
d. $\frac{\sqrt{7}}{8}$

31 The value of $\cos\frac{\pi}{14}\cos\frac{3\pi}{14}\cos\frac{5\pi}{14}$ is

a. $\frac{1}{4}$
b. $\frac{1}{8}$
c. $\frac{\sqrt{7}}{4}$
d. $\frac{\sqrt{7}}{8}$

32 The equation whose roots are $\tan^2\frac{\pi}{7}$, $\tan^2\frac{3\pi}{7}$ and $\tan^2\frac{5\pi}{7}$, is

a. $x^3 - 35x^2 + 7x - 21 = 0$
b. $x^3 - 35x^2 + 21x - 7 = 0$
c. $x^3 - 21x^2 + 35x - 7 = 0$
d. $x^3 - 21x^2 + 7x - 35 = 0$

33 The value of $\sum_{r=1}^{3} \tan^2\left(\frac{2r-1}{7}\right) \cdot \sum_{r=1}^{3} \cot^2\left(\frac{2r-1}{7}\right)$ is

a. 15
b. 105
c. 21
d. 147

JEE Main & AIEEE Archive

(Compilation of Last 13 Years' Questions)

1 If $f_k(x)=\frac{1}{k}(\sin^k x+\cos^k x)$, where $x\in R$, $k\geq 1$, then $f_4(x)-f_6(x)$ is equal to **(JEE Main 2014)**

a. $\frac{1}{6}$ b. $\frac{1}{3}$
c. $\frac{1}{4}$ d. $\frac{1}{12}$

2 The expression $\frac{\tan A}{1-\cot A}+\frac{\cot A}{1-\tan A}$ can be written as **(JEE Main 2013)**

a. $\sin A\cos A+1$ b. $\sec A\operatorname{cosec} A+1$
c. $\tan A+\cot A$ d. $\sec A+\operatorname{cosec} A$

3 If $A=\sin^2 x+\cos^4 x$, then for all real x **(AIEEE 2011)**

a. $\frac{13}{16}\leq A\leq 1$ b. $1\leq A\leq 2$
c. $\frac{3}{4}\leq A\leq\frac{13}{16}$ d. $\frac{3}{4}\leq A\leq 1$

4 Let $\cos(\alpha+\beta)=\frac{4}{5}$ and $\sin(\alpha-\beta)=\frac{5}{13}$, where $0\leq\alpha,\ \beta\leq\frac{\pi}{4}$. Then, $\tan 2\alpha$ is equal to **(AIEEE 2010)**

a. $\frac{25}{16}$ b. $\frac{56}{33}$
c. $\frac{19}{12}$ d. $\frac{20}{7}$

5 Let A and B denote the statements

A : $\cos\alpha+\cos\beta+\cos\gamma=0$

B : $\sin\alpha+\sin\beta+\sin\gamma=0$

If $\cos(\beta-\gamma)+\cos(\gamma-\alpha)+\cos(\alpha-\beta)=-\frac{3}{2}$, then **(AIEEE 2009)**

a. A is true and B is false
b. A is false and B is true
c. both A and B are true
d. both A and B are false

6 A triangular park is enclosed on two sides by a fence and on the third side by a straight river bank. The two sides having fence are of same length x. The maximum area enclosed by the park is **(AIEEE 2006)**

a. $\sqrt{\frac{x^3}{8}}$ b. $\frac{1}{2}x^2$
c. πx^2 d. $\frac{3}{2}x^2$

7 If $0<x<\pi$ and $\cos x+\sin x=\frac{1}{2}$, then $\tan x$ is **(AIEEE 2006)**

a. $\frac{(4-\sqrt{7})}{3}$ b. $-\frac{(4+\sqrt{7})}{3}$
c. $\frac{(1+\sqrt{7})}{4}$ d. $\frac{(1-\sqrt{7})}{4}$

8 In a ΔPQR, $\angle R=\frac{\pi}{2}$. If $\tan\left(\frac{P}{2}\right)$ and $\tan\left(\frac{Q}{2}\right)$ are the roots of $ax^2+bx+c=0, a\neq 0$, then **(AIEEE 2005)**

a. $b=a+c$ b. $b=c$
c. $c=a+b$ d. $a=b+c$

9 Let α and β be such that $\pi<\alpha-\beta<3\pi$. If $\sin\alpha+\sin\beta=-\frac{21}{65}$ and $\cos\alpha+\cos\beta=-\frac{27}{65}$, then the value of $\cos\left(\frac{\alpha-\beta}{2}\right)$ is **(AIEEE 2004)**

a. $-\frac{3}{\sqrt{130}}$ b. $\frac{3}{\sqrt{130}}$
c. $\frac{6}{65}$ d. $-\frac{6}{65}$

10 $\sin^2\theta=\frac{4xy}{(x+y)^2}$ is true, if and only if **(AIEEE 2002)**

a. $x-y\neq 0$ b. $x=-y$
c. $x+y\neq 0$ d. $x\neq 0, y\neq 0$

11 The value of $\frac{1-\tan^2 15°}{1+\tan^2 15°}$ is **(AIEEE 2002)**

a. 1 b. $\sqrt{3}$
c. $\frac{\sqrt{3}}{2}$ d. 2

12 If $\sin(\alpha+\beta)=1$ and $\sin(\alpha-\beta)=\frac{1}{2}$, then $\tan(\alpha+2\beta)\tan(2\alpha+\beta)$ is equal to **(AIEEE 2002)**

a. 1
b. −1
c. zero
d. None of the above

13 If $y=\sin^2\theta+\operatorname{cosec}^2\theta, \theta\neq 0$, then **(AIEEE 2002)**

a. $y=0$ b. $y\leq 2$
c. $y\geq -2$ d. $y\geq 2$

14 The equation $a\sin x+b\cos x=c$, where $|c|>\sqrt{a^2+b^2}$ has **(AIEEE 2002)**

a. a unique solution
b. infinite number of solutions
c. no solution
d. None of the above

15 If α is a root of $25\cos^2\theta+5\cos\theta-12=0$, $\frac{\pi}{2}<\alpha<\pi$, then $\sin 2\alpha$ is equal to **(AIEEE 2002)**

a. $\frac{24}{25}$ b. $-\frac{24}{25}$
c. $\frac{13}{18}$ d. $-\frac{13}{18}$

JEE Advanced & IIT JEE Archive

(Compilation of Last 10 Years' Questions)

1 Let $f:(-1,1) \to R$ be such that $f(\cos 4\theta) = \dfrac{2}{2-\sec^2\theta}$ for $\theta \in \left(0, \dfrac{\pi}{4}\right) \cup \left(\dfrac{\pi}{4}, \dfrac{\pi}{2}\right)$. Then, the value(s) of $f\left(\dfrac{1}{3}\right)$ is/are **(2012, One or More than One Options Correct Type)**

a. $1-\sqrt{\dfrac{3}{2}}$ b. $1+\sqrt{\dfrac{3}{2}}$

c. $1-\sqrt{\dfrac{2}{3}}$ d. $1+\sqrt{\dfrac{2}{3}}$

2 The number of all possible values of θ, where $0 < \theta < \pi$, for which the system of equations

$$(y+z)\cos 3\theta = (xyz)\sin 3\theta$$

$$x\sin 3\theta = \frac{2\cos 3\theta}{y} + \frac{2\sin 3\theta}{z}$$

and $(xyz)\sin 3\theta = (y+2z)\cos 3\theta + y\sin 3\theta$ have a solution (x_0, y_0, z_0) with $y_0 z_0 \neq 0$, is

(2010, Integer Type)

3 If $\dfrac{\sin^4 x}{2} + \dfrac{\cos^4 x}{3} = \dfrac{1}{5}$, then

(2009, One or More than One Options Correct Type)

a. $\tan^2 x = \dfrac{2}{3}$ b. $\dfrac{\sin^8 x}{8} + \dfrac{\cos^8 x}{27} = \dfrac{1}{125}$

c. $\tan^2 x = \dfrac{1}{3}$ d. $\dfrac{\sin^8 x}{8} + \dfrac{\cos^8 x}{27} = \dfrac{2}{125}$

4 For $0 < \theta < \dfrac{\pi}{2}$, the solution(s) of

$$\sum_{m=1}^{6} \text{cosec}\left(\theta + \frac{(m-1)\pi}{4}\right)\text{cosec}\left(\theta + \frac{m\pi}{4}\right) = 4\sqrt{2}$$

is/are **(2009, One or More than One Options Correct Type)**

a. $\dfrac{\pi}{4}$ b. $\dfrac{\pi}{6}$

c. $\dfrac{\pi}{12}$ d. $\dfrac{5\pi}{12}$

5 Let $\theta \in \left(0, \dfrac{\pi}{4}\right)$ and $t_1 = (\tan\theta)^{\tan\theta}$, $t_2 = (\tan\theta)^{\cot\theta}$, $t_3 = (\cot\theta)^{\tan\theta}$ and $t_4 = (\cot\theta)^{\cot\theta}$, then

(2006, Only One Option Correct Type)

a. $t_1 > t_2 > t_3 > t_4$

b. $t_4 > t_3 > t_1 > t_2$

c. $t_3 > t_1 > t_2 > t_4$

d. $t_2 > t_3 > t_1 > t_4$

6 The number of ordered pairs (α, β), where $\alpha, \beta \in (-\pi, \pi)$ satisfying $\cos(\alpha - \beta) = 1$ and $\cos(\alpha + \beta) = \dfrac{1}{e}$, is **(2005, Only One Option Correct Type)**

a. 0 b. 1

c. 2 d. 4

JEE Archive

2. Height and Distance

DPP-1 Height and Distance

Subjective Questions

Direction *(Q. Nos. 1-3) These questions are subjective in nature, need to be solved completely on notebook.*

Working Space

1 The shadow of a tower is found to be 60 m shorter, when the sun's altitude changes from 30° to 60°, then find the height of a tower from the ground.

2 The tower of height *b* subtends an angle at a point *O* on the level of the foot of a tower and at a distance '*a*' from foot of the tower. If a pole mounted on the tower also subtends an equal angle at *O*, then find the height of the pole.

3 If angles of elevation of the top of a tower from three points (collinear) *A*, *B* and *C* on a road leading to the foot of the tower are 30°, 45° and 60° respectively, then prove that $AB : BC = \sqrt{3}:1$.

Only One Option Correct Type

Direction *(Q. Nos. 4-13) This section contains 10 multiple choice questions. Each question has four choices (a), (b), (c) and (d), out of which ONLY ONE is correct.*

4 The shadow of a tower standing on a level ground is *x* m long when the sun's altitude is 30°, while it is *y* m long when the altitude is 60°. If the height of the tower is $\frac{45\sqrt{3}}{2}$ m, then $x - y$ is

a. 45 m
b. $45\sqrt{3}$ m
c. $\frac{45}{\sqrt{3}}$ m
d. $\frac{45\sqrt{3}}{2}$ m

5 For a man, the angle of elevation of the highest point of the temple situated East of him is 60°. On walking 240 m to North, the angle of elevation is reduced to 30°, then the height of temple is

a. $60\sqrt{6}$ m
b. 60 m
c. $50\sqrt{3}$ m
d. 360m

6 A balloon is observed simultaneously from three points *A*, *B* and *C* on a straight road directly under it. The angular elevation at *B* is twice and at *C* is thrice that of *A*. If the distance between *A* and *B* is 200 m and the distance between *B* and *C* is 100 m, then the height of balloon is

a. 50 m
b. $50\sqrt{3}$ m
c. $50\sqrt{2}$ m
d. None of these

7 The angle of elevation of the top of a tower from a point *A* due South of the tower is α and from a point *B* due East of the tower is β. If $AB = d$, then the height of tower is

a. $\frac{d}{\sqrt{\tan^2\alpha - \tan^2\beta}}$
b. $\frac{d}{\sqrt{\tan^2\alpha + \tan^2\beta}}$
c. $\frac{d}{\sqrt{\cot^2\alpha + \cot^2\beta}}$
d. $\frac{d}{\sqrt{\cot^2\alpha - \cot^2\beta}}$

Working Space

8 The angle of elevation of a stationary cloud from a point 2500 m above a lake is 15° and the angle of depression of its reflection in the lake is 45°. The height of cloud above the lake level is **[IIT JEE 1976]**

a. $2500\sqrt{3}$ m
b. 2500 m
c. $500\sqrt{3}$ m
d. None of these

9 A person standing on the bank of a river observes that the angle subtended by a tree on the opposite bank is 60° when he returns 40 m from the bank. He finds the angle to be 30°. The breadth of the river is **[IIT JEE 1975, AIEEE 2004]**

a. 20 m
b. 40 m
c. 30 m
d. 60 m

10 The angle of elevation of the top of a tower at a point on the ground is 30°. If on walking 20 m towards the tower, the angle of elevation becomes 60°, then the height of the tower is **[IIT JEE 1967]**

a. 10 m
b. $\frac{10}{\sqrt{3}}$ m
c. $10\sqrt{3}$ m
d. None of these

11 From the top of a light house 60 m high with its base at the sea level, the angle of depression of a boat is 15°. The distance of the boat from the foot of light house is **[IIT JEE 1983]**

a. $60\left(\frac{\sqrt{3}-1}{\sqrt{3}+1}\right)$ m
b. $60\left(\frac{\sqrt{3}+1}{\sqrt{3}-1}\right)$ m
c. $\left(\frac{\sqrt{3}+1}{\sqrt{3}-1}\right)$ m
d. None of these

12 An aeroplane flying horizontally 1 km above the ground is observed at an elevation of 60° and after 10 s the elevation is observed to be 30°. The uniform speed of aeroplane (in km/h) is **[IIT JEE 1965]**

a. 240
b. $240\sqrt{3}$
c. $60\sqrt{3}$
d. None of these

13 *AB* is vertical pole with *B* at the ground level and *A* at the top. A man finds that the angle of elevation of the point *A* from a certain point *C* on the ground is 60°. He moves away from the pole along the line *BC* to a point *D* such that *CD* = 7m. From *D*, the angle of elevation of the point *A* is 45°. Then, the height of the pole is **[AIEEE 2008]**

a. $\frac{7\sqrt{3}}{2}(\sqrt{3}+1)$m
b. $\frac{7\sqrt{3}}{2}(\sqrt{3}-1)$m
c. $\frac{7\sqrt{3}}{2}\left(\frac{1}{\sqrt{3}+1}\right)$m
d. $\frac{7\sqrt{3}}{2}\left(\frac{1}{\sqrt{3}-1}\right)$m

One or More than One Options Correct Type

Direction (Q. Nos. 14-16) *This section contains 3 multiple choice questions. Each question has four choices (a), (b), (c) and (d), out of which ONE or MORE THAN ONE are correct.*

14 A vertical pole consists of two parts, the lower part being one-third of the whole. At a point in the horizontal plane through the base of the pole and distance 20 m from it, the upper part of the pole subtends an angle whose tangent is $\frac{1}{2}$. The possible heights of the pole is/are

a. 20 m
b. 60 m
c. 40 m
d. None of these

Working Space

15 A ladder rests against a wall making an angle α with the horizontal. The foot of the ladder is pulled away from the wall through a distance x, so that it slides a distance y down the wall making an angle β with the horizontal. The correct relation is/are

a. $x = y\tan\left(\frac{\alpha+\beta}{2}\right)$

b. $y^2 - x^2 = (x^2 + y^2)\cos(\alpha+\beta)$

c. $(x^2 + y^2)\sin(\alpha+\beta) = 2xy$

d. None of the above

16 A tower AB leans towards West making an angle α with the vertical. The angular elevation B, the top most point of the tower is β as observed from a point C due East of A at a distance of d from A. If the angular elevation of B from a point D due East of C at a distance $2d$ from C is γ, then **[IIT JEE 1994]**

a. $2\tan\alpha = 3\cot\beta - 2\cot\gamma$

b. $2\tan\alpha = 3\cot\gamma - 2\cos\beta$

c. $2\tan\alpha = 3\cot\beta - \cot\gamma$

d. $2\tan\alpha = \cot\beta - 3\cot\gamma$

One Integer Value Correct Type

***Direction** (Q. Nos. 17-20) This section contains 4 questions. When worked out will result in an integer from 0 to 9 (both inclusive).*

17 The upper $\frac{3}{4}$th portion of a vertical pole subtends an angle $\tan^{-1}\left(\frac{3}{5}\right)$ at a point in the horizontal plane through its foot and at a distance 40 m from the foot. If possible height of the vertical pole is $h \times 10$, then find the value of h.

18 A man from the top of a 100 m high tower saw a car moving towards the tower at an angle of depression of 30°. After sometime, the angle of depression becomes 60°. If the distance travelled by the car during the time is $\frac{200}{\sqrt{\lambda}}$ m, then find the value of λ. **[IIT JEE 2001]**

19 The angular depressions of the top and the foot of a chimney as seen from the top of a second chimney, which is 150 m high and standing on the same level as the first are θ and ϕ respectively, then the distance between their tops when $\tan\theta = \frac{4}{3}$ and $\tan\phi = \frac{5}{2}$ is 25λ, then λ is equal to

20 At a point on the ground, the angle of elevation of a tower is such that its cotangent is $\frac{3}{5}$. On walking 32 m towards the tower, the cotangent of the angle of elevation is $\frac{2}{5}$. The height of the tower is 40λ, then λ is equal to

JEE Main & AIEEE Archive

(Compilation of Last 13 Years' Questions)

1 If the angles of elevation of the top of a tower from three collinear points A, B and C, on a line leading to the foot of the tower, are 30°, 45° and 60° respectively, then the ratio, $AB : BC$, is **(JEE Main 2015)**

a. $\sqrt{3}:1$
b. $\sqrt{3}:\sqrt{2}$
c. $1:\sqrt{3}$
d. $2:3$

2 A bird is sitting on the top of a vertical pole 20 m high and its elevation from a point O on the ground is 45°. It flies off horizontally straight away from the point O. After 1s, the elevation of the bird from O is reduced to 30°. Then, the speed (in ms^{-1}) of the bird is **(JEE Main 2014)**

a. $40(\sqrt{2}-1)$
b. $40(\sqrt{3}-\sqrt{2})$
c. $20\sqrt{2}$
d. $20(\sqrt{3}-1)$

3 $ABCD$ is a trapezium such that AB and CD are parallel and $BC \perp CD$. If $\angle ADB = \theta$, $BC = p$ and $CD = q$, then AB is equal to **(JEE Main 2013)**

a. $\dfrac{(p^2+q^2)\sin\theta}{p\cos\theta + q\sin\theta}$
b. $\dfrac{p^2+q^2\cos\theta}{p\cos\theta + q\sin\theta}$
c. $\dfrac{p^2+q^2}{p^2\cos\theta + q^2\sin\theta}$
d. $\dfrac{(p^2+q^2)\sin\theta}{(p\cos\theta + q\sin\theta)^2}$

4 AB is a vertical pole with B at the ground level and A at the top. A man finds that the angle of elevation of the point A from a certain point C on the ground is 60°. He moves away from the pole along the line BC to a point D such that $CD = 7$ m. From D, the angle of elevation of the point A is 45°. Then, the height of the pole is **(AIEEE 2008)**

a. $\dfrac{7\sqrt{3}}{2}\left(\dfrac{1}{\sqrt{3}+1}\right)$ m
b. $\dfrac{7\sqrt{3}}{2}\left(\dfrac{1}{\sqrt{3}-1}\right)$ m
c. $\dfrac{7\sqrt{3}}{2}(\sqrt{3}+1)$ m
d. $\dfrac{7\sqrt{3}}{2}(\sqrt{3}-1)$ m

5 A tower stands at the centre of a circular park. A and B are two points on the boundary of the park such that $AB (= a)$ subtends an angle of 60° at the foot of the tower and the angle of elevation of the top of the tower from A or B is 30°. The height of the tower is **(AIEEE 2007)**

a. $\dfrac{2a}{\sqrt{3}}$
b. $2a\sqrt{3}$
c. $\dfrac{a}{\sqrt{3}}$
d. $\sqrt{3}$

6 A person standing on the bank of a river, observes that the angle of elevation of the top of a tree on the opposite bank of the river is 60° and when he retires 40 m away from the tree, the angle of elevation becomes 30°. The breadth of the river is **(AIEEE 2004)**

a. 20 m
b. 30 m
c. 40 m
d. 60 m

7 The upper $\left(\dfrac{3}{4}\right)$th portion of a vertical pole subtends an angle $\tan^{-1}\left(\dfrac{3}{5}\right)$ at a point in the horizontal plane through its foot and at a distance 40 m from the foot. A possible height of the vertical pole is **(AIEEE 2003)**

a. 20 m
b. 40 m
c. 60 m
d. 80 m

3. Trigonometric Equations

DPP

DAILY PRACTICE PROBLEMS

DPP-1 For Trigonometric Equation with Conceptual Problems

True/False Statements

Working Space

Direction *(Q. Nos. 1- 4) This section contains 4 questions. Each question has statement which is TRUE or FALSE.*

1 A value of unknown angle which satisfies the given trigonometric equation is called a solution of equation.

2 Principal solution lies in $[0, 2\pi)$.

3 $\sin x + \cos x = 1$ is a trigonometric equation not trigonometric identity.

4 If $\tan\theta = \tan\alpha$, then $\theta = n\pi + \alpha$, $n \in I$, when θ and α are odd multiples of $\frac{\pi}{2}$.

Fill in the Blanks

Direction *(Q. Nos. 5-8) This section contains 4 sentences. Complete the given sentences filling the blanks with suitable words, informations, data, etc.*

5 General solution of $\sin\theta = \sin\alpha$ is $\theta = n\pi + (-1)^n \alpha$, $n \in I$ when $\alpha \in$ ....

6 If $\cos\theta = \cos\alpha \Leftrightarrow \theta = 2n\pi \pm \alpha$, $n \in I$ and $\alpha \in [a, b]$, then $a =$ ...... and $b =$ ....

7 The number of solutions of the trigonometric equation $\sec\theta = 0$ is ....

8 The number of solutions of $\sin x + \cos x = 3$ is ....

Only One Option Correct Type

Direction *(Q. Nos. 9-17) This section contains 9 multiple choice questions. Each question has four choices (a), (b), (c) and (d), out of which ONLY ONE is correct.*

9 Values of θ $(0° < \theta < 360°)$ satisfying $\operatorname{cosec}\theta + 2 = 0$ are

a. 210°, 300° b. 240°, 300°
c. 210°, 240° d. 210°, 330°

10 The smallest positive value of principal solutions of $\sin\theta = -\frac{1}{2}$ is

a. $\frac{\pi}{6}$ b. $\frac{7\pi}{6}$ c. $\frac{11\pi}{6}$ d. None of these

11 The principal solution of $\cot x = -\sqrt{3}$, when $x \in (\pi, 2\pi)$ is

a. $\frac{7\pi}{6}$ b. $\frac{9\pi}{6}$ c. $\frac{11\pi}{6}$ d. None of these

Working Space

12 The value of θ satisfying $\sin 5\theta = 0$ is

a. $\frac{n\pi}{5}, n \in I$ b. $\frac{n\pi}{3}, n \in I$ c. $\frac{2n\pi}{5}, n \in I$ d. None of these

13 The number of solutions of the equation $\operatorname{cosec}\theta = 0$ is/are

a. 0 b. 1 c. 2 d. 3

14 If $\sqrt{3}\sec 2\theta = 2$, then

a. $\theta = n\pi \pm \frac{\pi}{6}, n \in I$ b. $\theta = 2n\pi \pm \frac{\pi}{6}, n \in I$

c. $\theta = n\pi \pm \frac{\pi}{12}, n \in I$ d. None of these

15 The solution set for x, which satisfy the equation $\frac{\tan 3x - \tan 2x}{1 + \tan 3x \tan 2x} = 1$ is

a. $x = n\pi + \frac{\pi}{4}, n \in I$ b. $x = 2n\pi + \frac{\pi}{4}, n \in I$

c. Both (a) and (b) d. No such x exists

16 The equation $\sin x + \sin y + \sin z = -3$ for $0 \le x \le 2\pi$, $0 \le y \le 2\pi$, $0 \le z \le 2\pi$ has

a. one solution b. two sets of solutions

c. four sets of solutions d. no solution

17 The most general value of θ which will satisfy both the equations $\sin\theta = -\frac{1}{2}$ and $\tan\theta = \frac{1}{\sqrt{3}}$ is

a. $n\pi + (-1)^n \frac{\pi}{6}$ b. $n\pi + \frac{\pi}{6}$ c. $2n\pi \pm \frac{\pi}{6}$ d. None of these

One or More than One Options Correct Type

Direction *(Q. Nos. 18-20) This section contains 3 multiple choice questions. Each question has four choices (a), (b), (c) and (d), out of which ONE or MORE THAN ONE are correct.*

18 If $2\sin^2\theta = 3\cos\theta$, where $0 \le \theta \le 2\pi$, then θ is equal to **[IIT JEE 1963]**

a. $\frac{\pi}{3}$ b. $\frac{7\pi}{3}$ c. $\frac{5\pi}{3}$ d. $\frac{7\pi}{6}$

19 If $\cot\theta = \sin 2\theta$ ($\theta \ne n\pi$, n is an integer), then θ is equal to

a. 45° b. 60° c. Both (a) and (b) d. 90°

20 If $(2\cos x - 1)(3 + 2\cos x) = 0$, $0 \le x \le 2\pi$, then x is equal to

a. $\frac{\pi}{3}$ b. $\frac{5\pi}{3}$ c. $\cos^{-1}\left(\frac{-3}{2}\right)$ d. None of these

One Integer Value Correct Type

Direction *(Q. Nos. 21 and 22) This section contains 2 questions. When worked out will result in an integer from 0 to 9 (both inclusive).*

21 If $\cos A \sin\left(A - \frac{\pi}{6}\right)$ is maximum, when the value of A is equal to $\frac{\pi}{\lambda}$, then the value of λ is

22 The number of values of x in the interval $[0, 3\pi]$ satisfying the equation $2\sin^2 x + 5\sin x - 3 = 0$ is **[AIEEE 2006]**

DPP 2 Equations of the Form $\sin^2\theta = \sin^2\alpha$, $\cos^2\theta = \cos^2\alpha$, $\tan^2\theta = \tan^2\alpha$

Subjective Questions

Working Space

Direction *(Q. Nos. 1-3) These questions are subjective in nature, need to be solved completely on notebook.*

1 Solve : $3\tan\theta + \cot\theta = 5\,\text{cosec}\,\theta$

2 Solve : $7\cos^2\theta + 3\sin^2\theta = 4$

3 Solve: $\sqrt{3}\tan x + \sqrt{3}\tan 2x + \tan x \tan 2x = 1$

Only One Option Correct Type

Direction *(Q. Nos. 4-13) This section contains 10 multiple choice questions. Each question has four choices (a), (b), (c) and (d), out of which ONLY ONE is correct.*

4 If $\theta = k(4n+1)$ is the solution of $\tan\theta + \tan\left(\theta + \frac{\pi}{3}\right) + \tan\left(\theta + \frac{2\pi}{3}\right) = 3$, then k is

a. $\frac{\pi}{3}$ b. $\frac{\pi}{6}$ c. $\frac{\pi}{12}$ d. None of these

5 If $\sin x = \tan x$, then the value of x is

a. $n\pi, n \in I$ b. $(2n+1)\pi, n \in I$ c. $2n\pi, n \in I$ d. None of these

6 If $\sin x + \sin 3x + \sin 5x = 0$, then the solution is

a. $x = \frac{2n\pi}{3}, n \in I$ b. $\frac{n\pi}{3}, n \in I$ c. $(3n \pm 1)\frac{\pi}{3}, n \in I$ d. None of these

7 If $\tan x + \sec x = 2\cos x$, then

a. $x = n\pi + (-1)^n \frac{3\pi}{2}, n \in I$ b. $x = n\pi + (-1)^n \frac{\pi}{6}, n \in I$

c. $x = n\pi + (-1)^n \frac{\pi}{3}, n \in I$ d. None of these

8 If $x = n\pi + (-1)^n \alpha, n \in I$ and $x = n\pi + (-1)^n \beta$ are the roots of $4\cos x - 3\sec x = \tan x$, then $4(\sin\alpha + \sin\beta)$ is

a. -1 b. 1 c. 2 d. None of these

9 If $\sqrt{3}\cos x - \sin x = 1$, then x is

a. $2n\pi \pm \frac{\pi}{4} - \frac{\pi}{6}, n \in I$ b. $2n\pi \pm \frac{\pi}{6} - \frac{\pi}{3}, n \in I$

c. $2n\pi \pm \frac{\pi}{3} - \frac{\pi}{6}, n \in I$ d. None of these

Working Space

10 If $\sqrt{2}\sec x + \tan x = 1$, then the value of x is

a. $2n\pi + \frac{\pi}{3}$ b. $2n\pi - \frac{\pi}{4}$ c. $2n\pi + \frac{\pi}{6}$ d. $n\pi + \frac{\pi}{3}$

11 General solution of $\sin\alpha + \cos\alpha = \min_{a\in R}\{1, a^2 - 4a + 6\}$ is

a. $n\pi + (-1)^n \frac{\pi}{4} - \frac{\pi}{4}, \forall n \in I$ b. $n\pi + (-1)^n \frac{\pi}{4}$

c. $2n\pi \pm \frac{\pi}{4}, \forall n \in I$ d. None of these

12 If $\tan m\theta = \tan n\theta$ and general value of θ are in AP, then common difference is

a. $\frac{1}{m-n}$ b. $\frac{\pi}{m+n}$ c. $\frac{\pi}{m-n}$ d. None of these

13 The general solution of $\sin x - 3\sin 2x + \sin 3x = \cos x - 3\cos 2x + \cos 3x$ is

a. $n\pi + \frac{\pi}{8}$ b. $\frac{n\pi}{2} + \frac{\pi}{8}$ c. $(-1)^n \frac{n\pi}{2} + \frac{\pi}{8}$ d. $2n\pi + \cos^{-1}\left(\frac{3}{2}\right)$

One or More than One Options Correct Type

Direction (Q. Nos. 14-17) *This section contains 4 multiple choice questions. Each question has four choices (a), (b), (c) and (d), out of which ONE or MORE THAN ONE are correct.*

14 If $\cos\theta + \cos 3\theta - 2\cos 2\theta = 0$, then θ is equal to

a. $(2n+1)\frac{\pi}{4}, n \in I$ b. $2n\pi, n \in I$ c. $3n\pi, n \in I$ d. None of these

15 Let $P = \{\theta : \sin\theta - \cos\theta = \sqrt{2}\cos\theta\}$ and $Q = \{\theta: \sin\theta + \cos\theta = \sqrt{2}\sin\theta\}$ be two sets. Then, which of the following statement(s) is/are wrong?

a. $P \subset Q$ and $Q - P \neq \phi$ b. $Q \not\subset P$

c. $P \not\subset Q$ d. $P = Q$

16 The smallest positive values of x and y which satisfy $\tan(x - y) = 1$ and $\sec(x + y) = \frac{2}{\sqrt{3}}$ are

a. $x = \frac{25\pi}{24}, y = \frac{19\pi}{24}$ b. $x = \frac{37\pi}{24}, y = \frac{7\pi}{24}$

c. $x = \frac{\pi}{4}, y = \frac{\pi}{2}$ d. $x = \frac{\pi}{3}, y = \frac{7\pi}{12}$

17 If $R = \frac{1}{2}\sin^2\theta + \frac{1}{3}\cos^2\theta$ and $a \le R \le b$, then

a. $a = \frac{1}{3}$ b. $a = \frac{1}{2}$ c. $b = \frac{1}{3}$ d. $b = \frac{1}{2}$

One Integer Value Correct Type

Direction (Q. Nos. 18-20) *This section contains 3 questions. When worked out will result in an integer from 0 to 9 (both inclusive).*

18 The number of solutions which satisfy the equation $3\cos x + 4\sin x = 6$ is

19 The number of values of x in the interval $[0, 5\pi]$ satisfying the equation $3\sin^2 x - 7\sin x + 2 = 0$ is

20 If $\sin 2t = \cos 3t$ and t is an acute angle, then $\frac{t}{6}$ is equal to

DPP-3 Homogeneous, Extreme Values and T-Inequations

Subjective Questions

Direction (Q. No. 1) *This question is subjective in nature, need to be solved completely on notebook.*

Working Space

1 If $(1+\tan\alpha)(1+\tan\beta)=2$, then prove that $\alpha+\beta=45°$.

True/False Statements

Direction (Q. Nos. 2 and 3) *This section contains 2 questions. Each question has statement which is TRUE or FALSE.*

2 If $2^{\frac{1}{1-|\cos x|}}=4$, then the value of x is $n\pi\pm\frac{\pi}{3}$, $n\in I$.

3 If $\cos 40°=t$ and $\cos\theta=1-2t^2$, $0°<\theta<360°$, then the value of θ is 100° and 260°.

Only One Option Correct Type

Direction (Q. Nos. 4-13) *This section contains 10 multiple choice questions. Each question has four choices (a), (b), (c) and (d), out of which ONLY ONE is correct.*

4 The general solution of $\sin^2\theta\sec\theta+\sqrt{3}\tan\theta=0$ is

a. $\theta=\frac{n\pi}{3}, \forall n\in Z$

b. $\theta=n\pi, \forall n\in Z$

c. $\theta=n\pi+(-1)^n\sin^{-1}(-\sqrt{3}), n\in Z$

d. None of the above

5 If $\sin 3\alpha=4\sin\alpha\sin(x+\alpha)\sin(x-\alpha)$, then

a. $x=n\pi\pm\frac{\pi}{3}, n\in I$

b. $x=n\pi\pm\frac{\pi}{6}, n\in I$

c. $x=n\pi\pm\frac{\pi}{2}, n\in I$

d. None of these

6 If x is the positive root of equation $\sin x+\cos x-2\sqrt{2}\sin x\cos x=0$, then the value of x is

a. $2n\pi+\frac{\pi}{4}$

b. $2n\pi-\frac{\pi}{4}$

c. $n\pi+\frac{\pi}{4}$

d. None of these

Working Space

7 If $\sqrt{3}\cos^2\theta - 2\sqrt{3}\sin\theta\cos\theta - 3\sin^2\theta = 0$, then

a. $\theta = n\pi + \frac{\pi}{6}, n \in I$

b. $\theta = n\pi - \frac{\pi}{6}, n \in I$

c. $\theta = n\pi + \frac{\pi}{3}, n \in I$

d. None of the above

8 The number of solutions for $\sin 5\theta \cos 3\theta = \sin 9\theta \cos 7\theta$ in $\left[0, \frac{\pi}{2}\right]$ is

a. 0 b. 3
c. 6 d. 9

9 $\lambda \cos x - 3\sin x = \lambda + 1$ is solvable only, if

a. $\lambda \in [0, 5]$ b. $\lambda \in [4, 5]$
c. $\lambda \in (-\infty, 4]$ d. None of these

10 The number of distinct solutions of $\sqrt{3}\sec x - \sqrt{3}\tan x = 3$, $0 \le x \le 3\pi$ is

a. 0 b. 1
c. 2 d. 3

11 $\cos 2x - 3\cos x + 1 = \dfrac{1}{(\cot 2x - \cot x)\sin(x - \pi)}$ holds, if

a. $\cos x = 0$

b. $\cos x = 1$

c. $\cos x = \frac{5}{2}$

d. for no value of x

12 The number of solutions of $\sin\theta \tan 4\theta = \cos\theta$, $\theta \in (0, \pi)$ is

a. 4 b. 7
c. 8 d. None of these

13 The equation $\sin^4\theta + \cos^4\theta + 2\sin\theta \cdot \cos\theta + \alpha = 0$ is solvable for

a. $\alpha \in \left[-\frac{1}{2}, \frac{1}{2}\right]$ b. $\alpha \in [-3, 1]$

c. $\alpha \in \left[\frac{-3}{2}, \frac{1}{2}\right]$ d. $\alpha \in [-1, 1]$

One or More than One Options Correct Type

***Direction** (Q. Nos. 14-16) This section contains 3 multiple choice questions. Each question has four choices (a), (b), (c) and (d), out of which ONE or MORE THAN ONE are correct.*

14 If $\cos^2 x + \sin^2 y - 2\tan^2 z = 2$, then

a. $x = n\pi, n \in I$ b. $y = (2m + 1)\frac{\pi}{2}, m \in I$

c. $z = t\pi, t \in I$ d. None of these

15 If $x + y = \frac{4\pi}{3}$ and $\sin x = 2\sin y$, then

a. $x = n\pi + \frac{\pi}{2}, n \in I$ b. $y = \frac{5\pi}{6} - n\pi, n \in I$

c. Both (a) and (b) d. None of these

Working Space

16 If $\max_{\theta \in R} \{5\sin\theta + 3\sin(\theta - \alpha)\} = 7$, then

a. $\alpha = 2n\pi + \frac{\pi}{3}$
b. $\alpha = n\pi + \frac{\pi}{3}$
c. $\alpha = 2n\pi - \frac{\pi}{3}$
d. None of these

One Integer Value Correct Type

***Direction** (Q. Nos. 17 and 18) This section contains 2 questions. When worked out will result in an integer from 0 to 9 (both inclusive).*

17 If $|\sqrt{3}\cos x - \sin x| \geq 2$ for $x \in [0, 4\pi]$, then the number of solutions is

18 The number of values of t for which $\sin^4 t = 1 + \tan^4 t$ is

Comprehension Type

***Direction** (Q. Nos. 19 and 20) This section contains a paragraph, which describing theory, experiments, data, etc. Two questions related to the paragraph have been given. Each question has only one correct answer among the four given options (a), (b), (c) and (d).*

Passage

Let $f(x)$ and $g(x)$ be two distinct functions satisfying the equation $f(x) = g(x)$ for some values of x, then to find number of solutions, find number of intersection points of $f(x)$ and $g(x)$.

On the basis of above information, answer the following questions:

19 The number of solutions of $\sin x = \frac{x}{10}$ is

a. 2
b. 6
c. 7
d. 9

20 The total number of solutions of $\cos 2x = |\sin x|$, where $x \in \left(\frac{-\pi}{2}, \pi\right)$ is

a. 3
b. 4
c. 5
d. 6

SET - 1

Revisal Problems for JEE Main

Only One Option Correct Type

Working Space

Direction *(Q. Nos. 1-9) This section contains 9 multiple choice questions. Each question has four choices (a), (b), (c) and (d), out of which ONLY ONE is correct.*

1 If $\sqrt{3}\cos\theta+\sin\theta=\sqrt{2}$ and $\theta=n\pi+(-1)^n\alpha-\beta$, then the value of $\alpha+\beta$ is

a. $\frac{\pi}{12}$ b. $\frac{5\pi}{12}$

c. $\frac{7\pi}{12}$ d. None of these

2 If $1+\cot\alpha=\text{cosec}\,\alpha$, then the general value of α is

a. $n\pi+\frac{\pi}{2}, n\in I$ b. $2n\pi+\frac{\pi}{4}, n\in I$

c. $\frac{\pi}{2}, n\in I$ d. None of these

3 If $\sec x\cos 5x=-1$ and $0<x<\frac{\pi}{4}$, then x is equal to

a. $\frac{\pi}{6}$ b. $\frac{\pi}{3}$

c. $\frac{\pi}{4}$ d. None of these

4 The solution of equation $\cos^2\alpha+\sin\alpha+1=0$ lies in the interval $[0, 2\pi]$, when α is

a. $\frac{\pi}{2}$ b. $\frac{3\pi}{2}$

c. $\frac{3\pi}{4}$ d. None of these

5 The number of values of θ for which $\sin 2\theta+\cos 2\theta=2$ is

a. 0 b. 1

c. 2 d. infinite

6 If $2\tan^2\theta-5\sec\theta=1$ for exactly 7 distinct values of $\theta\in\left[0, \frac{n\pi}{2}\right], n\in N$, then the greatest value of n is

a. 6 b. 12

c. 13 d. 15

7 If $\sin^{100}\theta-\cos^{100}\theta=1$, then θ is

a. $2n\pi+\frac{\pi}{3}, n\in I$ b. $n\pi+\frac{\pi}{2}, n\in I$

c. $n\pi+\frac{\pi}{4}, n\in I$ d. $2n\pi-\frac{\pi}{3}, n\in I$

8 The number of solutions of the equation $\cos(\pi\sqrt{x-4})\cos(\pi\sqrt{x})=1$ is

a. 0 b. 1

c. 2 d. infinite

9 If $\sqrt{3}\sin x-\cos x=\min\limits_{\alpha\in R}\{2, e^2, \pi, \alpha^2-4\alpha+7\}$, then

a. $x=2n\pi, n\in I$ b. $x=2n\pi+\frac{2\pi}{3}, n\in I$

c. $x=n\pi+(-1)^n\frac{\pi}{4}+\frac{\pi}{6}, n\in I$ d. $x=n\pi+(-1)^{n+1}\frac{\pi}{4}-\frac{\pi}{3}, n\in I$

SET - 2

Revisal Problems
for JEE Advanced

Only One Option Correct Type

Direction (Q. Nos. 1-9) *This section contains 9 multiple choice questions. Each question has four choices (a), (b), (c) and (d), out of which ONLY ONE is correct.*

Working Space

1 The equation $(\cos\alpha - 1)t^2 + (\cos\alpha)t + \sin\alpha = 0$, where t is variable has real roots, then

a. $\alpha \in (0, \pi)$
b. $\alpha \in [0, \pi]$
c. $\alpha \in [0, 2\pi]$
d. None of these

2 The number of solutions of the equation $\cos 4x + 6 = 7\cos 2x$, when $x \in [315°, 317°]$ is

a. 0
b. 1
c. 2
d. 4

3 If $\cot\left(\frac{\pi}{3}\cos 2\pi x\right) = \sqrt{3}$, then general value of the equation is

a. $x = n \pm \frac{1}{6}, n \in I$
b. $x = n \pm \frac{1}{3}, n \in I$
c. $x = 2n\pi \pm \frac{\pi}{6}, n \in I$
d. None of these

4 The number of solutions of $\sin\theta + \sin 2\theta + \sin 3\theta + \sin 4\theta = 4$ in $[0, 10\pi]$ is

a. 1
b. 2
c. 3
d. None of these

5 If $F(t) = \sin t + \cos t$, then most general solution of $F(t) = \left[F\left(\frac{\pi}{10}\right)\right]$ is (where, $[t]$ denotes the greatest integer function)

a. $2n\pi + \frac{\pi}{2}, n \in I$
b. $n\pi, n \in I$
c. $2n\pi \pm \frac{2\pi}{3}, n \in I$
d. None of these

6 The solution of the equation $\log_{\cos\alpha} \sin\alpha + \log_{\sin\alpha} \cos\alpha = 2$ is given by

a. $\alpha = 2n\pi + \frac{\pi}{4}, n \in I$
b. $x = n\pi + \frac{\pi}{2}, n \in I$
c. $\alpha = n\pi + \frac{\pi}{8}, n \in I$
d. None of these

7 The equation $\cos^4 x - (p+2)\cos^2 x - (p+3) = 0$ possesses a solution, if

a. $p > -3$
b. $p < -2$
c. $-3 \le p \le -2$
d. $p \in I^+$

8 If $\cos\theta + \sin\theta = \sqrt{\left(x + \frac{1}{x}\right)}$, $\theta \in [0, \pi]$, then

a. $\theta = \frac{\pi}{4}, x = 1$
b. $x = 0$
c. $x = 2$
d. $\theta = \frac{3\pi}{4}$

9 The number of solutions of $\cot(5\pi\sin\theta) = \tan(5\pi\cos\theta)$, $\forall \theta \in (0, 2\pi)$ is

a. 7
b. 14
c. 21
d. 28

One or More than One Options Correct Type

Working Space

Direction *(Q. Nos. 10-18) This section contains 9 multiple choice questions. Each question has four choices (a), (b), (c) and (d), out of which ONE or MORE THAN ONE are correct.*

10 If $\alpha \in [0, \pi]$ and $81^{\sin^2 \alpha} + 81^{\cos^2 \alpha} = 30$, then α is equal to

a. $\frac{\pi}{6}$ b. $\frac{\pi}{3}$

c. $\frac{2\pi}{3}$ d. $\frac{\pi}{2}$

11 If $\cos(\sin\theta) = \frac{1}{\sqrt{2}}$, then θ must lie in the interval

a. $\left(\frac{\pi}{4}, \frac{\pi}{2}\right)$ b. $\left(\frac{-\pi}{4}, 0\right)$

c. $\left(\pi, \frac{3\pi}{2}\right)$ d. $\left(\frac{\pi}{2}, \pi\right)$

12 The solutions of the equation $\cos 2x + \sin 7x + 2 = 0$ are

a. $x = \frac{2n\pi}{7} + \frac{3\pi}{14}, \forall n \in I$

b. $x = n\pi + \frac{\pi}{4}, \forall n \in I$

c. $x = n\pi + \frac{\pi}{2}, \forall n \in I$

d. $x = n\pi + \frac{3\pi}{4}, \forall n \in I$

13 The number of solutions of the equations $y = \frac{1}{3}[\sin\theta + [\sin\theta + [\sin\theta]]]$ and $[y + [y]] = 2\cos\theta$ [where, $[\cdot]$ denotes the greatest integer function] is/are

a. 0 b. 1

c. 2 d. infinite

14 If $[\sin x] + [\sqrt{2}\cos x] = -3$, $x \in [0, 2\pi]$, (where, $[\cdot]$ denotes the greatest integer function), then

a. $x \in \left(\pi, \frac{5\pi}{4}\right)$ b. $x \in \left(\pi, \frac{7\pi}{6}\right)$

c. $x \in \left[\pi, \frac{5\pi}{4}\right]$ d. None of these

15 If $\sin\theta + \cos\theta + \tan\theta + \cot\theta + \sec\theta + \text{cosec}\,\theta = 7$ and $\sin 2\theta = a - b\sqrt{c}$, then the value of $a - b + 2c$ is not equal to

a. 0 b. 14

c. 28 d. 42

16 The equation $\sin x = [1 + \sin x] + [1 - \cos x]$ has (where, $[\cdot]$ denotes the greatest integer function)

a. no solution in $\left[\frac{-\pi}{2}, \frac{\pi}{2}\right]$

b. no solution in $\left[\frac{\pi}{2}, \pi\right]$

c. no solution in $\left[\pi, \frac{3\pi}{2}\right]$

d. no solution for $x \in R$

17 If $(1 - \tan\theta)(1 + \tan\theta)\sec^2\theta + 2^{\tan^2\theta} = 0$ and $\theta \in \left(\frac{-\pi}{2}, \frac{\pi}{2}\right)$, then

a. $\theta = 0$ b. $\theta = \frac{\pi}{3}$

c. $\theta = \frac{-\pi}{3}$ d. $\theta = \frac{\pi}{6}$

18 $\sqrt{\cos 2x} + \sqrt{1 + \sin 2x} = 2\sqrt{(\sin x + \cos x)}$, if

a. $\sin x + \cos x = 0$ b. $x = 2n\pi, n \in I$

c. $x = n\pi - \frac{\pi}{4}, n \in I$ d. $x = 2n\pi \pm \cos^{-1}\left(-\frac{1}{5}\right), n \in I$

Matching List Type

Working Space

Direction (Q. No. 19) *Choices for the correct combination of elements from Column I and Column II are given as options (a), (b), (c) and (d), out of which one is correct.*

19 Match the following columns and select the correct option from the codes given below.

	Column I		Column II
i.	If α, β are the solutions of $\sin x = -\frac{1}{2}$ in $[0, 2\pi]$ and α, γ are the solutions of $\cos x = \frac{-\sqrt{3}}{2}$ in $[0, 2\pi]$, then	p.	$\alpha - \beta = \pi$
ii.	If α, β are the solutions of $\cot x = -\sqrt{3}$ in $[0, 2\pi]$ and α, γ are the solutions of $\operatorname{cosec} x = -2$ in $[0, 2\pi]$, then	q.	$\beta - \gamma = \pi$
iii.	If α, β are the solutions of $\sin x = \frac{-1}{2}$ in $[0, 2\pi]$ and α, γ are the solutions of $\tan x = \frac{1}{\sqrt{3}}$ in $[0, 2\pi]$, then	r.	$\alpha - \gamma = \pi$
		s.	$\alpha + \beta = 3\pi$
		t.	$\beta + \gamma = 2\pi$

Codes

	i.	ii	iii		i.	ii	iii
a.	q,s	p,t	r,s,t	b.	p,q	r,s	p,t
c.	p,r	p,q	r,s	d.	p	q,s	t

Comprehension Type

Direction (Q. Nos. 20-24) *This section contains a paragraph, which describing theory, experiments, data, etc. Five questions related to the paragraph have been given. Each question has only one correct answer among the four given options (a), (b), (c) and (d).*

Passage

An equation of the form $f(\sin x \pm \cos x, \pm \sin x \cos x) = 0$ can be solved by changing variable.

Let $\sin x \pm \cos x = t \Rightarrow \sin^2 x + \cos^2 x \pm 2 \sin x \cos x = t^2$

$$\Rightarrow \pm 2 \sin x \cos x = t^2 - 1$$

$$\Rightarrow \pm \sin x \cos x = \frac{t^2 - 1}{2}$$

Hence, reduce the given equation into $f\left(t, \frac{t^2 - 1}{1}\right) = 0$.

On the basis of above information, answer the following questions:

20 If $1 - \sin 2x = \cos x - \sin x$, then x is

a. $2n\pi, 2n\pi + \frac{\pi}{2}, n \in I$
b. $2n\pi, n\pi + \frac{\pi}{4}, n \in I$
c. $2n\pi - \frac{\pi}{2}, n\pi + \frac{\pi}{4}, n \in I$
d. None of these

21 If $\sin x + \cos x = 1 + \sin x \cos x$, then x is

a. $2n\pi, 2n\pi + \frac{\pi}{2}, n \in I$
b. $2n\pi, n\pi + \frac{\pi}{4}, n \in I$
c. $2n\pi - \frac{\pi}{2}, n\pi + \frac{\pi}{4}, n \in I$
d. None of these

22 If $\sin^4 x + \cos^4 x = \sin x \cos x$, then x is

a. $n\pi, n \in I$
b. $(6n + 1)\frac{\pi}{6}, n \in I$
c. $(4n + 1)\frac{\pi}{4}, n \in I$
d. None of these

Working Space

23 If $1+\sin^3 x+\cos^3 x=\frac{3}{2}\sin 2x$, then x is

a. $(2n+1)\pi, \left(2n\pi+\frac{\pi}{4}\right), n\in I$
b. $(2n+1)\pi, \left(2n\pi-\frac{\pi}{2}\right), n\in I$
c. $\left(2n\pi+\frac{\pi}{4}\right)\left(2n\pi+\frac{\pi}{2}\right), n\in I$
d. None of these

24 If $(\sin x+\cos x)-2\sqrt{2}\sin x\cos x=0$, then x is

a. $2n\pi+\frac{\pi}{4}, n\in I$
b. $2n\pi-\frac{\pi}{4}, n\in I$
c. $n\pi+\frac{\pi}{4}, n\in I$
d. $n\pi-\frac{\pi}{4}, n\in I$

Statement Type

***Direction** (Q. Nos. 25 and 26) This section is based on Statement I and Statement II. Select the correct answer from the codes given below.*

a. Both Statement I and Statement II are correct and Statement II is the correct explanation of Statement I
b. Both Statement I and Statement II are correct but Statement II is not the correct explanation of Statement I
c. Statement I is correct but Statement II is incorrect
d. Statement II is correct but Statement I is incorrect

25 **Statement I** The most general solution of $2^{\sin\theta}+2^{\cos\theta}=2^{1-\frac{1}{\sqrt{2}}}$ is $\theta=n\pi+\frac{\pi}{4}, n\in I$.

Statement II As AM $\geq$ GM.

26 **Statement I** In $(0,\pi)$, the number of solutions of the equation $\tan\theta+\tan 2\theta+\tan 3\theta=\tan\theta\tan 2\theta\tan 3\theta$ is 2.

Statement II $\tan 6\theta$ is not defined at $(2n+1)\frac{\pi}{12}$

One Integer Value Correct Type

***Direction** (Q. Nos. 27-31) This section contains 5 questions. When worked out will result in an integer from 0 to 9 (both inclusive).*

27 The number of solutions of $\sum_{r=1}^{5}\cos rx=5$ in the interval $[0,2\pi]$ is

28 The sum of roots of the equation $\cos 4t+6=7\cos 2t$ over the interval $[0, 314]$ is $990\pi\lambda$, then the value of λ is

29 Let $p, q\in N$ and $q>p$, the number of solutions of the equation $q|\sin\theta|=p|\cos\theta|$ in the interval $[0,2\pi]$ is

30 The number of solutions of the equation $\tan^2 x-\sec^{10} x=-1$ in $(0, 10)$ is

31 The value of p for which the equation $4\operatorname{cosec}^2\{\pi(p+t)\}+p^2-4p=0$ has a real solution is

JEE Archive

(Compilation of Last 13 Years' Questions of JEE Main, AIEEE, JEE Advanced and IIT JEE)

1 The possible values of $\theta \in (0, \pi)$ such that $\sin(\theta) + \sin(4\theta) + \sin(7\theta) = 0$ are **(AIEEE 2011)**

a. $\frac{2\pi}{9}, \frac{\pi}{4}, \frac{4\pi}{9}, \frac{\pi}{2}, \frac{3\pi}{4}, \frac{8\pi}{9}$

b. $\frac{\pi}{4}, \frac{5\pi}{12}, \frac{\pi}{2}, \frac{2\pi}{3}, \frac{3\pi}{4}, \frac{8\pi}{9}$

c. $\frac{2\pi}{9}, \frac{\pi}{4}, \frac{\pi}{2}, \frac{2\pi}{3}, \frac{3\pi}{4}, \frac{35\pi}{36}$

d. $\frac{2\pi}{9}, \frac{\pi}{4}, \frac{\pi}{2}, \frac{2\pi}{3}, \frac{3\pi}{4}, \frac{8\pi}{9}$

2 The number of values of x in the interval $[0, 3\pi]$ satisfying the equation $2\sin^2 x + 5\sin x - 3 = 0$, is **(AIEEE 2006)**

a. 6 b. 1
c. 2 d. 4

3 The equation $a\sin x + b\cos x = c$, where $|c| > \sqrt{a^2 + b^2}$ has **(AIEEE 2002)**

a. a unique solution b. infinite number of solutions
c. no solution d. None of these

4 For $x \in (0, \pi)$, the equation $\sin x + 2\sin 2x - \sin 3x = 3$ has **(2014 Adv., Only One Option Correct Type)**

a. infinitely many solutions
b. three solutions
c. one solution
d. no solution

5 Let $\theta, \phi \in [0, 2\pi]$ be such that $2\cos\theta(1 - \sin\phi) = \sin^2\theta$ $\left(\tan\frac{\theta}{2} + \cot\frac{\theta}{2}\right)\cos\phi - 1$, $\tan(2\pi - \theta) > 0$ and $-1 < \sin\theta < -\frac{\sqrt{3}}{2}$. Then, ϕ cannot satisfy

(2012, One or More than One Options Correct Type)

a. $0 < \phi < \frac{\pi}{2}$ b. $\frac{\pi}{2} < \phi < \frac{4\pi}{3}$
c. $\frac{4\pi}{3} < \phi < \frac{3\pi}{2}$ d. $\frac{3\pi}{2} < \phi < 2\pi$

6 Let $P = \{\theta : \sin\theta - \cos\theta = \sqrt{2}\cos\theta\}$ and $Q = \{\theta : \sin\theta + \cos\theta = \sqrt{2}\sin\theta\}$ be two sets. Then, **(2011, Only One Option Correct Type)**

a. $P \subset Q$ and $Q - P \neq \phi$
b. $Q \not\subset P$
c. $P \not\subset Q$
d. $P = Q$

7 The positive integer value of $n > 3$ satisfying the equation $\frac{1}{\sin\left(\frac{\pi}{n}\right)} = \frac{1}{\sin\left(\frac{2\pi}{n}\right)} + \frac{1}{\sin\left(\frac{3\pi}{n}\right)}$ is

(2011, Integer Type)

8 The number of values of θ in the interval $\left(-\frac{\pi}{2}, \frac{\pi}{2}\right)$ such that $\theta \neq \frac{n\pi}{5}$ for $n = 0, \pm 1, \pm 2$ and $\tan\theta = \cot 5\theta$ as well as $\sin 2\theta = \cos 4\theta$ is **(2010, Integer Type)**

9 The number of solutions of the pair of equations $2\sin^2\theta - \cos 2\theta = 0$ and $2\cos^2\theta - 3\sin\theta = 0$ in the interval $[0, 2\pi]$, is **(2007, Only One Option Correct Type)**

a. 0 b. 1
c. 2 d. 4

10 The set of values of θ satisfying the inequation $2\sin^2\theta - 5\sin\theta + 2 > 0$, where $0 < \theta < 2\pi$, is

(2006, Only One Option Correct Type)

a. $\left(0, \frac{\pi}{6}\right) \cup \left(\frac{5\pi}{6}, 2\pi\right)$

b. $\left[0, \frac{\pi}{6}\right) \cup \left(\frac{5\pi}{6}, 2\pi\right]$

c. $\left[0, \frac{\pi}{3}\right] \cup \left[\frac{2\pi}{3}, 2\pi\right]$

d. None of the above

4. Solutions of Triangles

DPP
DAILY PRACTICE PROBLEMS

DPP-1 Sine Rule and Its Applications

Subjective Questions

Working Space

Direction *(Q. Nos. 1-4) These questions are subjective in nature, need to be solved completely on notebook.*

1 In ΔABC, $a = 3$, $b = 8$ and $\sin A = \frac{5}{13}$, then find the number of triangles.

2 In ΔPQR, $P : Q : R = 3 : 5 : 4$, then find the value of $p + q + \sqrt{2}r$.

3 If $A + B + C = \pi$, then prove that

$$a^3 \cos(B - C) + b^3 \cos(C - A) + c^3 \cos(A - B) = 3abc.$$

4 In ΔABC, prove that $\frac{\sin B}{\sin(A + B)} = \frac{b}{c}$.

Only One Option Correct Type

Direction *(Q. Nos. 5-14) This section contains 10 multiple choice questions. Each question has four choices (a), (b), (c) and (d), out of which ONLY ONE is correct.*

5 If $P + Q + R = 180°$, then the value of $p\sin(Q - R) + q\sin(R - P) + r\sin(P - Q)$ is

a. 0
b. 1
c. 2
d. 3

6 In ΔABC, if $(\sqrt{3} - 1)a = 2b$, $A = 3B$, then the value of C is

a. 45°
b. 60°
c. 120°
d. 135°

7 In ΔPQR, if $\sin P \sin Q = \frac{pq}{r^2}$, then R is equal to

a. 45°
b. 90°
c. 60°
d. 120°

8 In ΔABC, $\frac{\cos A}{a} = \frac{\cos B}{b} = \frac{\cos C}{c}$ and the side $a = 2$, then the area of triangle is

a. $2\sqrt{3}$ sq units
b. 2 sq units
c. $\sqrt{3}$ sq units
d. $4\sqrt{3}$ sq units

9 The value of $a^2 (\cos^2 B - \cos^2 C) + b^2 (\cos^2 C - \cos^2 A) + c^2 (\cos^2 A - \cos^2 B)$ is

a. 0
b. 2
c. abc
d. $a^2b^2c^2$

10 If in ΔABC, $AB = 2BC$, then $\tan\frac{B}{2} : \cot\left(\frac{C - A}{2}\right)$ is equal to

a. 1 : 3
b. 3 : 1
c. 2 : 3
d. None of these

Working Space

11 If the angles of a triangle are in the ratio 4 : 1 : 1, then the ratio of the largest side to the perimeter is

a. $\sqrt{3}:(2+\sqrt{3})$　　b. 1 : 6
c. $1:(2+\sqrt{3})$　　d. 2 : 3

12 If in ΔABC, $\dfrac{a^2-b^2}{a^2+b^2}=\dfrac{\sin(A-B)}{\sin(A+B)}$, then the triangle is

a. right angled　　b. isosceles
c. right angled or isosceles　　d. right angled isosceles

13 We are given b, c and $\sin B$ such that B is acute and $b < c\sin B$, then

a. no triangle is possible　　b. one triangle is possible
c. two triangles are possible　　d. a right angled triangle is possible

14 If in ΔPQR, $\sin P$, $\sin Q$, $\sin R$ are in AP, then

a. the altitudes are in AP　　b. the altitudes are in HP
c. the medians are in AP　　d. All of these

One or More than One Options Correct Type

***Direction** (Q. Nos. 15-17) This section contains 3 multiple choice questions. Each question has four choices (a), (b), (c) and (d), out of which ONE or MORE THAN ONE are correct.*

15 In ΔABC, $A = 15^\circ$, $b = 10\sqrt{2}$ cm, the value of a for which there will be unique triangle meeting these requirement is

a. $10\sqrt{2}$ cm　　b. 15cm
c. $5(\sqrt{3}+1)$cm　　d. $5(\sqrt{3}-1)$cm

16 If the angles of a triangle are in the ratio 2 : 3 : 7, then the sides opposite these angles are in the ratio

a. $\sqrt{2}:2:(\sqrt{3}+1)$　　b. $2:\sqrt{2}:(\sqrt{3}+1)$
c. $1:\sqrt{2}:\left(\dfrac{\sqrt{2}}{\sqrt{3}-1}\right)$　　d. $\dfrac{1}{\sqrt{2}}:1:\left(\dfrac{\sqrt{3}+1}{2}\right)$

17 If angles A, B and C of a ΔABC are in AP and $\dfrac{b}{c}=\dfrac{\sqrt{3}}{\sqrt{2}}$, then

a. $\angle A = 75^\circ$　　b. $\angle C = 45^\circ$
c. $\angle B = 60^\circ$　　d. None of these

Comprehension Type

***Direction** (Q. Nos. 18-20) This section contains a paragraph, which describing theory, experiments, data, etc. Three questions related to the paragraph have been given. Each question has only one correct answer among the four given options (a), (b), (c) and (d).*

Passage

Consider ΔXYZ whose sides are x, y and z opposite to angular points X, Y and Z are in GP.

On the basis of above information, answer the following questions :

18 If r is the common ratio of GP, then

a. $\dfrac{\sqrt{5}-1}{2}<r<\dfrac{\sqrt{5}+1}{2}$　　b. $\dfrac{\sqrt{5}-2}{2}<r<\dfrac{\sqrt{5}+2}{2}$
c. $\dfrac{\sqrt{5}-1}{3}<r<\dfrac{\sqrt{5}+1}{3}$　　d. $\dfrac{\sqrt{5}-2}{3}<r<\dfrac{\sqrt{5}+2}{3}$

19 The integral value of $\dfrac{\sin Y}{\sin X}$ is

a. prime only
b. even
c. composite
d. odd

Working Space

20 The maximum value of $\dfrac{\sin Z}{\sin Y}$ is

a. irrational number
b. rational number but not integer
c. integer
d. not defined

One Integer Value Correct Type

Direction *(Q. Nos. 21-23) This section contains 3 questions. When worked out will result in an integer from 0 to 9 (both inclusive).*

21 A student standing on the bank of a river observes that the angle subtended by pole on the opposite bank is 60°, when he retreat 20 m from the bank, he finds the angle to be 30°. If the height of the pole is $10\sqrt{\lambda}$ and breadth of the river is μ, then $\mu - 2\lambda$ is equal to

22 An object is observed from three points *A*, *B* and *C* in the same horizontal line passing through the base of the object. The angle of elevation at *B* is twice and at *C* thrice that at *A*. If $AB = 4$, $BC = 3$ and height of the object is $\dfrac{\lambda}{3}\sqrt{35}$, then find the value of λ.

23 In ΔABC, the value of $(b - c)\cot\dfrac{A}{2} + (c - a)\cot\dfrac{B}{2} + (a - b)\cot\dfrac{C}{2}$ is

DPP-2 Cosine Rule and Projection Formulae

Subjective Questions

Working Space

Direction *(Q. Nos. 1-3) These questions are subjective in nature, need to be solved completely on notebook.*

1 In ΔABC, find the value of $(b-c)^2 \cos^2 \frac{A}{2} + (b+c)^2 \sin^2 \frac{A}{2}$.

2 In ΔABC, prove that

$$a(\cos B + \cos C - 1) + b(\cos C + \cos A - 1) + c(\cos A + \cos B - 1) = 0.$$

3 Two poles, A and B are on the same side of a river. From a point C in the river, the distance of poles A and B are 25 m and 30 m respectively, then find the distance between the poles when $\angle C = 45°$. (Given, $\sqrt{2} = 1.44$)

Only One Option Correct Type

Direction *(Q. Nos. 4-14) This section contains 11 multiple choice questions. Each question has four choices (a), (b), (c) and (d), out of which ONLY ONE is correct.*

4 In ΔABC, if

$$\frac{\cos A}{b\cos C + c\cos B} + \frac{\cos B}{c\cos A + a\cos C} + \frac{\cos C}{a\cos B + b\cos A} = \frac{a^2+b^2+c^2}{\lambda abc},$$

then the value of $\lambda + \lambda + \lambda + \ldots$ upto 20 terms is

a. 10 b. 20

c. 30 d. 40

5 If the angles of ΔABC are in AP, then

a. $c^2 = a^2 + b^2 - ab$ b. $b^2 = a^2 + c^2 - ac$

c. $a^2 = b^2 + c^2 - ac$ d. $b^2 = a^2 + c^2$

6 In ΔABC, if $a^2 + b^2 = 3c^2$, then $\cot A + \cot B - \cot C$ is equal to

a. 1 b. 2

c. 0 d. None of these

7 In ΔABC, if $(a+b+c)(a-b+c) = 3ac$, then

a. $\angle B = 60°$ b. $\angle B = 30°$

c. $\angle C = 60°$ d. $\angle A = \angle C = 90°$

8 If in a ΔABC, $\angle C = 60°$, then $\frac{1}{a+c} + \frac{1}{b+c}$ is equal to

a. $\frac{1}{a+b+c}$ b. $\frac{2}{a+b+c}$

c. $\frac{3}{a+b+c}$ d. None of these

Working Space

9 In ΔABC, $(b^2 - c^2)\cot A + (c^2 - a^2)\cot B + (a^2 - b^2)\cot C$ is equal to

a. 0
b. $a^2 + b^2 + c^2$
c. $2(a^2 + b^2 + c^2)$
d. $\frac{1}{2abc}$

10 In ΔABC, $\frac{2\cos A}{a} + \frac{\cos B}{b} + \frac{2\cos C}{c} = \frac{a}{bc} + \frac{b}{ca}$, then the value of angle A is **[IIT JEE 1993]**

a. 45°
b. 30°
c. 90°
d. 60°

11 In ΔABC, if $A + C = 2B$, then $\frac{a + c}{\sqrt{a^2 + c^2 - ac}}$ is equal to

a. $2\cos\left(\frac{A - C}{2}\right)$
b. $\sin\left(\frac{A + C}{2}\right)$
c. $\sin\frac{A}{2}$
d. None of these

12 In ΔABC, the value of $2ab\sin\left(\frac{A - C + B}{2}\right)$ is equal to

a. $a^2 + b^2 - c^2$
b. $b^2 + c^2 - a^2$
c. $c^2 + a^2 - b^2$
d. None of these

13 If in a ΔABC, $\cos A + \cos B + \cos C = \frac{3}{2}$, then the triangle is

a. isosceles
b. equilateral
c. right angled
d. None of these

14 If the line segment joining the points $A(a, b)$ and $B(c, d)$ subtends an angle θ at the origin, then $\cos\theta$ is

a. $\frac{ab + cd}{\sqrt{(a^2 + b^2)(c^2 + d^2)}}$
b. $\frac{ac + bd}{\sqrt{(a^2 + b^2)(c^2 + d^2)}}$
c. $\frac{ac + bd}{\sqrt{(a^2 + c^2)(b^2 + d^2)}}$
d. None of these

One or More than One Options Correct Type

Direction *(Q. Nos. 15-18) This section contains 4 multiple choice questions. Each question has four choices (a), (b), (c) and (d), out of which ONE or MORE THAN ONE are correct.*

15 If a, b, and c are the sides of a ΔABC such that $\frac{a + b}{11} = \frac{b + c}{13} = \frac{c + a}{12}$ and $\frac{\cos A}{p} = \frac{\cos B}{q} = \frac{\cos C}{r}$, then

a. $p = 25$
b. $q = 19$
c. $r = 7$
d. None of these

16 If in ΔABC, $\angle B = 60°$, then

a. $(a - b)^2 = c^2 - ab$
b. $(b - c)^2 = a^2 - bc$
c. $(c - a)^2 = b^2 - ac$
d. $a^2 + b^2 + c^2 = 2b^2 + ac$

Working Space

17 If there exists a ΔABC, then

a. $\tan A + \tan B + \tan C = 0$

b. $\dfrac{\sin A}{2} = \dfrac{\sin B}{3} = \dfrac{\sin C}{7}$

c. $(a+b)^2 = c^2 + ab$ and $\sqrt{2}(\sin A + \cos A) = \sqrt{3}$

d. $\sin A + \sin B = \dfrac{\sqrt{3}+1}{2}$, $\cos A \cos B = \dfrac{\sqrt{3}}{4} = \sin A \sin B$

18 The lengths of the sides of a triangle are $a = \alpha - \beta$, $b = \alpha + \beta$ and $c = \sqrt{3\alpha^2 + \beta^2}$ $(\alpha > \beta > 0)$, then

a. $C = \dfrac{2\pi}{3}$

b. $A + B = \dfrac{\pi}{3}$

c. $C = \dfrac{3\pi}{4}$

d. None of these

One Integer Value Correct Type

Direction (Q. Nos. 19 and 20) *This section contains 2 questions. When worked out will result in an integer from 0 to 9 (both inclusive).*

19 In ΔABC, $AB = 2$, $BC = 4$, $CA = 3$ and D is the mid-point of BC, then $2AD^2 = \ldots\ldots$.

20 If in a ΔABC, $BC = 5$, $CA = 4$, $AB = 3$ and D, E are the points on BC such that $BD = DE = EC$, then $16\tan(\angle CAE) = \ldots$.

Matching List Type

Direction (Q. No. 21) *Choices for the correct combination of elements from Column I and Column II are given as options (a), (b), (c) and (d), out of which one is correct.*

21 Match the Column I with Column II and select the correct option from the codes given below.

	Column I		Column II
i.	In ΔABC, if $a^4 - 2(b^2 + c^2)a^2 + b^4 + b^2c^2 + c^4 = 0$, then $\angle A$ is	p.	30°
ii.	In ΔABC, if $a^4 + b^4 + c^4 = a^2b^2 + 2b^2c^2 + 2c^2a^2$, then $\angle C$ is	q.	60°
iii.	In ΔABC, if $a^4 + b^4 + c^4 + 2a^2c^2 = 2a^2b^2 + 2b^2c^2$, then $\angle B$ is	r.	90°
		s.	120°
		t.	150°

Codes

	i	ii	iii
a.	q,s	p,t	r
b.	p	q,r	t
c.	p,q	r,s	p,t
d.	q	p	r,s

DPP-3 Napier's Analogy, Area of Triangle (Heron's Formula), Half-Angle Formulae

Subjective Questions

Direction (*Q. Nos. 1-3*) *These questions are subjective in nature, need to be solved completely on notebook.*

Working Space

1 In ΔABC, prove that $\left(\cot\frac{B}{2}+\cot\frac{C}{2}\right)\left(a\sin^2\frac{B}{2}+b\sin^2\frac{A}{2}\right)=a\cot\frac{C}{2}$.

2 In ΔABC, if $\cos B+\cos C=4\sin^2\frac{A}{2}$, then prove that b, a, c are in AP.

3 In ΔABC, if $\tan\frac{A}{2}\tan\frac{C}{2}=\frac{1}{2}$, then prove that $3b=a+c$.

True/False Statements

Direction (*Q. Nos. 4 and 5*) *This section contains 2 questions. Each question has statement which is TRUE or FALSE.*

4 In ΔABC, $(s-b)(s-c)=s(s-a)$, then $\angle A$ is 90°.

5 In ΔABC, $(s-a)(s-b)=\lambda\sin^2\frac{C}{2}$, then λ is bc.

Only One Option Correct Type

Direction (*Q. Nos. 6-15*) *This section contains 10 multiple choice questions. Each question has four choices (a), (b), (c) and (d), out of which ONLY ONE is correct.*

6 In ΔABC, if a, b, c are in AP, then the value of $\dfrac{\sin\frac{A}{2}\sin\frac{C}{2}}{\sin\frac{B}{2}}$ is

a. 1 b. $\frac{1}{2}$ c. 2 d. −1

7 In ΔABC, the value of $1-\tan\frac{B}{2}\tan\frac{C}{2}$ is

a. $\frac{2a}{a+b+c}$ b. $\frac{a}{a+b+c}$ c. $\frac{2}{a+b+c}$ d. $\frac{4a}{a+b+c}$

8 In ΔABC, if $\sin\frac{B}{2}\cdot\sin\frac{C}{2}=\frac{b+c-a}{a}$, then the triangle is

a. equilateral b. isosceles c. right angled d. not possible

Working Space

9 If in a ΔABC, $a=\sqrt{3}$, $b=1$ and $A-B=90°$, then $\angle C$ is

a. 30°
b. 45°
c. 75°
d. 15°

10 If in a ΔABC, $a\cos^2\frac{B}{2}+b\cos^2\frac{A}{2}=\frac{3c}{2}$, then its sides are in

a. AP
b. GP
c. HP
d. None of these

11 In ΔABC, $\sin\frac{A}{2}\sin\frac{B}{2}=\sin\frac{C}{2}$ and $2s$ is the perimeter of the triangle, then s is equal to

a. $2c$
b. c
c. $3c$
d. $4b$

12 In ΔABC, $\dfrac{\tan\frac{B}{2}-\tan\frac{C}{2}}{\tan\frac{B}{2}+\tan\frac{C}{2}}$ is equal to

a. $\dfrac{b-c}{b+c}$
b. $\dfrac{b-c}{a}$
c. $\dfrac{b-c}{a+b+c}$
d. $\dfrac{a}{b+c}$

13 Observe the following statements :

I. In a ΔABC,

$$b\cos^2\frac{C}{2}+c\cos^2\frac{B}{2}=s$$

II. In a ΔABC,

$$\cot\frac{A}{2}=\frac{b+c}{a}\quad\Rightarrow\quad\angle B=90°$$

Which of the following is correct?

a. Both I and II are true
b. I is true and II is false
c. I is false and II is true
d. Both I and II are false

14 In a ΔABC, if $\cot\frac{A}{2}\cot\frac{B}{2}=c$, $\cot\frac{B}{2}\cot\frac{C}{2}=a$, $\cot\frac{C}{2}\cot\frac{A}{2}=b$. Then, the value of $\dfrac{1}{s-a}+\dfrac{1}{s-b}+\dfrac{1}{s-c}$ is

a. 1
b. 2
c. 3
d. None of these

15 In a ΔABC, $\tan\frac{A}{2}=\frac{5}{6}$ and $\tan\frac{C}{2}=\frac{2}{5}$, then

a. a, c, b are in AP
b. a, b, c are in AP
c. b, a, c are in AP
d. a, b, c are in GP

One or More than One Options Correct Type

Direction *(Q. Nos. 16-18) This section contains 3 multiple choice questions. Each question has four choices (a), (b), (c) and (d), out of which ONE or MORE THAN ONE are correct.*

16 There exists a ΔABC satisfying the conditions **[IIT JEE 1986]**

a. $b\sin A=a$, $A<\frac{\pi}{2}$
b. $b\sin A>a$, $A>\frac{\pi}{2}$
c. $b\sin A>a$, $A<\frac{\pi}{2}$
d. $b\sin A<a$, $A<\frac{\pi}{2}$, $b>a$

Working Space

17 In a ΔABC with fixed base BC, the vertex moves such that $\cos B + \cos C = 4\sin^2 \frac{A}{2}$. If a, b and c denote the lengths of sides of the triangle opposite to the angles A, B and C respectively, then **[IIT JEE 2009]**

a. $b + c = 4a$
b. $b + c = 2a$
c. locus of point A is an ellipse
d. locus of point A is a pair of straight line

18 Which of the following is/are not true in a ΔABC? **[IIT JEE 2005]**

a. $(b+c)\sin\left(\frac{B-C}{2}\right) = 2a\cos\frac{A}{2}$
b. $(b+c)\cos\frac{A}{2} = 2a\sin\left(\frac{B-C}{2}\right)$
c. $(b-c)\cos\frac{A}{2} = a\sin\left(\frac{B-C}{2}\right)$
d. $(b-c)\sin\left(\frac{B-C}{2}\right) = 2a\cos\frac{A}{2}$

Comprehension Type

Direction *(Q. Nos. 19-21) This section contains a paragraph, which describing theory, experiments, data, etc. Three questions related to the paragraph have been given. Each question has only one correct answer among the four given options (a), (b), (c) and (d).*

Passage

G is centroid of ΔABC. Perpendiculars from vertices A, B, C meet on the sides BC, CA and AB at D, E and F, respectively. P, Q and R are feet of the perpendiculars from G on sides BC, CA and AB, respectively. L, M and N are the mid-points of sides BC, CA and AB respectively.

On the basis of above information, answer the following questions :

19 Length of the side PG is

a. $\frac{1}{2}b\sin C$
b. $\frac{1}{2}c\sin C$
c. $\frac{2}{3}b\sin C$
d. $\frac{1}{3}c\sin B$

20 The ratio of area of ΔGPL to the area of ΔALD is equal to

a. $\frac{1}{3}$
b. $\frac{1}{9}$
c. $\frac{2}{3}$
d. $\frac{4}{9}$

21 Area of ΔPQR is

a. $\frac{1}{9}(a^2+b^2+c^2)\sin A\sin B\sin C$
b. $\frac{1}{18}(a^2+b^2+c^2)\sin A\sin B\sin C$
c. $\frac{2}{9}(a^2+b^2+c^2)\sin A\sin B\sin C$
d. $\frac{1}{3}(a^2+b^2+c^2)\sin A\sin B\sin C$

One Integer Value Correct Type

Direction *(Q. Nos. 22-24) This section contains 3 questions. When worked out will result in an integer from 0 to 9 (both inclusive).*

22 If area of triangle is given by $\Delta = a^2 - (b-c)^2$, then $\tan\frac{A}{2} = \frac{1}{\lambda}$, thus λ is

23 In ΔABC, if $25 = a + b + c$, $\frac{s-a}{11} = \frac{s-b}{12} = \frac{s-c}{13}$ and $33\mu\tan^2\frac{A}{2} = 65$, then μ is

24 Let PQR be a triangle of area Δ sq units with $a = 2$, $b = \frac{7}{2}$ and $c = \frac{5}{2}$, where a, b, c are the lengths of the sides of the triangle opposite to the angles at P,Q and R, respectively and $\frac{2\sin P - \sin 2P}{2\sin P + \sin 2P} = \frac{\lambda}{16\Delta^2}$, then λ is

DPP-4 Half-Angle Formulae using 's', Heron's Formulae, Area of Triangle using Sides and Angles

Only One Option Correct Type

Working Space

Direction *(Q. Nos. 1-12) This section contains 12 multiple choice questions. Each question has four choices (a), (b), (c) and (d), out of which ONLY ONE is correct.*

1 In ΔABC, side $a = 4$cm, $\angle B = 30°$ and $\angle C = 45°$, then the area of the triangle is

a. $\left(\dfrac{\sqrt{3}+1}{3}\right)$ cm^2
b. $4(\sqrt{3}-1)$ cm^2
c. $\left(\dfrac{\sqrt{3}+1}{2\sqrt{2}}\right)$ cm^2
d. None of these

2 The area of isosceles triangle is 9 cm^2. If the equal sides are 6 cm, then the angle between them is

a. 60°
b. 30°
c. 90°
d. 45°

3 The area of ΔABC, in which $a = 1, b = 2, \angle C = 60°$, is

a. $\dfrac{1}{2}$ sq units
b. $\sqrt{3}$ sq units
c. $\dfrac{\sqrt{3}}{2}$ sq units
d. $\dfrac{3}{2}$ sq units

4 If in ΔABC, the altitudes p_1, p_2, p_3 are in HP which are drawn from the vertices A, B, C on opposite sides, then $\sin A$, $\sin B$, $\sin C$ are in

a. AP
b. GP
c. HP
d. None of these

5 If the area of a ΔABC is Δ, then $a^2 \sin 2B + b^2 \sin 2A$ is equal to

a. 3Δ
b. 2Δ
c. 4Δ
d. None of these

6 The area of ΔABC is $20\sqrt{3}$ cm^2, $\angle C = 60°$ and its perimeter is 20 cm, then the side c will be

a. 4 cm
b. 7 cm
c. 8 cm
d. None of these

Working Space

7 If $\Delta = a^2 - (b-c)^2$, where Δ is the area of the ΔABC, then $\tan A$ is equal to

a. $\frac{15}{18}$ b. $\frac{8}{15}$

c. $\frac{8}{17}$ d. $\frac{1}{2}$

8 In ΔABC and ΔPQR, $AB = PQ$, $AC = QR$ and $\angle A = 2\angle Q$. Two triangles will have the same area, if $\angle A$ is equal to

a. $\frac{\pi}{3}$ b. $\frac{\pi}{2}$

c. $\frac{2\pi}{3}$ d. $\frac{5\pi}{6}$

9 If in a ΔABC, $a = 6$, $b = 3$ and $\cos(A+B) = \frac{4}{5}$, then its area will be

a. 5.4 sq units b. 5 sq units

c. 3.2 sq units d. None of these

10 In ΔABC, $a^2 + b^2 = 1999c^2$, then $999\left(\frac{\cot A + \cot B}{\cot C}\right)$ is equal to

a. 0 b. 1

c. 2 d. None of these

11 If Δ is the area of triangle and $2s$ is the sum of three sides of a triangle such that $\Delta \le \frac{s^2}{\lambda\sqrt{3}}$, then λ is equal to

a. 1 b. 2

c. 3 d. None of these

12 In ΔABC, if $\tan\frac{A}{2} = \frac{5}{6}$ and $\tan\frac{B}{2} = \frac{20}{37}$, then the sides a, b, c of triangle are in

a. AP b. GP

c. HP d. None of these

One or More than One Options Correct Type

***Direction** (Q. Nos. 13-15) This section contains 3 multiple choice questions. Each question has four choices (a), (b), (c) and (d), out of which ONE or MORE THAN ONE are correct.*

13 If the angles of a triangle are 30° and 45° and the included side is $\sqrt{3} + 1$, then area of triangle **[IIT JEE 1988]**

a. $\frac{1+\sqrt{3}}{2}$ b. $\sqrt{2}\cos 15°$

c. $\sqrt{2}\sin 75°$ d. None of these

14 If p_1, p_2, p_3 are the altitudes of a triangle from the vertices A, B, C and Δ is the area of triangle, then $\frac{1}{p_1} + \frac{1}{p_2} - \frac{1}{p_3}$ is equal to **[IIT JEE 1978]**

a. $\frac{s-c}{\Delta}$

b. $\frac{a+b-c}{2\Delta}$

c. $\frac{2ab}{(a+b+c)\Delta}\cos^2\frac{C}{2}$

d. None of the above

Working Space

15 Internal bisector of $\angle A$ of ΔABC meets side BC at D. A line drawn through D perpendicular to AD intersects the side AC at E and side AB at F. If a, b, c represent the sides of ΔABC, then **[IIT JEE 2006]**

a. AE is HM of b and c

b. $AD = \dfrac{2bc}{b+c} \cos \dfrac{A}{2}$

c. $EF = \dfrac{4bc}{b+c} \sin \dfrac{A}{2}$

d. the ΔAEF is isosceles

Comprehension Type

Direction *(Q. Nos. 16-18) This section contains a paragraph, which describing theory, experiments, data, etc. Three questions related to the paragraph have been given. Each question has only one correct answer among the four given options (a), (b), (c) and (d).*

Passage

In a triangle, if sum of two sides is p and product is q ($p \geq 2\sqrt{q}$) such that $(p+r)(p-r) = q$, where r is the third side of a triangle.

On the basis of above information, answer the following questions :

16 The greatest angle of triangle is

a. 105°

b. 120°

c. 135°

d. 150°

17 The area of triangle is

a. $\dfrac{\sqrt{3}}{4} q$ sq units

b. $\dfrac{\sqrt{3}}{4} p$ sq units

c. $\dfrac{\sqrt{3}}{4} r$ sq units

d. None of these

18 The sides of triangle are

a. $\dfrac{p \pm \sqrt{p^2 - 4q}}{2}, r$

b. $\dfrac{q \pm \sqrt{q^2 - 4r}}{2}, r$

c. $\dfrac{r \pm \sqrt{r^2 - 4p}}{2}, r$

d. None of these

One Integer Value Correct Type

Direction *(Q. Nos. 19 and 20) This section contains 2 questions. When worked out will result in an integer from 0 to 9 (both inclusive).*

19 The sides of a cyclic quadrilateral are in AP, the shortest side is 6 and the difference of the longest side and the shortest side is also 6. Let A be the area of quadrilateral such that $A^2 = 1152\lambda$, then λ is

20 In a ΔABC, if $\cot \dfrac{A}{2} \cot \dfrac{B}{2} = c$, $\cot \dfrac{B}{2} \cot \dfrac{C}{2} = a$ and $\cot \dfrac{C}{2} \cot \dfrac{A}{2} = b$, then the value of $\dfrac{1}{s-a} + \dfrac{1}{s-b} + \dfrac{1}{s-c}$ is

DPP-5 Properties of Triangles

Subjective Questions

Working Space

Direction (Q. Nos. 1-3) *These questions are subjective in nature, need to be solved completely on notebook.*

1 In ΔABC, prove that $\dfrac{a\cos A + b\cos B + c\cos C}{a+b+c} = \dfrac{r}{R}$.

2 In ΔABC, prove that $2R^2(\sin 2A + \sin 2B + \sin 2C) = 4\Delta$.

3 In ΔABC, if $r + r_1 + r_3 - r_2 = \lambda R\cos B$, then find the value of λ.

True/False Statements

Direction (Q. Nos. 4-7) *This section contains 4 questions. Each question has statement which is TRUE or FALSE.*

4 $\tan\dfrac{A}{2}\tan\dfrac{B}{2} = \dfrac{s-c}{s}$

5 $\cot\dfrac{A}{2} + \cot\dfrac{B}{2} = \dfrac{c}{s-c}\cot\dfrac{C}{2}$

6 $\sin C = \dfrac{2\Delta}{ab}$

7 $r = (s-c)\tan\dfrac{A}{2}$, where r is radius.

Fill in the Blanks

Direction (Q. Nos. 8 and 9) *This section contains 2 sentences. Complete the given sentences filling the blanks with suitable words, information, data, etc.*

8 If $r_1 = \lambda R\sin\dfrac{A}{2}\cos\dfrac{B}{2}\cos\dfrac{C}{2}$ and λ is any real constant, then λ = ...

9 If $\sin C = \dfrac{\sqrt{s(s-a)(s-b)(s-c)}}{ab}\cdot\lambda$, then λ = ...

Only One Option Correct Type

Direction (Q. Nos. 10-19) *This section contains 10 multiple choice questions. Each question has four choices (a), (b), (c) and (d), out of which ONLY ONE is correct.*

10 In right angled ΔABC, if $\angle B = 90°$, a = 6cm and the radius of the circumcircle is 5 cm, then the area of ΔABC is

a. 24 cm^2 b. 30cm^2

c. 36 cm^2 d. None of these

Working Space

11 If 3, 4, 5 be the sides of a ΔABC, then its inradius is

a. $\sqrt{\frac{8}{7}}$ b. $\sqrt{8}$

c. $\sqrt{7}$ d. None of these

12 In ΔABC, if $a = 5, b = 12$ and $\angle C = 90°$, then R is

a. $\frac{a}{2\sin B}$ b. $\frac{b}{3\sin B}$

c. $\frac{c}{\sin C}$ d. None of these

13 In an equilateral triangle, the inradius and circumradius are connected by

a. $r = 4R$ b. $R = 2r$

c. $R = 3r$ d. None of these

14 In ΔABC, $2R^2 \sin A \sin B \sin C$ is equal to

a. S^2 b. $ab + bc + ca$

c. Δ d. None of these

15 In any ΔABC, the value of $a \cot A + b \cot B + c \cot C$ is

a. $r + R$ b. $r - R$

c. $2(r + R)$ d. $2(r - R)$

16 If the radius of circumcircle of an isosceles ΔABC is equal to $AB = (AC)$, then the $\angle A$ is

a. $\frac{\pi}{6}$ b. $\frac{\pi}{3}$ c. $\frac{\pi}{2}$ d. $\frac{2\pi}{3}$

17 Let p, q, r be the sides of a ΔPQR such that $p : q : r = 4 : 5 : 6$, then the ratio of radius of circumcircle to that of incircle is

a. 16 : 7 b. 7 : 16

c. 5 : 16 d. None of these

18 In ΔABC, $\angle C = \frac{\pi}{2}$, $a = 4$ and $b = 3$, then the value of $\frac{R}{r}$ is

a. $\frac{3}{5}$ b. $\frac{5}{2}$ c. $\frac{5}{3}$ d. $\frac{2}{5}$

19 The sum of radii of inscribed and circumscribed circles for an n-sided regular polygon of side a, is

a. $\frac{a}{2}\cot\frac{\pi}{2n}$ b. $\frac{a}{2}\tan\frac{\pi}{2n}$

c. $a\cot\frac{\pi}{2n}$ d. None of these

One or More than One Options Correct Type

Direction (Q. Nos. 20-22) *This section contains 3 multiple choice questions. Each question has four choices (a), (b), (c) and (d), out of which ONE or MORE THAN ONE are correct.*

20 If in ΔABC, $r_1 = 2r_2 = 3r_3$, then

a. $\frac{a}{b} = \frac{4}{5}$ b. $\frac{a}{b} = \frac{5}{4}$

c. $\frac{a}{c} = \frac{3}{5}$ d. $\frac{a}{c} = \frac{5}{3}$

Working Space

21 If in ΔABC, a, b, c are in AP and p_1, p_2, p_3 are the altitudes from the vertices A, B, C respectively, then

a. p_1, p_2, p_3 are in AP
b. p_1, p_2, p_3 are in HP
c. $p_1 + p_2 + p_3 \le \frac{3R}{\Delta}$
d. $\frac{1}{p_1} + \frac{1}{p_2} + \frac{1}{p_3} \le \frac{3R}{\Delta}$

22 Given an isosceles triangle with equal sides of length b, base angle $\alpha < \frac{\pi}{4}$ and R, r the radii and O, I the centres of the circumcircle and incircle respectively. Then,

a. $R = \frac{1}{2} b \operatorname{cosec} \alpha$
b. $\Delta = 2b^2 \sin 2\alpha$
c. $r = \frac{b \sin 2\alpha}{2(1 + \cos \alpha)}$
d. $OI = \left| \frac{b \cos \frac{3\alpha}{2}}{2 \sin \alpha \cos \frac{\alpha}{2}} \right|$

Comprehension Type

Direction (Q. Nos. 23-29) *This section contains 2 paragraphs, each describing theory, experiments, data, etc. Seven questions related to the paragraphs have been given. Each question has only one correct answer among the four given options (a), (b), (c) and (d).*

Passage I

Consider the circle $x^2 + y^2 = 9$ and the parabola $y^2 = 8x$. They intersect at P and Q in the first and fourth quadrants respectively. Tangents to the circle at P and Q intersect the X-axis at R and tangents to the parabola at P and Q intersect the X-axis at S. **[IIT JEE 2007]**

On the basis of above information, answer the following questions :

23 The ratio of the areas of ΔPQS and ΔPQR is

a. $1:\sqrt{2}$
b. $1:2$
c. $1:4$
d. $1:8$

24 The radius of circumcircle of the ΔPRS is

a. 5
b. $3\sqrt{3}$
c. $3\sqrt{2}$
d. $2\sqrt{3}$

25 The radius of incircle of the ΔPQR is

a. 4
b. 3
c. $\frac{8}{3}$
d. 2

Passage II

The Δ formed by joining the three excentres I_1, I_2 and I_3 of ΔABC is called the excentral or excentric triangle and in this case internal angle bisector of ΔABC are the altitudes of $\Delta I_1 I_2 I_3$.

On the basis of above information, answer the following questions :

26 The incentre I of ΔABC is the .... of the excentral $\Delta I_1 I_2 I_3$.

a. circumcentre
b. orthocentre
c. centroid
d. None of these

Working Space

27 Angles of the $\Delta I_1I_2I_3$ are

a. $\frac{\pi}{2}-\frac{A}{2}, \frac{\pi}{2}-\frac{B}{2}, \frac{\pi}{2}-\frac{C}{2}$

b. $\frac{\pi}{2}+\frac{A}{2}, \frac{\pi}{2}+\frac{B}{2}, \frac{\pi}{2}+\frac{C}{2}$

c. $\frac{\pi}{2}-A, \frac{\pi}{2}-B, \frac{\pi}{2}-C$

d. None of these

28 Sides of the $\Delta I_1I_2I_3$ are

a. $R\cos\frac{A}{2}, R\cos\frac{B}{2}, R\cos\frac{C}{2}$

b. $4R\cos\frac{A}{2}, 4R\cos\frac{B}{2}, 4R\cos\frac{C}{2}$

c. $2R\cos\frac{A}{2}, 2R\cos\frac{B}{2}, 2R\cos\frac{C}{2}$

d. None of the above

29 Value of $II_1^2 + I_2I_3^2 = II_2^2 + I_1I_3^2 = II_3^2 + I_1I_2^2$

a. $4R^2$

b. $16R^2$

c. $32R^2$

d. $64R^2$

One Integer Value Correct Type

Direction *(Q. Nos. 30 and 31) This section contains 2 questions. When worked out will result in an integer from 0 to 9 (both inclusive).*

30 If the area of equilateral ΔABC which contains three coins of unit radius, is $6 + 4\sqrt{p}$. Then, the value of p is

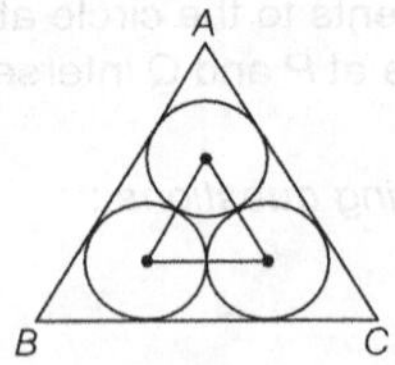

31 In ΔABC, if $\frac{r_1}{6} = \frac{r_2}{3} = \frac{r_3}{2}$ and $\frac{a}{b} + \frac{b}{c} + \frac{c}{a} = \frac{191}{20\lambda}$, then the value of λ is

SET - 1

Revisal Problems for JEE Main

Only One Option Correct Type

Working Space

Direction *(Q. Nos. 1-10) This section contains 10 multiple choice questions. Each question has four choices (a), (b), (c) and (d), out of which ONLY ONE is correct.*

1 If in ΔABC, $(a+b+c)(b+c-a)=\lambda bc$, then

a. $\lambda<0$
b. $\lambda>6$
c. $0<\lambda<4$
d. $\lambda>4$

2 In ΔABC, if $r=r_1+r_3-r_2$ and $\angle B>\frac{\pi}{3}$, then the range of $\frac{s}{r}$ is

a. $\left(\frac{1}{2}, 2\right)$
b. $\left(\frac{1}{2}, \infty\right)$
c. $\left(\frac{1}{2}, 3\right)$
d. $(3, \infty)$

3 In an equilateral triangle, $R:r:r_2$ is equal to

a. 1 : 1 : 1
b. 1 : 2 : 3
c. 2 : 1 : 3
d. 3 : 2 : 4

4 If p, q, r are the length of internal angle bisectors of a ΔABC, then $\frac{1}{p}\cos\frac{A}{2}+\frac{1}{q}\cos\frac{B}{2}+\frac{1}{r}\cos\frac{C}{2}$ is equal to

a. $a^{-1}+b^{-1}-c^{-1}$
b. $a^{-1}-b^{-1}+c^{-1}$
c. $a^{-1}+b^{-1}+c^{-1}$
d. None of these

5 If a, b, c are the sides of a ΔABC such that $\cos A\cos B+\sin A\sin B\sin C=1$, then $a:b:c$ is equal to

a. $1:1:\sqrt{2}$
b. $1:\sqrt{2}:1$
c. $\sqrt{2}:1:1$
d. None of these

6 In ΔABC, if $\sin A=x$, $\cos B=y$, then the value of $\cos^2 C$ is

a. $x^2-2xy\sin C+y^2$
b. $x^2+2xy\sin C+y^2$
c. $x^2-2xy\sin C-y^2$
d. None of these

7 In ΔABC, $2a^2+4b^2+c^2=4ab+2ac$, then the numerical value of $\cos B$ is

a. $\frac{5}{8}$
b. $\frac{3}{8}$
c. $\frac{7}{8}$
d. None of these

8 If a, b, c are the sides of a ΔABC and $\lambda=\frac{(a+b+c)^2}{ab+bc+ca}$, then $\lambda\in$

a. [1, 2]
b. [3, 4]
c. [3, 4)
d. None of these

9 Circle with radii 3, 4, 5 touch each other externally. If P is the point of intersection of tangents to these circles at their point of contact the distance of P from the point of contanct is

a. $\sqrt{2}$
b. $\sqrt{3}$
c. $\sqrt{5}$
d. None of these

10 If a, b, c are the sides of ΔABC and if roots of the equation $a(b-c)x^2+b(c-a)x+c(a-b)=0$ are equal, then $\sin^2\frac{A}{2}, \sin^2\frac{B}{2}, \sin^2\frac{C}{2}$ are in

a. AP
b. GP
c. HP
d. None of these

SET - 2

Revisal Problems for JEE Advanced

Only One Option Correct Type

Working Space

Direction (Q. Nos. 1-11) *This section contains 11 multiple choice questions. Each question has four choices (a), (b), (c) and (d), out of which ONLY ONE is correct.*

1 If in ΔABC, $b=c$ and $\angle A=20°$, then the value of a^3+b^3 is

a. $3a^2b$ b. $3b^2c$

c. $3c^2a$ d. abc

2 In ΔABC, $a=3$, $b=5$ and a, b, c are in AP, then the maximum value of $\tan A \tan B$ is

a. $\frac{1}{2}$ b. $\frac{1}{3}$

c. $\frac{1}{4}$ d. $\frac{1}{5}$

3 In ΔABC, $\angle A=60°$, $\angle B=45°$ and let D divides AB internally in 1 : 3.

Then, $\frac{\sqrt{6}\sin(\angle ACD)}{\sin(\angle BCD)}$ is equal to

a. $\frac{1}{\sqrt{6}}$ b. $\sqrt{6}$

c. 1 d. None of these

4 If two adjacent sides of a cyclic quadrilateral are 2 and 5 and the angle between them is 60°. If the area of the quadrilateral is $4\sqrt{3}$ sq units, then the remaining two sides are

a. 2, 3 b. 1, 2

c. 3, 4 d. 2, 2

5 If a, b, c, d are sides of a quadrilateral, then the value of $\frac{a^2+b^2+c^2}{d^2}$ is greater than

a. 1 b. $\frac{1}{2}$

c. $\frac{1}{3}$ d. $\frac{1}{4}$

6 If p is the product of sines of angles of a triangle and q be the product of their cosines, then the tangents of the angle are roots of the equation

a. $qx^3-px^2+(1+q)x-p=0$

b. $px^3-qx^2+(1+p)x-q=0$

c. $(1+q)x^3-px^2+qx-q=0$

d. None of the above

7 Let $F(x+y)=F(x)\cdot F(y), \forall x, y$ and $F(1)=2$. If in a ΔABC, $a=F(3)$, $b=F(1)+F(3)$ and $c=F(2)+F(3)$, then $2A$ is

a. C b. $2C$

c. $3C$ d. $4C$

8 In ΔABC, if r_1, r_2, r_3 be the exradii and $\frac{bc}{r_1}+\frac{ca}{r_2}+\frac{ab}{r_3}=\lambda\left[s\Sigma\frac{1}{a}-3\right]$, then λ is equal to

a. $\frac{abc}{\Delta}$ b. $\frac{2abc}{\Delta}$

c. abc d. None of these

Working Space

9 In ΔABC, if $\cos A + \cos B + \cos C = \frac{7}{4}$, then $\frac{27R}{r}$ is equal to

a. 26
b. 36
c. 18
d. None of these

10 The cosine of the obtuse angle formed by medians drawn from the vertices of acute angles of an isosceles right angled triangle is

a. $\frac{4}{5}$
b. $\frac{3}{5}$
c. $\frac{-4}{5}$
d. None of these

11 A triangle is inscribed in a circle. The vertices of the triangle divide the circle into three arcs of lengths 3, 4 and 5 units, then the area of triangle is

a. $\frac{9\sqrt{3}(1+\sqrt{3})}{\pi^2}$ sq units
b. $\frac{8\sqrt{3}(1+\sqrt{3})}{\pi^2}$ sq units
c. $\frac{9\sqrt{3}(1+\sqrt{3})}{\pi}$ sq units
d. None of these

One or More than One Options Correct Type

Direction *(Q. Nos. 12-18) This section contains 7 multiple choice questions. Each question has four choices (a), (b), (c) and (d), out of which ONE or MORE THAN ONE are correct.*

12 In ΔABC, if $\frac{a\cos A + b\cos B + c\cos C}{a\sin B + b\sin C + c\sin A} = \frac{a+b+c}{9R}$, then

a. $\angle A = 60°$
b. $\angle B = 60°$
c. $\angle C = 60°$
d. None of these

13 Let ABC be an isosceles triangle with base BC. If r is the radius of the circle inscribed in ΔABC and r_1 is the radius of the circle escribed opposite to the $\angle A$, then the product $r_1 \cdot r$ can be equal to

a. $R^2 \sin^2 A$
b. $R^2 \sin^2 2B$
c. $\frac{1}{2}a^2$
d. $\frac{a^2}{4}$

14 Lengths of the tangent from A, B and C to the incircle are in AP, then

a. r_1, r_2, r_3 are in HP
b. r_1, r_2, r_3 are in AP
c. a, b, c are in AP
d. $\cos A = \frac{4c - 3b}{2c}$

15 If p is the length of median from the vertex A to side BC of a ΔABC, then

a. $4p^2 = 2b^2 + 2c^2 - a^2$
b. $4p^2 = b^2 + c^2 + 2bc\cos A$
c. $4p^2 = a^2 + 4bc\cos A$
d. $4p^2 = (2s-a)^2 - 4bc\sin^2 \frac{A}{2}$

16 If a, b, A be given in a ΔABC and c_1 and c_2 be two possible values of third side such that $c_1^2 + c_1c_2 + c_2^2 = a^2$, then $\angle A$ is

a. 30°
b. 60°
c. 90°
d. 120°

17 In a ΔABC, if the radii of ex-circles r_1, r_2, r_3 are given by $r_1 = 8, r_2 = 12, r_3 = 24$, then

a. $r = 8$
b. $a = 12$
c. $b = 16$
d. $c = 20$

18 If A, A_1, A_2, A_3 be the areas of the inscribed and escribed of a ΔABC, then

a. $\sqrt{A_1} + \sqrt{A_2} + \sqrt{A_3} = \sqrt{\pi}(r_1 + r_2 + r_3)$
b. $\frac{1}{\sqrt{A_1}} + \frac{1}{\sqrt{A_2}} + \frac{1}{\sqrt{A_3}} = \frac{1}{\sqrt{A}}$
c. $\frac{1}{\sqrt{A_1}} + \frac{1}{\sqrt{A_2}} + \frac{1}{\sqrt{A_3}} = \frac{s^2}{\sqrt{\pi}r_1r_2r_3}$
d. $\sqrt{A_1} + \sqrt{A_2} + \sqrt{A_3} = \sqrt{\pi}(4R + r)$

Working Space

Statement Type

Direction *(Q. Nos. 19 and 20) This section is based on Statement I and Statement II. Select the correct answer from the codes given below.*

a. Both Statement I and Statement II are correct and Statement II is the correct explanation of Statement I

b. Both Statement I and Statement II are correct but Statement II is not the correct explanation of Statement I

c. Statement I is correct but Statement II is incorrect

d. Statement II is correct but Statement I is incorrect

19 **Statement I** In any ΔABC, the minimum value $\frac{r_1 + r_2 + r_3}{r}$ is 9.

Statement II AM $\geq$ GM

20 **Statement I** If A, B, C, D are the angles of a cyclic quadrilateral, then $\Sigma \sin A = 0$.

Statement II If A, B, C, D are the angles of a cyclic quadrilateral, then $\Sigma \cos A = 0$.

One Integer Value Correct Type

Direction *(Q. Nos. 21-24) This section contains 4 questions. When worked out will result in an integer from 0 to 9 (both inclusive).*

21 Let $A_1, A_2, A_3, \cdots, A_n$ be the vertices of an n-sided regular polygon such that $\frac{1}{A_1A_2} = \frac{1}{A_1A_3} + \frac{1}{A_1A_4}$, then the value of n is

22 Let ABC and ABC' be two non-congruent triangles with sides $AB = 4$, $AC = AC' = 2\sqrt{2}$ and $\angle B = 30°$. The absolute value of the difference between the areas of there triangles is

23 If r and R be respectively the radii of inscribed and circumscribed circles of a regular polygon of n-sides such that $\frac{R}{r} = \sqrt{5} - 1$, then n is equal to

24 Consider a ΔABC and let a, b, c denote the length of sides opposite to vertices A, B and C respectively. Suppose $a = 6, b = 10$ and the area of triangle is $15\sqrt{3}$.If $\angle ACB$ is obtuse and r denotes the radius of incircle of the triangle, then r^2 is equal to **[IIT JEE 2010]**

Comprehension Type

Direction *(Q. Nos. 25-36) This section contains 4 paragraphs, each describing theory, experiments, data, etc. Twelve questions related to the paragraphs have been given. Each question has only one correct answer among the four given options (a), (b), (c) and (d).*

Passage I

A polygon has n-sides. If all sides and all angles are same, then the polygon is called regular polygon. Let $A_1, A_2, A_3, \ldots, A_n$ are the vertices of a regular polygon of n-sides. R be the radius of circumscribed circle of regular polygon and r be the radius of inscribed circle of regular polygon and $A_1A_2 = A_2A_3 = A_3A_4 = \ldots = A_nA_1 = a$.

On the basis of above information, answer the following questions :

25 The value $r + R$ is

a. $\frac{a}{2}\cot\left(\frac{\pi}{n}\right)$

b. $a\cot\left(\frac{\pi}{2n}\right)$

c. $\frac{a}{4}\cot\left(\frac{\pi}{2n}\right)$

d. $\frac{a}{2}\cot\left(\frac{\pi}{2n}\right)$

26 The area of a regular polygon of n-sides is

a. $\frac{1}{2}nR^2\sin\left(\frac{2\pi}{n}\right)$

b. $nR^2\tan\left(\frac{\pi}{n}\right)$

c. $\frac{nr^2}{2}\sin\left(\frac{2\pi}{n}\right)$

d. $nr^2\tan\left(\frac{2\pi}{n}\right)$

Working Space

Passage II

Let $ABCD$ be a cyclic quadrilateral inscribed in a circle of radius R.

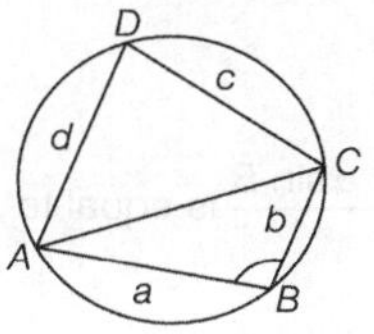

Then, $$\cos B = \frac{a^2 + b^2 - c^2 - d^2}{2ab + 2cd}$$

and $$\text{area of quadrilateral} = \frac{1}{2}(ab + cd)\sin B = \sqrt{(s-a)(s-b)(s-c)(s-d)}$$

when, $s = \dfrac{a+b+c+d}{2}$

Also, $$AC^2 \cdot BD^2 = (ac + bd)^2$$

i.e. $$AC \cdot BD = AB \cdot CD + BC \cdot AD$$

and $$R = \frac{AC}{2\sin B}$$

On the basis of above information, answer the following questions :

27 The sides of quadrilateral which can be inscribed in a circle are 6, 6, 8 and 8 cm. Then, radius of circumcircle is

a. $\frac{5}{12}$ cm b. $\frac{24}{7}$ cm

c. $\frac{11}{7}$ cm d. None of these

28 The sides of a quadrilateral which can be inscribed in a circle are 5, 5, 12 and 12 cm. Then, the radius of incircle of the quadrilateral is

a. $\frac{15}{17}$ cm b. $\frac{30}{17}$ cm

c. $\frac{60}{17}$ cm d. incircle does not exist

29 A quadrilateral $ABCD$ is such that one circle can be inscribed in it and another circle circumscribed about it, then $\tan^2 \frac{A}{2}$ is

a. $\frac{bc}{ad}$ b. $\frac{ab}{cd}$

c. $\frac{ad - bc}{ad + bc}$ d. $\frac{a + c}{b + d}$

30 If a quadrilateral with side of lengths a, b, c, d can be inscribed in one circle and circumscribed about another circle, then its area is

a. $\sqrt{abcd}$ b. $\sqrt{2abcd}$

c. $2\sqrt{abcd}$ d. None of these

Passage III

Consider a ΔABC, where x, y, z are the length of perpendicular drawn from the vertices of the triangle to the opposite side a, b, c respectively and let the letters R, r, s, Δ denote the circumradius, inradius, semi-perimeter and area of the triangle respectively.

On the basis of above information, answer the following questions :

31 If $\frac{bx}{c} + \frac{cy}{a} + \frac{az}{b} = \frac{a^2 + b^2 + c^2}{k}$, then the value of k is

a. R b. s

c. $2R$ d. $\frac{3}{2}R$

Working Space

32 If $\cot A + \cot B + \cot C = k\left(\frac{1}{x^2} + \frac{1}{y^2} + \frac{1}{z^2}\right)$, then the value of k is

a. R^2
b. Rr
c. Δ
d. $a^2 + b^2 + c^2$

33 The value of $\frac{c \sin B + b \sin C}{x} + \frac{a \sin C + c \sin A}{y} + \frac{b \sin A + a \sin B}{z}$ is equal to

a. $\frac{R}{r}$
b. $\frac{s}{R}$
c. 2
d. 6

Passage IV

In ΔABC with internal angles A, B and C, points O, I and H respectively denote circumcentre, incentre and orthocentre. Given that the points A, H, I, O and B are concyclic.

On the basis of above information, answer the following questions :

34 Angle C is equal to

a. $\frac{\pi}{6}$
b. $\frac{\pi}{4}$
c. $\frac{\pi}{3}$
d. $\frac{5\pi}{12}$

35 The ratio $HI : IO$ is equal to

a. $\frac{1}{2}$
b. 1
c. $\sqrt{2}$
d. $\sqrt{3}$

36 If $AH = HI$, then the angles A, C and B are in AP. The common difference of the AP is

a. $\frac{\pi}{18}$
b. $\frac{\pi}{12}$
c. $\frac{\pi}{9}$
d. $\frac{\pi}{10}$

JEE Main & AIEEE Archive

(Compilation of Last 13 Years' Questions)

1 $ABCD$ is a trapezium such that AB and CD are parallel and $BC \perp CD$. If $\angle ADB = \theta$, $BC = p$ and $CD = q$, then AB is equal to **(JEE Main 2013)**

a. $\dfrac{(p^2+q^2)\sin\theta}{p\cos\theta + q\sin\theta}$

b. $\dfrac{p^2+q^2\cos\theta}{p\cos\theta + q\sin\theta}$

c. $\dfrac{p^2+q^2}{p^2\cos\theta + q^2\sin\theta}$

d. $\dfrac{(p^2+q^2)\sin\theta}{(p\cos\theta + q\sin\theta)^2}$

2 In a ΔPQR, if $3\sin P + 4\cos Q = 6$ and $4\sin Q + 3\cos P = 1$, then the angle R is equal to **(AIEEE 2012)**

a. $\dfrac{5\pi}{6}$ b. $\dfrac{\pi}{6}$

c. $\dfrac{\pi}{4}$ d. $\dfrac{3\pi}{4}$

3 In a ΔPQR, $\angle R = \dfrac{\pi}{2}$. If $\tan\left(\dfrac{P}{2}\right)$ and $\tan\left(\dfrac{Q}{2}\right)$ are the roots of $ax^2 + bx + c = 0, a \neq 0$, then **(AIEEE 2005)**

a. $b = a + c$ b. $b = c$

c. $c = a + b$ d. $a = b + c$

4 In a ΔABC, let $\angle C = \dfrac{\pi}{2}$, if r is the inradius and R is the circumradius of the ΔABC, then $2(r+R)$ is equal to **(AIEEE 2005)**

a. $c + a$ b. $a + b + c$

c. $a + b$ d. $b + c$

5 If $\cos^{-1} x - \cos^{-1}\dfrac{y}{2} = \alpha$, then $4x^2 - 4xy\cos\alpha + y^2$ is equal to **(AIEEE 2005)**

a. $-4\sin^2\alpha$ b. $4\sin^2\alpha$

c. 4 d. $2\sin 2\alpha$

6 If in a ΔABC, the altitudes from the vertices A, B and C on opposite sides are in HP, then $\sin A, \sin B, \sin C$ are in **(AIEEE 2005)**

a. HP b. AGP

c. AP d. GP

7 The sides of a triangle are $\sin\alpha, \cos\alpha$ and $\sqrt{1+\sin\alpha\cos\alpha}$ for some $0° < \alpha < \dfrac{\pi}{2}$. Then, the greatest angle of the triangle is **(AIEEE 2004)**

a. 60° b. 90°

c. 120° d. 150°

8 If in a ΔABC, $a\cos^2\left(\dfrac{C}{2}\right) + c\cos^2\left(\dfrac{A}{2}\right) = \dfrac{3b}{2}$, then the sides a, b, c **(AIEEE 2003)**

a. are in AP

b. are in GP

c. are in HP

d. satisfy $a + b = c$

9 In a ΔABC, medians AD and BE are drawn. If $AD = 4$, $\angle DAB = \dfrac{\pi}{6}$ and $\angle ABE = \dfrac{\pi}{3}$, then the area of the ΔABC is **(AIEEE 2003)**

a. $\dfrac{8}{3}$ sq units b. $\dfrac{16}{3}$ sq units

c. $\dfrac{32}{3\sqrt{3}}$ sq units d. $\dfrac{64}{3}$ sq units

10 In a ΔABC, $2ca\sin\left(\dfrac{A-B+C}{2}\right)$ is equal to **(AIEEE 2002)**

a. $a^2 + b^2 - c^2$

b. $c^2 + a^2 - b^2$

c. $b^2 - c^2 - a^2$

d. $c^2 - a^2 - b^2$

11 In a ΔABC, $a = 4, b = 3$ and $\angle A = 60°$, then c is the root of the equation **(AIEEE 2002)**

a. $c^2 - 3c - 7 = 0$

b. $c^2 + 3c + 7 = 0$

c. $c^2 - 3c + 7 = 0$

d. $c^2 + 3c - 7 = 0$

12 In a ΔABC, $\tan\dfrac{A}{2} = \dfrac{5}{6}$, $\tan\dfrac{C}{2} = \dfrac{2}{5}$, then **(AIEEE 2002)**

a. a, c, b are in AP

b. a, b, c are in AP

c. b, a, c are in AP

d. a, b, c are in GP

JEE Advanced & IIT JEE Archive

(Compilation of Last 10 Years' Questions)

1 In a triangle, the sum of two sides is x and the product of same two sides is y. If $x^2 - c^2 = y$, where c is the third side of triangle, then the ratio of the inradius to the circumradius of triangle is **(2014 Adv., Only One Option Correct Type)**

a. $\frac{3y}{2x(x+c)}$ b. $\frac{3y}{2c(x+c)}$

c. $\frac{3y}{4x(x+c)}$ d. $\frac{3y}{4c(x+c)}$

2 In a ΔPQR, P is the largest angle and $\cos P = \frac{1}{3}$. Further incircle of the triangle touches the sides PQ, QR and RP at N, L and M respectively, such that the lengths of PN, QL and RM are consecutive even integers. Then, possible length(s) of the side(s) of the triangle is(are) **(2013 Adv., One or More than One Options Correct Type)**

a. 16 b. 18

c. 24 d. 22

3 Let PQR be a triangle of area Δ with $a = 2$, $b = \frac{7}{2}$ and $c = \frac{5}{2}$, where a, b and c are lengths of the sides of triangle opposite to the angles at P, Q and R, respectively. Then, $\frac{2\sin P - \sin 2P}{2\sin P + \sin 2P}$ is equal to **(2012, Only One Option Correct Type)**

a. $\frac{3}{4\Delta}$ b. $\frac{45}{4\Delta}$

c. $\left(\frac{3}{4\Delta}\right)^2$ d. $\left(\frac{45}{4\Delta}\right)^5$

4 If the angles A, B and C of a triangle are in an arithmetic progression and a, b and c denote lengths of the sides opposite to A, B and C respectively, then the value of the expression $\frac{a}{c}\sin 2C + \frac{c}{a}\sin 2A$ is **(2010, Only One Option Correct Type)**

a. $\frac{1}{2}$ b. $\frac{\sqrt{3}}{2}$

c. 1 d. $\sqrt{3}$

5 Consider a ΔABC and let a, b and c denote lengths of the sides opposite to vertices A, B and C respectively. Suppose $a = 6$, $b = 10$ and the area of triangle is $15\sqrt{3}$. If $\angle ACB$ is obtuse and r denotes radius of the incircle of triangle, then r^2 is equal to **(2010, Integer Type)**

6 Let ABC be a triangle such that $\angle ACB = \frac{\pi}{6}$ and let a, b and c denote lengths of the sides opposite to A, B and C respectively. The value(s) of x for which $a = x^2 + x + 1$, $b = x^2 - 1$ and $c = 2x + 1$ is(are) **(2010, One or More than One Options Correct Type)**

a. $-(2+\sqrt{3})$ b. $1+\sqrt{3}$

c. $2+\sqrt{3}$ d. $4\sqrt{3}$

7 In a ΔABC with fixed base BC, the vertex A moves such that $\cos B + \cos C = 4\sin^2\frac{A}{2}$. If a, b and c denote lengths of the sides of triangle opposite to the angles A, B and C respectively, then **(2009, One or More than One Options Correct Type)**

a. $b + c = 4a$

b. $b + c = 2a$

c. locus of point A is an ellipse

d. locus of point A is a pair of straight lines

8 Let ABC and ABC' be two non-congruent triangles with sides $AB = 4$, $AC = AC' = 2\sqrt{2}$ and $\angle B = 30°$. The absolute value of difference between the areas of these triangles is **(2009, Integer Type)**

9 A straight line through the vertex P of a ΔPQR intersects the side QR at point S and the circumcircle of ΔPQR at point T. If S is not the centre of circumcircle, then **(2008, One or More than One Options Correct Type)**

a. $\frac{1}{PS} + \frac{1}{ST} < \frac{2}{\sqrt{QS \times SR}}$

b. $\frac{1}{PS} + \frac{1}{ST} > \frac{2}{\sqrt{QS \times SR}}$

c. $\frac{1}{PS} + \frac{1}{ST} < \frac{4}{QR}$

d. $\frac{1}{PS} + \frac{1}{ST} > \frac{4}{QR}$

Passage for Q. Nos. (10-12)

Consider the circle $x^2 + y^2 = 9$ and the parabola $y^2 = 8x$. They intersect at P and Q in the first and fourth quadrants, respectively. Tangents to the circle at P and Q intersect the X-axis at R and tangents to the parabola at P and Q intersect the X-axis at S. **(2007, Comprehension Type)**

10 The ratio of areas of the ΔPQS and ΔPQR is

a. $1:\sqrt{2}$ b. $1:2$

c. $1:4$ d. $1:8$

11 The radius of circumcircle of the ΔPRS is

a. 5 b. $3\sqrt{3}$
c. $3\sqrt{2}$ d. $2\sqrt{3}$

12 The radius of incircle of the ΔPQR is

a. 4 b. 3
c. $\frac{8}{3}$ d. 2

13 Internal bisector of $\angle A$ of ΔABC meets side BC at D. A line drawn through D perpendicular to AD intersects the side AC at E and side AB at F. If a, b, c represent sides of ΔABC, then

(2006, One or More than One Options Correct Type)

a. AE is HM of b and c

b. $AD = \frac{2bc}{b+c} \cos \frac{A}{2}$

c. $EF = \frac{4bc}{b+c} \sin \frac{A}{2}$

d. the ΔAEF is isosceles

14 Inradius of a circle which is inscribed in an isosceles triangle one of whose angle is $2\pi/3$, is $\sqrt{3}$, then area of triangle is

(2006, Only One Option Correct Type)

a. $4\sqrt{3}$ sq units
b. $(12 - 7\sqrt{3})$ sq units
c. $(12 + 7\sqrt{3})$ sq units
d. None of the above

15 In a ΔABC, among the following which one is true? **(2005, Only One Option Correct Type)**

a. $(b+c)\cos \frac{A}{2} = a \sin\left(\frac{B+C}{2}\right)$

b. $(b+c)\cos\left(\frac{B+C}{2}\right) = a \sin \frac{A}{2}$

c. $(b-c)\cos\left(\frac{B-C}{2}\right) = a \cos \frac{A}{2}$

d. $(b-c)\cos \frac{A}{2} = a \sin\left(\frac{B-C}{2}\right)$

Answers *with* Explanations

Answers

1. Trigonometric Ratios and Identities

DPP-1 Measurement of Angles

1. (i) 300° (ii) –37° 30′ (iii) –229° 5′ 27″ **2.** (i) $\frac{7\pi}{12}$ rad (ii) $\frac{-19\pi}{72}$ rad (iii) $\left(\frac{3}{4}\right)^c$ **3.** 72°,18°

4. $\frac{\pi}{2}$cm **5.** 25°12′ **6.** *a* **7.** *a* **8.** *c* **9.** *c* **10.** *b* **11.** *b* **12.** *d* **13.** *b*
14. *d* **15.** *b* **16.** *a,c* **17.** *b,c* **18.** *a* **19.** 2 **20.** 7 **21.** *c* **22.** *a*

DPP-2 Trigonometric Functions using Identities

5. *d* **6.** *a* **7.** *b* **8.** *c* **9.** *c* **10.** *c* **11.** *a* **12.** *b* **13.** *c* **14.** *d*
15. *a* **16.** *a,c,d* **17.** *a,b* **18.** *a,b,c* **19.** *a* **20.** *b* **21.** 4 **22.** 4

DPP-3 Sign of Trigonometric Ratios and Allied Angles

4. *b* **5.** *d* **6.** *b* **7.** *b* **8.** *d* **9.** *d* **10.** *c* **11.** *b* **12.** *c* **13.** *a*
14. *a,b* **15.** *a,b* **16.** *b,d* **17.** *c* **18.** *a* **19.** 5 **20.** 7

DPP-4 Sign of T-Ratios and T-Ratios of Standard Angles

1. $\sin\theta=-\frac{5}{13}, \operatorname{cosec}\theta=-\frac{13}{5}, \cos\theta=-\frac{12}{13}, \sec\theta=-\frac{13}{12}, \tan\theta=\frac{5}{12}$ **3.** $-\frac{7}{2}$ **4.** *b* **5.** *c* **6.** *a*
7. *c* **8.** *d* **9.** *d* **10.** *d* **11.** *a* **12.** *c* **13.** *b* **14.** *a,b* **15.** *a* **16.** *b,c*
17. 2 **18.** 0 **19.** 3 **20.** True

DPP-5 T-Ratios using Addition and Subtraction Formulae

1. (1,–1) **2.** I and IV **3.** 2 **7.** *b* **8.** *a* **9.** *a* **10.** *b* **11.** *a* **12.** *b* **13.** *b*
14. *b* **15.** *b* **16.** *a* **17.** *c* **18.** *a* **19.** *a,b* **20.** *a,b,c* **21.** *a,b* **22.** 4 **23.** 1
24. 2

DPP-6 Trigonometric Functions using Greatest and Least Values of T-Ratios

2. 0 **3.** 1 **5.** *b* **6.** *d* **7.** *c* **8.** *a* **9.** *b* **10.** *b* **11.** *b* **12.** *c*
13. *b* **14.** *c* **15.** *b,c* **16.** *b* **17.** *a,c* **18.** *b* **19.** *b* **20.** *d* **21.** *c*
22. –1,1 **23.** IInd **24.** $\sqrt{a^2+b^2}, -\sqrt{a^2+b^2}$

DPP-7 Addition and Subtraction Formulae using Sign of T-Ratios and Allied Angles

4. *a* **5.** *a* **6.** *b* **7.** *d* **8.** *c* **9.** *c* **10.** *a* **11.** *a* **12.** *c* **13.** *a*
14. *c* **15.** *b* **16.** *c,d* **17.** *b,c* **18.** *a,b* **19.** *a* **20.** 6 **21.** *a*

DPP-8 Transformation Formulae-I

5. *b* **6.** *c* **7.** *b* **8.** *b* **9.** *b* **10.** *b* **11.** *a* **12.** *b* **13.** *b* **14.** *b*
15. *a,c* **16.** *a,b,c* **17.** *a,b* **18.** 0 **19.** 0 **20.** 2

DPP-9 Transformation Formulae-II

4. $\tan 5x - \tan 3x - \tan 2x$ **5.** *b* **6.** *b* **7.** *a* **8.** *c* **9.** *c* **10.** *b* **11.** *d*
12. *d* **13.** *c* **14.** *b* **15.** *a* **16.** *a,b,c* **17.** *a,c* **18.** *a,b,c* **19.** 3 **20.** 4 **21.** 3

DPP-10 Trigonometric Ratios using Half-Angle Formulae

1. 2 2. –1 3. μ 4. True 5. True 6. False 7. tanθ 8. $\frac{2}{\sqrt{5}}$ 10. $\alpha=-1, \beta=3$
11. *d* 12. *d* 13. *c* 14. *c* 15. *d* 16. *d* 17. *b* 18. *a* 19. *d* 20. *a*
21. *b* 22. *b,c* 23. *b,c* 24. *a,c* 25. *d* 26. *a* 27. *d* 28. *d*

DPP-11 Trigonometric Ratios of Multiple Angles

1. ± 2 4. 3 5. –1 6. 3 7. True 8. True 9. False 10. *a* 11. *c* 12. *b*
13. *c* 14. *b* 15. *b* 16. *b* 17. *b* 18. *a* 19. *a* 20. *a,c* 21. *a,b,c* 22. *c,d*
23. 2 24. 1

DPP-12 T-Ratios of Sub-Multiple Angles

1. True 2. False 3. True 7. *a* 8. *b* 9. *c* 10. *b* 11. *c* 12. *d* 13. *c*
14. *c* 15. *b* 16. *a,b,c,d* 17. *a,c* 18. *a,b,c* 19. *a,c,d* 20. *a* 21. 8 22. 6

DPP-13 Sum of Three or More Angles

4. *b* 5. *b* 6. *a* 7. *b* 8. *c* 9. *c* 10. *d* 11. *c* 12. *d* 13. *d*
14. *a* 15. *a,c,d* 16. *a,b,c* 17. *a,b* 18. *b* 19. *b* 20. *c* 21. *c*

Revisal Problems (JEE Main)

1. *a* 2. *a* 3. *c* 4. *a* 5. *a* 6. *b* 7. *b*

Revisal Problems (JEE Advanced)

1. *a* 2. *b* 3. *a* 4. *c* 5. *d* 6. *a* 7. *c* 8. *a,b,c,d* 9. *b,c*
10. *a,b,c* 11. *a,b* 12. *b,d* 13. *a,b,c* 14. *b,c,d* 15. *a,b,c,d* 16. *b,c,d* 17. 8 18. 3 19. 3
20. 7 21. 1 22. 1 23. 3 24. 0 25. 5 26. 2 27. *a* 28. *a* 29. *a*
30. *b* 31. *d* 32. *c* 33. *b*

JEE Main & AIEEE Archive

1. *d* 2. *b* 3. *d* 4. *b* 5. *c* 6. *b* 7. *b* 8. *c* 9. *a* 10. *c*
11. *c* 12. *a* 13. *d* 14. *c* 15. *b*

JEE Advanced & IIT JEE Archive

1. *a,b* 2. 3 3. *a,b* 4. *c,d* 5. *b* 6. *d*

2. Height and Distance

DPP-1 Height and Distance

1. $30\sqrt{3}$ m 2. $\frac{b(a^2+b^2)}{a^2-b^2}$ 4. *a* 5. *a* 6. *d* 7. *c* 8. *a* 9. *a*
10. *c* 11. *b* 12. *b* 13. *a* 14. *a,b* 15. *a,b,c* 16. *c* 17. 4 18. 3 19. 4
20. 4

JEE Main & AIEEE Archive

1. *a* 2. *d* 3. *a* 4. *c* 5. *c* 6. *a* 7. *b*

3. Trigonometric Equations

DPP-1 For Trigonometric Equation with Conceptual Problems

1. True 2. True 3. True 4. False 5. $\left[\frac{-\pi}{2}, \frac{\pi}{2}\right]$ 6. $a=0, b=\pi$ 7. 0 8. 0
9. *d* 10. *b* 11. *c* 12. *a* 13. *a* 14. *c* 15. *d* 16. *a* 17. *d* 18. *a,c*
19. *a,d* 20. *a,b* 21. 3 22. 4

DPP-2 Equations of the Form $\sin^2\theta = \sin^2\alpha$, $\cos^2\theta = \cos^2\alpha$, $\tan^2\theta = \tan^2\alpha$

1. $\theta = 2n\pi \pm \frac{\pi}{3}, \forall n \in I$ 2. $\theta = n\pi \pm \frac{\pi}{3}, \forall n \in I$ 3. $x = \frac{n\pi}{3} + \frac{\pi}{18}, \forall n \in I$ 4. *c* 5. *a* 6. *b*
7. *b* 8. *a* 9. *c* 10. *b* 11. *a* 12. *c* 13. *b* 14. *a,b* 15. *a,b,c* 16. *a,b*
17. *a,d* 18. 0 19. 6 20. 3

DPP-3 Homogeneous, Extreme Values and T-Inequations

2. True 3. True 4. *b* 5. *a* 6. *a* 7. *d* 8. *d* 9. *c* 10. *d* 11. *a*
12. *d* 13. *c* 14. *a,b,c* 15. *c* 16. *a,c* 17. 4 18. 0 19. *c* 20. *a*

Revisal Problems (JEE Main)

1. *c* 2. *d* 3. *a* 4. *b* 5. *a* 6. *d* 7. *b* 8. *b* 9. *b*

Revisal Problems (JEE Advanced)

1. *b* 2. *a* 3. *a* 4. *d* 5. *a* 6. *a* 7. *c* 8. *a* 9. *b*
10. *a,b,c* 11. *a,c,d* 12. *a,c* 13. *a* 14. *a* 15. *a,b,d* 16. *a,b,c,d* 17. *b,c*
18. *a,b,c* 19. *a* 20. *d* 21. *a* 22. *c* 23. *b* 24. *a* 25. *a* 26. *b*
27. 2 28. 5 29. 4 30. 3 31. 2

JEE Archive

1. *a* 2. *d* 3. *c* 4. *d* 5. *a,c,d* 6. *d* 7. 7 8. 3 9. *c* 10. *b*

4. Solutions of Triangles

DPP-1 Sine Rule and Its Applications

1. 0 2. $3q$ 5. *a* 6. *c* 7. *b* 8. *c* 9. *a* 10. *a* 11. *a* 12. *c*
13. *a* 14. *b* 15. *a,b,c,d* 16. *a,c,d* 17. *a,b,c* 18. *a* 19. *d* 20. *d* 21. 4
22. 2 23. 0

DPP-2 Cosine Rule and Projection Formulae

1. a^2 3. 21.09 m 4. *d* 5. *b* 6. *c* 7. *a* 8. *c* 9. *a* 10. *c* 11. *a*
12. *a* 13. *b* 14. *b* 15. *a,b,c* 16. *c,d* 17. *c,d* 18. *a,b* 19. 5 20. 6 21. *a*

DPP-3 Napier's Analogy, Area of Triangle (Heron's Formula), Half-Angle Formulae

4. True 5. False 6. *b* 7. *a* 8. *d* 9. *a* 10. *a* 11. *a* 12. *b* 13. *a*
14. *b* 15. *b* 16. *a,d* 17. *b,c* 18. *a,b,d* 19. *d* 20. *b* 21. *b* 22. 4 23. 5
24. 9

DPP-4 Half-Angle Formulae using 's', Heron's Formulae, Area of Triangle using Sides and Angles

1. *b* 2. *b* 3. *c* 4. *a* 5. *c* 6. *a* 7. *b* 8. *c* 9. *a* 10. *b*
11. *c* 12. *a* 13. *a,b,c* 14. *a,b,c* 15. *a,b,c,d* 16. *b* 17. *a* 18. *a* 19. 5 20. 2

DPP-5 Properties of Triangles

3. 4 4. True 5. True 6. True 7. False 8. 4 9. 2 10. *a* 11. *d* 12. *d*
13. *b* 14. *c* 15. *c* 16. *d* 17. *a* 18. *b* 19. *a* 20. *b,d* 21. *b,d*
22. *a,c,d* 23. *c* 24. *b* 25. *d* 26. *b* 27. *a* 28. *b* 29. *b* 30. 3 31. 3

Revisal Problems (JEE Main)

1. *c* 2. *a* 3. *c* 4. *c* 5. *a* 6. *a* 7. *c* 8. *c* 9. *c* 10. *c*

Revisal Problems (JEE Advanced)

1. *c* 2. *b* 3. *c* 4. *a* 5. *c* 6. *a* 7. *a* 8. *a* 9. *b* 10. *c*
11. *a* 12. *a,b,c* 13. *a,b,d* 14. *a,c,d* 15. *a,b,c* 16. *b,d* 17. *b,c,d* 18. *a,b,c,d* 19. *b*
20. *d* 21. 7 22. 4 23. 5 24. 3 25. *d* 26. *a* 27. *d* 28. *c* 29. *a*
30. *a* 31. *c* 32. *c* 33. *d* 34. *c* 35. *b* 36. *c*

JEE Main & AIEEE Archive

1. *a* 2. *b* 3. *c* 4. *c* 5. *b* 6. *c* 7. *c* 8. *a* 9. *c* 10. *b*
11. *a* 12. *b*

JEE Advanced & IIT JEE Archive

1. *b* 2. *b,d* 3. *c* 4. *d* 5. 3 6. *b* 7. *b,c* 8. 4 9. *b,d* 10. *c*
11. *b* 12. *d* 13. *a,b,c,d* 14. *c* 15. *d*

1. Trigonometric Ratios and Identities

DPP-1 Measurement of Angles

1. $\because 1^C = \left(\frac{180}{\pi}\right)^\circ$

(i) $\left(\frac{5\pi}{3}\right)^C = \left(\frac{5\pi}{3} \times \frac{180}{\pi}\right)^\circ = 300^\circ$

(ii) $\left(-\frac{5\pi}{24}\right)^C = \left(-\frac{5\pi}{24} \times \frac{180}{\pi}\right)^\circ$

$= -\left(\frac{75}{2}\right)^\circ = -37^\circ\left(\frac{1}{2} \times 60\right)'$

$= -37^\circ 30'$

(iii) $(-4)^C = -\left(4 \times \frac{180}{\pi}\right)^\circ$

$= -\left(4 \times \frac{180 \times 7}{22}\right)^\circ = -\left(\frac{2520}{11}\right)^\circ$

$= -229^\circ\left(\frac{1}{11} \times 60\right)'$

$= -229^\circ 5'\left(\frac{5}{11} \times 60\right)'' = -229^\circ 5' 27''$

2. $\because 1^\circ = \left(\frac{\pi}{180}\right)^C$

(i) $105^\circ = \left(105 \times \frac{\pi}{180}\right)^C = \frac{7\pi}{12}$ rad

(ii) $-47^\circ 30' = -\left[47^\circ + \left(\frac{30}{60}\right)^\circ\right]$

$= -\left(47\frac{1}{2}\right)^\circ = -\left(\frac{95}{2}\right)^\circ = -\left(\frac{95}{2} \times \frac{\pi}{180}\right)^C$

$= -\frac{19\pi}{72}$ rad

(iii) $\because \quad 16'' = \left(\frac{16}{60}\right)' = \left(\frac{16}{60 \times 60}\right)^\circ = \left(\frac{1}{225}\right)^\circ$

$57' = \left(\frac{57}{60}\right)^\circ$

$\therefore \; 42^\circ 57' 16'' = \left(42 + \frac{57}{60} + \frac{1}{225}\right)^\circ$

$= \left(\frac{38659}{900}\right)^\circ = \left(\frac{38659}{900} \times \frac{\pi}{180}\right)^C = \left(\frac{3}{4}\right)^C$

3. Let A and B be two acute angles of a ΔABC in which $\angle ACB = 90^\circ$

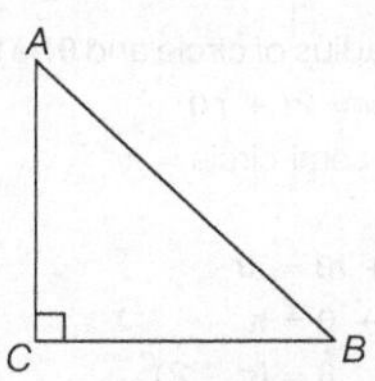

$A + B + C = 180^\circ$

$A + B = 180^\circ - C$

$A + B = 180^\circ - 90^\circ$

$\Rightarrow \quad A + B = 90^\circ$ ...(i)

According to the question,

$A - B = \frac{3\pi}{10} = \frac{3}{10} \times 180^\circ = 54^\circ$

$\therefore \quad A - B = 54^\circ$ ...(ii)

On solving Eqs. (i) and (ii), we get

$A = 72^\circ$ and $B = 18^\circ$

4. Let S be the length of an arc subtending an angle θ^C at the centre of circle with radius r.

Given that,

$\theta = 15^\circ = \left(15 \times \frac{\pi}{180}\right)^C = \frac{\pi}{12}$ rad

$r = 6$ cm

We know that, $\theta = \frac{S}{r} \Rightarrow S = r\theta$

$S = 6 \times \frac{\pi}{12} = \frac{\pi}{2}$ cm

5. Let θ^C be the angle subtending at centre of circle

$\because \quad r = \frac{100}{2} = 50$ cm

$S = 22$ cm

$\therefore \quad \theta = \frac{S}{r} = \frac{22}{50}$ rad $= \frac{11}{25}$ rad

$= \frac{11}{25} \times \frac{180}{\pi} = \frac{11}{25} \times \frac{180 \times 7}{22}$

$= 25^\circ\left(\frac{1}{5} \times 60\right)' = 25^\circ 12'$

6. (a) $\because$ Angle traced by minute hand in $60' = 360^\circ$

Angle traced by minute hand in $1' = \frac{360^\circ}{60} = 6^\circ$

Angle traced by minute hand in $30' = 30 \times 6 = 180^\circ \times \frac{\pi}{180} = \pi^C$

Length of minute hand $(r) = 14$ cm

Let distance moved by minute hand be S in $30'$.

$\because \quad S = r\theta = 14 \times \pi = 14 \times \frac{22}{7} = 44$ cm

7. (a) Angle traced by hour hand in 12 h $= 360^\circ$

Angle traced by hour hand in 1 h $= \frac{360^\circ}{12} = 30^\circ$

Angle traced by hour hand in 4 h $= 30^\circ \times 4 = 120^\circ \times \frac{\pi}{180} = \frac{2\pi}{3}$

8. (c) Let $3x$, $4x$, $5x$ be angles of a triangle, then

$3x + 4x + 5x = 180^\circ$

$12x = 180^\circ \Rightarrow x = \frac{180^\circ}{12}$

$x = 15^\circ$

$\therefore$ Greatest angle $= 5x = 5 \times 15^\circ = 75^\circ$

$= 75^\circ \times \frac{\pi}{180} = \frac{5\pi}{12}$ rad

9. (c) Let P be the post and PA be the length of rope in tight position, i.e. r (let).

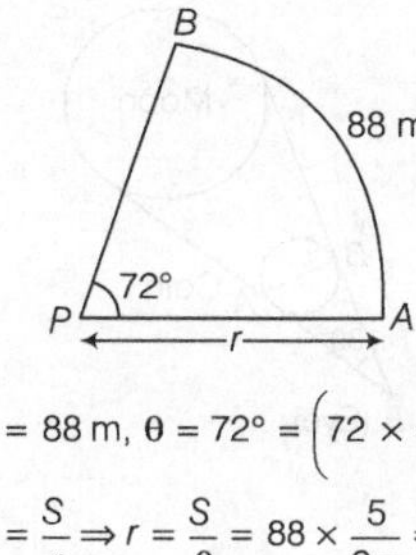

$S = 88$ m, $\theta = 72^\circ = \left(72 \times \frac{\pi}{180}\right)^C = \left(\frac{2\pi}{5}\right)^C$

$\because \quad \theta = \frac{S}{r} \Rightarrow r = \frac{S}{\theta} = 88 \times \frac{5}{2\pi} = 70$ m

10. (b) We know that, angle traced by hour hand in 1 h = 30°

Angle traced by hour hand in $7\frac{1}{2}$ h $= 30° \times \frac{15}{2} = 225°$

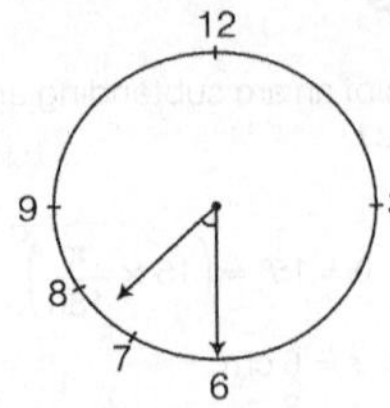

$\because$ Angle traced by minute hand in 1 min = 6°

Angle traced by minute hand in 30 min

$$= 30 \times 6° = 180°$$

$\therefore$ The angle between two hands $= 225° - 180° = 45°$

11. (b) $\because r = \frac{120}{2} = 60$ cm

$$AB = 60 \text{ cm} \qquad \text{(given)}$$

$$\therefore \quad OA = OB = AB$$

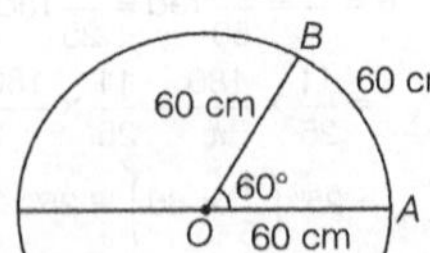

Then, ΔOAB is an equilateral triangle.

Now, $\theta = \angle AOB = 60° = \left(60 \times \frac{\pi}{180}\right)^C = \left(\frac{\pi}{3}\right)^C$

Let S be the length of an arc of the circle, then

$$S = r\theta = 60 \times \frac{\pi}{3} = 20\pi \text{ cm}$$

12. (d) Angle traced by OQ in 1 revolution $= 2\pi^C$

Angle traced by OQ in 4 revolutions $= 4 \times 2\pi = 8\pi^C$

$\therefore$ Angular velocity in rad/sec $= 8\pi$

13. (b) Radius of the curve (r) = 1500 m

Distance covered in 1 s

$$= 66 \times \frac{1000}{3600} = \frac{55}{3} \text{ m}$$

Distance covered in 10 s

$$= 10 \times \frac{55}{3} = \frac{550}{3} \text{ m} = S \text{ (let)}$$

Let θ be the required angle.

$$\because \quad \theta = \frac{S}{r} = \frac{550}{3} \times \frac{1}{1500} = \left(\frac{11}{90}\right)^C$$

14. (d) Let r be the required distance of a coin from the eye to conceal the full moon.

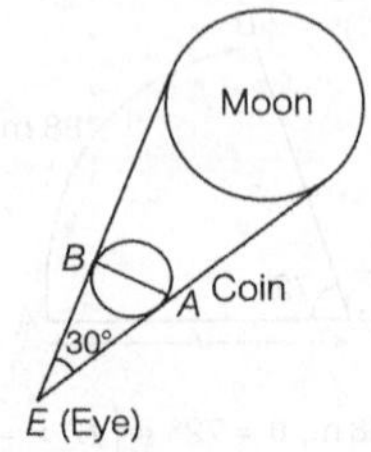

$$\therefore \quad \theta = 30' = \left(\frac{30}{60}\right)^\circ$$

$$= \left(\frac{1}{2} \times \frac{\pi}{180}\right)^C$$

$$= \left(\frac{\pi}{360}\right)^C$$

Let arc AB = diameter AB of coin = 2.2 cm

$$\because \quad r = \frac{S}{\theta} = \frac{2.2 \times 360}{\pi} \text{ cm}$$

$$r = 252 \text{ cm}$$

15. (b) Let radius of wheel (r) $= \frac{40}{2} = 20$ cm

$\therefore$ Circumference of the wheel $= 2\pi r$

$$= 2 \times \pi \times 20$$

$$= 40\pi \text{ cm}$$

Number of revolutions made in 1 s = 7

Distance covered in 1s $= 40\pi \times 7$

$$= 40 \times \frac{22}{7} \times 7$$

$$= 880 \text{ cm}$$

$\therefore$ Speed of the train = 880 cm/s

16. (a,c) Let r_1 and r_2 be two radii of two circles with same arcs S (let)

$$\theta_1 = 65°$$

and $\theta_2 = 110°$

$$\frac{S}{r_1} = \left(65 \times \frac{\pi}{180}\right)^C$$

and $\frac{S}{r_2} = \left(110 \times \frac{\pi}{180}\right)^C$

$$\frac{S}{r_1} = \left(\frac{65\pi}{180}\right)^C \qquad \text{...(i)}$$

and $\frac{S}{r_2} = \left(\frac{110\pi}{180}\right)^C \qquad \text{...(ii)}$

On dividing Eq. (ii) by Eq. (i), we get

$$\frac{S}{r_2} \div \frac{S}{r_1} = \frac{110\pi}{180} \div \frac{65\pi}{180}$$

$$\frac{r_1}{r_2} = \frac{22}{13}$$

$$\therefore \quad r_1 : r_2 = 22 : 13$$

Now, from option (c),

$$\frac{r_1}{r_2} = \frac{x}{y} = \frac{44}{26} = \frac{22}{13}$$

17. (b,c) Let r be the radius of circle and θ be the angle of sector. Then,

Perimeter of sector $= 2r + r\theta$

Length of arc of a semi-circle $= \pi r$

It is given that

$$2r + r\theta = \pi r$$

$$2 + \theta = \pi$$

$$\therefore \quad \theta = (\pi - 2)^C \qquad \text{[option (c)]}$$

$$\theta = \left[(\pi - 2) \times \frac{180}{\pi}\right]^\circ$$

$$\therefore \quad \theta = 180° - \left(\frac{360}{\pi}\right)^\circ$$

$$\theta = 180° - 114°32'44''$$

$$\theta = 65°27'16'' \qquad \text{[option (b)]}$$

18. (a) Let h be the height of letters.
h may be considered as the arc of a circle of radius 12 m, i.e. $S = h$.

Now, $\theta = 5' = \left(\frac{5}{60}\right)^\circ$

$= \left(\frac{5}{60} \times \frac{\pi}{180}\right)^C$

$= \frac{\pi}{12 \times 180}$ rad

$\because$ $S = r\theta$

$\therefore$ $h = 12 \times \frac{\pi}{12 \times 180} = \frac{\pi}{180}$ m = 1.7 cm

19. (2) Let $a - d, a, a + d$ are angles in degree of a triangle in AP, then

$a - d + a + a + d = 180°$

$3a = 180°$

$\Rightarrow a = 60°$

Greatest angle in radian = $(a + d) \times \frac{\pi}{180}$ rad

According to the question,

$$\frac{a-d}{\frac{\pi}{180}(a+d)} = \frac{60}{\pi}$$

or $3(a - d) = a + d$

$\Rightarrow d = \frac{1}{2}a = \frac{1}{2} \times 60° = 30°$

$\therefore$ Angles are $60° - 30°, 60°, 60° + 30° = 30°, 60°, 90°$

$\Rightarrow \frac{\pi}{6}, \frac{\pi}{3}, \frac{\pi}{2}$

According to the question, $\frac{\pi}{2} - \frac{\pi}{6} = \frac{\lambda\pi}{6}$

$\Rightarrow \frac{2\pi}{6} = \frac{\lambda\pi}{6}$

$\therefore \lambda = 2$

20. (7) The radius of circle $C_1 = 1, C_2 = 2, C_3 = 3, \ldots$. It starts from $A_1(1, 0)$ on C_1 moves a distance of 1 cm on C_1 to come to B_1. The angle subtended by A_1B_1 at the centre will be $\frac{S}{r} = \theta$ radians, i.e. 1 rad.
From B_1, it moves along radius OB_1 and comes to A_2 on circle C_2 of radius 2.
From A_2, it moves on C_2 a distance 2 cm and comes to B_2. The angle subtended by A_2B_2 is again as before 1 rad. The total angle subtended at the centre is 2 rad. The process continues in order to cross the X-axis again it must describe 2π rad, i.e.
$2 \times \frac{22}{7} = 6.7$ rad
Hence, it must be moving on circle C_7.

$\therefore n = 7$

21. (c) We have, $1^C = 57° < 60°$

$\Rightarrow \cos 1^C > \cos 60°$

$\Rightarrow \cos 1^C > \frac{1}{2}$

22. (a) Let number of sides of two regular polygons are 5λ and 4λ.
Now, according to the question,

$$\left(\pi - \frac{2\pi}{5\lambda}\right) - \left(\pi - \frac{2\pi}{4\lambda}\right) = \frac{\pi}{40}$$

$\Rightarrow \frac{2\pi}{4\lambda} - \frac{2\pi}{5\lambda} = \frac{\pi}{40}$

$\Rightarrow \frac{2\pi}{\lambda}\left[\frac{1}{4} - \frac{1}{5}\right] = \frac{\pi}{40}$

$\Rightarrow \frac{2}{\lambda}\left(\frac{1}{20}\right) = \frac{1}{40}$

$\Rightarrow \lambda = \frac{40}{10}$

$\therefore \lambda = 4$

Thus, number of sides are 20 and 16.

DPP-2 Trigonometric Functions using Identities

1. LHS = $(\text{cosec }\theta - \sin\theta)(\sec\theta - \cos\theta)(\tan\theta + \cot\theta)$

$= \left(\frac{1}{\sin\theta} - \sin\theta\right)\left(\frac{1}{\cos\theta} - \cos\theta\right)\left(\frac{\sin\theta}{\cos\theta} + \frac{\cos\theta}{\sin\theta}\right)$

$= \frac{1-\sin^2\theta}{\sin\theta} \times \frac{1-\cos^2\theta}{\cos\theta} \times \frac{\sin^2\theta + \cos^2\theta}{\sin\theta\cos\theta}$

$= \frac{\cos^2\theta}{\sin\theta} \times \frac{\sin^2\theta}{\cos\theta} \times \frac{1}{\sin\theta\cos\theta}$ **Hence proved.**

$= 1 = $ RHS

2. LHS = $\frac{\sin^3\theta + \cos^3\theta}{\sin\theta + \cos\theta} + \frac{\sin^3\theta - \cos^3\theta}{\sin\theta - \cos\theta}$

$= \frac{(\sin\theta + \cos\theta)(\sin^2\theta + \cos^2\theta - \sin\theta\cos\theta)}{(\sin\theta + \cos\theta)}$

$+ \frac{(\sin\theta - \cos\theta)(\sin^2\theta + \cos^2\theta + \sin\theta\cos\theta)}{(\sin\theta - \cos\theta)}$

$[\because a^3 + b^3 = (a+b)(a^2 + b^2 - ab)$

and $a^3 - b^3 = (a-b)(a^2 + b^2 + ab)]$

$= 1 - \sin\theta\cos\theta + 1 + \sin\theta\cos\theta$

$= 2 = $ RHS **Hence proved.**

3. We have,

$x = a\sec\theta\cos\phi$

$y = b\sec\theta\sin\phi$

and $z = c\tan\theta$

$\Rightarrow \frac{x}{a} = \frac{\cos\phi}{\cos\theta}$

$\frac{y}{b} = \frac{\sin\phi}{\cos\theta}$

and $\frac{z}{c} = \frac{\sin\theta}{\cos\theta}$

Now, LHS = $\frac{x^2}{a^2} + \frac{y^2}{b^2} - \frac{z^2}{c^2}$

$= \frac{\cos^2\phi}{\cos^2\theta} + \frac{\sin^2\phi}{\cos^2\theta} - \frac{\sin^2\theta}{\cos^2\theta}$

$= \frac{(\cos^2\phi + \sin^2\phi) - \sin^2\theta}{\cos^2\theta}$

$= \frac{1 - \sin^2\theta}{\cos^2\theta}$ $(\because \sin^2 A + \cos^2 A = 1)$

$= \frac{\cos^2\theta}{\cos^2\theta}$

$= 1 = $ RHS **Hence proved.**

4. We have,

$$\cos^2\alpha - \sin^2\alpha = \tan^2\beta$$

On dividing both sides by $\cos^2\alpha$, we get

$$1 - \tan^2\alpha = \frac{\tan^2\beta}{\cos^2\alpha} = \tan^2\beta \sec^2\alpha$$

$$1 - \tan^2\alpha = \tan^2\beta\,(1 + \tan^2\alpha)$$

$$1 - \tan^2\alpha = \tan^2\beta + \tan^2\alpha \tan^2\beta$$

$$1 - \tan^2\beta = \tan^2\alpha\,(1 + \tan^2\beta)$$

$$\frac{\cos^2\beta - \sin^2\beta}{\cos^2\beta} = \frac{\tan^2\alpha}{\cos^2\beta}$$

$\therefore \quad \tan^2\alpha = \cos^2\beta - \sin^2\beta$ **Hence proved.**

Alternate Method

We have,

$$\cos^2\alpha - \sin^2\alpha = \tan^2\beta$$

$$\Rightarrow \quad \frac{\cos^2\alpha - \sin^2\alpha}{\cos^2\alpha + \sin^2\alpha} = \frac{\sin^2\beta}{\cos^2\beta} \qquad (\because \cos^2\theta + \sin^2\theta = 1)$$

Now, applying componendo and dividendo, we get

$$\frac{\cos^2\alpha - \sin^2\alpha + \cos^2\alpha + \sin^2\alpha}{\cos^2\alpha - \sin^2\alpha - \cos^2\alpha - \sin^2\alpha} = \frac{\sin^2\beta + \cos^2\beta}{\sin^2\beta - \cos^2\beta}$$

$$\frac{2\cos^2\alpha}{-2\sin^2\alpha} = \frac{1}{-(\cos^2\beta - \sin^2\beta)}$$

$$\frac{1}{\tan^2\alpha} = \frac{1}{\cos^2\beta - \sin^2\beta}$$

$\therefore \quad \tan^2\alpha = \cos^2\beta - \sin^2\beta$ **Hence proved.**

5. (d) Given,

$$\sin^6\theta + \cos^6\theta - 1 = \lambda \sin^2\theta\cos^2\theta$$

$$\Rightarrow \quad (\sin^2\theta)^3 + (\cos^2\theta)^3 - 1 = \lambda\sin^2\theta\cos^2\theta$$

$$\Rightarrow (\sin^2\theta + \cos^2\theta)(\sin^4\theta + \cos^4\theta - \sin^2\theta\cos^2\theta) - 1 = \lambda\sin^2\theta\cos^2\theta$$

$$\Rightarrow [(\sin^2\theta + \cos^2\theta)^2 - 3\sin^2\theta\cos^2\theta] - 1 = \lambda\sin^2\theta\cos^2\theta$$

$$\Rightarrow \quad [1 - 3\sin^2\theta\cos^2\theta] - 1 = \lambda\sin^2\theta\cos^2\theta$$

$$\Rightarrow \quad -3\sin^2\theta\cos^2\theta = \lambda\sin^2\theta\cos^2\theta$$

$$\therefore \quad \lambda = -3$$

6. (a) LHS $= (1 + \tan\alpha\tan\beta)^2 + (\tan\alpha - \tan\beta)^2$

$$= 1 + \tan^2\alpha\tan^2\beta + 2\tan\alpha\tan\beta + \tan^2\alpha + \tan^2\beta - 2\tan\alpha\tan\beta$$

$$= (1 + \tan^2\alpha) + (\tan^2\beta + \tan^2\alpha\tan^2\beta)$$

$$= (1 + \tan^2\alpha) + \tan^2\beta\,(1 + \tan^2\alpha)$$

$$= (1 + \tan^2\alpha)(1 + \tan^2\beta)$$

$$= \sec^2\alpha\sec^2\beta$$

7. (b) We have,

$$\frac{2\sin\theta\cos\theta - \cos\theta}{1 - \sin\theta + \sin^2\theta - \cos^2\theta}$$

$$= \frac{\cos\theta(2\sin\theta - 1)}{1 - \cos^2\theta + \sin^2\theta - \sin\theta}$$

$$= \frac{\cos\theta\,(2\sin\theta - 1)}{\sin^2\theta + \sin^2\theta - \sin\theta}$$

$$= \frac{\cos\theta\,(2\sin\theta - 1)}{2\sin^2\theta - \sin\theta}$$

$$= \frac{\cos\theta\,(2\sin\theta - 1)}{\sin\theta\,(2\sin\theta - 1)}$$

$$= \frac{\cos\theta}{\sin\theta} = \cot\theta$$

8. (c) Let $\quad a\sin\theta + b\cos\theta = p \quad$ ...(i)

$a\cos\theta - b\sin\theta = c \quad$ (given) ...(ii)

On squaring and adding Eqs. (i) and (ii), we get

$$(a\sin\theta + b\cos\theta)^2 + (a\cos\theta - b\sin\theta)^2 = p^2 + c^2$$

$$a^2\sin^2\theta + b^2\cos^2\theta + 2ab\sin\theta\cdot\cos\theta + a^2\cos^2\theta + b^2\sin^2\theta - 2ab\sin\theta\cdot\cos\theta = p^2 + c^2$$

$$a^2(\sin^2\theta + \cos^2\theta) + b^2(\sin^2\theta + \cos^2\theta) = p^2 + c^2$$

$$\Rightarrow \quad p^2 = a^2 + b^2 - c^2$$

$$\Rightarrow \quad p = \pm\sqrt{a^2 + b^2 - c^2}$$

$\therefore \quad a\sin\theta + b\cos\theta = \pm\sqrt{a^2 + b^2 - c^2} \quad$ [from Eq. (i)]

9. (c) $3(\sin x - \cos x)^4 + 6(\sin x + \cos x)^2 + 4(\sin^6 x + \cos^6 x)$

$$= 3(\sin^2 x + \cos^2 x - 2\sin x\cos x)^2 + 6(\sin^2 x + \cos^2 x + 2\sin x\cos x) + 4(\sin^2 x + \cos^2 x)(\sin^4 x + \cos^4 x - \sin^2 x\cos^2 x)$$

$$= 3(1 - 2\sin x\cos x)^2 + 6(1 + 2\sin x\cos x) + 4(\sin^4 x + \cos^4 x) - 4\sin^2 x\cos^2 x$$

$$= 3(1 + 4\sin^2 x\cos^2 x - 4\sin x\cos x) + 6 + 12\sin x\cos x + 4(\sin^4 x + \cos^4 x) - 4\sin^2 x\cos^2 x$$

$$= 9 + 4(\sin^4 x + \cos^4 x + 2\sin^2 x\cos^2 x)$$

$$= 9 + 4(\sin^2 x + \cos^2 x)^2 = 9 + 4 = 13$$

10. (c) $\because \quad \cos\theta + \sin\theta = \sqrt{2}\cos\theta \quad$ (given) ...(i)

and $\quad \cos\theta - \sin\theta = p$ (let) ...(ii)

On squaring and adding Eqs. (i) and (ii), we get

$$(\cos\theta + \sin\theta)^2 + (\cos\theta - \sin\theta)^2 = 2\cos^2\theta + p^2$$

$$\cos^2\theta + \sin^2\theta + 2\sin\theta\cos\theta + \cos^2\theta + \sin^2\theta - 2\sin\theta\cos\theta = 2\cos^2\theta + p^2$$

$$1 + 1 = 2\cos^2\theta + p^2$$

$$\therefore \quad p^2 = 2 - 2\cos^2\theta = 2(1 - \cos^2\theta)$$

$$p^2 = 2\sin^2\theta$$

$$\therefore \quad p = \sqrt{2}\sin\theta$$

$\Rightarrow \quad \cos\theta - \sin\theta = \sqrt{2}\sin\theta \quad$ ...(iii)

But $\quad \cos\theta - \sin\theta = k\sin\theta \quad$ (given) ...(iv)

From Eqs. (iii) and (iv), we get

$$k = \sqrt{2}$$

11. (a) We know that, $\quad \sec^2\theta \ge 1 \quad$ ...(i)

and $\quad$ AM $\ge$ GM

$$\frac{x + y}{2} \ge \sqrt{xy}$$

$$\frac{(x + y)^2}{4} \ge xy$$

$\dfrac{4xy}{(x + y)^2} \le 1 \quad$ ...(ii)

From Eqs. (i) and (ii), we get

$$\frac{4xy}{(x + y)^2} = 1$$

$$(x + y)^2 = 4xy$$

$$(x - y)^2 = 0$$

$$\Rightarrow \quad x = y$$

$$\therefore \quad \frac{x}{y} = 1, \text{ i.e. } x : y = 1 : 1$$

12. (b) Given, $\sin\theta + \operatorname{cosec}\theta = 2$

$$\Rightarrow \sin\theta + \frac{1}{\sin\theta} = 2$$

$$\Rightarrow \sin^2\theta + 1 - 2\sin\theta = 0$$

$$\Rightarrow (\sin\theta - 1)^2 = 0$$

$$\Rightarrow \sin\theta = 1, \operatorname{cosec}\theta = 1$$

Now, $\sin^{20}\theta + \operatorname{cosec}^{20}\theta = (1)^{20} + (1)^{20}$

$$= 2$$

Remember

1. If $\sin\theta + \operatorname{cosec}\theta = 2$, *then*

$\sin\theta = 1$ *and* $\operatorname{cosec}\theta = 1$

2. If $\tan\theta + \cot\theta = 2$, *then*

$\tan\theta = 1$ *and* $\cot\theta = 1$

3. If $\cos\theta + \sec\theta = 2$, *then*

$\cos\theta = 1$ *and* $\sec\theta = 1$

13. (c) $\because \sec\theta + \tan\theta = \frac{3}{2}$ ...(i)

We know that,

$$\sec^2\theta - \tan^2\theta = 1$$

$$(\sec\theta + \tan\theta)(\sec\theta - \tan\theta) = 1$$

$$\sec\theta - \tan\theta = \frac{1}{\sec\theta + \tan\theta} = \frac{1}{3/2} \quad \text{[from Eq. (i)]}$$

$$\Rightarrow \sec\theta - \tan\theta = \frac{2}{3} \quad \text{...(ii)}$$

On adding Eqs. (i) and (ii), we get

$$2\sec\theta = \frac{13}{6}$$

$$\Rightarrow \sec\theta = \frac{13}{12}$$

On subtracting Eq. (ii) from Eq. (i), we get

$$2\tan\theta = \frac{3}{2} - \frac{2}{3} = \frac{9-4}{6} = \frac{5}{6}$$

$$\therefore \tan\theta = \frac{5}{6} \times \frac{1}{2} = \frac{5}{12}$$

$$\therefore \sec\theta = \frac{13}{12}, \tan\theta = \frac{5}{12}$$

So, $a = \frac{13}{12}$ and $b = \frac{5}{12}$

14. (d) $\because F_k(x) = \frac{1}{k}(\sin^k x + \cos^k x)$

Now, $F_4(x) = \frac{1}{4}(\sin^4 x + \cos^4 x)$

$$= \frac{1}{4}[(\sin^2 x + \cos^2 x)^2 - 2\sin^2 x\cos^2 x]$$

$$= \frac{1}{4}[1 - 2\sin^2 x\cos^2 x]$$

and $F_6(x) = \frac{1}{6}[\sin^6 x + \cos^6 x]$

$$= \frac{1}{6}[(\sin^2 x)^3 + (\cos^2 x)^3]$$

$$= \frac{1}{6}[(\sin^2 x + \cos^2 x)(\sin^4 x + \cos^4 x - \sin^2 x\cos^2 x)]$$

$$= \frac{1}{6}[1(1 - 3\sin^2 x\cos^2 x)]$$

$$\therefore F_4(x) - F_6(x) = \frac{1}{4}(1 - 2\sin^2 x\cos^2 x) - \frac{1}{6}(1 - 3\sin^2 x\cos^2 x)$$

$$= \frac{1}{12}[3 - 6\sin^2 x\cos^2 x - 2 + 6\sin^2 x\cos^2 x]$$

$$= \frac{1}{12}$$

15. (a) Given, $m = a\cos^3\theta + 3a\cos\theta\sin^2\theta$

$$n = a\sin^3\theta + 3a\cos^2\theta\sin\theta$$

Now, $m + n = a(\sin^3\theta + \cos^3\theta) + 3a\sin\theta\cos\theta(\sin\theta + \cos\theta)$

$$= a(\sin\theta + \cos\theta)(\sin^2\theta + \cos^2\theta - \sin\theta\cos\theta) + 3a\sin\theta\cos\theta(\sin\theta + \cos\theta)$$

$$= a(\sin\theta + \cos\theta)[\sin^2\theta + \cos^2\theta - \sin\theta\cos\theta + 3\sin\theta\cos\theta]$$

$$= a(\sin\theta + \cos\theta)[\sin^2\theta + \cos^2\theta + 2\sin\theta\cos\theta]$$

$$= a(\sin\theta + \cos\theta)(\sin\theta + \cos\theta)^2$$

$$m + n = a(\sin\theta + \cos\theta)^3$$

$$(m+n)^{1/3} = a^{1/3}(\sin\theta + \cos\theta)$$

$$(m+n)^{2/3} = a^{2/3}(\sin\theta + \cos\theta)^2 \quad \text{...(i)}$$

Similarly,

$$(m-n)^{2/3} = a^{2/3}(\sin\theta - \cos\theta)^2 \quad \text{...(ii)}$$

On adding Eqs. (i) and (ii), we get

$(m+n)^{2/3} + (m-n)^{2/3}$

$$= a^{2/3}[\sin^2\theta + \cos^2\theta + 2\sin\theta\cos\theta + \sin^2\theta + \cos^2\theta - 2\sin\theta\cos\theta]$$

$$= a^{2/3}(1+1) = 2a^{2/3}$$

16. (a,c,d) $10\sin^4\alpha + 15\cos^4\alpha = 6$

On dividing both sides by $\cos^4\alpha$, we get

$$\frac{10\sin^4\alpha}{\cos^4\alpha} + 15 = \frac{6}{\cos^4\alpha}$$

$$10\tan^4\alpha + 15 = 6\sec^4\alpha = 6(1 + \tan^2\alpha)^2$$

$$10\tan^4\alpha + 15 = 6[1 + \tan^4\alpha + 2\tan^2\alpha]$$

$$10\tan^4\alpha + 15 - 6\tan^4\alpha - 12\tan^2\alpha - 6 = 0$$

$$4\tan^4\alpha - 12\tan^2\alpha + 9 = 0$$

$$\therefore (2\tan^2\alpha - 3)^2 = 0$$

$$\Rightarrow 2\tan^2\alpha - 3 = 0$$

$\Rightarrow \tan^2\alpha = \frac{3}{2}$, then $\cot^2\alpha = \frac{2}{3}$

$$\therefore 27\operatorname{cosec}^6\alpha + 8\sec^6\alpha = 27(\operatorname{cosec}^2\alpha)^3 + 8(\sec^2\alpha)^3$$

$$= 27(1 + \cot^2\alpha)^3 + 8(1 + \tan^2\alpha)^3$$

$$= 27\left(1 + \frac{2}{3}\right)^3 + 8\left(1 + \frac{3}{2}\right)^3$$

$$= 27 \times \frac{125}{27} + 8 \times \frac{125}{8} = 250$$

17. (a,b) We have,

$$\frac{\sin^4 x}{2} + \frac{\cos^4 x}{3} = \frac{1}{5}$$

$$\Rightarrow \frac{\sin^4 x}{2} + \frac{(1 - \sin^2 x)^2}{3} = \frac{1}{5}$$

$$\Rightarrow \frac{3\sin^4 x + 2(1 + \sin^4 x - 2\sin^2 x)}{6} = \frac{1}{5}$$

$$\Rightarrow 25\sin^4 x - 20\sin^2 x + 4 = 0$$

$$\Rightarrow (5\sin^2 x - 2)^2 = 0$$

$$\Rightarrow \sin^2 x = \frac{2}{5}$$

$$\Rightarrow \cos^2 x = \frac{3}{5}$$

$$\therefore \tan^2 x = \frac{2}{3} \quad \text{[option (a)]}$$

Now, $\frac{\sin^8 x}{8} + \frac{\cos^8 x}{27} = \frac{16}{625}\cdot\frac{1}{8} + \frac{81}{625}\cdot\frac{1}{27}$

$$= \frac{2}{625} + \frac{3}{625} = \frac{5}{625} = \frac{1}{125} \quad \text{[option (b)]}$$

18. (a,b,c) We have, $\cot\theta + \tan\theta = x$

$\Rightarrow \dfrac{\cos\theta}{\sin\theta} + \dfrac{\sin\theta}{\cos\theta} = x$

$\Rightarrow \dfrac{\cos^2\theta + \sin^2\theta}{\sin\theta\cos\theta} = x$

$\Rightarrow \dfrac{1}{\sin\theta\cos\theta} = x$

$\therefore \sin\theta\cos\theta = \dfrac{1}{x}$ [option (a)]

and $\sec\theta - \cos\theta = y$

$\Rightarrow \dfrac{1-\cos^2\theta}{\cos\theta} = y \Rightarrow \dfrac{\sin^2\theta}{\cos\theta} = y$

$\Rightarrow \sin\theta\cdot\dfrac{\sin\theta}{\cos\theta} = y$

$\therefore \sin\theta\tan\theta = y$ [option (b)]

Now, $(x^2y)^{2/3} - (xy^2)^{2/3}$

$= \left(\dfrac{1}{\sin^2\theta\cos^2\theta}\cdot\sin\theta\tan\theta\right)^{2/3} - \left(\dfrac{1}{\sin\theta\cos\theta}\cdot\sin^2\theta\tan^2\theta\right)^{2/3}$

$= \left(\dfrac{1}{\cos^3\theta}\right)^{2/3} - \left(\dfrac{\sin^3\theta}{\cos^3\theta}\right)^{2/3}$

$= \sec^2\theta - \tan^2\theta$

$= 1$ [option (c)]

and $(x^2y)^{1/3} + (xy^2)^{1/3} = \left(\dfrac{1}{\cos^3\theta}\right)^{1/3} + (\tan^3\theta)^{1/3}$

$= \sec\theta + \tan\theta \neq 1$

[option (d) is not correct]

19. (a) $\because \sin A, \cos A, \tan A$ are in GP, then using given information.

$\cos^2 A = \sin A \tan A$

$\cos^2 A = \sin A\cdot\dfrac{\sin A}{\cos A}$

$\dfrac{\cos^2 A}{\sin^2 A} = \dfrac{1}{\cos A}$

$\cot^2 A = \sec A$

On squaring both sides, we get

$\cot^4 A = \sec^2 A = 1 + \tan^2 A$

$\cot^4 A = 1 + \dfrac{1}{\cot^2 A}$

$\cot^6 A = 1 + \cot^2 A$

$\cot^6 A - \cot^2 A = 1$

20. (b) $\because \tan A, 2\tan A + 2, 3\tan A + 3$ are in GP, then using given information.

$(2\tan A + 2)^2 = \tan A(3\tan A + 3)$

$\Rightarrow 4\tan^2 A + 4 + 8\tan A = 3\tan^2 A + 3\tan A$

$\Rightarrow \tan^2 A + 5\tan A + 4 = 0$

$\Rightarrow (\tan A + 1)(\tan A + 4) = 0$

$\Rightarrow \tan A = -4, -1$ (neglecting $\tan A = -1$)

$\therefore \tan A = -4$

21. (4) We have,

$\tan\theta + \sin\theta = m$

and $\tan\theta - \sin\theta = n$

Now, $m + n = 2\tan\theta$

and $m - n = 2\sin\theta$

and $mn = \tan^2\theta - \sin^2\theta$

$= \dfrac{\sin^2\theta}{\cos^2\theta} - \sin^2\theta$

$\Rightarrow mn = \dfrac{\sin^2\theta(1-\cos^2\theta)}{\cos^2\theta}$

$\Rightarrow mn = \tan^2\theta\sin^2\theta$

$\Rightarrow 16mn = (2\tan\theta\cdot 2\sin\theta)^2$

$\Rightarrow 16mn = [(m+n)(m-n)]^2$

$\Rightarrow (m^2 - n^2)^2 = 16mn$

$\Rightarrow 4\lambda mn = 16mn$ (given)

$\Rightarrow 4\lambda = 16$

$\therefore \lambda = 4$

22. (4) We have,

$\sin A + \sin^2 A + \sin^3 A = 1$

$\sin A + \sin^3 A = 1 - \sin^2 A$

$\sin A(1 + \sin^2 A) = \cos^2 A$

$\sin A(2 - \cos^2 A) = \cos^2 A$

On squaring both sides, we get

$\sin^2 A(4 + \cos^4 A - 4\cos^2 A) = \cos^4 A$

$\Rightarrow (1 - \cos^2 A)(4 + \cos^4 A - 4\cos^2 A) = \cos^4 A$

$\Rightarrow 4 + \cos^4 A - 4\cos^2 A - 4\cos^2 A - \cos^6 A + 4\cos^4 A - \cos^4 A = 0$

$\Rightarrow -\cos^6 A + 4\cos^4 A - 8\cos^2 A + 4 = 0$

$\therefore \cos^6 A - 4\cos^4 A + 8\cos^2 A = 4$

DPP-3 Sign of Trigonometric Ratios and Allied Angles

1. LHS

$= \left(1 + \cos\dfrac{\pi}{4}\right)\left(1 + \cos\dfrac{3\pi}{4}\right)\left(1 + \cos\dfrac{5\pi}{4}\right)\left(1 + \cos\dfrac{7\pi}{4}\right)$

$= \left(1 + \cos\dfrac{\pi}{4}\right)\left(1 + \cos\dfrac{3\pi}{4}\right)\left[1 + \cos\left(\pi + \dfrac{\pi}{4}\right)\right]\left[1 + \cos\left(\pi + \dfrac{3\pi}{4}\right)\right]$

$= \left(1 + \cos\dfrac{\pi}{4}\right)\left(1 + \cos\dfrac{3\pi}{4}\right)\left(1 - \cos\dfrac{\pi}{4}\right)\left(1 - \cos\dfrac{3\pi}{4}\right)$

$= \left(1 - \cos^2\dfrac{\pi}{4}\right)\left(1 - \cos^2\dfrac{3\pi}{4}\right)$

$= \sin^2\dfrac{\pi}{4}\cdot\sin^2\dfrac{3\pi}{4}$

$= \sin^2\dfrac{\pi}{4}\sin^2\left(\pi - \dfrac{\pi}{4}\right)$

$= \sin^2\dfrac{\pi}{4}\cdot\sin^2\dfrac{\pi}{4} = \left(\dfrac{1}{\sqrt{2}}\right)^2\left(\dfrac{1}{\sqrt{2}}\right)^2$

$= \dfrac{1}{2}\times\dfrac{1}{2} = \dfrac{1}{4} = \text{RHS}$ **Hence proved.**

2. Given, $x = y\cos\dfrac{2\pi}{3} = z\cos\dfrac{4\pi}{3} = k$

$\Rightarrow x = -\dfrac{y}{2} = -\dfrac{z}{2} = k$

$\Rightarrow x = k, y = -2k, z = -2k$

Now, LHS $= xy + yz + zx$

$= k(-2k) + (-2k)\cdot(-2k) + k(-2k)$

$= -2k^2 + 4k^2 - 2k^2$

$= 0 = \text{RHS}$ **Hence proved.**

3. We have,

$$\sec^2\theta + \text{cosec}^2\theta = 1 + \tan^2\theta + 1 + \cot^2\theta$$

$$\sec^2\theta + \text{cosec}^2\theta = 2 + \tan^2\theta + \cot^2\theta \quad \text{...(i)}$$

$\because$ AM $\geq$ GM, then

$$\frac{\tan^2\theta + \cot^2\theta}{2} \geq \sqrt{\tan^2\theta \cot^2\theta}$$

$$\tan^2\theta + \cot^2\theta \geq 2$$

$$2 + \tan^2\theta + \cot^2\theta \geq 2 + 2$$

$$2 + \tan^2\theta + \cot^2\theta \geq 4$$

$\therefore \quad \sec^2\theta + \text{cosec}^2\theta \geq 4$ [from Eq. (i)]

Hence proved.

4. (b) From option (a),

$$4\sin^2\theta = 5$$

$\Rightarrow \quad \sin^2\theta = \frac{5}{4}$, which is not possible. $(\because 0 \leq \sin^2\theta \leq 1)$

From option (b),

$$(a^2 + b^2)\cos\theta = 2ab$$

$$\cos\theta = \frac{2ab}{a^2 + b^2}$$

$\because$ AM $\geq$ GM, then $\quad \frac{a^2 + b^2}{2} \geq \sqrt{a^2b^2}$

$$a^2 + b^2 \geq 2ab$$

$\Rightarrow \quad \frac{2ab}{a^2 + b^2} \leq 1$

$\cos\theta \leq 1$, which is true.

From option (c),

$$(m^2 + n^2)\,\text{cosec}\,\theta = m^2 - n^2$$

$$\text{cosec}\,\theta = \frac{m^2 - n^2}{m^2 + n^2} = \frac{(m^2 + n^2) - 2n^2}{m^2 + n^2}$$

$= 1 - \frac{2n^2}{m^2 + n^2} < 1$, which is wrong. $(\because \text{cosec}\,\theta \geq 1)$

From option (d),

$\sin\theta = 2.375$, which is wrong. $(\because -1 \leq \sin\theta \leq 1)$

5. (d) Let $\quad y = e^{\log_{10}\tan 1° + \log_{10}\tan 2° + \log_{10}\tan 3° + \ldots + \log_{10}\tan 89°}$

$$y = e^{\log_{10}(\tan 1° \cdot \tan 2° \cdot \tan 3° \cdot \ldots \cdot \tan 89°)} \quad \text{...(i)}$$

$(\because \log m + \log n = \log m \cdot n)$

$\because \quad \tan 89° = \tan(90° - 1°) = \cot 1°$

$\tan 88° = \tan(90° - 2°) = \cot 2°$

....

$\therefore \quad \tan 1° \cdot \tan 89° = \tan 1° \cdot \cot 1° = 1$

and $\quad \tan 2° \cdot \tan 88° = \tan 2° \cdot \cot 2° = 1$

Now, from Eq. (i),

$$y = e^{\log_{10}(\tan 1° \cdot \tan 89°)(\tan 2° \cdot \tan 88°) \ldots \tan 45°}$$

$$= e^{\log_{10}(1 \cdot 1 \cdot 1 \ldots 1)}$$

$$= e^{\log_{10} 1} = e^0 = 1$$

6. (b) LHS $= \sqrt{\frac{1 - \sin\alpha}{1 + \sin\alpha}} = \frac{\sqrt{1 - \sin\alpha}}{\sqrt{1 + \sin\alpha}} \times \frac{\sqrt{1 - \sin\alpha}}{\sqrt{1 - \sin\alpha}}$

$$= \frac{1 - \sin\alpha}{\sqrt{1 - \sin^2\alpha}} = \frac{1 - \sin\alpha}{|\cos\alpha|}$$

and $\quad$ RHS $= \sec\alpha - \tan\alpha$

$$= \frac{1}{\cos\alpha} - \frac{\sin\alpha}{\cos\alpha} = \frac{1 - \sin\alpha}{\cos\alpha}$$

$\because \quad \frac{1 - \sin\alpha}{|\cos\alpha|} = \frac{1 - \sin\alpha}{\cos\alpha}$

If $|\cos\alpha| = \cos\alpha$, it is possible only when $-\frac{\pi}{2} < \alpha < \frac{\pi}{2}$.

Hence, $\sqrt{\frac{1 - \sin\alpha}{1 + \sin\alpha}} = \sec\alpha - \tan\alpha$, when $\alpha \in \left(-\frac{\pi}{2}, \frac{\pi}{2}\right)$.

7. (b) $\because \cot(2n - 1)\alpha = \cot(2n\alpha - \alpha)$

$$= \cot\left(\frac{\pi}{2} - \alpha\right) \quad \left(\because 4n\alpha = \pi \therefore 2n\alpha = \frac{\pi}{2}\right)$$

$$\cot(2n - 1)\alpha = \tan\alpha$$

and $\cot(2n - 2)\alpha = \cot(2n\alpha - 2\alpha)$

$$= \cot\left(\frac{\pi}{2} - 2\alpha\right) = \tan 2\alpha$$

...... and so on.

Now, $\cot\alpha \cdot \cot 2\alpha \cdot \cot 3\alpha \cdots \cot(2n - 2)\alpha \cdot \cot(2n - 1)\alpha$

$= [\cot\alpha \cdot \cot(2n - 1)\alpha][\cot 2\alpha \cdot \cot(2n - 2)\alpha] \ldots \cot(n\alpha)$

$= (\cot\alpha \cdot \tan\alpha)(\cot 2\alpha \cdot \tan 2\alpha) \ldots \cot\frac{\pi}{4}$

$= 1 \cdot 1 \ldots 1 = 1$

8. (d) $\because \quad \cos^7 x + \sin^4 x = 1$

It is possible only when

(i) $\quad \cos^7 x = 0$ and $\sin^4 x = 1$

or $\quad \cos x = 0$ and $\sin x = \pm 1$

$\therefore$ In $[-\pi, \pi]$, $\quad x = \frac{\pi}{2}, -\frac{\pi}{2}$

(ii) $\quad \cos^7 x = 1$ and $\sin^4 x = 0$

$\cos x = 1$ and $\sin x = 0$

$\therefore$ In $[-\pi, \pi]$, $\quad x = 0$

So, $\quad x = 0, -\frac{\pi}{2}, \frac{\pi}{2}$

$\therefore$ The number of real solutions is 3.

9. (d) We know that,

$$\cos 90° = 0$$

$$\cos 85° = \cos(90° - 5°) = \sin 5°$$

$$\cos 80° = \cos(90° - 10°) = \sin 10°$$

....

....

$\therefore \quad (\cos^2 5° + \cos^2 85°) + (\cos^2 10° + \cos^2 80°)$

$+ \ldots + \cos^2 45° + \cos^2 90°$

$= (\cos^2 5° + \sin^2 5°) + (\cos^2 10° + \sin^2 10°)$

$+ \ldots + (\cos^2 40° + \sin^2 40°) + \left(\frac{1}{\sqrt{2}}\right)^2 + 0$

$= 1 + 1 + \ldots$ upto 8 terms $+ \frac{1}{2}$

$= 8 + \frac{1}{2} = 8.5$

10. (c) We know that,

$$-1 \leq \cos\theta \leq 1 \quad \text{...(i)}$$

and $\quad$ AM $\geq$ GM

$$\frac{x + \frac{1}{x}}{2} \geq \sqrt{x \cdot \frac{1}{x}}$$

$\Rightarrow \quad x + \frac{1}{x} \geq 2 \times 1$

$$x + \frac{1}{x} \geq 2 \quad \text{...(ii)}$$

From Eqs. (i) and (ii), we get

$$\cos\theta \neq x + \frac{1}{x}$$

11. (b) $3\left[\sin^4\left(\frac{3\pi}{2}-\alpha\right)+\sin^4(3\pi+\alpha)\right]$

$$-2\left[\sin^6\left(\frac{\pi}{2}+\alpha\right)+\sin^6(5\pi-\alpha)\right]$$

$= 3[\cos^4\alpha+\sin^4\alpha]-2[\cos^6\alpha+\sin^6\alpha]$

$= 3[(\sin^2\alpha+\cos^2\alpha)^2-2\sin^2\alpha\cos^2\alpha]$

$\quad -2[(\sin^2\alpha+\cos^2\alpha)(\sin^4\alpha+\cos^4\alpha-\sin^2\alpha\cos^2\alpha)]$

$= 3[1-2\sin^2\alpha\cos^2\alpha]-2[1(\sin^2\alpha+\cos^2\alpha)^2-3\sin^2\alpha\cos^2\alpha]$

$= 3-6\sin^2\alpha\cos^2\alpha-2+6\sin^2\alpha\cos^2\alpha = 3-2=1$

12. (c) $\because \frac{\sin^3\theta-\cos^3\theta}{\sin\theta-\cos\theta}$

$$=\frac{(\sin\theta-\cos\theta)(\sin^2\theta+\cos^2\theta+\sin\theta\cos\theta)}{(\sin\theta-\cos\theta)}$$

$$=1+\sin\theta\cos\theta,\text{ when }\theta\neq\frac{\pi}{4},\frac{5\pi}{4}$$

and $-\frac{\cos\theta}{\sqrt{1+\cot^2\theta}}=-\frac{\cos\theta}{|\operatorname{cosec}\theta|}$

$=-\sin\theta\cos\theta,$ when $\theta\in(0,\pi)$

and $-2\tan\theta\cot\theta=-2,$ when $\theta\neq 0,\frac{\pi}{2}$

Now, $\frac{\sin^3\theta-\cos^3\theta}{\sin\theta-\cos\theta}-\frac{\cos\theta}{\sqrt{1+\cot^2\theta}}-2\tan\theta\cot\theta=-1$

When, $\theta\in(0,\pi)$ and $\theta\neq 0,\frac{\pi}{4},\frac{5\pi}{4},\frac{\pi}{2}$

$\therefore \quad \theta\in(0,\pi)-\left\{\frac{\pi}{4},\frac{\pi}{2}\right\}$

13. (a) $\because e^{\sin x}-e^{-\sin x}=4$

$$e^{\sin x}-\frac{1}{e^{\sin x}}=4$$

$$(e^{\sin x})^2-4e^{\sin x}-1=0$$

Let $e^{\sin x}=y$, then

$$y^2-4y-1=0$$

$$y=\frac{4\pm\sqrt{16+4\times1\times1}}{2}=\frac{4\pm2\cdot\sqrt{5}}{2}$$

$$y=2\pm\sqrt{5}$$

$\therefore \quad e^{\sin x}=2+\sqrt{5}$ or $2-\sqrt{5}$

$e^{\sin x}=2+2.2$ or $2-2.2$

$e^{\sin x}=4.2$ or -0.2

We know that, $-1\le\sin x\le1$

$0<e^{\sin x}\le e$

$0<e^{\sin x}\le 2.7 \quad (\because e\approx2.7)$...(ii)

From Eqs. (i) and (ii), we get

$e^{\sin x}=2\pm\sqrt{5}$ is not possible.

Hence, there is no solution exists.

14. (a,b) We have,

$$81^{\sin^2x}+81^{\cos^2x}=30$$

$\Rightarrow \quad 3^{4\sin^2x}+(3^4)^{1-\sin^2x}=30$

$\Rightarrow \quad 3^{4\sin^2x}+\frac{3^4}{3^{4\sin^2x}}=30$

Let $\quad 3^{4\sin^2x}=y$

$\Rightarrow \quad y+\frac{81}{y}=30$

$\Rightarrow \quad y^2-30y+81=0$

$\Rightarrow \quad (y-3)(y-27)=0$

$$y=3,27$$

Now, $\quad 3^{4\sin^2x}=3$ or $3^{4\sin^2x}=27$

$$\sin^2x=\frac{1}{4},\sin^2x=\frac{3}{4}$$

$$\sin x=\frac{1}{2},\sin x=\frac{\sqrt{3}}{2}$$

$\therefore \quad x=\frac{\pi}{6},\frac{\pi}{3} \quad (\because 0\le x\le\pi)$

15. (a,b) We have,

$$\sin(\alpha+\beta)=1\text{ and }\sin(\alpha-\beta)=\frac{1}{2}$$

where, $\alpha,\beta\in\left[0,\frac{\pi}{2}\right]$

Now, $0\le\alpha+\beta\le\pi$

and $-\frac{\pi}{2}\le\alpha-\beta\le\frac{\pi}{2}$

Then, $\sin(\alpha+\beta)=1=\sin\frac{\pi}{2}$

$\alpha+\beta=\frac{\pi}{2}$ ...(i)

and $\sin(\alpha-\beta)=\frac{1}{2}=\sin\frac{\pi}{6}$

$\therefore \quad \alpha-\beta=\frac{\pi}{6}$ ...(ii)

On solving Eqs. (i) and (ii), we get

$$\alpha=\frac{\pi}{3}\text{ and }\beta=\frac{\pi}{6}$$

$\tan(\alpha+\beta)=\tan\left(\frac{\pi}{3}+\frac{\pi}{6}\right)=\tan\frac{\pi}{2}=\infty$ [option (c) is wrong]

Now, $\tan(\alpha+2\beta)=\tan\left(\frac{\pi}{3}+2\cdot\frac{\pi}{6}\right)=\tan\left(\frac{2\pi}{3}\right)$

$=\tan\left(\pi-\frac{\pi}{3}\right)=-\tan\frac{\pi}{3}=-\sqrt{3}$ [option (a)]

and $\tan(2\alpha+\beta)=\tan\left(\frac{2\pi}{3}+\frac{\pi}{6}\right)=\tan\frac{5\pi}{6}$

$=\tan\left(\pi-\frac{\pi}{6}\right)=-\tan\frac{\pi}{6}=-\frac{1}{\sqrt{3}}$ [option (b)]

16. (b,d) We have,

$$\frac{4x^2+1}{64x^2-96x\sin\alpha+5}<\frac{1}{32}$$

$$64x^2+96x\sin\alpha+27<0$$

For real x, $D\ge0$

$$(96\sin\alpha)^2-4\cdot64\cdot27\ge0$$

$$\sin^2\alpha\ge\frac{3}{4}$$

$\therefore \quad \alpha\in\left(\frac{\pi}{3},\frac{2\pi}{3}\right)$ and $\left(\frac{4\pi}{3},\frac{5\pi}{3}\right)$

17. (c) We have,

$$0<i<\pi, i=1,2,3,\dots,n$$

$\therefore \quad \sin\theta_i>0, i=1,2,3,\dots,n$

We know that,

$$\left(\frac{\sin\theta_1+\sin\theta_2+\sin\theta_3+\dots+\sin\theta_n}{n}\right)$$

$$\le\sin\left(\frac{\theta_1+\theta_2+\theta_3+\dots+\theta_n}{n}\right)=\sin\left(\frac{\pi}{n}\right)$$

Thus, $\sin\theta_1+\sin\theta_2+\sin\theta_3+\dots+\sin\theta_n\le n\sin\left(\frac{\pi}{n}\right)$

18. (a)

(i) $\tan\theta(\cot\theta\cos\theta + \sin\theta)$

$$= \tan\theta\left(\frac{\cos^2\theta}{\sin\theta} + \sin\theta\right)$$

$$= \frac{\sin\theta}{\cos\theta}\left(\frac{\cos^2\theta + \sin^2\theta}{\sin\theta}\right)$$

$$= \frac{1}{\cos\theta} = \sec\theta$$

Thus, **(i) → (r)**

(ii) $$\frac{\tan\theta\cdot\operatorname{cosec}^2\theta}{1+\tan^2\theta} = \frac{\frac{\sin\theta}{\cos\theta}\cdot\frac{1}{\sin^2\theta}}{\sec^2\theta}$$

$$= \frac{\sin\theta}{\cos\theta}\cdot\frac{1}{\sin^2\theta}\cdot\cos^2\theta$$

$$= \frac{\cos\theta}{\sin\theta} = \cot\theta$$

Thus, **(ii) → (p)**

(iii) $$\frac{\operatorname{cosec}^2\theta - 1}{\operatorname{cosec}\theta\cos^2\theta} = \frac{\cot^2\theta}{\frac{1}{\sin\theta}\cdot\cos^2\theta}$$

$$= \frac{\cos^2\theta}{\sin^2\theta}\cdot\frac{\sin\theta}{\cos^2\theta} = \frac{1}{\sin\theta} = \operatorname{cosec}\theta$$

Thus, **(iii) → (s)**

(iv) $$\frac{\cot\theta\sec^2\theta}{1+\cot^2\theta} = \frac{\frac{\cos\theta}{\sin\theta}\cdot\frac{1}{\cos^2\theta}}{\operatorname{cosec}^2\theta}$$

$$= \frac{1}{\sin\theta\cos\theta}\times\sin^2\theta$$

$$= \frac{\sin\theta}{\cos\theta} = \tan\theta$$

Thus, **(iv) → (q)**

19. (5) $\because \sec A\tan B + \tan A\sec B = 2$ ...(i)

Now, $(\sec A\sec B + \tan A\tan B)^2 - (\sec A\tan B + \tan A\sec B)^2$

$$= \left(\frac{1}{\cos A\cos B} + \frac{\sin A}{\cos A}\cdot\frac{\sin B}{\cos B}\right)^2 - \left(\frac{1}{\cos A}\cdot\frac{\sin B}{\cos B} + \frac{\sin A}{\cos A}\cdot\frac{1}{\cos B}\right)^2$$

$$= \frac{(1+\sin A\sin B)^2}{\cos^2 A\cos^2 B} - \frac{(\sin A+\sin B)^2}{\cos^2 A\cos^2 B}$$

$$= \frac{1+\sin^2 A\sin^2 B + 2\sin A\sin B - \sin^2 A - \sin^2 B - 2\sin A\sin B}{\cos^2 A\cos^2 B}$$

$$= \frac{(1-\sin^2 A) - \sin^2 B + \sin^2 A\sin^2 B}{\cos^2 A\cos^2 B}$$

$$= \frac{(1-\sin^2 A) - \sin^2 B(1-\sin^2 A)}{\cos^2 A\cos^2 B}$$

$$= \frac{(1-\sin^2 A)(1-\sin^2 B)}{\cos^2 A\cos^2 B}$$

$$= \frac{\cos^2 A\cos^2 B}{\cos^2 A\cos^2 B} = 1$$

$\therefore$ $(\sec A\sec B + \tan A\tan B)^2$

$= 1 + (\sec A\tan B + \tan A\sec B)^2$

$= 1 + (2)^2 = 1 + 4 = 5$

20. (7) Given equation is

$$4x^2 - 3x + \alpha = 0$$

$\sin A$ and $\cos A$ are its roots, then

$$\sin A + \cos A = \frac{3}{4}$$

$$\sin A\cos A = \frac{\alpha}{4}$$

Now, $\sin A + \cos A + \tan A + \cot A + \sec A + \operatorname{cosec} A = 7$

$$(\sin A + \cos A) + \frac{\sin A}{\cos A} + \frac{\cos A}{\sin A} + \frac{1}{\cos A} + \frac{1}{\sin A} = 7$$

$$(\sin A + \cos A) + \frac{\sin^2 A + \cos^2 A}{\sin A\cos A} + \frac{\sin A + \cos A}{\sin A\cos A} = 7$$

$$\Rightarrow \quad \frac{3}{4} + \frac{1 + (\sin A + \cos A)}{\sin A\cos A} = 7$$

$$\Rightarrow \quad \frac{1+\frac{3}{4}}{\frac{\alpha}{4}} = 7 - \frac{3}{4}$$

$$= \frac{28-3}{4}$$

$$\Rightarrow \quad \frac{7}{\alpha} = \frac{25}{4}$$

$$\therefore \quad \frac{25\alpha}{4} = 7$$

DPP-4 Sign of T-Ratios and T-Ratios of Standard Angles

1. $\because \cot\theta = \frac{12}{5}$, θ lies in III quadrant, then

$$\tan\theta = \frac{5}{12}$$

$$\because \quad \operatorname{cosec}^2\theta = 1 + \cot^2\theta = 1 + \left(\frac{12}{5}\right)^2$$

$$= \frac{25+144}{25} = \frac{169}{25}$$

$$\therefore \quad \operatorname{cosec}\theta = \pm\sqrt{\frac{169}{25}} = \pm\frac{13}{5}$$

$\because \theta$ lies in III quadrant, then

$$\operatorname{cosec}\theta = -\frac{13}{5}$$

$$\sin\theta = \frac{1}{\operatorname{cosec}\theta} = \frac{1}{-\frac{13}{5}}$$

$$\therefore \quad \sin\theta = \frac{-5}{13}$$

Now, $$\sec^2\theta = 1 + \tan^2\theta$$

$$= 1 + \left(\frac{5}{12}\right)^2 = \frac{144+25}{144} = \frac{169}{144}$$

$$\Rightarrow \quad \sec\theta = \pm\frac{13}{12}$$

$\because \theta \in$ III quadrant, then $\sec\theta = -\frac{13}{12}$

and $\cos\theta = -\frac{12}{13}$

2. $\because \sin\theta = \frac{12}{13}$ and θ lies in II quadrant.

We know that,

$$\cos^2\theta = 1 - \sin^2\theta$$

$$= 1 - \left(\frac{12}{13}\right)^2 = 1 - \frac{144}{169} = \frac{25}{169}$$

$$\Rightarrow \quad \cos\theta = \pm\frac{5}{13}$$

$$\Rightarrow \quad \cos\theta = -\frac{5}{13} \quad (\because \theta \text{ lies in II quadrant})$$

$$\therefore \quad \sec\theta = -\frac{13}{5}$$

and $\tan\theta = \frac{\sin\theta}{\cos\theta} = \frac{\frac{12}{13}}{-\frac{5}{13}} = -\frac{12}{5}$

Now, LHS $= \sec\theta + \tan\theta = -\frac{13}{5} - \frac{12}{5}$

$= -\frac{25}{5} = -5 =$ RHS **Hence proved.**

3. $\because \sin\theta = \frac{3}{5}$

We know that,

$$\cos^2\theta = 1 - \sin^2\theta = 1 - \left(\frac{3}{5}\right)^2$$

$$\Rightarrow \quad \cos^2\theta = 1 - \frac{9}{25} = \frac{16}{25}$$

$$\Rightarrow \quad \cos\theta = \pm\frac{4}{5}$$

$$\therefore \quad \cos\theta = -\frac{4}{5} \quad \left(\because \frac{\pi}{2} < \theta < \pi\right)$$

and $\tan\theta = \frac{\sin\theta}{\cos\theta} = \frac{\frac{3}{5}}{-\frac{4}{5}} = -\frac{3}{4}$

$$\Rightarrow \quad \tan\theta = -\frac{3}{4} \quad \text{...(i)}$$

$$\because \quad \tan\phi = \frac{1}{2} \quad \text{(given)}$$

We know that,

$$\sec^2\phi = 1 + \tan^2\phi = 1 + \left(\frac{1}{2}\right)^2$$

$$= 1 + \frac{1}{4} = \frac{5}{4}$$

$$\Rightarrow \quad \sec\phi = \pm\frac{\sqrt{5}}{2}$$

$$\therefore \quad \sec\phi = -\frac{\sqrt{5}}{2} \quad \left(\because \pi < \phi < \frac{3\pi}{2}\right) \text{...(ii)}$$

Now, $8\tan\theta - \sqrt{5}\sec\phi = 8 \times \left(-\frac{3}{4}\right) - \sqrt{5} \times \left(-\frac{\sqrt{5}}{2}\right)$

[from Eqs. (i) and (ii)]

$$= -6 + \frac{5}{2}$$

$$= \frac{-12+5}{2} = -\frac{7}{2}$$

4. (b) We have, $\sin^2\frac{\pi}{18} + \sin^2\frac{\pi}{9} + \sin^2\frac{7\pi}{18} + \sin^2\frac{4\pi}{9}$

$= \sin^2\frac{\pi}{18} + \cos^2\left(\frac{\pi}{2} - \frac{\pi}{9}\right) + \sin^2\frac{7\pi}{18} + \cos^2\left(\frac{\pi}{2} - \frac{4\pi}{9}\right)$

$= \left(\sin^2\frac{\pi}{18} + \cos^2\frac{\pi}{18}\right) + \left(\cos^2\frac{7\pi}{18} + \sin^2\frac{7\pi}{18}\right)$

$= 1 + 1 = 2$ $(\because \sin^2\theta + \cos^2\theta = 1)$

5. (c) We have, $\sin\theta = \frac{2t}{1+t^2}$ and θ lies in II quadrant.

We know that,

$$\cos\theta = \sqrt{1 - \sin^2\theta} = \sqrt{1 - \frac{4t^2}{(1+t^2)^2}}$$

$$= \sqrt{\frac{(1+t^2)^2 - 4t^2}{(1+t^2)^2}} = \sqrt{\frac{(1-t^2)^2}{(1+t^2)^2}}$$

$$= \frac{-|1-t^2|}{1+t^2} \quad (\because \cos\theta < 0 \text{ in II quadrant})$$

6. (a) $\because \tan\theta = x - \frac{1}{4x}$ (given)

Let $\quad \sec\theta - \tan\theta = p$ ...(i)

$\quad \sec\theta + \tan\theta = \frac{1}{p}$ ...(ii)

On subtracting Eq. (ii) from Eq. (i), we get

$$-2\tan\theta = p - \frac{1}{p}$$

$$-2\left(x - \frac{1}{4x}\right) = p - \frac{1}{p}$$

$$-2x + \frac{1}{2x} = p - \frac{1}{p} \quad \text{...(iii)}$$

$\because$ Option (a) satisfies the Eq. (iii), then

$$\sec\theta - \tan\theta = -2x, \frac{1}{2x}$$

7. (c) $\because \quad \tan\theta + \sec\theta = \sqrt{3}$ ...(i)

Then, $\quad \sec\theta - \tan\theta = \frac{1}{\sqrt{3}}$ ...(ii)

On adding Eqs. (i) and (ii), we get

$$2\sec\theta = \sqrt{3} + \frac{1}{\sqrt{3}}$$

$$2\sec\theta = \frac{3+1}{\sqrt{3}}$$

$$\sec\theta = \frac{4}{\sqrt{3}} \times \frac{1}{2}$$

$$\sec\theta = \frac{2}{\sqrt{3}}$$

Now, $\quad \cos\theta = \frac{\sqrt{3}}{2} = \cos\frac{\pi}{6} \quad (\because 0 < \theta < \pi)$

$$\therefore \quad \theta = \frac{\pi}{6}$$

8. (d) $\because \frac{3\pi}{4} < \alpha < \pi$, then we have

$$\sqrt{2\cot\alpha + \frac{1}{\sin^2\alpha}} = \sqrt{2\cot\alpha + \text{cosec}^2\alpha}$$

$$= \sqrt{2\cot\alpha + 1 + \cot^2\alpha}$$

$$= \sqrt{(\cot\alpha + 1)^2}$$

$$= \pm(1 + \cot\alpha)$$

$$= -(1 + \cot\alpha) \quad \left(\because \frac{3\pi}{4} < \alpha < \pi\right)$$

9. (d) $\because \frac{99\pi}{2} < \theta < 50\pi \Rightarrow \theta$ lies in IV quadrant, then

$$|\sin\theta| = -\sin\theta \text{ and } |\cos\theta| = \cos\theta$$

Now, $x = \sin\theta|\sin\theta|$ and $y = \cos\theta|\cos\theta|$

$$x = -\sin^2\theta \text{ and } y = \cos^2\theta$$

Therefore, $\quad y - x = \cos^2\theta - (-\sin^2\theta)$

$$\Rightarrow \quad y - x = \cos^2\theta + \sin^2\theta$$

$$\therefore \quad y - x = 1$$

10. (d) We have,

$$x \sin 45° \cos^2 60° = \frac{\tan^2 60° \operatorname{cosec} 30°}{\sec 45° \cot^2 30°}$$

$$\Rightarrow \quad x \cdot \frac{1}{\sqrt{2}} \cdot \frac{1}{4} = \frac{(\sqrt{3})^2\, 2}{\sqrt{2}\, (\sqrt{3})^2}$$

$$\Rightarrow \quad \frac{x}{4\sqrt{2}} = \frac{6}{3\sqrt{2}}$$

$$\Rightarrow \quad x = \frac{6}{3\sqrt{2}} \times 4\sqrt{2}$$

$$\therefore \quad x = 8$$

11. (a) We have,

$$\frac{\sin A}{\sin B} = \sqrt{2} \quad \text{...(i)}$$

$$\sin A = \sqrt{2} \sin B$$

$$\sin^2 A = 2 \sin^2 B$$

$$1 - \cos^2 A = 2(1 - \cos^2 B)$$

$$1 - \cos^2 A = 2 - 2\cos^2 B$$

$$\cos^2 A = 2\cos^2 B - 1 \quad \text{...(ii)}$$

and

$$\frac{\tan A}{\tan B} = \sqrt{3}$$

$$\frac{\sin A}{\sin B} \cdot \frac{\cos B}{\cos A} = \sqrt{3}$$

$$\sqrt{2} \cdot \frac{\cos B}{\cos A} = \sqrt{3} \quad \text{[from Eq. (i)]}$$

$$2\cos^2 B = 3\cos^2 A$$

$$2\cos^2 B = 3(2\cos^2 B - 1) \quad \text{[from Eq. (ii)]}$$

$$2\cos^2 B = 6\cos^2 B - 3$$

$$4\cos^2 B = 3$$

$$\cos B = \pm \frac{\sqrt{3}}{2}$$

$$\cos B = \frac{\sqrt{3}}{2} = \cos 30° \quad \left(\because 0 \le B < \frac{\pi}{2}\right)$$

$$B = 30°$$

$B = 30°$ put in Eq. (ii)

$$\therefore \quad \cos^2 A = 2\cos^2 30° - 1 = 2 \times \frac{3}{4} - 1$$

$$\cos^2 A = \frac{1}{2}$$

$$\cos A = \pm \frac{1}{\sqrt{2}} = \frac{1}{\sqrt{2}} \quad \left(\because 0 < A < \frac{\pi}{2}\right)$$

$$\cos A = \cos 45°$$

$$\therefore \quad A = 45° \text{ and } B = 30°$$

12. (c) $\because \cot(\alpha + \beta) = 0 = \cot \frac{\pi}{2}$

$$\Rightarrow \quad \alpha + \beta = \frac{\pi}{2} \quad \text{...(i)}$$

Now, if we take $\alpha = \frac{\pi}{2} - \beta$, then

$$\sin(\alpha + 2\beta) = \sin\left(\frac{\pi}{2} - \beta + 2\beta\right)$$

$$= \sin\left(\frac{\pi}{2} + \beta\right) = \cos \beta$$

If we take $\beta = \frac{\pi}{2} - \alpha$, then

$$\sin(\alpha + 2\beta) = \sin(\alpha + \pi - 2\alpha)$$

$$= \sin(\pi - \alpha)$$

$$= \sin \alpha$$

$$\therefore \quad \sin(\alpha + 2\beta) = \sin\alpha \text{ or } \cos\beta$$

13. (b) We have, $\sqrt{\frac{1 - \sin\theta}{1 + \sin\theta}} + \sqrt{\frac{1 + \sin\theta}{1 - \sin\theta}} = \frac{1 - \sin\theta + 1 + \sin\theta}{\sqrt{1 - \sin^2\theta}}$

$$= \frac{2}{\sqrt{\cos^2\theta}} = \frac{2}{|\cos\theta|}$$

$$= \frac{2}{-\cos\theta} \quad (\because \theta \text{ lies in II quadrant})$$

$$= -2\sec\theta$$

14. (a,b) We have,

$$\sqrt{\frac{1 + \cos\theta}{1 - \cos\theta}} = \sqrt{\frac{(1 + \cos\theta)^2}{1 - \cos^2\theta}} = \frac{|1 + \cos\theta|}{|\sin\theta|} = \frac{(1 + \cos\theta)}{|\sin\theta|}$$

$$(\because 1 + \cos\theta \ge 0, \forall \theta)$$

When $0 < \theta < \pi$, then

$$\sqrt{\frac{1 + \cos\theta}{1 - \cos\theta}} = \frac{(1 + \cos\theta)}{\sin\theta} = \operatorname{cosec}\theta + \cot\theta$$

When $\pi < \theta < 2\pi$, then

$$\sqrt{\frac{1 + \cos\theta}{1 - \cos\theta}} = \frac{(1 + \cos\theta)}{-\sin\theta} = -(\operatorname{cosec}\theta + \cot\theta)$$

15. (a) $\because \sec\alpha = -\frac{5}{4}$ and $\frac{\pi}{2} < \alpha < \pi$

We know that, $\tan^2\alpha = \sec^2\alpha - 1$

$$= \left(\frac{-5}{4}\right)^2 - 1 = \frac{25}{16} - 1 = \frac{9}{16}$$

$$\tan\alpha = \pm\frac{3}{4}$$

$$\tan\alpha = \frac{-3}{4} \quad \left(\because \frac{\pi}{2} < \alpha < \pi\right)$$

and

$$\cot\alpha = \frac{-4}{3}$$

Then, $\frac{7 - 5\cot\alpha}{9 - 4\sqrt{\sec^2\alpha - 1}} = \frac{7 - 5\cot\alpha}{9 - 4|\tan\alpha|}$

$$= \frac{7 - 5\cot\alpha}{9 - 4(-\tan\alpha)} = \frac{7 - 5\cot\alpha}{9 + 4\tan\alpha}$$

$$= \frac{7 + 5 \times \frac{4}{3}}{9 - 4 \times \frac{3}{4}} = \frac{\frac{21 + 20}{3}}{9 - 3} = \frac{41}{3 \times 6} = \frac{41}{18}$$

16. (b,c) Given, $x = p\cos\theta + q\sin\theta$

and $y = p\sin\theta - q\cos\theta, 0 < \theta < \frac{\pi}{2}$

Such that $x^2 + 4xy + y^2 = ap^2 + bq^2$

Where, a and b are constants.

Now, $x^2 + 4xy + y^2$

$$= p^2(\sin^2\theta + \cos^2\theta) + q^2(\sin^2\theta + \cos^2\theta) + 4(p^2\sin\theta\cos\theta - pq\cos^2\theta + pq\sin^2\theta - q^2\cos\theta\sin\theta)$$

$$= p^2 + q^2 + 4\sin\theta\cos\theta(p^2 - q^2) + 4pq(\sin^2\theta - \cos^2\theta)$$

$$= p^2 + q^2 + 2(p^2 - q^2)\sin 2\theta - 4pq\cos 2\theta$$

$$= p^2(1 + 2\sin 2\theta) + q^2(1 - 2\sin 2\theta) - 4pq\cos 2\theta$$

$$x^2 + 4xy + y^2 = p^2(1 + 2\sin 2\theta) + q^2(1 - 2\sin 2\theta) \quad \text{...(i)}$$

When, $\cos 2\theta = 0 = \cos\frac{\pi}{2}$

$$2\theta = \frac{\pi}{2}$$

$$\Rightarrow \quad \theta = \frac{\pi}{4} = 45°$$

From Eq. (i), $x^2 + 4xy + y^2 = 3p^2 - q^2$ ...(ii)

Comparing with $x^2 + 4xy + y^2 = ap^2 + bq^2$, we have

$$a = 3 \text{ and } b = -1$$

17. (2) We have, $2k - 4 = 0$

$$\therefore \quad k = 2$$

Now, $\quad \sin x + \operatorname{cosec} x = 2$

$$\sin x + \frac{1}{\sin x} = 2$$

$$\sin^2 x + 1 = 2\sin x$$

$$\sin^2 x - 2\sin x + 1 = 0$$

$$(\sin x - 1)^2 = 0$$

$$\sin x - 1 = 0$$

$$\therefore \quad \sin x = 1$$

Then, $\quad \operatorname{cosec} x = 1 \quad \left(\because 0 < x \le \frac{\pi}{2}\right)$

So, $\quad \sin^n x + \operatorname{cosec}^n x = (1)^n + (1)^n = 2 \quad (\because n \in N)$

18. (0) We have,

$$0 \le x \le \frac{\pi}{2}$$

and $\quad \cos x + \cos 2x + \cos 3x = 3$

It is possible only when

$$\cos x = \cos 2x = \cos 3x = 1$$

because $\quad -1 \le \cos\theta \le 1$

Now, if $\cos x = 1 = \cos 0°$, then $x = 0°$

$$\therefore \quad \sin x + \sin 2x + \sin 3x = 0$$

19. (3) We have, $\quad (y + z)\cos 3\theta = xyz \sin 3\theta$

$$\Rightarrow \quad x\sin 3\theta - \frac{1}{y}\cos 3\theta - \frac{1}{z}\cos 3\theta = 0 \quad \text{...(i)}$$

and $\quad x \sin 3\theta = \dfrac{2\cos 3\theta}{y} + \dfrac{2\sin 3\theta}{z}$

$$\Rightarrow \quad x\sin 3\theta - \frac{2}{y}\cos 3\theta - \frac{2}{z}\sin 3\theta = 0 \quad \text{...(ii)}$$

and $\quad xyz\sin 3\theta = (y + 2z)\cos 3\theta + y\sin 3\theta$

$$\Rightarrow \quad x\sin 3\theta - \frac{2}{y}\cos 3\theta - \frac{1}{z}(\sin 3\theta + \cos 3\theta) = 0 \quad \text{...(iii)}$$

On solving Eqs. (ii) and (iii), we get

$$2\sin 3\theta = \cos 3\theta + \sin 3\theta$$

$$\Rightarrow \quad \sin 3\theta = \cos 3\theta$$

$$\Rightarrow \quad \tan 3\theta = 1 = \tan\frac{\pi}{4}, \tan\frac{5\pi}{4}, \tan\frac{9\pi}{4}$$

$$3\theta = \frac{\pi}{4}, \frac{5\pi}{4}, \frac{9\pi}{4}$$

$$\theta = \frac{\pi}{12}, \frac{5\pi}{12}, \frac{9\pi}{12} \quad (\because 0 < \theta < \pi)$$

$\therefore$ Number of possible values is 3.

20. True, we know that,

$$0 \le \sin^2\theta \le 1$$

and $\quad 0 \le \cos^2\theta \le 1$

$$\Rightarrow \quad 0 \le \sin^8\theta \le \sin^2\theta$$

and $\quad 0 \le \cos^{14}\theta \le \cos^2\theta$

$$\Rightarrow \quad 0 \le \sin^8\theta + \cos^{14}\theta \le \sin^2\theta + \cos^2\theta$$

$$\Rightarrow \quad 0 \le \sin^8\theta + \cos^{14}\theta \le 1$$

$$\therefore \quad 0 \le A \le 1, \forall \theta$$

DPP-5 T-Ratios using Addition and Subtraction Formulae

1. We have,

$$\cos(A + B) = \alpha\cos A\cos B + \beta\sin A\sin B \quad \text{...(i)}$$

and we know that,

$$\cos(A + B) = \cos A\cos B - \sin A\sin B \quad \text{...(ii)}$$

From Eqs. (i) and (ii), we get

$$\alpha = 1, \beta = -1$$

$$\therefore \quad (\alpha, \beta) = (1, -1)$$

2. $\because \alpha$ lies in II quadrant, i.e.

$$90° < \alpha < 180°$$

and β lies in III quadrant, i.e.

$$180° < \beta < 270°$$

Now, $\quad 270° < \alpha + \beta < 450°$

i.e. $\alpha + \beta$ lies in I and IV quadrants.

$\therefore$ $(\alpha + \beta)$ lies in I quadrant. $\quad [\because \tan(\alpha + \beta) > 0]$

3. We have, $\tan A = \frac{1}{2}$, $\tan B = \frac{1}{3}$

We know that,

$$\tan(A + B) = \frac{\tan A + \tan B}{1 - \tan A\tan B} = \frac{\frac{1}{2} + \frac{1}{3}}{1 - \frac{1}{2}\cdot\frac{1}{3}} = \frac{\frac{5}{6}}{\frac{5}{6}} = 1 = \tan\frac{\pi}{4}$$

$$\therefore \quad A + B = \frac{\pi}{4}$$

$$A = \frac{\pi}{4} - B$$

Now, $\quad \cos 2A = \cos 2\left(\frac{\pi}{4} - B\right) = \cos\left(\frac{\pi}{2} - 2B\right) = \sin 2B$

But $\quad \cos 2A = \sin\lambda B$

$$\therefore \quad \lambda = 2$$

4. $\text{LHS} = \cos\left(\frac{3\pi}{4} - A\right)\cos\left(\frac{3\pi}{4} - B\right) - \sin\left(\frac{3\pi}{4} - A\right)\sin\left(\frac{3\pi}{4} - B\right)$

$$= \cos\left[\left(\frac{3\pi}{4} - A\right) + \left(\frac{3\pi}{4} - B\right)\right]$$

$[\because \cos(A + B) = \cos A\cos B - \sin A\sin B]$

$$= \cos\left[\frac{3\pi}{2} - (A + B)\right]$$

$$= -\sin(A + B) = \text{RHS}$$

Hence proved.

5. $\because \cos A = \frac{1}{7}$

$$\therefore \quad \sin^2 A = 1 - \cos^2 A$$

$$= 1 - \left(\frac{1}{7}\right)^2 = 1 - \frac{1}{49} = \frac{48}{49}$$

$$\Rightarrow \quad \sin A = \pm\frac{4\sqrt{3}}{7}$$

$$\therefore \quad \sin A = \frac{4\sqrt{3}}{7} \quad (\because A \text{ is an acute angle})$$

and $\quad \cos B = \frac{13}{14}$

$$\therefore \quad \sin^2 B = 1 - \cos^2 B$$

$$= 1 - \left(\frac{13}{14}\right)^2$$

$$= 1 - \frac{169}{196} = \frac{27}{196}$$

$$\Rightarrow \quad \sin B = \pm\frac{3\sqrt{3}}{14}$$

$$\therefore \quad \sin B = \frac{3\sqrt{3}}{14} \quad (\because B \text{ is an acute angle})$$

Now, we know that,

$\sin(A-B) = \sin A\cos B - \cos A\sin B$

$$= \frac{4\sqrt{3}}{7}\times\frac{13}{14} - \frac{1}{7}\times\frac{3\sqrt{3}}{14} = \frac{52\sqrt{3}-3\sqrt{3}}{14\times 7}$$

$$= \frac{49\sqrt{3}}{14\times 7} = \frac{\sqrt{3}}{2} = \sin\frac{\pi}{3}$$

$\Rightarrow \quad \sin(A-B) = \sin\frac{\pi}{3}$

$\therefore \quad A - B = \frac{\pi}{3}$ **Hence proved.**

6. $\because \quad 3\tan A\tan B = 1$

$\Rightarrow \quad \tan A\tan B = \frac{1}{3}$ (given)

$$\text{LHS} = \frac{\cos(A-B)}{\cos(A+B)} = \frac{\cos A\cos B + \sin A\sin B}{\cos A\cos B - \sin A\sin B}$$

$$= \frac{1+\tan A\tan B}{1-\tan A\tan B}$$

(dividing numerator and denominator by $\cos A\cos B$)

$$= \frac{1+\frac{1}{3}}{1-\frac{1}{3}} = \frac{4}{3}\times\frac{3}{2} = 2 = \text{RHS}$$

Hence proved.

7. (b) $\because \tan\alpha = \frac{m}{m+1}$ and $\tan\beta = \frac{1}{2m+1}$

Now, $\tan(\alpha+\beta) = \frac{\tan\alpha+\tan\beta}{1-\tan\alpha\cdot\tan\beta} = \frac{\frac{m}{m+1}+\frac{1}{2m+1}}{1-\frac{m}{m+1}\times\frac{1}{2m+1}}$

$$= \frac{m(2m+1)+(m+1)}{(m+1)(2m+1)-m}$$

$$= \frac{2m^2+m+m+1}{2m^2+m+2m+1-m}$$

$$= \frac{2m^2+2m+1}{2m^2+2m+1} = 1$$

$\Rightarrow \quad \tan(\alpha+\beta) = \tan\frac{\pi}{4}$

$\therefore \quad \alpha+\beta = \frac{\pi}{4}$

8. (a) $\because \cot\alpha = \frac{1}{2}$

$\tan\alpha = 2$

and $\sec\beta = -\frac{5}{3}$

We know that, $\tan^2\beta = \sec^2\beta - 1$

$$= \left(-\frac{5}{3}\right)^2 - 1 = \frac{25}{9} - 1 = \frac{16}{9}$$

$\Rightarrow \quad \tan\beta = \pm\frac{4}{3}$

$\therefore \quad \tan\beta = -\frac{4}{3} \quad \left(\because \frac{\pi}{2} < \beta < \pi\right)$

Now, $\tan(\alpha+\beta) = \frac{\tan\alpha+\tan\beta}{1-\tan\alpha\tan\beta} = \frac{2-\frac{4}{3}}{1-2\times\left(-\frac{4}{3}\right)}$

$$= \frac{\frac{6-4}{3}}{1+\frac{8}{3}} = \frac{2}{3}\times\frac{3}{11} = \frac{2}{11}$$

$\therefore \quad \tan(\alpha+\beta) = \frac{2}{11}$ ...(i)

$\because \quad \pi < \alpha < \frac{3\pi}{2}$

and $\frac{\pi}{2} < \beta < \pi$

$\therefore \quad \frac{3\pi}{2} < \alpha+\beta < \frac{5\pi}{2}$

i.e. I and IV quadrants.

We know that, $\tan\theta$ is positive in I and III quadrants and from Eq. (i),

$$\tan(\alpha+\beta) = \frac{2}{11} > 0$$

So, $\alpha+\beta$ lies in I quadrant.

9. (a) $\because 0 < \alpha < \frac{\pi}{4}, 0 < \beta < \frac{\pi}{4}$

$\Rightarrow \quad -\frac{\pi}{4} < -\beta < 0$

$0 < (\alpha+\beta) < \frac{\pi}{2}$

and $-\frac{\pi}{4} < (\alpha-\beta) < \frac{\pi}{4}$

Now, $\cos(\alpha+\beta) = \frac{4}{5}$

$\sec(\alpha+\beta) = \frac{5}{4}$

$\sqrt{1+\tan^2(\alpha+\beta)} = \frac{5}{4}$

$1+\tan^2(\alpha+\beta) = \frac{25}{16}$

$\tan(\alpha+\beta) = \sqrt{\frac{25}{16}-1} = \pm\frac{3}{4}$

$\therefore \quad \tan(\alpha+\beta) = \frac{3}{4} \quad \left(\because 0 < \alpha+\beta < \frac{\pi}{2}\right)$

and $\sin(\alpha-\beta) = \frac{5}{13}$ (given)

$\Rightarrow \quad \cos^2(\alpha-\beta) = 1-\sin^2(\alpha-\beta)$

$$= 1-\left(\frac{5}{13}\right)^2 = 1-\frac{25}{169} = \frac{144}{169}$$

$\therefore \quad \cos(\alpha-\beta) = \pm\frac{12}{13}$

$\cos(\alpha-\beta) = \frac{12}{13}$

($\because \alpha-\beta$ lies in I and IV quadrants)

$\therefore \quad \tan(\alpha-\beta) = \frac{\sin(\alpha-\beta)}{\cos(\alpha-\beta)} = \frac{5}{12}$

Now, $\tan 2\alpha = \tan[(\alpha+\beta)+(\alpha-\beta)]$

$$= \frac{\tan(\alpha+\beta)+\tan(\alpha-\beta)}{1-\tan(\alpha+\beta)\tan(\alpha-\beta)}$$

$$= \frac{\frac{3}{4}+\frac{5}{12}}{1-\frac{3}{4}\times\frac{5}{12}} = \frac{56}{33}$$

10. (b) $\because \quad \tan 70° = \tan(50° + 20°)$

$$\tan 70° = \frac{\tan 50° + \tan 20°}{1-\tan 50°\tan 20°}$$

$\tan 70° - \tan 70°\tan 50°\tan 20° = \tan 50° + \tan 20°$

$\tan 70° - \tan(90° - 20°)\tan 50°\tan 20° = \tan 50° + \tan 20°$

$\tan 70° - \cot 20°\tan 50°\tan 20° = \tan 50° + \tan 20°$

$\tan 70° - \tan 50° = \tan 50° + \tan 20°$

$\tan 70° = 2\tan 50° + \tan 20°$

But $\tan 70° = \lambda\tan 50° + \tan 20°$

$\therefore \quad \lambda = 2$

11. (a) $\because 3\sin(2A+B) = \sin B$

$$\frac{\sin(2A+B)}{\sin B} = \frac{1}{3}$$

$$\frac{\sin[(A+B)+A]}{\sin[(A+B)-A]} = \frac{1}{3}$$

$$\Rightarrow \frac{\sin(A+B)\cos A + \cos(A+B)\sin A}{\sin(A+B)\cos A - \cos(A+B)\sin A} = \frac{1}{3}$$

On applying componendo and dividendo, we get

$$\Rightarrow \frac{2\sin(A+B)\cos A}{2\cos(A+B)\sin A} = \frac{4}{-2}$$

$$\Rightarrow \frac{\tan(A+B)}{\tan A} = -2$$

$$\Rightarrow \tan(A+B) = -2\tan A$$

$$\therefore 2\tan A + \tan(A+B) = 0$$

12. (b) Given, $\tan\beta = \dfrac{n\sin\alpha\cos\alpha}{1-n\sin^2\alpha}$ ...(i)

and $\tan(\alpha-\beta) = k\tan\alpha$

Now, $\dfrac{\tan\alpha - \tan\beta}{1+\tan\alpha\tan\beta} = k\tan\alpha$

$$\frac{\dfrac{\sin\alpha}{\cos\alpha} - \dfrac{n\sin\alpha\cos\alpha}{1-n\sin^2\alpha}}{1+\dfrac{\sin\alpha}{\cos\alpha}\cdot\dfrac{n\sin\alpha\cos\alpha}{1-n\sin^2\alpha}} = k\tan\alpha \quad \text{[from Eq. (i)]}$$

$$\frac{\dfrac{\sin\alpha[1-n\sin^2\alpha] - n\sin\alpha\cos^2\alpha}{\cos\alpha(1-n\sin^2\alpha)}}{\dfrac{(1-n\sin^2\alpha)\cos\alpha + n\sin^2\alpha\cos\alpha}{\cos\alpha(1-n\sin^2\alpha)}} = k\tan\alpha$$

$$\Rightarrow \frac{\sin\alpha[1-n\sin^2\alpha - n\cos^2\alpha]}{\cos\alpha[1-n\sin^2\alpha + n\sin^2\alpha]} = k\tan\alpha$$

$$\Rightarrow \frac{\sin\alpha}{\cos\alpha}\cdot\frac{[1-n(\sin^2\alpha+\cos^2\alpha)]}{1} = k\tan\alpha$$

$$\Rightarrow (1-n)\tan\alpha = k\tan\alpha$$

$$\therefore k = (1-n)$$

13. (b) Given that,

$$2\tan\beta + \cot\beta = \tan\alpha \quad \text{...(i)}$$

Now, $\cot\beta = n\tan(\alpha-\beta) = n\cdot\dfrac{\tan\alpha - \tan\beta}{1+\tan\alpha\cdot\tan\beta}$

$$= n\cdot\frac{2\tan\beta + \cot\beta - \tan\beta}{1+(2\tan\beta+\cot\beta)\tan\beta} \quad \text{[from Eq. (i)]}$$

$$= n\cdot\frac{\tan\beta+\cot\beta}{1+2\tan^2\beta+1} = n\cdot\frac{\tan\beta + \dfrac{1}{\tan\beta}}{2+2\tan^2\beta}$$

$$= n\cdot\frac{\tan^2\beta+1}{\tan\beta}\times\frac{1}{2(1+\tan^2\beta)} = \frac{n}{2}\cdot\frac{1}{\tan\beta}$$

$$\cot\beta = \frac{n}{2}\cdot\cot\beta$$

$$\therefore \frac{n}{2} = 1 \Rightarrow n = 2$$

14. (b) We have, $A+B+C = 180°$

$$A+B = 180° - C$$

$$\tan(A+B) = \tan(180° - C)$$

$$\frac{\tan A + \tan B}{1-\tan A\tan B} = -\tan C \quad [\because \tan(180° - \theta) = -\tan\theta]$$

$$\tan A + \tan B = -\tan C + \tan A\tan B\tan C$$

$$\tan A + \tan B + \tan C = \tan A\tan B\tan C$$

But $\tan A + \tan B + \tan C = \tan A\tan B\tan C$

$$\therefore \mu = 1$$

15. (b) $\dfrac{\sin(x+y)}{\sin(x-y)} = k\left(\dfrac{\tan x + \tan y}{\tan x - \tan y}\right)$

$$\Rightarrow \frac{\sin(x+y)}{\sin(x-y)} = k\left(\frac{\dfrac{\sin(x+y)}{\cos x\cos y}}{\dfrac{\sin(x-y)}{\cos x\cos y}}\right)$$

$$\left[\because \tan A \pm \tan B = \frac{\sin(A\pm B)}{\cos A\cos B}\right]$$

$$\Rightarrow \frac{\sin(x+y)}{\sin(x-y)} = k\frac{\sin(x+y)}{\sin(x-y)}$$

$$\therefore k = 1$$

16. (a) We know that, $\sin^2 A - \sin^2 B = \sin(A+B)\sin(A-B)$

Then, we have

$$\cos 2\alpha\cos 2\beta + \sin^2(\alpha-\beta) - \sin^2(\alpha+\beta) = \cos\lambda(\alpha+\beta)$$

$$\Rightarrow \cos 2\alpha\cos 2\beta + \sin(\alpha-\beta+\alpha+\beta)\sin(\alpha-\beta-\alpha-\beta) = \cos\lambda(\alpha+\beta)$$

$$\cos 2\alpha\cos 2\beta + \sin 2\alpha\sin(-2\beta) = \cos\lambda(\alpha+\beta)$$

$$\cos 2\alpha\cos 2\beta - \sin 2\alpha\sin 2\beta = \cos\lambda(\alpha+\beta)$$

$$\Rightarrow \cos(2\alpha+2\beta) = \cos\lambda(\alpha+\beta)$$

$$\cos 2(\alpha+\beta) = \cos\lambda(\alpha+\beta)$$

$$\therefore \lambda = 2$$

17. (c) We have, $\alpha+\beta = \dfrac{\pi}{2} \Rightarrow \alpha = \dfrac{\pi}{2} - \beta$

$$\Rightarrow \tan\alpha = \tan\left(\frac{\pi}{2} - \beta\right)$$

$$\Rightarrow \tan\alpha = \cot\beta$$

$$\Rightarrow \tan\alpha\cdot\tan\beta = 1 \quad \text{...(i)}$$

$$\Rightarrow \beta + \gamma = \alpha$$

$$\Rightarrow \gamma = \alpha - \beta$$

$$\Rightarrow \tan\gamma = \tan(\alpha-\beta)$$

$$\Rightarrow \tan\gamma = \frac{\tan\alpha - \tan\beta}{1+\tan\alpha\cdot\tan\beta}$$

and $\tan\gamma = \dfrac{\tan\alpha - \tan\beta}{1+1}$ [from Eq. (i)]

$$2\tan\gamma = \tan\alpha - \tan\beta$$

$$\therefore \tan\alpha = \tan\beta + 2\tan\gamma$$

18. (a) We have, θ and ϕ are acute angles and

$$\sin\theta = \frac{1}{2}, \cos\phi = \frac{1}{3}$$

$$\Rightarrow \sin\theta = \sin\frac{\pi}{6} \text{ and } 0 < \cos\phi < \frac{1}{2}$$

$$\Rightarrow \theta = \frac{\pi}{6} \text{ and } \cos\frac{\pi}{2} < \cos\phi < \cos\frac{\pi}{3}$$

and $\dfrac{\pi}{3} < \phi < \dfrac{\pi}{2}$

Now, $\dfrac{\pi}{6} + \dfrac{\pi}{3} < \theta + \phi < \dfrac{\pi}{2} + \dfrac{\pi}{6}$

$$\Rightarrow \frac{\pi}{2} < \theta + \phi < \frac{2\pi}{3}$$

$$\therefore \theta + \phi \in \left(\frac{\pi}{2}, \frac{2\pi}{3}\right)$$

19. (a,b) $\because \tan\alpha, \tan\beta$ are roots of the equation $x^2 + px + q = 0$.

$$\tan\alpha + \tan\beta = -p$$

$$\tan\alpha\tan\beta = q$$

We know that, $\tan(\alpha+\beta) = \dfrac{\tan\alpha + \tan\beta}{1-\tan\alpha\tan\beta} = \dfrac{-p}{1-q} = \dfrac{p}{q-1}$

Hence, option (b) is correct.

Now, in option (a)

LHS $= \sin^2(\alpha+\beta) + p\sin(\alpha+\beta)\cos(\alpha+\beta) + q\cos^2(\alpha+\beta)$

$$= \cos^2(\alpha+\beta)\,[\tan^2(\alpha+\beta) + p\tan(\alpha+\beta) + q]$$

$$= \frac{\left(\frac{p}{q-1}\right)^2 + p\left(\frac{p}{q-1}\right) + q}{1 + \left(\frac{p}{q-1}\right)^2} \quad \text{[from Eq. (i)]}$$

$$= \frac{p^2 + p^2(q-1) + q(q-1)^2}{(q-1)^2 + p^2} = \frac{p^2(1+q-1) + q(q-1)^2}{(q-1)^2 + p^2}$$

$$= \frac{[p^2 + (q-1)^2]q}{(q-1)^2 + p^2} = q = \text{RHS}$$

$\therefore$ Option (a) is correct.

Now, we know that,

$$\sec^2(\alpha+\beta) = 1 + \tan^2(\alpha+\beta)$$

$$= 1 + \frac{p^2}{(q-1)^2} = \frac{(q-1)^2 + p^2}{(q-1)^2} \quad \text{[from Eq. (i)]}$$

$$\Rightarrow \cos(\alpha+\beta) = \frac{q-1}{\sqrt{(q-1)^2 + p^2}}$$

$\therefore$ Option (c) is wrong.

We know that,

$$\sin^2(\alpha+\beta) = 1 - \cos^2(\alpha+\beta)$$

$$= 1 - \frac{(q-1)^2}{p^2 + (q-1)^2}$$

$$= \frac{p^2 + (q-1)^2 - (q-1)^2}{p^2 + (q-1)^2}$$

$$\Rightarrow \sin(\alpha+\beta) = \frac{p}{\sqrt{p^2 + (q-1)^2}}$$

$\therefore$ Option (d) is wrong.

20. (a,b,c) We have,

$$\frac{\tan A}{1} = \frac{\tan B}{2} = \frac{\tan C}{3} = \lambda \text{ (let)}$$

$\therefore$ $\tan A = \lambda,\ \tan B = 2\lambda,\ \tan C = 3\lambda$

$\because A + B + C = \pi$, then we know that

$$\tan A + \tan B + \tan C = \tan A \tan B \tan C$$

$\Rightarrow$ $\lambda + 2\lambda + 3\lambda = \lambda \cdot 2\lambda \cdot 3\lambda$

$\Rightarrow$ $6\lambda = 6\lambda^3$

$\Rightarrow$ $\lambda^3 - \lambda = 0$

$\Rightarrow$ $\lambda(\lambda^2 - 1) = 0$

$\Rightarrow$ $\lambda^2 - 1 = 0 \quad (\because \lambda \neq 0)$

$\Rightarrow$ $\lambda = \pm 1$

$\therefore$ $\tan A = \pm 1,\ \tan B = \pm 2,\ \tan C = \pm 3$ [option (a)]

Now, $\tan A + \tan B + \tan C = \pm 6$

$\therefore$ Maximum value of $\tan A + \tan B + \tan C = 6$ [option (c)]

Minimum value of $\tan A + \tan B + \tan C = -6$ [option (b)]

21. (a,b) $\because$ In ΔABC, $A + B + C = 180°$, then

$$\tan A + \tan B + \tan C = \tan A \tan B \tan C$$

$$6 = 2\tan C \quad \text{(given)}$$

$\therefore$ $\tan C = 3$ ...(i)

$\because$ $\tan A + \tan B + \tan C = 6$

$\tan A + \tan B = 3$ [from Eq. (i)]

Now, $\tan A + \tan B = 3$ and $\tan A \tan B = 2$

Then, either $\tan A = 1$ and $\tan B = 2$

or $\tan A = 2$ and $\tan B = 1$

$\therefore$ $\tan A$, $\tan B$ and $\tan C$ are 1, 2, 3 or 2, 1, 3.

22. (4) Given, $\tan(\pi\cos\theta) = \cot(\pi\sin\theta)$

$$\tan(\pi\cos\theta) = \tan\left[\pm\frac{\pi}{2} - \pi\sin\theta\right]$$

$$\pi\cos\theta = \pm\frac{\pi}{2} - \pi\sin\theta$$

$$\cos\theta = \pm\frac{1}{2} - \sin\theta$$

$$\sin\theta + \cos\theta = \pm\frac{1}{2}$$

$$\frac{1}{\sqrt{2}}\sin\theta + \frac{1}{\sqrt{2}}\cos\theta = \pm\frac{1}{2\sqrt{2}}$$

$$\cos\theta\cos\frac{\pi}{4} + \sin\theta\sin\frac{\pi}{4} = \pm\frac{1}{2\sqrt{2}}$$

$$\cos\left(\theta - \frac{\pi}{4}\right) = \pm\frac{1}{2\sqrt{2}}$$

$$\pm\frac{1}{\lambda\sqrt{\mu}} = \pm\frac{1}{2\sqrt{2}} \quad \text{(given)}$$

$\Rightarrow$ $\lambda = 2,\ \mu = 2$

$\therefore$ $\lambda + \mu = 4$

23. (1) Given, equation is

$$x^4 - x^3\sin 2\beta + x^2\cos 2\beta - x\cos\beta - \sin\beta = 0$$

Now,

$$S_1 = \tan\theta_1 + \tan\theta_2 + \tan\theta_3 + \tan\theta_4 = \sin 2\beta$$

$$S_2 = \Sigma\tan\theta_1\tan\theta_2 = \cos 2\beta$$

$$S_3 = \Sigma\tan\theta_1\tan\theta_2\tan\theta_3 = \cos\beta$$

$$S_4 = \tan\theta_1\tan\theta_2\tan\theta_3\tan\theta_4 = -\sin\beta$$

We know that,

$$\tan(\theta_1 + \theta_2 + \theta_3 + \theta_4) = \frac{S_1 - S_3}{1 - S_2 + S_4}$$

$$= \frac{\sin 2\beta - \cos\beta}{1 - \cos 2\beta - \sin\beta} = \frac{2\sin\beta\cos\beta - \cos\beta}{2\sin^2\beta - \sin\beta}$$

$$= \frac{\cos\beta\,[2\sin\beta - 1]}{\sin\beta\,[2\sin\beta - 1]} = \frac{\cos\beta}{\sin\beta} = \cot\beta$$

But $\tan(\theta_1 + \theta_2 + \theta_3 + \theta_4) = \cot(n\beta)$

$\therefore$ $n = 1$

24. (2) $\because$ $\sec 290° = \sec(270° + 20°) = \text{cosec } 20°$

and $\text{cosec } 250° = \text{cosec}(270° - 20°) = -\sec 20°$

Now, $\sqrt{3}\lambda = \sqrt{3}\sec 290° + \text{cosec } 250°$

$$= \sqrt{3}\,\text{cosec } 20° - \sec 20°$$

$$= \frac{\sqrt{3}}{\sin 20°} - \frac{1}{\cos 20°} = \frac{\sqrt{3}\cos 20° - \sin 20°}{\sin 20°\cos 20°}$$

$$= \frac{\frac{\sqrt{3}}{2}\cdot\cos 20° - \frac{1}{2}\cdot\sin 20°}{\sin 20°\cos 20°} \times 2$$

$$= \frac{\sin 60°\cos 20° - \cos 60°\sin 20°}{\sin 20°\cos 20°} \times 2$$

$$\Rightarrow \lambda = \frac{\sin(60° - 20°)}{\frac{\sqrt{3}}{2}\cdot\sin 20°\cos 20°}$$

[$\because \sin(A - B) = \sin A\cos B - \cos A\sin B$ and $\sin 2A = 2\sin A\cos A$]

$$= 2\cdot\frac{\sin 40°}{\frac{\sqrt{3}}{2}\cdot\sin 40°} = \frac{4}{\sqrt{3}}$$

$$\Rightarrow \lambda^2 = \frac{16}{3}$$

$$\therefore 9\lambda^4 + 3\lambda^2 - 270 = 9\cdot\frac{256}{9} + 3\cdot\frac{16}{3} - 270$$

$$= 256 + 16 - 270 = 2$$

DPP-6 Trigonometric Functions using Greatest and Least Values of T-Ratios

1. We know that,

$$-\sqrt{a^2+b^2}+c \le a\sin\theta + b\cos\theta + c \le \sqrt{a^2+b^2}+c$$

$\therefore$ Minimum value of $3\cos x + 4\sin x + 5$

$$= -\sqrt{4^2+3^2}+5 = -5+5 = 0$$

Hence proved.

2. $\because \sin\theta_1 + \sin\theta_2 + \sin\theta_3 = 3$

We know that, $-1 \le \sin A \le 1$, then this equation is possible only when

$$\sin\theta_1 = \sin\theta_2 = \sin\theta_3 = 1$$

$$\therefore \quad \theta_1 = \theta_2 = \theta_3 = \frac{\pi}{2}$$

Now, $\cos\theta_1 = \cos\theta_2 = \cos\theta_3 = \cos\frac{\pi}{2} = 0$

$$\therefore \quad \cos\theta_1 + \cos\theta_2 + \cos\theta_3 = 0+0+0 = 0$$

3. Given, $\sin x + \sin^2 x = 1$

$\Rightarrow \sin x = 1 - \sin^2 x$

$\Rightarrow \sin x = \cos^2 x$

On squaring both sides, we get

$\Rightarrow \sin^2 x = \cos^4 x$

$\Rightarrow 1 - \cos^2 x = \cos^4 x$

$\Rightarrow \cos^4 x + \cos^2 x = 1$

Again, on squaring both sides, we get

$$(\cos^4 x + \cos^2 x)^2 = (1)^2$$

$$\therefore \quad \cos^8 x + \cos^4 x + 2\cos^6 x = 1$$

4. $\because x = r\sin\theta\cos\phi$, $y = r\sin\theta\sin\phi$ and $z = r\cos\theta$

$\therefore x^2+y^2+z^2 = r^2\sin^2\theta\cos^2\phi + r^2\sin^2\theta\sin^2\phi + r^2\cos^2\theta$

$= r^2\sin^2\theta(\cos^2\phi+\sin^2\phi) + r^2\cos^2\theta$

$= r^2\sin^2\theta + r^2\cos^2\theta$

$= r^2(\sin^2\theta+\cos^2\theta) = r^2$

$\therefore x^2+y^2+z^2$ is independent of θ and ϕ. **Hence proved.**

5. (b) $\because 2\sin^2\theta + 3\cos^2\theta = 2\sin^2\theta + 2\cos^2\theta + \cos^2\theta$

$= 2(\sin^2\theta+\cos^2\theta) + \cos^2\theta$

$= 2 + \cos^2\theta$

We know that,

$$0 \le \cos^2\theta \le 1$$

$\Rightarrow 2 < 2 + \cos^2\theta \le 2+1$

$\Rightarrow 2 \le 2+\cos^2\theta \le 3$

$\therefore$ Minimum value of $2\sin^2\theta + 3\cos^2\theta = 2$.

6. (d) $\because \cos x + \sec x = -2$

$\Rightarrow \cos x + \dfrac{1}{\cos x} = -2$

$\Rightarrow \cos^2 x + 1 = -2\cos x$

$\Rightarrow \cos^2 x + 1 + 2\cos x = 0$

$\Rightarrow (\cos x + 1)^2 = 0$

$\Rightarrow \cos x + 1 = 0$

$\Rightarrow \cos x = -1$ and $\sec x = -1$

$\therefore \cos^n x + \sec^n x = (-1)^n + (-1)^n$

$= 2(-1)^n$

(i) If n is even, then

$$\cos^n x + \sec^n x = 2(-1)^{\text{even}} = 2\times 1 = 2$$

(ii) If n is odd, then

$$\cos^n x + \sec^n x = 2(-1)^{\text{odd}} = 2\times(-1) = -2$$

7. (c) We know that,

$\sin\alpha + \sin\beta + \sin\gamma > \sin(\alpha+\beta+\gamma)$

$\sin\alpha + \sin\beta + \sin\gamma > \sin\pi$

$\sin\alpha + \sin\beta + \sin\gamma > 0$ (positive)

8. (a) We know that,

if $A + B + C = 180°$, then

$$\tan A + \tan B + \tan C = \tan A\tan B\tan C \quad \text{...(i)}$$

and AM $\ge$ GM

$\Rightarrow \dfrac{\tan A + \tan B + \tan C}{3} \ge (\tan A\tan B\tan C)^{1/3}$

$\Rightarrow \tan A\tan B\tan C \ge 3(\tan A\tan B\tan C)^{1/3}$ [from Eq. (i)]

$\Rightarrow (\tan A\tan B\tan C)^{2/3} \ge 3$

$\Rightarrow \tan A\tan B\tan C \ge 3^{3/2}$

$\Rightarrow \dfrac{1}{\lambda} \ge 3^{3/2}$ $(\because \cot A\cot B\cot C = \lambda)$

$\therefore \lambda \le \dfrac{1}{3\sqrt{3}}$

9. (b) $\because A = \sin^2\theta + \cos^4\theta$

$= 1 - \cos^2\theta + \cos^4\theta$

$= \cos^4\theta - \cos^2\theta + \left(\frac{1}{2}\right)^2 - \frac{1}{4} + 1$

$= \left(\cos^2\theta - \frac{1}{2}\right)^2 + \frac{3}{4} \ge \frac{3}{4}$ $(\because x^2 \ge 0)$

and $A \le \left(\cos^2\theta - \frac{1}{2}\right)^2 + \frac{3}{4} \le \left(1 - \frac{1}{2}\right)^2 + \frac{3}{4}$ $(\because \cos^2\theta \le 1)$

≤ 1

$\therefore \dfrac{3}{4} \le A \le 1$

10. (b) $\because \sec^2\theta + \operatorname{cosec}^2\theta - 4 = 1 + \tan^2\theta + 1 + \cot^2\theta - 4$

$= \tan^2\theta + \cot^2\theta - 2$

$= (\tan\theta - \cot\theta)^2 \ge 0$

$\therefore \sec^2\theta + \operatorname{cosec}^2\theta \ge 4$

11. (b) As, $1^c \simeq 57°$

$\therefore \sin 1^c > \sin 1°$

12. (c) $\because 3\sin\theta + 5\cos\theta = 5$ ...(i)

Let $5\sin\theta - 3\cos\theta = p$ ...(ii)

On squaring and adding Eqs. (i) and (ii), we get

$(3\sin\theta + 5\cos\theta)^2 + (5\sin\theta - 3\cos\theta)^2 = 25 + p^2$

$\Rightarrow 9\sin^2\theta + 25\cos^2\theta + 30\sin\theta\cdot\cos\theta + 25\sin^2\theta + 9\cos^2\theta - 30\sin\theta\cdot\cos\theta = 25 + p^2$

$\Rightarrow 34\sin^2\theta + 34\cos^2\theta = 25 + p^2$

$\Rightarrow 34(\sin^2\theta + \cos^2\theta) = 25 + p^2$

$\Rightarrow 34 - 25 = p^2$

$\Rightarrow 9 = p^2$

$\Rightarrow p = \pm 3$

$\therefore 5\sin\theta - 3\cos\theta = \pm 3$

13. (b) $\because$ $\sin^2 x = 1 - \sin x$

$\sin x = 1 - \sin^2 x$

$\sin x = \cos^2 x$

$\sin^2 x = \cos^4 x$ ...(i)

Now, $\cos^{12} x + 3\cos^{10} x + 3\cos^8 x + \cos^6 x$

$= \cos^6 x(\cos^6 x + 3\cos^4 x + 3\cos^2 x + 1)$

$= \cos^6 x[(\cos^2 x)^3 + (1)^3 + 3\cos^2 x(\cos^2 x + 1)]$

$= (\cos^2 x)^3 [\cos^2 x + 1]^3$

$= [\cos^2 x(\cos^2 x + 1)]^3 = (\cos^4 x + \cos^2 x)^3$

$= (\sin^2 x + \cos^2 x)^3$ [from Eq. (i)]

$= (1)^3 = 1$

14. (c) We have, $P = \cos(\cos x) + \sin(\cos x)$, then

$-\sqrt{1^2 + 1^2} \le P \le \sqrt{1^2 + 1^2}$

$-\sqrt{2} \le P \le \sqrt{2}$

$\therefore$ $P_{\max} = \sqrt{2}$

and $P_{\min} = -\sqrt{2}$

15. (b,c) We have, $x = \operatorname{cosec}^2\theta$, $y = \sec^2\theta$ and $z = \dfrac{1}{1 - \sin^2\theta\cos^2\theta}$

$$xyz = \frac{1}{\sin^2\theta\cos^2\theta(1 - \sin^2\theta\cos^2\theta)}$$

$$= \frac{(1 - \sin^2\theta\cos^2\theta) + \sin^2\theta\cos^2\theta}{\sin^2\theta\cos^2\theta(1 - \sin^2\theta\cos^2\theta)}$$

$$= \frac{1}{\sin^2\theta\cos^2\theta} + \frac{1}{1 - \sin^2\theta\cos^2\theta}$$

$$= \operatorname{cosec}^2\theta\sec^2\theta + \frac{1}{1 - \sin^2\theta\cos^2\theta}$$

$xyz = xy + z$ [option (b)]

Now, $$x + y + z = \frac{1}{\sin^2\theta} + \frac{1}{\cos^2\theta} + \frac{1}{1 - \sin^2\theta\cos^2\theta}$$

$$= \frac{\sin^2\theta + \cos^2\theta}{\sin^2\theta\cos^2\theta} + \frac{1}{1 - \sin^2\theta\cos^2\theta}$$

$$= \frac{1 - \sin^2\theta\cos^2\theta + \sin^2\theta\cos^2\theta}{\sin^2\theta\cos^2\theta(1 - \sin^2\theta\cos^2\theta)}$$

$$= \operatorname{cosec}^2\theta\sec^2\theta \cdot \frac{1}{1 - \sin^2\theta\cos^2\theta}$$

$\therefore$ $x + y + z = xyz$ [option (c)]

16. (b) We have, $0 < \theta < \dfrac{\pi}{4}$, then

$0 < \tan\theta < 1$ and $\cot\theta > 1$

Now, $t_1 = (\tan\theta)^{\tan\theta} < 1$ and $t_2 = (\tan\theta)^{\cot\theta} < 1$

$t_3 = (\cot\theta)^{\tan\theta} > 1$ and $t_4 = (\cot\theta)^{\cot\theta} > 1$

$\Rightarrow$ $t_1 > t_2$ and $t_4 > t_3$

$\therefore$ $t_4 > t_3 > t_1 > t_2$

17. (a,c) Let $P = 2 - \cos x + \sin^2 x$

$= 3 - \cos^2 x - \cos x - \dfrac{1}{4} + \dfrac{1}{4}$

$= \dfrac{13}{4} - \left(\cos x + \dfrac{1}{2}\right)^2$

Now, if $\left(\cos x + \dfrac{1}{2}\right)^2$ is minimum, then P is maximum.

i.e. $P_{\max} = \dfrac{13}{4} - 0 = \dfrac{13}{4}$ $\left[\because \left(\cos x + \dfrac{1}{2}\right)^2 \ge 0\right]$

If $\left(\cos x + \dfrac{1}{2}\right)^2$ is maximum, then P is minimum.

$\therefore$ $P_{\min} = \dfrac{13}{4} - \left(1 + \dfrac{1}{2}\right)^2$ $(\because \cos x \ge 1)$

$= \dfrac{13}{4} - \dfrac{9}{4} = 1$

$\therefore$ Required ratio = 13 : 4 or 26 : 8

18. (b) $\because P_n = \sin^n\theta + \cos^n\theta$

$P_{n-2} = \sin^{n-2}\theta + \cos^{n-2}\theta$

and $P_{n-4} = \sin^{n-4}\theta + \cos^{n-4}\theta$ ...(i)

Now, $P_n - P_{n-2} = \sin^n\theta + \cos^n\theta - \sin^{n-2}\theta - \cos^{n-2}\theta$

$= \sin^{n-2}\theta(\sin^2\theta - 1) + \cos^{n-2}\theta(\cos^2\theta - 1)$

$= -\sin^{n-2}\theta\cos^2\theta - \cos^{n-2}\theta\sin^2\theta$

$= -\sin^2\theta\cos^2\theta(\sin^{n-4}\theta + \cos^{n-4}\theta)$

$\therefore$ $P_n - P_{n-2} = -\sin^2\theta\cos^2\theta\, P_{n-4}$ [from Eq. (i)] ...(ii)

It is given that $P_n - P_{n-2} = kP_{n-4}$ ...(iii)

From Eqs. (ii) and (iii), we get

$k = -\sin^2\theta\cos^2\theta$

19. (b) Given, $P_1 = m$, $\sin\theta + \cos\theta = m$

$\sin^2\theta + \cos^2\theta + 2\sin\theta\cos\theta = m^2$

$1 + 2\sin\theta\cos\theta = m^2$

$\sin\theta \cdot \cos\theta = \dfrac{m^2 - 1}{2}$ ...(i)

and $P_6 = \sin^6\theta + \cos^6\theta = (\sin^2\theta)^3 + (\cos^2\theta)^3$

$= (\sin^2\theta + \cos^2\theta)(\sin^4\theta + \cos^4\theta - \sin^2\theta \cdot \cos^2\theta)$

$= 1[(\sin^2\theta + \cos^2\theta)^2 - 2\sin^2\theta\cos^2\theta - \sin^2\theta\cos^2\theta]$

$= 1 - 3\sin^2\theta\cos^2\theta$

Now, $4(1 - P_6) = 4[1 - (1 - 3\sin^2\theta\cos^2\theta)]$

$= 4[1 - 1 + 3(\sin\theta\cos\theta)^2]$

$= 4\left[3 \cdot \left(\dfrac{m^2 - 1}{2}\right)^2\right]$ [from Eq. (i)]

$= 4 \cdot \dfrac{3}{4} \cdot (m^2 - 1)^2$

$\therefore$ $4(1 - P_6) = 3(m^2 - 1)^2$

20. (d) From Q. 18, we have,

$P_n - P_{n-2} = -\sin^2\theta\cos^2\theta\, P_{n-4}$

$\Rightarrow$ $P_{n-2} - P_n = \sin^2\theta\cos^2\theta\, P_{n-4}$ ...(i)

Now, $P_{n-2} - P_n = \sin^2\theta\cos^2\theta\, P_\lambda$ ...(ii)

On comparing Eqs. (i) and (ii), we get

$\lambda = n - 4$

21. (c) $\because P_n = \sin^n\theta + \cos^n\theta$

$\therefore$ $P_6 = \sin^6\theta + \cos^6\theta = 1 - 3\sin^2\theta\cos^2\theta$ [from Q. 19]

$P_4 = \sin^4\theta + \cos^4\theta = (\sin^2\theta + \cos^2\theta)^2 - 2\sin^2\theta\cos^2\theta$

$= 1 - 2\sin^2\theta\cos^2\theta$

Now, $2P_6 - 3P_4 + 10$

$= 2(1 - 3\sin^2\theta\cos^2\theta) - 3(1 - 2\sin^2\theta\cos^2\theta) + 10$

$= 2 - 6\sin^2\theta\cos^2\theta - 3 + 6\sin^2\theta\cos^2\theta + 10$

$\therefore$ $2P_6 - 3P_4 + 10 = 9$

22. –1, 1

23. IInd quadrant

24. $\sqrt{a^2 + b^2}, -\sqrt{a^2 + b^2}$

DPP-7 Addition and Subtraction Formulae using Sign of T-Ratios and Allied Angles

1. $\because$ $\tan A - \tan B = x$

and $\cot B - \cot A = y$

$$\frac{1}{\cot A} - \frac{1}{\cot B} = x$$

$$\Rightarrow \frac{\cot B - \cot A}{\cot A \cot B} = x$$

$$\Rightarrow \frac{\cot A \cot B}{\cot B - \cot A} = \frac{1}{x}$$

and $\frac{1}{\cot B - \cot A} = \frac{1}{y}$

$$\text{RHS} = \frac{1}{x} + \frac{1}{y}$$

$$= \frac{\cot A \cot B}{\cot B - \cot A} + \frac{1}{\cot B - \cot A}$$

$$= \frac{\cot A \cot B + 1}{\cot B - \cot A}$$

$$= \cot (A - B) = \text{LHS}$$

Hence proved.

2. $\because m \tan (\theta - 30°) = n \tan (\theta + 120°)$

$$\frac{m}{n} = \frac{\tan (\theta + 120°)}{\tan (\theta - 30°)}$$

$$\frac{m}{n} = \frac{\sin (\theta + 120°) \cos (\theta - 30°)}{\cos (\theta + 120°) \sin (\theta - 30°)}$$

On applying componendo and dividendo, we get

$$\frac{m+n}{m-n} = \frac{\sin(\theta + 120°)\cos(\theta - 30°) + \cos(\theta + 120°)\sin(\theta - 30°)}{\sin(\theta + 120°)\cos(\theta - 30°) - \cos(\theta + 120°)\sin(\theta - 30°)}$$

$$\Rightarrow \frac{m+n}{m-n} = \frac{\sin (\theta + 120° + \theta - 30°)}{\sin (\theta + 120° - \theta + 30°)}$$

$[\because \sin(A \pm B) = \sin A \cos B \pm \cos A \sin B]$

$$\Rightarrow \frac{m+n}{m-n} = \frac{\sin (2\theta + 90°)}{\sin 150°} = \frac{\cos 2\theta}{\sin (180° - 30°)}$$

$$\Rightarrow \frac{m+n}{m-n} = \frac{\cos 2\theta}{\sin 30°}$$

$$\therefore \frac{m+n}{m-n} = 2 \cos 2\theta$$

Hence proved.

3. We know that,

$$\tan 60° = \tan (40° + 20°)$$

$$\sqrt{3} = \frac{\tan 40° + \tan 20°}{1 - \tan 40° \tan 20°}$$

$$\sqrt{3} - \sqrt{3} \tan 40° \tan 20° = \tan 40° + \tan 20°$$

$$\therefore \tan 40° + \tan 20° + \sqrt{3} \tan 40° \tan 20° = \sqrt{3}$$

Hence proved.

4. (a) We have, $\tan \theta = \frac{\sin \alpha - \cos \alpha}{\sin \alpha + \cos \alpha}$

$$\tan \theta = \frac{-\cos \alpha (1 - \tan \alpha)}{\cos \alpha (1 + \tan \alpha)}$$

$$\tan \theta = -\tan\left(\frac{\pi}{4} - \alpha\right) \quad \left[\because \frac{1 - \tan \theta}{1 + \tan \theta} = \tan\left(\frac{\pi}{4} - \theta\right)\right]$$

$$\tan \theta = \tan\left(\alpha - \frac{\pi}{4}\right) \quad [\because \tan(-\theta) = -\tan \theta]$$

$$\theta = \alpha - \frac{\pi}{4}$$

$$\alpha = \theta + \frac{\pi}{4} \quad \text{...(i)}$$

Now, $\sin \alpha + \cos \alpha = \sin\left(\theta + \frac{\pi}{4}\right) + \cos\left(\theta + \frac{\pi}{4}\right)$ [from Eq. (i)]

$$= \sin \theta \cos \frac{\pi}{4} + \cos \theta \sin \frac{\pi}{4} + \cos \theta \cos \frac{\pi}{4} - \sin \theta \sin \frac{\pi}{4}$$

$$= \frac{1}{\sqrt{2}} \sin \theta + \frac{1}{\sqrt{2}} \cos \theta + \frac{1}{\sqrt{2}} \cos \theta - \frac{1}{\sqrt{2}} \sin \theta$$

$$= \frac{2}{\sqrt{2}} \cos \theta = \sqrt{2} \cos \theta$$

Again, $\sin \alpha - \cos \alpha = \sin\left(\theta + \frac{\pi}{4}\right) - \cos\left(\theta + \frac{\pi}{4}\right)$ [from Eq. (i)]

$$= \frac{1}{\sqrt{2}} \sin \theta + \frac{1}{\sqrt{2}} \cos \theta - \frac{1}{\sqrt{2}} \cos \theta + \frac{1}{\sqrt{2}} \sin \theta$$

$$= \frac{2}{\sqrt{2}} \sin \theta = \sqrt{2} \sin \theta$$

5. (a) $\because \tan A \tan B = 2$

$$\frac{\sin A}{\cos A} \cdot \frac{\sin B}{\cos B} = 2$$

$$\sin A \sin B = 2 \cos A \cos B \quad \text{...(i)}$$

Now, we know that,

$$\cos (A - B) = \cos A \cos B + \sin A \sin B$$

$$\frac{3}{5} = \cos A \cos B + 2 \cos A \cos B \quad \text{[from Eq. (i)]}$$

$$3 \cos A \cos B = \frac{3}{5}$$

$$\cos A \cos B = \frac{1}{5}$$

6. (b) $\because$ $A + B + C = 180° \quad \text{...(i)}$

Now, $\sec A (\cos B \cos C - \sin B \sin C) = \frac{1}{\cos A} \cdot \cos (B + C)$

$$= \frac{1}{\cos A} \cdot \cos (\pi - A) \quad \text{[from Eq. (i)]}$$

$$= \frac{1}{\cos A} \cdot (-\cos A) = -1$$

7. (d) $\because$ $\sin (3\pi \cos \theta) = \cos (3\pi \sin \theta)$

$$\sin (3\pi \cos \theta) = \sin\left[\frac{\pi}{2} \pm 3\pi \sin \theta\right]$$

$$3\pi \cos \theta = \frac{\pi}{2} \pm 3\pi \sin \theta$$

$$\cos \theta = \frac{1}{3} \times \frac{1}{2} \pm \sin \theta$$

$$\cos \theta \mp \sin \theta = \frac{1}{6}$$

On squaring both sides, we get

$$\cos^2 \theta + \sin^2 \theta \mp 2 \sin \theta \cos \theta = \frac{1}{36}$$

$$1 \mp \sin 2\theta = \frac{1}{36}$$

$$\pm \sin 2\theta = 1 - \frac{1}{36} = \frac{35}{36}$$

$$\therefore \sin 2\theta = \pm \frac{35}{36}$$

8. (c) $\because \tan 69° + \tan 66° - \tan 69° \tan 66° = 2k$

$$\tan 69° + \tan 66° = 2k + \tan 69° \tan 66°$$

$$\frac{\tan 69° + \tan 66°}{2k + \tan 69° \tan 66°} = 1$$

Now, we will verify the given options.

If we put $k = -\frac{1}{2}$, then we get

$$\frac{\tan 69° + \tan 66°}{2 \times \left(-\frac{1}{2}\right) + \tan 69° \tan 66°} = 1$$

$$\frac{\tan 69° + \tan 66°}{-(1 - \tan 69° \tan 66°)} = 1$$

$$-\tan (69° + 66°) = 1$$

$$-\tan (135°) = 1$$

$$-\tan (180° - 45°) = 1$$

$$\tan 45° = 1$$

$$1 = 1$$

9. (c) We have, $\text{cosec}10° - \sqrt{3}\sec 10°$

$$= \frac{1}{\sin 10°} - \frac{\sqrt{3}}{\cos 10°}$$

$$= \frac{\cos 10° - \sqrt{3}\sin 10°}{\sin 10° \cos 10°}$$

$$= 2 \cdot \frac{\frac{1}{2}\cos 10° - \frac{\sqrt{3}}{2}\sin 10°}{\frac{1}{2} \cdot 2 \sin 10° \cos 10°}$$

$$= 4 \cdot \frac{\cos 60° \cos 10° - \sin 60° \sin 10°}{\sin 20°}$$

$$= 4 \cdot \frac{\cos (60° + 10°)}{\sin 20°}$$

$$= 4 \cdot \frac{\cos 70°}{\sin 20°} = 4 \cdot \frac{\cos (90° - 20°)}{\sin 20°} = 4 \cdot \frac{\sin 20°}{\sin 20°}$$

$$= 4$$

10. (a) $\because \tan \frac{\pi}{9}, x, \tan \frac{5\pi}{18}$ are in AP.

$\Rightarrow \tan 20°, x, \tan 50°$ are in AP, then

$$2x = \tan 50° + \tan 20°$$

$$= \frac{\sin (50° + 20°)}{\cos 50° \cos 20°}$$

$$\left[\because \tan A + \tan B = \frac{\sin (A + B)}{\cos A \cos B}\right]$$

$$= \frac{\sin 70°}{\cos 50° \cos 20°}$$

$$= \frac{\sin (90° - 20°)}{\cos 50° \cos 20°}$$

$$= \frac{\cos 20°}{\cos 50° \cos 20°}$$

$\therefore \quad 2x = \frac{1}{\cos 50°}$ ...(i)

Again, $\tan \frac{\pi}{9}, y, \tan \frac{7\pi}{18}$ are in AP.

$\Rightarrow \tan 20°, y, \tan 70°$ are in AP, then

$$2y = \tan 70° + \tan 20°$$

$$= \frac{\sin (70° + 20°)}{\cos 70° \cos 20°}$$

$$= \frac{\sin 90°}{\cos (90° - 20°) \cos 20°}$$

$$= \frac{1 \times 2}{2 \sin 20° \cos 20°} = \frac{2}{\sin 40°}$$

$$= \frac{2}{\sin (90° - 50°)} = \frac{2}{\cos 50°}$$

$\therefore \quad y = \frac{1}{\cos 50°} = 2x$ [from Eq. (i)]

Hence, $2x = y$

11. (a) $\because ABCD$ is a convex quadrilateral, then

$$\frac{A}{2} < A < \pi$$

and $4 \sec A + 5 = 0$ (given)

$$\sec A = -\frac{5}{4}$$

We know that,

$$\tan^2 A = \sec^2 A - 1$$

$$= \left(-\frac{5}{4}\right)^2 - 1 = \frac{25}{16} - 1$$

$$\tan^2 A = \frac{9}{16}$$

$\Rightarrow \quad \tan A = \pm \frac{3}{4}$

$\therefore \quad \tan A = -\frac{3}{4} \quad \left(\because \frac{\pi}{2} < A < \pi\right)$

Let $\alpha = \tan A = \frac{-3}{4}$

$$\cot A = -\frac{4}{3}$$

Now, $\text{cosec}^2 A = 1 + \cot^2 A$

$$= 1 + \left(-\frac{4}{3}\right)^2$$

$$= 1 + \frac{16}{9} = \frac{25}{9}$$

$$\text{cosec } A = \pm \frac{5}{3}$$

$\text{cosec } A = \frac{5}{3}$ ($\because A$ lies in IInd quadrant)

Let $\beta = \text{cosec } A = \frac{5}{3}$

$\because \quad \alpha + \beta = \tan A + \text{cosec } A$

$$= -\frac{3}{4} + \frac{5}{3}$$

$$= \frac{-9 + 20}{12} = \frac{11}{12}$$

$$\alpha \cdot \beta = -\frac{3}{4} \cdot \frac{5}{3} = -\frac{5}{4}$$

Now, the quadratic equation is

$$x^2 - (\alpha + \beta)x + \alpha\beta = 0$$

$$x^2 - \frac{11}{12}x - \frac{5}{4} = 0$$

$$12x^2 - 11x - 15 = 0$$

12. (c) It is given that

$$3 \sin^2 x - 7 \sin x + 2 = 0$$

$$3 \sin^2 x - 6 \sin x - \sin x + 2 = 0$$

$$3 \sin x (\sin x - 2) - 1 (\sin x - 2) = 0$$

$$(\sin x - 2)(3 \sin x - 1) = 0$$

Either $\sin x - 2 = 0$

$\Rightarrow \quad \sin x = 2$ (not possible)

or $3 \sin x - 1 = 0$

$$\sin x = \frac{1}{3} > 0$$

$\because \sin x > 0$ in I and II quadrants, when $0 < x < 2\pi$.

Thus, 2 values in $[0, 2\pi]$

2 values in $[2\pi, 4\pi]$

2 values in $[4\pi, 5\pi]$

$\therefore$ Total number of solutions in $[0, 5\pi] = 6$

13. (a) Here, $\dfrac{\sqrt{x^2+1}+\dfrac{\tan^2\alpha}{\sqrt{x^2+1}}}{2} \geq \sqrt{\sqrt{x^2+1}\cdot\dfrac{\tan^2\alpha}{\sqrt{x^2+1}}}$ $(\because AM \geq GM)$

$$\frac{\sqrt{x^2+1}+\dfrac{\tan^2\alpha}{\sqrt{x^2+1}}}{2} \geq |\tan\alpha|$$

$$\frac{\sqrt{x^2+1}+\dfrac{\tan^2\alpha}{\sqrt{x^2+1}}}{2} \geq \tan\alpha \qquad \left(\because 0<\alpha<\frac{\pi}{2}\right)$$

$$\therefore \quad \sqrt{x^2+1}+\frac{\tan^2\alpha}{\sqrt{x^2+1}} \geq 2\tan\alpha$$

14. (c) We have,

$$\cos(\beta-\gamma)+\cos(\gamma-\alpha)+\cos(\alpha-\beta) = -\frac{3}{2}$$

$$\Rightarrow 2[\cos\beta\cos\gamma+\sin\beta\sin\gamma+\cos\gamma\cos\alpha+\sin\gamma\sin\alpha + \cos\alpha\cos\beta+\sin\alpha\sin\beta]+3=0$$

$$\Rightarrow 2\cos\beta\cos\gamma+2\sin\beta\sin\gamma+2\cos\gamma\cos\alpha + 2\sin\gamma\sin\alpha+2\cos\alpha\cos\beta+2\sin\alpha\sin\beta + (\sin^2\alpha+\cos^2\alpha)+(\sin^2\beta+\cos^2\beta)+(\sin^2\gamma+\cos^2\gamma)=0$$

$$\Rightarrow (\sin^2\alpha+\sin^2\beta+\sin^2\gamma+2\sin\alpha\sin\beta+2\sin\beta\sin\gamma + 2\sin\gamma\sin\alpha)+(\cos^2\alpha+\cos^2\beta+\cos^2\gamma + 2\cos\alpha\cos\beta+2\cos\beta\cos\gamma+2\cos\gamma\cos\alpha)=0$$

$$\Rightarrow \quad (\sin\alpha+\sin\beta+\sin\gamma)^2+(\cos\alpha+\cos\beta+\cos\gamma)^2=0$$

It is possible only when,

$$\sin\alpha+\sin\beta+\sin\gamma=0$$

and $$\cos\alpha+\cos\beta+\cos\gamma=0$$

15. (b) We have,

$$3\sin P+4\cos Q=6 \quad \text{...(i)}$$

and $$4\sin Q+3\cos P=1 \quad \text{...(ii)}$$

On squaring and adding Eqs. (i) and (ii), we get

$$\Rightarrow 9\sin^2 P+16\cos^2 Q+24\sin P\cos Q + 16\sin^2 Q+9\cos^2 P+24\sin Q\cos P=37$$

$$\Rightarrow 9(\sin^2 P+\cos^2 P)+16(\cos^2 Q+\sin^2 Q) + 24(\sin P\cos Q+\cos P\sin Q)=37$$

$$24\sin(P+Q)=37-25$$

$$\Rightarrow \quad \sin(P+Q)=\frac{1}{2}=\sin\frac{\pi}{6} \text{ or } \sin\frac{5\pi}{6}$$

$$\therefore \quad P+Q=\frac{\pi}{6} \text{ or } \frac{5\pi}{6}$$

Now, if $P+Q=\dfrac{\pi}{6}$, then $R=\dfrac{5\pi}{6}$

If $\quad P+Q=\dfrac{5\pi}{6}$, then $R=\dfrac{\pi}{6}$

If we take $R=\dfrac{5\pi}{6}$, then $\quad 0<P, Q<\dfrac{\pi}{6}$

$$\Rightarrow \quad \cos Q<1$$

and $$\sin P<\frac{1}{2}$$

Then, $\quad 3\sin P+4\cos Q<\dfrac{3}{2}+4=\dfrac{11}{2}$

But given that $\quad 3\sin P+4\cos Q=6$

Thus, $R=\dfrac{5\pi}{6}$ is not possible.

Hence, $\angle R=\dfrac{\pi}{6}$

16. (c,d) $\because \dfrac{\cot 76^\circ\cot 16^\circ+3}{\cot 76^\circ+\cot 16^\circ} = \dfrac{\dfrac{\cot 76^\circ}{\sin 76^\circ}\cdot\dfrac{\cos 16^\circ}{\sin 16^\circ}+3}{\dfrac{\cos 76^\circ}{\sin 76^\circ}+\dfrac{\cos 16^\circ}{\sin 16^\circ}}$

$$=\frac{(\cos 76^\circ\cos 16^\circ+\sin 76^\circ\sin 16^\circ)+2\sin 76^\circ\sin 16^\circ}{\cos 76^\circ\sin 16^\circ+\sin 76^\circ\cos 16^\circ}$$

$$=\frac{\cos(76^\circ-16^\circ)+\cos(76^\circ-16^\circ)-\cos(76^\circ+16^\circ)}{\sin(76^\circ+16^\circ)}$$

$$=\frac{\cos 60^\circ+\cos 60^\circ-\cos 92^\circ}{\sin 92^\circ}$$

$$=\frac{\frac{1}{2}+\frac{1}{2}-\cos 92^\circ}{\sin 92^\circ}=\frac{1-\cos 92^\circ}{\sin 92^\circ}=\tan 46^\circ \quad \text{[option (c)]}$$

$$=\cot(90^\circ-46^\circ)=\cot 44^\circ \quad \text{[option (d)]}$$

17. (b,c) Given that,

$$\tan A+\tan B+\tan C=6 \quad \text{...(i)}$$

and $$\tan A\tan B=2 \quad \text{...(ii)}$$

$$\tan A\tan B\tan C=6$$

$$(\because \tan A+\tan B+\tan C=\tan A\cdot\tan B\cdot\tan C)$$

$$2\tan C=6$$

$$\tan C=3$$

Put in Eq. (i), we get

$$\tan A+\tan B+3=6$$

$$\tan A+\tan B=3$$

$$\tan B=3-\tan A \quad \text{...(iii)}$$

Put in Eq. (ii), we get

Now, $$\tan A(3-\tan A)=2$$

$$3\tan A-\tan^2 A=2$$

$$\tan^2 A-3\tan A+2=0$$

$$(\tan A-1)(\tan A-2)=0$$

$\therefore$ $\tan A=1$ or 2 put in Eq. (iii), we get

$$\tan B=3-1=2$$

$$\tan B=3-2=1$$

We know that, $\sin^2\theta=\dfrac{\tan^2\theta}{1+\tan^2\theta}$

If $\tan A=1$, then

$$\sin^2 A=\frac{1^2}{1+1^2}$$

$$\Rightarrow \quad \sin^2 A=\frac{1}{2}$$

If $\tan B=2$, then

$$\sin^2 B=\frac{2^2}{1+2^2}$$

$$\Rightarrow \quad \sin^2 B=\frac{4}{5}$$

and if $\tan C=3$, then

$$\sin^2 C=\frac{3^2}{1+3^2}$$

$$\Rightarrow \quad \sin^2 C=\frac{9}{10}$$

$$\therefore \quad \sin^2 A:\sin^2 B:\sin^2 C=\frac{1}{2}:\frac{4}{5}:\frac{9}{10}$$

$$=5:8:9 \quad \text{[option (c)]}$$

Now, if $\tan A=2$, then

$$\sin^2 A=\frac{2^2}{1+2^2}$$

$$\Rightarrow \quad \sin^2 A=\frac{4}{5}$$

If $\tan B = 1$, then

$$\sin^2 B = \frac{1^2}{1+1^2}$$

$$\Rightarrow \quad \sin^2 B = \frac{1}{2}$$

and if $\tan C = 3$, then

$$\sin^2 C = \frac{3^2}{1+3^2}$$

$$\Rightarrow \quad \sin^2 C = \frac{9}{10}$$

$$\therefore \quad \sin^2 A : \sin^2 B : \sin^2 C = \frac{4}{5} : \frac{1}{2} : \frac{9}{10}$$

$$= 8 : 5 : 9 \qquad \text{[option (b)]}$$

18. (a,b) $\because \quad \sec\theta + \tan\theta = 1 \qquad \text{...(i)}$

Then, $\quad \sec\theta - \tan\theta = 1 \qquad \text{...(ii)}$

On solving Eqs. (i) and (ii), we get

$$\sec\theta = \tan\theta = 1$$

Now, given that

$$a(b-c)x^2 + b(c-a)x + c(a-b) = 0 \qquad \text{...(iii)}$$

On putting $x = 1$, we get

$$ab - ac + bc - ab + ca - bc = 0$$

$$0 = 0$$

$\therefore$ $x = 1$ is the root of Eq. (iii) but $\sec\theta = \tan\theta = 1$

So, $\tan\theta$ and $\sec\theta$ are the roots of Eq. (iii).

19. (a) $\because k_1 = \dfrac{\sin\theta}{\cos 3\theta} + \dfrac{\sin 3\theta}{\cos 9\theta} + \dfrac{\sin 9\theta}{\cos 27\theta}$

$$= \frac{1}{2}\left[\frac{2\sin\theta\cos\theta}{\cos 3\theta\cos\theta} + \frac{2\sin 3\theta\cos 3\theta}{\cos 9\theta\cos 3\theta} + \frac{2\sin 9\theta\cos 9\theta}{\cos 27\theta\cos 9\theta}\right]$$

$$= \frac{1}{2}\left[\frac{\sin 2\theta}{\cos\theta\cos 3\theta} + \frac{\sin 6\theta}{\cos 9\theta\cos 3\theta} + \frac{\sin 18\theta}{\cos 27\theta\cos 9\theta}\right]$$

$$= \frac{1}{2}\left[\frac{\sin(3\theta-\theta)}{\cos 3\theta\cos\theta} + \frac{\sin(9\theta-3\theta)}{\cos 9\theta\cos 3\theta} + \frac{\sin(27\theta-9\theta)}{\cos 27\theta\cos 9\theta}\right]$$

$$= \frac{1}{2}[\tan 3\theta - \tan\theta + \tan 9\theta - \tan 3\theta + \tan 27\theta - \tan 9\theta]$$

$$= \frac{1}{2}(\tan 27\theta - \tan\theta) \qquad \left[\because \tan A - \tan B = \frac{\sin(A-B)}{\cos A\cos B}\right]$$

$$\Rightarrow \quad k_1 = \frac{1}{2}k_2 \qquad \text{(given)}$$

$$\therefore \quad k_2 = 2k_1$$

20. (6) We know that, AM $\geq$ GM, then for all +ve $\sec A$, $\sec B$, $\sec C$

$$\frac{\sec A + \sec B + \sec C}{3} \geq 3\sqrt{\sec A\sec B\sec C} \qquad \text{...(i)}$$

Now, $\cos A\cos B\cos C = \dfrac{1}{2}\cos A\,[2\cos B\cos C]$

$$= \frac{1}{2}\cos A\,[\cos(B+C) + \cos(B-C)]$$

$$= \frac{1}{2}\cos A\,[\cos(\pi - A) + \cos(B-C)]$$

$$\Rightarrow \cos A\cos B\cos C = \frac{1}{2}\cos A\,[\cos(B-C) - \cos A]$$

$$\because \quad \cos(B-C) \leq 1$$

$$\therefore \quad \cos(B-C) \leq \frac{1}{2}\cos A\,(1-\cos A)$$

$$= \frac{1}{2}\left[\cos A - \cos^2 A - \frac{1}{4} + \frac{1}{4}\right]$$

$$= \frac{1}{8} - \frac{1}{2}\left(\cos A - \frac{1}{2}\right)^2$$

$$\therefore \quad \cos A\cos B\cos C \leq \frac{1}{8}$$

$$\sec A\sec B\sec C \geq 8$$

$$\sqrt[3]{\sec A\sec B\sec C} \geq 2$$

Now, from Eq. (i),

$$\frac{\sec A + \sec B + \sec C}{3} \geq 2$$

$$\sec A + \sec B + \sec C \geq 6$$

$\therefore$ Least value of $\sec A + \sec B + \sec C$ is 6.

21. (a)

(i) Let $P = 3\sin 2\theta + 4\cos 2\theta + 3$

We know that,

$$-\sqrt{3^2+4^2} + 3 \leq 3\sin 2\theta + 4\cos 2\theta + 3 \leq \sqrt{3^2+4^2} + 3$$

$$-2 \leq P \leq 8$$

But given that $\quad \mu \leq P \leq \lambda$

$$\therefore \quad \mu = -2, \lambda = 8$$

Now, $\quad \lambda + \mu = 6$

$$\lambda - \mu = 10$$

Thus, **(i) → (r, s)**

(ii) Let $\quad P = 5\cos\theta + 3\cos\left(\theta + \dfrac{\pi}{3}\right) + 3$

$$= 5\cos\theta + 3\left[\cos\theta\cos\frac{\pi}{3} - \sin\theta\sin\frac{\pi}{3}\right] + 3$$

$$= 5\cos\theta + \frac{3}{2}\cos\theta - \frac{3\sqrt{3}}{2}\sin\theta + 3$$

$$= \frac{13}{2}\cos\theta - \frac{3\sqrt{3}}{2}\sin\theta + 3$$

$$\therefore -\sqrt{\left(\frac{13}{2}\right)^2 + \left(\frac{-3\sqrt{3}}{2}\right)^2} + 3 \leq \frac{13}{2}\cos\theta - \frac{3\sqrt{3}}{2}\sin\theta + 3$$

$$\leq \sqrt{\left(\frac{13}{2}\right)^2 + \left(\frac{-3\sqrt{3}}{2}\right)^2} + 3$$

$$\Rightarrow \quad -7 + 3 \leq P \leq 7 + 3$$

$$\Rightarrow \quad -4 \leq P \leq 10$$

But $\quad \mu \leq P \leq \lambda$

$$\therefore \quad \lambda = 10 \text{ and } \mu = -4$$

Now, $\quad \lambda + \mu = 10 - 4 = 6$

$$\lambda - \mu = 10 + 4 = 14$$

Thus, **(ii) → (r, t)**

(iii) Let $\quad P = 1 + \sin\left(\dfrac{\pi}{4} + \theta\right) + 2\cos\left(\dfrac{\pi}{4} - \theta\right)$

$$= 1 + \left(\sin\frac{\pi}{4}\cos\theta + \cos\frac{\pi}{4}\sin\theta\right)$$

$$+ 2\left(\cos\frac{\pi}{4}\cos\theta + \sin\frac{\pi}{4}\sin\theta\right)$$

$$= 1 + \frac{1}{\sqrt{2}}(\sin\theta + \cos\theta) + \frac{2}{\sqrt{2}}(\sin\theta + \cos\theta)$$

$$= 1 + \frac{3}{\sqrt{2}}(\sin\theta + \cos\theta)$$

Now, $\quad -\sqrt{2} \leq \sin\theta + \cos\theta \leq \sqrt{2}$

$$1 - \sqrt{2}\cdot\frac{3}{\sqrt{2}} \leq 1 + \frac{3}{\sqrt{2}}(\sin\theta + \cos\theta) \leq 1 + \frac{3}{\sqrt{2}}\sqrt{2}$$

$$-2 \leq P \leq 4$$

But given that $\mu \leq P \leq \lambda$, then

$$\lambda = 4 \text{ and } \mu = -2$$

$$\therefore \quad \lambda + \mu = 4 - 2 = 2$$

$$\lambda - \mu = 4 + 2 = 6$$

Thus, **(iii) → (p, q)**

DPP-8 Transformation Formulae-I

1. LHS $= 4\sin\theta\sin\left(\frac{\pi}{3}+\theta\right)\sin\left(\frac{2\pi}{3}+\theta\right)$

$= 4\sin\theta\sin(60^\circ+\theta)\sin(120^\circ+\theta)$

$= 4\sin\theta\sin(60^\circ+\theta)\sin[180^\circ-(\theta+120^\circ)]$

$= 4\sin\theta\sin(60^\circ+\theta)\sin(60^\circ-\theta)$

$= 4\sin\theta(\sin^2 60^\circ - \sin^2\theta)$

$[\because \sin^2 A - \sin^2 B = \sin(A+B)\sin(A-B)]$

$= 4\left(\frac{3\sin\theta - 4\sin^3\theta}{4}\right)$

$= 4\times\frac{1}{4}\sin 3\theta \quad [\sin 3\theta = 3\sin\theta - 4\sin^3\theta]$

$= \sin 3\theta =$ RHS

Hence proved.

2. LHS $= \sin(B-C)\cos(A-D) + \sin(C-A)\cos(B-D) + \sin(A-B)\cos(C-D)$

$= \frac{1}{2}[2\sin(B-C)\cos(A-D) + 2\sin(C-A)\cos(B-D) + 2\sin(A-B)\cos(C-D)]$

$= \frac{1}{2}[\sin(B-C+A-D) + \sin(B-C-A+D) + \sin(C-A+B-D) + \sin(C-A-B+D) + \sin(A-B+C-D) + \sin(A-B-C+D)]$

$= \frac{1}{2}[\sin(A+B-C-D) + \sin(-A+B-C+D) + \sin(-A+B+C-D) + \sin\{-(A+B-C-D)\} + \sin\{-(-A+B-C+D)\} + \sin\{-(-A+B+C-D)\}]$

$= \frac{1}{2}[\sin(A+B-C-D) + \sin(-A+B-C+D) + \sin(-A+B+C-D) - \sin(A+B-C-D) - \sin(-A+B-C+D) - \sin(-A+B+C-D)]$

$[\because \sin(-\theta) = -\sin\theta]$

$= \frac{1}{2}\times 0 = 0 =$ RHS **Hence proved.**

3. LHS $= \frac{1+\cos 2x + \cos 4x + \cos 6x}{\cos 3x\cos 2x\cos x}$

$= \frac{(1+\cos 6x) + (\cos 4x + \cos 2x)}{\cos 3x\cos 2x\cos x}$

$= \frac{2\cos^2 3x + 2\cos\left(\frac{4x+2x}{2}\right)\cos\left(\frac{4x-2x}{2}\right)}{\cos 3x\cos 2x\cos x}$

$= \frac{2\cos^2 3x + 2\cos 3x\cos x}{\cos 3x\cos 2x\cos x} = \frac{2\cos 3x(\cos 3x + \cos x)}{\cos 3x\cos 2x\cos x}$

$= \frac{2\cdot 2\cos\left(\frac{3x+x}{2}\right)\cos\left(\frac{3x-x}{2}\right)}{\cos 2x\cos x}$

$= \frac{4\cdot\cos 2x\cos x}{\cos 2x\cos x} = 4 =$ RHS **Hence proved.**

4. LHS $= \sin\alpha + \sin\left(\alpha+\frac{2\pi}{3}\right) + \sin\left(\alpha+\frac{4\pi}{3}\right)$

$= \sin\alpha + \sin(\alpha+120^\circ) + \sin(\alpha+240^\circ)$

$= \sin\alpha + \sin(\alpha+120^\circ) + \sin[180^\circ+(60^\circ+\alpha)]$

$= 2\sin\left(\frac{\alpha+\alpha+120^\circ}{2}\right)\cos\left(\frac{\alpha-\alpha-120^\circ}{2}\right) - \sin(60^\circ+\alpha)$

$= 2\sin(\alpha+60^\circ)\cos 60^\circ - \sin(60^\circ+\alpha)$

$= 2\times\frac{1}{2}\sin(60^\circ+\alpha) - \sin(60^\circ+\alpha)$

$= \sin(60^\circ+\alpha) - \sin(60^\circ+\alpha) = 0 =$ RHS **Hence proved.**

5. (b) $\because \cos 20^\circ\cos 40^\circ\cos 60^\circ\cos 80^\circ = p$

$\Rightarrow \frac{1}{2}\cdot\cos 20^\circ\cos(60^\circ-20^\circ)\cos(60^\circ+20^\circ) = p$

$\Rightarrow p = \frac{1}{2}\cdot\frac{1}{4}\cos(3\times 20^\circ) = \frac{1}{8}\cdot\cos 60^\circ = \frac{1}{8}\times\frac{1}{2} = \frac{1}{16}$

$\therefore p = \frac{1}{16}$ ...(i)

$\left[\text{using } \cos\theta\cos(60^\circ-\theta)\cos(60^\circ+\theta) = \frac{1}{4}\cdot\cos 3\theta\right]$

Now, $\sin 10^\circ\sin 30^\circ\sin 50^\circ\sin 70^\circ = q$

$\frac{1}{2}\cdot\sin 10^\circ\sin(60^\circ-10^\circ)\sin(60^\circ+10^\circ) = q$

$\Rightarrow \frac{1}{2}\cdot\frac{1}{4}\cdot\sin(3\times 10)^\circ = q$

$\left[\text{using } \sin\theta\sin(60^\circ-\theta)\sin(60^\circ+\theta) = \frac{1}{4}\cdot\sin 3\theta\right]$

$\Rightarrow \frac{1}{8}\times\sin 30^\circ = q \Rightarrow \frac{1}{8}\times\frac{1}{2} = q$

$\therefore q = \frac{1}{16}$ ...(ii)

From Eqs. (i) and (ii), we get $p = q$

$\therefore \frac{p}{q} = 1$

6. (c) It is given that, $\tan 20^\circ\tan 40^\circ\tan 80^\circ = \sqrt{k}$

$\tan 20^\circ\tan(60^\circ-20^\circ)\tan(60^\circ+20^\circ) = \sqrt{k}$

[using $\tan\theta\tan(60^\circ-\theta)\tan(60^\circ+\theta) = \tan 3\theta$]

$\Rightarrow \tan 3\times 20^\circ = \sqrt{k}$

$\Rightarrow \tan 60^\circ = \sqrt{k} \Rightarrow \sqrt{3} = \sqrt{k}$

$\therefore k = 3$

7. (b) Given that,

$\frac{\sin(A-C) + 2\sin A + \sin(A+C)}{\sin(B-C) + 2\sin B + \sin(B+C)} = k\sin A\operatorname{cosec} B$

$\Rightarrow \frac{[\sin(A+C) + \sin(A-C)] + 2\sin A}{[\sin(B+C) + \sin(B-C)] + 2\sin B} = k\sin A\operatorname{cosec} B$

[using $\sin(A+B) + \sin(A-B) = 2\sin A\cos B$]

$\Rightarrow \frac{2\sin A\cos C + 2\sin A}{2\sin B\cos C + 2\sin B} = k\sin A\operatorname{cosec} B$

$\Rightarrow \frac{2\sin A(\cos C+1)}{2\sin B(\cos C+1)} = k\sin A\operatorname{cosec} B$

$\Rightarrow \sin A\cdot\operatorname{cosec} B = k\sin A\operatorname{cosec} B$

$\therefore k = 1$

8. (b) $\because \sin 2\theta + \sin 2\phi = \frac{1}{2}$ ...(i)

$\cos 2\theta + \cos 2\phi = \frac{3}{2}$ ...(ii)

On squaring and adding Eqs. (i) and (ii), we get

$(\sin^2 2\theta + \cos^2 2\theta) + (\sin^2 2\phi + \cos^2 2\phi) + 2\sin 2\theta\sin 2\phi + 2\cos 2\theta\cos 2\phi = \frac{1}{4} + \frac{9}{4}$

$\Rightarrow 1 + 1 + 2(\cos 2\theta\cos 2\phi + \sin 2\theta\sin 2\phi) = \frac{5}{2}$

$2 + 2\cos(2\theta - 2\phi) = \frac{5}{2}$

$[\because \cos(A-B) = \cos A\cos B + \sin A\sin B]$

$2[1 + \cos(2\theta - 2\phi)] = \frac{5}{2}$

$2\cdot 2\cos^2(\theta-\phi) = \frac{5}{2} \quad (\because 1 + \cos 2A = 2\cos^2 A)$

$\cos^2(\theta-\phi) = \frac{5}{8}$

9. (b) $\because \sin 78° - \sin 66° - \sin 42° + \sin 6°$

$$= (\sin 78° - \sin 42°) - (\sin 66° - \sin 6°)$$

$$= 2\cos\left(\frac{78° + 42°}{2}\right)\sin\left(\frac{78° - 42°}{2}\right) - 2\cos\left(\frac{66° + 6°}{2}\right)\sin\left(\frac{66° - 6°}{2}\right)$$

$$= 2\cos 60° \sin 18° - 2\cos 36° \sin 30°$$

$$\left[\because \sin C - \sin D = 2\cos\left(\frac{C+D}{2}\right)\sin\left(\frac{C-D}{2}\right)\right]$$

$$= 2 \times \frac{1}{2}\cdot \sin 18° - 2\cos 36° \times \frac{1}{2}$$

$$= \sin 18° - \cos 36° \quad \left(\because \sin 18° = \frac{\sqrt{5}-1}{4} \text{ and } \cos 36° = \frac{\sqrt{5}+1}{4}\right)$$

$$= \frac{\sqrt{5}-1}{4} - \frac{\sqrt{5}+1}{4} = -\frac{1}{2}$$

10. (b) It is given that, $\sin x + \sin y = \sqrt{3}(\cos y - \cos x)$

$$\Rightarrow \frac{\sin x + \sin y}{\cos y - \cos x} = \sqrt{3}$$

$$\Rightarrow \frac{2\sin\left(\frac{x+y}{2}\right)\cos\left(\frac{x-y}{2}\right)}{2\sin\left(\frac{x+y}{2}\right)\sin\left(\frac{x-y}{2}\right)} = \sqrt{3}$$

$$\Rightarrow \cot\left(\frac{x-y}{2}\right) = \sqrt{3} \Rightarrow \cot\left(\frac{x-y}{2}\right) = \cot 30°$$

$$\therefore \frac{x-y}{2} = 30° \quad \text{...(i)}$$

Now, $\sin 3x + \sin 3y = 2\sin 3\left(\frac{x+y}{2}\right)\cos 3\left(\frac{x-y}{2}\right)$

$$= 2\sin 3\left(\frac{x+y}{2}\right)\cos 3 \times 30° \quad \text{[from Eq. (i)]}$$

$$= 2\sin 3\left(\frac{x+y}{2}\right) \times 0 = 0$$

11. (a) Here, $\sin(y + z - x)$, $\sin(z + x - y)$, $\sin(x + y - z)$ are in AP.

Then $\sin(z + x - y) - \sin(y + z - x)$

$$= \sin(x + y - z) - \sin(z + x - y)$$

$$\Rightarrow 2\cos z \sin(x - y) = 2\cos x \sin(y - z)$$

$$\Rightarrow \cos z(\sin x \cos y - \cos x \sin y) = \cos x(\sin y \cos z - \cos y \sin z)$$

$$\Rightarrow \sin x \cos y \cos z - \cos x \sin y \cos z = \cos x \sin y \cos z - \cos x \cos y \sin z$$

On dividing both sides by $\cos x \cos y \cos z$, we get

$$\frac{\sin x}{\cos x} - \frac{\sin y}{\cos y} = \frac{\sin y}{\cos y} - \frac{\sin z}{\cos z}$$

$$\tan x - \tan y = \tan y - \tan z$$

$$\tan x + \tan z = 2\tan y$$

$\therefore$ $\tan x, \tan y, \tan z$ are in AP.

12. (b) $\because \cot\left(\frac{\pi}{4} + \theta\right)\cot\left(\frac{\pi}{4} - \theta\right) = \frac{1}{\tan\left(\frac{\pi}{4} + \theta\right)\tan\left(\frac{\pi}{4} - \theta\right)}$

$$= \frac{1}{\frac{1 + \tan\theta}{1 - \tan\theta} \times \frac{1 - \tan\theta}{1 + \tan\theta}} = 1 \quad \left(\because \theta \neq \pm\frac{\pi}{4}\right)$$

13. (b) For maximum value

$$2a^2 - 1 - \cos^2 x = 0 \Rightarrow \cos^2 x = 2a^2 - 1$$

$$\Rightarrow 1 - \sin^2 x = 2a^2 - 1 \Rightarrow \sin^2 x = 2 - 2a^2$$

$$\sin^2 x + 2a^2 = 2$$

$\therefore$ Maximum value of

$$\left|\sqrt{\sin^2 x + 2a^2} - \sqrt{2a^2 - 1 - \cos^2 x}\right| = |\sqrt{2} - 0| = \sqrt{2}$$

14. (b) $\because \sin\theta + \cos\theta = a$ ...(i)

On squaring both sides, we get

$$\sin^2\theta + \cos^2\theta + 2\sin\theta\cos\theta = a^2$$

$$1 + 2\sin\theta\cos\theta = a^2$$

$$\sin\theta\cos\theta = \frac{a^2 - 1}{2} \quad \text{...(ii)}$$

and $\sin^3\theta + \cos^3\theta = b$

$$(\sin\theta + \cos\theta)^3 - 3\sin\theta\cos\theta(\sin\theta + \cos\theta) = b$$

$$a^3 - 3\left(\frac{a^2 - 1}{2}\right)a = b \quad \text{[from Eqs. (i) and (ii)]}$$

$$2a^3 - 3a^3 + 3a = 2b$$

$$2a^3 - 3a^3 + 3a - 2b = 0$$

$$-a^3 + 3a - 2b = 0$$

$$a^3 - 3a + 2b = 0 \quad \text{...(iii)}$$

But given that, $\lambda a^3 + \mu b + \nu c = 0$ ...(iv)

From Eqs. (iii) and (iv), we get

$$\lambda = 1, \mu = -3 \text{ and } \nu = 2$$

Now, $\lambda^3 + \mu^3 + \nu^3 = 1^3 + (-3)^3 + (2)^3 = 1 - 27 + 8 = -18$

15. (a,c) Here, $\left(\frac{\cos A + \cos B}{\sin A - \sin B}\right)^n + \left(\frac{\sin A + \sin B}{\cos A - \cos B}\right)^n$

$$= \left[\frac{2\cos\left(\frac{A+B}{2}\right)\cos\left(\frac{A-B}{2}\right)}{2\cos\left(\frac{A+B}{2}\right)\sin\left(\frac{A-B}{2}\right)}\right]^n + \left[\frac{2\sin\left(\frac{A+B}{2}\right)\cos\left(\frac{A-B}{2}\right)}{-2\sin\left(\frac{A+B}{2}\right)\sin\left(\frac{A-B}{2}\right)}\right]^n$$

$$= \cot^n\left(\frac{A-B}{2}\right) + (-1)^n\cot^n\left(\frac{A-B}{2}\right)$$

Now, if n is even, then

$$= \cot^n\left(\frac{A-B}{2}\right) + \cot^n\left(\frac{A-B}{2}\right) \quad [\because (-1)^{\text{even}} = 1]$$

$$= 2\cot^n\left(\frac{A-B}{2}\right)$$

If n is odd, then

$$= \cot^n\left(\frac{A-B}{2}\right) - \cot^n\left(\frac{A-B}{2}\right) \quad [\because (-1)^{\text{odd}} = -1]$$

$$= 0$$

16. (a,b,c) We have, $\frac{x}{y} = \frac{\cos A}{\cos B} \Rightarrow \frac{x}{\cos A} = \frac{y}{\cos B} = k$ (let)

$$\Rightarrow x = k\cos A, \; y = k\cos B$$

Now,

(a) $\text{RHS} = \frac{x\tan A + y\tan B}{x + y}$

$$= \frac{k\cos A \cdot \frac{\sin A}{\cos A} + k\cos B \cdot \frac{\sin B}{\cos B}}{k\cos A + k\cos B}$$

$$= \frac{\sin A + \sin B}{\cos A + \cos B}$$

$$= \frac{2\sin\left(\frac{A+B}{2}\right)\cos\left(\frac{A-B}{2}\right)}{2\cos\left(\frac{A+B}{2}\right)\cos\left(\frac{A-B}{2}\right)} = \tan\left(\frac{A+B}{2}\right) = \text{LHS}$$

$\therefore$ Option (a) is correct.

(b) RHS $= \dfrac{x \tan A - y \tan B}{x + y}$

$$= \frac{k \cos A \cdot \dfrac{\sin A}{\cos A} - k \cos B \cdot \dfrac{\sin B}{\cos B}}{k \cos A + k \cos B}$$

$$= \frac{\sin A - \sin B}{\cos A + \cos B} = \frac{2 \cos\left(\dfrac{A+B}{2}\right) \sin\left(\dfrac{A-B}{2}\right)}{2 \cos\left(\dfrac{A+B}{2}\right) \cos\left(\dfrac{A-B}{2}\right)}$$

$$= \tan\left(\frac{A-B}{2}\right) = \text{LHS}$$

$\therefore$ Option (b) is correct.

(c) RHS $= \dfrac{y \sin A + x \sin B}{y \sin A - x \sin B}$

$$= \frac{k \cos B \sin A + k \cos A \sin B}{k \cos B \sin A - k \cos A \sin B}$$

$$= \frac{\sin (A + B)}{\sin (A - B)} = \text{LHS}$$

$\therefore$ Option (c) is correct.

(d) LHS $= x \cos A + y \cos B$

$= k \cos A \cos A + k \cos B \cos B$

$= k(\cos^2 A + \cos^2 B) \neq 0$

$\therefore$ Option (d) is incorrect.

17. (a,b) We know that AM $\geq$ GM, then

$$\frac{2^{\sin \theta} + 2^{-\cos \theta}}{2} \geq \sqrt{2^{\sin \theta} \cdot 2^{-\cos \theta}}$$

$$\frac{2^{\sin \theta} + 2^{-\cos \theta}}{2} \geq (2^{\sin \theta - \cos \theta})^{\frac{1}{2}}$$

$$2^{\sin \theta} + 2^{-\cos \theta} \geq 2^{1 + \frac{1}{2}(\sin \theta - \cos \theta)}$$

$$= 2^{1 + \frac{\sqrt{2}}{2}\left[\frac{1}{\sqrt{2}} \sin \theta - \frac{1}{\sqrt{2}} \cos \theta\right]}$$

$$= 2^{1 + \frac{1}{\sqrt{2}} \sin\left(\theta - \frac{\pi}{4}\right)}$$

Now, for minimum value,

$$\sin\left(\theta - \frac{\pi}{4}\right) = -1$$

$$\sin\left(\theta - \frac{\pi}{4}\right) = \sin \frac{3\pi}{2}$$

$$\theta = 2n\pi + \frac{3\pi}{2} + \frac{\pi}{4}$$

$$\theta = 2n\pi + \frac{7\pi}{4}, n \in I$$

When $n = 0$, then $\theta = \dfrac{7\pi}{4}$

When $n = 1$, then $\theta = \dfrac{15\pi}{4}$

and $2^{\sin \theta} + 2^{-\cos \theta} \geq 2^{1 + \frac{1}{\sqrt{2}}(-1)}$

$$2^{\sin \theta} + 2^{-\cos \theta} \geq 2^{1 - \frac{1}{\sqrt{2}}}$$

$\therefore$ $\lambda = 2^{1 - \frac{1}{\sqrt{2}}}$

18. (0) $\because a \sin \theta = b \sin\left(\theta + \dfrac{2\pi}{3}\right) = c \sin\left(\theta + \dfrac{4\pi}{3}\right) = \dfrac{1}{k}$ (let)

Then, $\dfrac{1}{a} = k \sin \theta$, $\dfrac{1}{b} = k \sin\left(\theta + \dfrac{2\pi}{3}\right)$, $\dfrac{1}{c} = k \sin\left(\theta + \dfrac{4\pi}{3}\right)$

$\therefore$ $\dfrac{1}{a} + \dfrac{1}{b} + \dfrac{1}{c} = k\left[\sin \theta + \sin\left(\theta + \dfrac{2\pi}{3}\right) + \sin\left(\theta + \dfrac{4\pi}{3}\right)\right]$

$\Rightarrow$ $\dfrac{bc + ca + ab}{abc} = k\left[\sin \theta + 2 \sin (\theta + \pi) \cos \dfrac{\pi}{3}\right]$

$\Rightarrow$ $\dfrac{ab + bc + ca}{abc} = k\left[\sin \theta - 2 \sin \theta \cdot \dfrac{1}{2}\right]$

$\Rightarrow$ $\dfrac{ab + bc + ca}{abc} = k(\sin \theta - \sin \theta) = 0$

$\therefore$ $ab + bc + ca = 0$

19. (0) We have,

$$\frac{\tan (\theta + \alpha)}{a} = \frac{\tan (\theta + \beta)}{b} = \frac{\tan (\theta + \gamma)}{c}$$

Now, by taking first two members, we get

$$\frac{\tan (\theta + \alpha)}{a} = \frac{\tan (\theta + \beta)}{b}$$

$\Rightarrow$ $\dfrac{a}{b} = \dfrac{\tan (\theta + \alpha)}{\tan (\theta + \beta)}$

On applying componendo and dividendo both sides, we get

$$\frac{a+b}{a-b} = \frac{\tan (\theta + \alpha) + \tan (\theta + \beta)}{\tan (\theta + \alpha) - \tan (\theta + \beta)}$$

$$= \frac{\dfrac{\sin (\theta + \alpha + \theta + \beta)}{\cos (\theta + \alpha) \cos (\theta + \beta)}}{\dfrac{\sin (\theta + \alpha - \theta - \beta)}{\cos (\theta + \alpha) \cos (\theta + \beta)}}$$

$\therefore$ $\dfrac{a+b}{a-b} = \dfrac{\sin (2\theta + \alpha + \beta)}{\sin (\alpha - \beta)}$

$\Rightarrow$ $\dfrac{a+b}{a-b} = \dfrac{\sin (2\theta + \alpha + \beta) \sin (\alpha - \beta)}{\sin^2(\alpha - \beta)}$

$\Rightarrow$ $\left(\dfrac{a+b}{a-b}\right) \sin^2(\alpha - \beta) = \dfrac{1}{2} [2 \sin (2\theta + \alpha + \beta) \sin (\alpha - \beta)]$

$\Rightarrow$ $\left(\dfrac{a+b}{a-b}\right) \sin^2(\alpha - \beta)$

$$= \frac{1}{2} [\cos (2\theta + \alpha + \beta - \alpha + \beta) - \cos (2\theta + \alpha + \beta + \alpha - \beta)]$$

$\Rightarrow$ $\left(\dfrac{a+b}{a-b}\right) \sin^2(\alpha - \beta) = \dfrac{1}{2} [\cos (2\theta + 2\beta) - \cos (2\theta + 2\alpha)]$ ...(i)

Similarly, by taking 2nd and 3rd members and 1st and 3rd members, we get

$\Rightarrow$ $\left(\dfrac{b+c}{b-c}\right) \sin^2(\beta - \gamma) = \dfrac{1}{2} [\cos (2\theta + 2\gamma) - \cos (2\theta + 2\beta)]$ ...(ii)

and $\left(\dfrac{c+a}{c-a}\right) \sin^2(\gamma - \alpha) = \dfrac{1}{2} [\cos (2\theta + 2\alpha) - \cos (2\theta + 2\gamma)]$ ...(iii)

On adding Eqs. (i), (ii) and (iii), we get

$$\left(\frac{a+b}{a-b}\right) \sin^2(\alpha - \beta) + \left(\frac{b+c}{b-c}\right) \sin^2(\beta - \gamma) + \left(\frac{c+a}{c-a}\right) \sin^2(\gamma - \alpha)$$

$$= \frac{1}{2} [\cos (2\theta + 2\beta) + \cos (2\theta + 2\alpha) + \cos (2\theta + 2\gamma) - \cos (2\theta + 2\beta) - \cos (2\theta + 2\alpha) - \cos (2\theta + 2\gamma)] = 0$$

20. (2) LHS $= \dfrac{\cos 6x + 6 \cos 4x + 15 \cos 2x + 10}{\cos 5x + 5 \cos 3x + 10 \cos x}$

$$= \frac{(\cos 6x + \cos 4x) + 5 \cos 4x + 5 \cos 2x + 10 \cos 2x + 10}{\cos 5x + 5 \cos 3x + 10 \cos x}$$

$$= \frac{2 \cos 5x \cdot \cos x + 5(\cos 4x + \cos 2x) + 10(1 + \cos 2x)}{\cos 5x + 5 \cos 3x + 10 \cos x}$$

$$= \frac{2 \cos 5x \cdot \cos x + 10 \cos 3x \cdot \cos x + 20 \cos^2 x}{\cos 5x + 5 \cos 3x + 10 \cos x}$$

$$= \frac{2 \cos x [\cos 5x + 5 \cos 3x + 10 \cos x]}{(\cos 5x + 5 \cos 3x + 10 \cos x)} = 2 \cos x$$

But RHS $= \lambda \cos x$

$\therefore$ $\lambda = 2$

DPP-9 Transformation Formulae-II

1. $\because \quad \cos x + \cos y + \cos z = 0$

$\Rightarrow \quad \cos x + \cos y = -\cos z$

$\Rightarrow \quad 2\cos\left(\frac{x+y}{2}\right)\cos\left(\frac{x-y}{2}\right) = -\cos z \quad \ldots(i)$

and $\quad \sin x + \sin y + \sin z = 0$

$\sin x + \sin y = -\sin z$

$\Rightarrow \quad 2\sin\left(\frac{x+y}{2}\right)\cos\left(\frac{x-y}{2}\right) = -\sin z \quad \ldots(ii)$

On dividing Eq. (i) by Eq. (ii), we get

$$\Rightarrow \frac{2\cos\left(\frac{x+y}{2}\right)\cos\left(\frac{x-y}{2}\right)}{2\sin\left(\frac{x+y}{2}\right)\cos\left(\frac{x-y}{2}\right)} = \frac{-\cos z}{-\sin z}$$

$\therefore \quad \cot\left(\frac{x+y}{2}\right) = \cot z$ **Hence proved.**

2. We have, $\tan A = 2\tan B + \cot B \quad \ldots(i)$

LHS $= 2\tan(A-B)$

$$= 2\left(\frac{\tan A - \tan B}{1 + \tan A \tan B}\right) = 2\left(\frac{2\tan B + \cot B - \tan B}{1 + (2\tan B + \cot B)\tan B}\right)$$

[from Eq. (i)]

$$= 2\left(\frac{\tan B + \cot B}{1 + 2\tan^2 B + 1}\right) = 2\left(\frac{\frac{1+\tan^2 B}{\tan B}}{2(1+\tan^2 B)}\right)$$

$$= 2\cdot\frac{1+\tan^2 B}{\tan B} \times \frac{1}{2(1+\tan^2 B)}$$

$= \frac{1}{\tan B} = \cot B =$ RHS **Hence proved.**

3. LHS $= \frac{\tan 70° - \tan 20°}{\tan 50°} = \frac{\frac{\sin(70° - 20°)}{\cos 70° \cos 20°}}{\frac{\sin 50°}{\cos 50°}}$

$\left[\because \tan A - \tan B = \frac{\sin(A-B)}{\cos A \cos B}\right]$

$$= \frac{\sin 50°}{\cos 70° \cos 20°} \times \frac{\cos 50°}{\sin 50°} = \frac{2\cos 50°}{2\cos 70° \cos 20°}$$

$$= \frac{2\cos 50°}{\cos 90° + \cos 50°} = \frac{2\cos 50°}{\cos 50°}$$

$= 2 =$ RHS **Hence proved.**

4. We know that, $\quad \tan 5x = \tan(3x + 2x)$

$$\tan 5x = \frac{\tan 3x + \tan 2x}{1 - \tan 3x \tan 2x}$$

$\Rightarrow \tan 5x - \tan 5x \tan 3x \tan 2x = \tan 3x + \tan 2x$

$\therefore \quad \tan 5x \tan 3x \tan 2x = \tan 5x - \tan 3x - \tan 2x$

5. (b) We have, $\cos(A+B)\sin(C-D) = \cos(A-B)\sin(C+D)$

$$\Rightarrow \frac{\cos(A+B)}{\cos(A-B)} = \frac{\sin(C+D)}{\sin(C-D)}$$

On applying componendo and dividendo, we get

$$\frac{\cos(A+B) + \cos(A-B)}{\cos(A+B) - \cos(A-B)} = \frac{\sin(C+D) + \sin(C-D)}{\sin(C+D) - \sin(C-D)}$$

$$\Rightarrow \frac{2\cos A \cos B}{-2\sin A \sin B} = \frac{2\sin C \cos D}{2\cos C \sin D}$$

$$\Rightarrow \frac{-1}{\tan A \tan B} = \frac{\tan C}{\tan D}$$

$\therefore \quad \tan A \tan B \tan C = -\tan D$

6. (b) We have, A, B, C are in AP, then

$2B = A + C \quad \ldots(i)$

Now, $$\frac{\sin A - \sin C}{\cos C - \cos A} = \frac{2\cos\left(\frac{A+C}{2}\right)\sin\left(\frac{A-C}{2}\right)}{2\sin\left(\frac{A+C}{2}\right)\sin\left(\frac{A-C}{2}\right)}$$

$= \cot\left(\frac{A+C}{2}\right) = \cot\left(\frac{2B}{2}\right)$ [from Eq. (i)]

$= \cot B$

7. (a) We have, $A + B = C$, then

$\tan(A+B) = \tan C$

$$\Rightarrow \frac{\tan A + \tan B}{1 - \tan A \tan B} = \tan C$$

$\Rightarrow \quad \tan A + \tan B = \tan C - \tan A \tan B \tan C$

$\therefore \quad \tan A \tan B \tan C = \tan C - \tan A - \tan B$

8. (c) $\because \quad \sin 2\theta = \lambda \sin 2\phi$

$$\Rightarrow \frac{\sin 2\theta}{\sin 2\phi} = \frac{\lambda}{1}$$

On applying componendo and dividendo, we get

$$\frac{\sin 2\theta + \sin 2\phi}{\sin 2\theta - \sin 2\phi} = \frac{\lambda + 1}{\lambda - 1}$$

$$\Rightarrow \frac{2\sin(\theta+\phi)\cos(\theta-\phi)}{2\cos(\theta+\phi)\sin(\theta-\phi)} = \frac{\lambda+1}{\lambda-1}$$

$$\therefore \frac{\tan(\theta+\phi)}{\tan(\theta-\phi)} = \frac{\lambda+1}{\lambda-1}$$

9. (c) $\quad \sin A + \sin B = a$

and $\quad \cos A + \cos B = b$

On dividing, we get

$$\frac{\sin A + \sin B}{\cos A + \cos B} = \frac{a}{b}$$

$$\Rightarrow \frac{2\sin\left(\frac{A+B}{2}\right)\cdot\cos\left(\frac{A-B}{2}\right)}{2\cos\left(\frac{A+B}{2}\right)\cdot\cos\left(\frac{A-B}{2}\right)} = \frac{a}{b}$$

$$\Rightarrow \tan\left(\frac{A+B}{2}\right) = \frac{a}{b}$$

$$\therefore \cos(A+B) = \frac{1 - \tan^2\left(\frac{A+B}{2}\right)}{1 + \tan^2\left(\frac{A+B}{2}\right)} = \frac{1 - \frac{a^2}{b^2}}{1 + \frac{a^2}{b^2}} = \frac{b^2 - a^2}{b^2 + a^2}$$

10. (b) $\because A + B = \frac{\pi}{4}$, then $\tan(A+B) = \tan\frac{\pi}{4}$

$$\Rightarrow \frac{\tan A + \tan B}{1 - \tan A \tan B} = 1$$

$\Rightarrow \quad \tan A + \tan B = 1 - \tan A \tan B$

$\Rightarrow \quad 1 + \tan A + \tan B + \tan A \tan B = 2$

$\Rightarrow \quad (1 + \tan A) + \tan B(1 + \tan A) = 2$

$\therefore \quad (1 + \tan A)(1 + \tan B) = 2$

11. (d) We have, $\sin y \cos\left(\frac{\pi}{2} - x\right) - \cos\left(\frac{\pi}{2} - y\right)\cos x$

$+ \cos x \sin\left(\frac{\pi}{2} - y\right) + \cos y \cos\left(\frac{\pi}{2} - x\right) = 0$

$\Rightarrow \sin y \sin x - \cos x \sin y + \cos x \cos y + \sin x \cos y = 0$

$\Rightarrow \cos x \cos y + \sin x \sin y + \sin x \cos y - \cos x \sin y = 0$

$\Rightarrow \quad \cos(x-y) + \sin(x-y) = 0$

$$\Rightarrow \quad \frac{\sin(x-y)}{\cos(x-y)} + 1 = 0 \quad \text{[dividing both sides by } \cos(x-y)]$$

$$\Rightarrow \quad \tan(x-y) = -1 = \tan\frac{3\pi}{4}$$

$$\Rightarrow \quad x - y = \frac{3\pi}{4}$$

$$\therefore \quad x - \frac{3\pi}{4} = y$$

12. (d) $\because \tan A = \frac{1}{2}$ and $\tan B = \frac{1}{3}$, then

$$\tan(A+B) = \frac{\tan A + \tan B}{1 - \tan A \tan B} = \frac{\frac{1}{2}+\frac{1}{3}}{1-\frac{1}{2}\cdot\frac{1}{3}} = \frac{\frac{5}{6}}{\frac{5}{6}} = 1$$

$$\Rightarrow \quad \tan(A+B) = \tan\frac{\pi}{4}$$

$$\therefore \quad A + B = \frac{\pi}{4}$$

13. (c) $\because \quad A + B + C = \pi \quad$ ...(i)

Now, $\sin A \sin(B-C) = \sin(\pi - \overline{B+C})\sin(B-C)$ [from Eq. (i)]

$= \sin(B+C)\sin(B-C) = \sin^2 B - \sin^2 C$

Similarly, $\sin B \sin(C-A) = \sin^2 C - \sin^2 A$

and $\sin C \sin(A-B) = \sin^2 A - \sin^2 B$

$\therefore \sin A \sin(B-C) + \sin B \sin(C-A) + \sin C \sin(A-B)$

$= \sin^2 B - \sin^2 C + \sin^2 C - \sin^2 A + \sin^2 A - \sin^2 B = 0$

14. (b) $\because k = \dfrac{\cos 18° - \sin 18°}{\sin 27°} = \dfrac{\cos 18° - \sin(90° - 72°)}{\sin 27°}$

$$= \frac{\cos 18° - \cos 72°}{\sin 27°} = \frac{2\sin\left(\frac{18°+72°}{2}\right)\sin\left(\frac{72°-18°}{2}\right)}{\sin 27°}$$

$$= \frac{2\sin 45° \sin 27°}{\sin 27°} = 2\sin 45° = 2 \times \frac{1}{\sqrt{2}} = \sqrt{2}$$

$\therefore \quad k = \sqrt{2}$

15. (a) $\because \sin\theta = n\sin(\theta + 2\alpha)$

$$\Rightarrow \quad \frac{\sin(\theta+2\alpha)}{\sin\theta} = \frac{1}{n}$$

On applying componendo and dividendo, we get

$$\frac{\sin(\theta+2\alpha)+\sin\theta}{\sin(\theta+2\alpha)-\sin\theta} = \frac{1+n}{1-n}$$

$$\Rightarrow \quad \frac{2\sin(\theta+\alpha)\cos\alpha}{2\cos(\theta+\alpha)\sin\alpha} = \frac{1+n}{1-n}$$

$$\therefore \quad \frac{\tan(\theta+\alpha)}{\tan\alpha} = \frac{1+n}{1-n}$$

16. (a,b,c) $\because \cos\alpha = \frac{3}{5}$

$$\Rightarrow \quad \sin\alpha = \sqrt{1-\cos^2\alpha}$$

$$\Rightarrow \quad \sin\alpha = \sqrt{1-\left(\frac{3}{5}\right)^2} = \frac{4}{5}$$

and $\quad \cos\beta = \frac{5}{13}$

$$\Rightarrow \quad \sin\beta = \sqrt{1-\cos^2\beta}$$

$$\Rightarrow \quad \sin\beta = \sqrt{1-\frac{25}{169}} = \frac{12}{13}$$

$$\therefore \quad \sin\alpha = \frac{4}{5} \text{ and } \sin\beta = \frac{12}{13}$$

Now, (a) $\cos(\alpha+\beta) = \cos\alpha\cos\beta - \sin\alpha\sin\beta$

$$= \frac{3}{5}\cdot\frac{5}{13} - \frac{4}{5}\cdot\frac{12}{13} = \frac{15-48}{65} = \frac{-33}{65}$$

$\therefore$ Option (a) is correct.

(b) $\sin(\alpha+\beta) = \sin\alpha\cos\beta + \cos\alpha\sin\beta$

$$= \frac{4}{5}\cdot\frac{5}{13} + \frac{3}{5}\cdot\frac{12}{13} = \frac{20+36}{65} = \frac{56}{65}$$

$\therefore$ Option (b) is correct.

(c) $\cos(\alpha-\beta) = \cos\alpha\cos\beta + \sin\alpha\sin\beta$

$$= \frac{3}{5}\cdot\frac{5}{13} + \frac{4}{5}\cdot\frac{12}{13} = \frac{15+48}{65} = \frac{63}{65}$$

$\therefore$ Option (c) is correct.

17. (a,c) We have, $\quad \sin\alpha = \sin\beta$ and $\cos\alpha = \cos\beta$

$\sin\alpha - \sin\beta = 0$ and $\cos\alpha - \cos\beta = 0$

$$2\cos\left(\frac{\alpha+\beta}{2}\right)\sin\left(\frac{\alpha-\beta}{2}\right) = 0 \quad \text{...(i)}$$

and $$-2\sin\left(\frac{\alpha+\beta}{2}\right)\sin\left(\frac{\alpha-\beta}{2}\right) = 0 \quad \text{...(ii)}$$

From Eqs. (i) and (ii), we get

$$2\cos\left(\frac{\alpha+\beta}{2}\right)\sin\left(\frac{\alpha-\beta}{2}\right) = -2\sin\left(\frac{\alpha+\beta}{2}\right)\sin\left(\frac{\alpha-\beta}{2}\right)$$

$$\Rightarrow \quad \cos\left(\frac{\alpha+\beta}{2}\right)\sin\left(\frac{\alpha-\beta}{2}\right) + \sin\left(\frac{\alpha+\beta}{2}\right)\sin\left(\frac{\alpha-\beta}{2}\right) = 0$$

$$\Rightarrow \quad \sin\left(\frac{\alpha-\beta}{2}\right)\left[\sin\left(\frac{\alpha+\beta}{2}\right) + \cos\left(\frac{\alpha+\beta}{2}\right)\right] = 0$$

$$\sin\left(\frac{\alpha-\beta}{2}\right) = 0, \text{ then } \cos\left(\frac{\alpha-\beta}{2}\right) = 1$$

18. (a,b,c) We have,

$$a\sin^2 x + b\cos^2 x = c$$

$$a\tan^2 x + b = c\sec^2 x$$

$$a\tan^2 x + b = c + c\tan^2 x$$

$$a\tan^2 x - c\tan^2 x = c - b$$

$$\tan^2 x = \frac{c-b}{a-c} \quad \text{[option (a)]}$$

and $$b\sin^2 y + a\cos^2 y = d$$

$$b\tan^2 y + a = d\sec^2 y$$

$$b\tan^2 y + a = d + d\tan^2 y$$

$$(b-d)\tan^2 y = d - a \quad \text{[option (b)]}$$

$$\tan^2 y = \frac{d-a}{b-d}$$

and $$a\tan x = b\tan y$$

$$\frac{a^2}{b^2} = \frac{\tan^2 y}{\tan^2 x}$$

$$\frac{a^2}{b^2} = \frac{(d-a)}{(b-d)} \times \frac{(a-c)}{(c-b)}$$

$$\frac{a^2}{b^2} = \frac{(a-d)(c-a)}{(b-c)(d-b)} \quad \text{[option (c)]}$$

19. (3) We have, $\quad A - B = \frac{\pi}{4}$

$$\Rightarrow \quad \tan(A-B) = \tan\frac{\pi}{4} = 1$$

$$\Rightarrow \quad \frac{\tan A - \tan B}{1 + \tan A \tan B} = 1$$

$$\Rightarrow \quad \tan A - \tan B = 1 + \tan A \tan B$$

$$\Rightarrow \quad 1 + \tan A - \tan B - \tan A \tan B = 2$$

$$\Rightarrow \quad (1 + \tan A) - \tan B(1 + \tan A) = 2$$

$$\Rightarrow \quad (1 + \tan A)(1 - \tan B) = 2 \quad \text{...(i)}$$

But $\quad y = (\tan A + 1)(1 - \tan B)$

$\therefore \quad y = 2 \quad$ [from Eq. (i)]

Now, $\quad (y+1)^{y-1} = (2+1)^{2-1} = 3$

20. (4) $\because \lambda \cos\left(\frac{\alpha+\beta}{2}\right)\cos\left(\frac{\beta+\gamma}{2}\right)\cos\left(\frac{\gamma+\alpha}{2}\right)$

$= \cos\alpha + \cos\beta + \cos\gamma + \cos(\alpha+\beta+\gamma)$

$= 2\cos\left(\frac{\alpha+\beta}{2}\right)\cos\left(\frac{\alpha-\beta}{2}\right) + 2\cos\left(\frac{\alpha+\beta+2\gamma}{2}\right)\cos\left(\frac{\alpha+\beta}{2}\right)$

$= 2\cos\left(\frac{\alpha+\beta}{2}\right)\left[\cos\left(\frac{\alpha-\beta}{2}\right) + \cos\left(\frac{\alpha+\beta+2\gamma}{2}\right)\right]$

$= 2\cos\left(\frac{\alpha+\beta}{2}\right)\left[2\cos\left(\frac{2\alpha+2\gamma}{4}\right)\cos\left(\frac{2\beta+2\gamma}{4}\right)\right]$

$= 4\cos\left(\frac{\alpha+\beta}{2}\right)\cos\left(\frac{\alpha+\gamma}{2}\right)\cos\left(\frac{\beta+\gamma}{2}\right)$

$\therefore \quad \lambda = 4$

21. (3) We have,

$\sin\left(\frac{A+B}{2}\right)\cos\left(\frac{A-B}{2}\right) + \sin\left(\frac{C+D}{2}\right)\cos\left(\frac{C-D}{2}\right) = 2$

$\Rightarrow 2\sin\left(\frac{A+B}{2}\right)\cos\left(\frac{A-B}{2}\right) + 2\sin\left(\frac{C+D}{2}\right)\cos\left(\frac{C-D}{2}\right) = 4$

$\Rightarrow \quad \sin A + \sin B + \sin C + \sin D = 4$

It is possible only when,

$\sin A = \sin B = \sin C = \sin D = 1 \quad (\because -1 \le \sin\theta \le 1)$

Then, $\quad A = B = C = D = \frac{\pi}{2}$

Now,

$\Sigma \cos\frac{A}{2}\cos\frac{B}{2} = \cos\frac{A}{2}\cos\frac{B}{2} + \cos\frac{A}{2}\cos\frac{C}{2} + \cos\frac{A}{2}\cos\frac{D}{2}$

$+ \cos\frac{B}{2}\cos\frac{C}{2} + \cos\frac{B}{2}\cos\frac{D}{2} + \cos\frac{C}{2}\cos\frac{D}{2}$

$= 6\times\left(\cos\frac{\pi}{4}\cos\frac{\pi}{4}\right) \quad \left(\because A = B = C = D = \frac{\pi}{2}\right)$

$= 6\times\frac{1}{\sqrt{2}}\times\frac{1}{\sqrt{2}} = 3$

DPP-10 **Trigonometric Ratios using Half-Angle Formulae**

1. $\because \quad \cos 2\theta = \frac{1-\tan^2\theta}{1+\tan^2\theta}$

$\therefore \quad m = 2$

2. $\because \frac{1-\cos 2A}{2} = \sin^2 A$

$\therefore \quad \lambda = -1$

3. $\lambda\sin x + \mu\cos x = \frac{\lambda\cdot 2\tan\frac{x}{2}}{1+\tan^2\frac{x}{2}} + \mu\left(\frac{1-\tan^2\frac{x}{2}}{1+\tan^2\frac{x}{2}}\right)$

$= \frac{2\lambda\cdot\tan\frac{x}{2} + \mu\left(1-\tan^2\frac{x}{2}\right)}{1+\tan^2\frac{x}{2}}$

$= \frac{2\lambda\cdot\frac{\lambda}{\mu} + \mu\left(1-\frac{\lambda^2}{\mu^2}\right)}{1+\frac{\lambda^2}{\mu^2}} \quad \left(\because \tan\frac{x}{2} = \frac{\lambda}{\mu}\right)$

$= \frac{2\lambda^2+\mu^2-\lambda^2}{\mu}\times\frac{\mu^2}{\lambda^2+\mu^2} = \frac{\lambda^2+\mu^2}{\mu}\times\frac{\mu^2}{\lambda^2+\mu^2} = \mu$

4. True, LHS $= \tan\theta\sin\left(\frac{\pi}{2}+\theta\right)\cos\left(\frac{\pi}{2}-\theta\right) - k\cos^2\theta$

$= \frac{\sin\theta}{\cos\theta}\cos\theta\sin\theta - k\cos^2\theta$

$= \sin^2\theta - k\cos^2\theta$

It will be 1 only, when $k = -1$ $\quad (\because \sin^2\theta + \cos^2\theta = 1)$

5. True, we have, $\cos\frac{\pi}{7}\cos\frac{5\pi}{7}\cos\frac{3\pi}{7}$

$= \cos\frac{\pi}{7}\cos\left(\pi-\frac{2\pi}{7}\right)\cos\left(\pi-\frac{4\pi}{7}\right)$

$= \cos\frac{\pi}{7}\cos\frac{2\pi}{7}\cos\frac{4\pi}{7} = \frac{\sin\frac{8\pi}{7}}{2^3\sin\frac{\pi}{7}}$

$= \frac{\sin\left(\pi+\frac{\pi}{7}\right)}{8\sin\frac{\pi}{7}} = \frac{-\sin\frac{\pi}{7}}{8\sin\frac{\pi}{7}} = -\frac{1}{8}$

6. False, we know that, $\sin\theta > \cos\theta$, when $\frac{\pi}{4} < \theta < \frac{\pi}{2}$

Then, $\sqrt{1-\sin 2\theta} = \sqrt{(\sin\theta-\cos\theta)^2} = |\sin\theta - \cos\theta|$

$= \sin\theta - \cos\theta \quad \left(\because \frac{\pi}{4} < \theta < \frac{\pi}{2}\right)$

7. We have, $\sqrt{\frac{1-\cos 2\theta}{1+\cos 2\theta}} = \sqrt{\frac{2\sin^2\theta}{2\cos^2\theta}} = \pm\tan\theta$

$= \tan\theta \quad (\because \theta$ lies in III quadrant)

8. We have, $\tan x = -\frac{4}{3}$ and x lies in II quadrant.

$\frac{\pi}{2} < x < \pi$

$\frac{\pi}{4} < \frac{x}{2} < \frac{\pi}{2}$

$\because \quad \sec^2 x = 1 + \tan^2 x$

$= 1 + \left(-\frac{4}{3}\right)^2 = \frac{25}{9}$

$\sec x = \pm\frac{5}{3} = -\frac{5}{3} \quad \left(\because \frac{\pi}{2} < x < \pi\right)$

$\Rightarrow \quad \cos x = -\frac{3}{5}$

$\Rightarrow \quad 1 - 2\sin^2\frac{x}{2} = -\frac{3}{5}$

$\Rightarrow \quad 2\sin^2\frac{x}{2} = 1 + \frac{3}{5}$

$\Rightarrow \quad \sin\frac{x}{2} = \sqrt{\frac{4}{5}} = \pm\frac{2}{\sqrt{5}}$

$\therefore \quad \sin\frac{x}{2} = \frac{2}{\sqrt{5}} \quad \left(\because \frac{x}{2}\text{ lies in I quadrant}\right)$

9. LHS $= \sin^4\frac{\pi}{8} + \sin^4\frac{3\pi}{8} + \sin^4\frac{5\pi}{8} + \sin^4\frac{7\pi}{8}$

$= \sin^4\frac{\pi}{8} + \sin^4\frac{3\pi}{8} + \sin^4\left(\pi-\frac{3\pi}{8}\right) + \sin^4\left(\pi-\frac{\pi}{8}\right)$

$= 2\sin^4\frac{\pi}{8} + 2\sin^4\frac{3\pi}{8}$

$= \frac{1}{2}\left[\left(2\sin^2\frac{\pi}{8}\right)^2 + \left(2\sin^2\frac{3\pi}{8}\right)^2\right]$

$$= \frac{1}{2}\left[\left(1 - \cos\frac{\pi}{4}\right)^2 + \left(1 - \cos\frac{3\pi}{4}\right)^2\right]$$

$$= \frac{1}{2}\left[\left(1 - \frac{1}{\sqrt{2}}\right)^2 + \left(1 + \frac{1}{\sqrt{2}}\right)^2\right]$$

$$= \frac{1}{2}\left[1 + \frac{1}{2} - 2\cdot\frac{1}{\sqrt{2}} + 1 + \frac{1}{2} + 2\cdot\frac{1}{\sqrt{2}}\right]$$

$$= \frac{1}{2}\cdot 2\cdot\frac{3}{2} = \frac{3}{2} = \text{RHS}$$ **Hence proved.**

10. We have, $A = 2\sin^2\theta - \cos 2\theta$

$= 1 - \cos 2\theta - \cos 2\theta$

$A = 1 - 2\cos 2\theta$ ...(i)

We know that,

$$-1 \le \cos 2\theta \le 1$$
$$-2 \le -2\cos 2\theta \le 2$$
$$-1 \le 1 - 2\cos 2\theta \le 3$$

$\therefore$ $-1 \le A \le 3$ [from Eq. (i)]

i.e. $A \in [-1, 3]$

But $A \in [\alpha, \beta]$

$\therefore$ $\alpha = -1$ and $\beta = 3$

11. (d) We have, $\sqrt{3}\operatorname{cosec} 20° - \sec 20° = \frac{\sqrt{3}}{\sin 20°} - \frac{1}{\cos 20°}$

$$= \frac{\sqrt{3}\cos 20° - \sin 20°}{\sin 20° \cos 20°}$$

$$= \frac{4\cdot\left(\frac{\sqrt{3}}{2}\cos 20° - \frac{1}{2}\sin 20°\right)}{2\sin 20° \cos 20°}$$

$$= \frac{4\cdot\sin 60° \cos 20° - \cos 60° \sin 20°}{\sin 40°}$$

$$= \frac{4\cdot\sin(60° - 20°)}{\sin 40°} = \frac{4\cdot\sin 40°}{\sin 40°} = 4$$

12. (d) We have, $\left(\cot\frac{x}{2} - \tan\frac{x}{2}\right)^2 (1 - 2\tan x \cot 2x)$

$$= \left(\frac{1}{\tan\frac{x}{2}} - \tan\frac{x}{2}\right)^2 \left(1 - \frac{2\tan x}{\tan 2x}\right)$$

$$= \left(\frac{1 - \tan^2\frac{x}{2}}{\tan\frac{x}{2}}\right)^2 \left(1 - \frac{2\tan x}{\frac{2\tan x}{1 - \tan^2 x}}\right)$$

$$= \left(\frac{2}{\frac{2\tan\frac{x}{2}}{1 - \tan^2\frac{x}{2}}}\right)^2 (1 - 1 + \tan^2 x)$$

$$= \left(\frac{2}{\tan x}\right)^2 \tan^2 x$$

$$= \frac{4}{\tan^2 x}\cdot\tan^2 x = 4$$

13. (c) We have,

$$\frac{\sin 68°}{k\cos 83°} = \cos 7° \cos 14° \cos 28° \cos 56°$$

$$= \frac{(2\sin 7° \cos 7°)\cos 14° \cos 28° \cos 56°}{2\sin 7°}$$

$$= \frac{(2\sin 14° \cos 14°)\cos 28° \cos 56°}{2^2 \sin 7°}$$

$$= \frac{(2\times\sin 28° \cos 28°)\cos 56°}{2^3 \sin 7°}$$

$$= \frac{2\times\sin 56° \cos 56°}{2^4 \sin 7°}$$

$$= \frac{\sin 112°}{2^4 \sin 7°} = \frac{\sin(180° - 68°)}{16\sin(90° - 83°)}$$

$$\Rightarrow \quad \frac{\sin 68°}{k\cos 83°} = \frac{\sin 68°}{16\cos 83°}$$

$\therefore$ $k = 16$

14. (c) $\because A = 125° \Rightarrow \frac{A}{2} = 62\frac{1}{2}°$ (I quadrant)

Now, $\sqrt{1 + \sin A} = \pm\left(\sin\frac{A}{2} + \cos\frac{A}{2}\right)$

$= \sin\frac{A}{2} + \cos\frac{A}{2}$ ...(i)

and $\sqrt{1 - \sin A} = \pm\left(\sin\frac{A}{2} - \cos\frac{A}{2}\right)$

$= \sin\frac{A}{2} - \cos\frac{A}{2}$ ...(ii)

$\left(\because \frac{A}{2} = 62\frac{1}{2}°, \text{ then } \cos\frac{A}{2} < \sin\frac{A}{2}\right)$

On subtracting Eq. (ii) from Eq. (i), we get

$$2\cos\frac{A}{2} = \sqrt{1 + \sin A} - \sqrt{1 - \sin A}$$

15. (d) $\because \sec 2\theta = p + \tan 2\theta$

$\sec 2\theta - \tan 2\theta = p$ ...(i)

$\sec 2\theta + \tan 2\theta = \frac{1}{p}$ ...(ii)

On adding Eqs. (i) and (ii), we get

$$2\sec 2\theta = p + \frac{1}{p} = \frac{p^2 + 1}{p}$$

$$\sec 2\theta = \frac{p^2 + 1}{2p}$$

$$\cos 2\theta = \frac{2p}{p^2 + 1}$$

$$1 - 2\sin^2\theta = \frac{2p}{p^2 + 1}$$

$$2\sin^2\theta = 1 - \frac{2p}{p^2 + 1} = \frac{p^2 + 1 - 2p}{p^2 + 1}$$

$\therefore$ $\sin^2\theta = \frac{(p - 1)^2}{2(p^2 + 1)}$

16. (d) We have, $\sin x + \cos x = \frac{1}{5}$

On squaring both sides, we get

$$1 + \sin 2x = \frac{1}{25}$$

$$\sin 2x = -\frac{24}{25}$$

We know that, $\cos 2x = \sqrt{1 - \sin^2 2x}$

$$= \sqrt{1 - \left(-\frac{24}{25}\right)^2}$$

$$= \sqrt{1 - \frac{576}{625}} = \sqrt{\frac{49}{625}} = \pm\frac{7}{25}$$

$\therefore$ $\tan 2x = \frac{\sin 2x}{\cos 2x} = \frac{24}{7}$

17. (b) $\because 0 < x < \frac{\pi}{4}$

$$\therefore \quad \sec 2x - \tan 2x = \frac{1 - \sin 2x}{\cos 2x} = \frac{1 - \cos\left(\frac{\pi}{2} - 2x\right)}{\sin\left(\frac{\pi}{2} - 2x\right)}$$

$$= \frac{2\sin^2\left(\frac{\pi}{4} - x\right)}{2\sin\left(\frac{\pi}{4} - x\right)\cos\left(\frac{\pi}{4} - x\right)}$$

$$= \frac{\sin\left(\frac{\pi}{4} - x\right)}{\cos\left(\frac{\pi}{4} - x\right)} = \tan\left(\frac{\pi}{4} - x\right)$$

Alternate Method

$$\because \sec 2x - \tan 2x = \frac{1}{\cos 2x} - \frac{\sin 2x}{\cos 2x}$$

$$= \frac{1 - \sin 2x}{\cos 2x} = \frac{(\cos x - \sin x)^2}{\cos^2 x - \sin^2 x}$$

$$= \frac{\cos x - \sin x}{\cos x + \sin x} = \frac{1 - \tan x}{1 + \tan x}$$

(dividing numerator and denominator by $\cos x$)

$$= \tan\left(\frac{\pi}{4} - x\right)$$

18. (a) We have, $\cos(\theta - \alpha)$, $\cos\theta$, $\cos(\theta + \alpha)$ are in HP, then

$\frac{1}{\cos(\theta - \alpha)}, \frac{1}{\cos\theta}, \frac{1}{\cos(\theta + \alpha)}$ are in AP.

$$\therefore \quad \frac{2}{\cos\theta} = \frac{1}{\cos(\theta - \alpha)} + \frac{1}{\cos(\theta + \alpha)}$$

$$= \frac{\cos(\theta + \alpha) + \cos(\theta - \alpha)}{\cos(\theta - \alpha)\cos(\theta + \alpha)}$$

$$\Rightarrow \quad \frac{2}{\cos\theta} = \frac{2\cos\theta \cdot \cos\alpha}{\cos^2\theta - \sin^2\alpha}$$

$$\left[\begin{array}{l}\because \cos(A + B) + \cos(A - B) = 2\cos A\cos B \\ \text{and} \cos(A + B)\cos(A - B) = \cos^2 A - \sin^2 B\end{array}\right]$$

$$\Rightarrow \quad \cos^2\theta - \sin^2\alpha = \cos^2\theta\cos\alpha$$

$$\Rightarrow \quad \cos^2\theta - \cos^2\theta\cos\alpha = \sin^2\alpha$$

$$\Rightarrow \quad \cos^2\theta(1 - \cos\alpha) = \sin^2\alpha$$

$$\Rightarrow \quad \cos^2\theta \cdot 2\sin^2\frac{\alpha}{2} = \left(2\sin\frac{\alpha}{2}\cos\frac{\alpha}{2}\right)^2$$

$$\Rightarrow \quad \cos^2\theta \cdot 2\sin^2\frac{\alpha}{2} = 4\sin^2\frac{\alpha}{2}\cos^2\frac{\alpha}{2}$$

$$\Rightarrow \quad \cos^2\theta = 2\cos^2\frac{\alpha}{2}$$

$$\Rightarrow \quad \frac{\cos^2\theta}{\cos^2\frac{\alpha}{2}} = 2$$

$$\therefore \quad \cos\theta \cdot \sec\frac{\alpha}{2} = \pm\sqrt{2}$$

19. (d) We know that, AM > GM

Then, $\frac{\cos^2 x + \sec^2 x}{2} \geq \sqrt{\cos^2 x \sec^2 x}$

$\cos^2 x + \sec^2 x \geq 2$ (given)

$\therefore \quad F(x) \geq 2$

20. (a) $\because 450° < \alpha < 540°$ (II quadrant)

$225° < \frac{\alpha}{2} < 270°$ (III quadrant)

$\because \quad \sin\alpha = \frac{4}{5}$

Then, $\cos\alpha = \sqrt{1 - \sin^2\alpha} = \sqrt{1 - \left(\frac{4}{5}\right)^2} = \pm\frac{3}{5}$

$\cos\alpha = -\frac{3}{5}$ ($\because \alpha$ lies in II quadrant)

We know that,

$$2\sin^2\frac{\alpha}{2} = 1 - \cos\alpha = 1 + \frac{3}{5} = \frac{8}{5}$$

$$\sin^2\frac{\alpha}{2} = \frac{4}{5}$$

$$\sin\frac{\alpha}{2} = \pm\frac{2}{\sqrt{5}}$$

$\sin\frac{\alpha}{2} = -\frac{2}{\sqrt{5}}$ ($\because \frac{\alpha}{2}$ lies in III quadrant)

21. (b) $\because \quad 0 < \alpha < \frac{\pi}{6}$

and $\quad \sin\alpha + \cos\alpha = \frac{\sqrt{7}}{2}$

$$\therefore \quad 0 < \frac{\alpha}{2} < \frac{\pi}{12}$$

and $\quad \frac{2\tan\frac{\alpha}{2}}{1 + \tan^2\frac{\alpha}{2}} + \frac{1 - \tan^2\frac{\alpha}{2}}{1 + \tan^2\frac{\alpha}{2}} = \frac{\sqrt{7}}{2}$

$$\Rightarrow \quad \frac{2\tan\frac{\alpha}{2} + 1 - \tan^2\frac{\alpha}{2}}{1 + \tan^2\frac{\alpha}{2}} = \frac{\sqrt{7}}{2}$$

$$\Rightarrow \quad 4\tan\frac{\alpha}{2} + 2 - 2\tan^2\frac{\alpha}{2} = \sqrt{7} + \sqrt{7}\tan^2\frac{\alpha}{2}$$

$$\Rightarrow \quad (\sqrt{7} + 2)\tan^2\frac{\alpha}{2} - 4\tan\frac{\alpha}{2} + (\sqrt{7} - 2) = 0$$

$$\therefore \quad \tan\frac{\alpha}{2} = \frac{4 \pm \sqrt{(4)^2 - 4(\sqrt{7} + 2)(\sqrt{7} - 2)}}{2(\sqrt{7} + 2)}$$

$$= \frac{4 \pm \sqrt{16 - 12}}{2(\sqrt{7} + 2)} = \frac{4 \pm 2}{2(\sqrt{7} + 2)} = \frac{2 \pm 1}{\sqrt{7} + 2}$$

$$= \frac{3}{\sqrt{7} + 2} \text{ or } \frac{1}{\sqrt{7} + 2} = (\sqrt{7} - 2) \text{ or } \frac{\sqrt{7} - 2}{3}$$

$$\Rightarrow \quad \tan\frac{\alpha}{2} = \frac{\sqrt{7} - 2}{3} = \frac{\sqrt{k} - 2}{3} \quad \left(\because 0 < \frac{\alpha}{2} < \frac{\pi}{12}\right)$$

$$\therefore \quad k = 7$$

22. (b,c) We have, $A + B = \frac{\pi}{3}$ and $\cos A + \cos B = 1$

$$2\cos\left(\frac{A + B}{2}\right)\cos\left(\frac{A - B}{2}\right) = 1$$

$$\Rightarrow \quad 2\cos\left(\frac{\pi}{6}\right)\cos\left(\frac{A - B}{2}\right) = 1$$

$$\Rightarrow \quad \cos\left(\frac{A - B}{2}\right) = \frac{1}{\sqrt{3}}$$

$$\Rightarrow \quad \cos^2\left(\frac{A - B}{2}\right) = \frac{1}{3}$$

$$\Rightarrow \quad 2\cos^2\left(\frac{A - B}{2}\right) - 1 = \frac{2}{3} - 1 = -\frac{1}{3}$$

$\therefore \quad \cos(A - B) = -\frac{1}{3}$ [option (c)]

Now, $(\cos A - \cos B) = 2\sin\left(\frac{A + B}{2}\right)\sin\left(\frac{B - A}{2}\right)$

$$= 2\sin\frac{\pi}{6}\sqrt{1 + \cos^2\left(\frac{A - B}{2}\right)}$$

$= 2 \cdot \frac{1}{2} \cdot \sqrt{1 - \frac{1}{3}} = \sqrt{\frac{2}{3}}$ [option (b)]

23. (b,c) We have,

$$\tan x = \frac{2b}{a-c}, (a \neq c)$$

$$y = a\cos^2 x + 2b \sin x \cos x + c \sin^2 x$$

and $$z = a\sin^2 x - 2b \sin x \cos x + c\cos^2 x$$

Now, $y + z = a(\sin^2 x + \cos^2 x) + c(\sin^2 x + \cos^2 x)$

$$y + z = a + c \quad \text{[option (b)]}$$

Now,

$y - z = a(\cos^2 x - \sin^2 x) + 2b \sin 2x - c\cos 2x$

$$= a\cos 2x + 2b\sin 2x - c\cos 2x$$

$$= (a-c)\cos 2x + 2b \sin 2x$$

$$= (a-c)\left(\frac{1-\tan^2 x}{1+\tan^2 x}\right) + 2b\left(\frac{2\tan x}{1+\tan^2 x}\right)$$

$$= (a-c)\left(\frac{1-\frac{4b^2}{(a-c)^2}}{1+\frac{4b^2}{(a-c)^2}}\right) + 4b\left(\frac{\frac{2b}{a-c}}{1+\frac{4b^2}{(a-c)^2}}\right)$$

$$= (a-c)\left[\frac{(a-c)^2-4b^2}{(a-c)^2+4b^2}\right] + \frac{8b^2(a-c)}{(a-c)^2+4b^2}$$

$$= \frac{(a-c)[(a-c)^2-4b^2+8b^2]}{(a-c)^2+4b^2}$$

$$= \frac{(a-c)[(a-c)^2+4b^2]}{[(a-c)^2+4b^2]}$$

$$\therefore \quad y - z = a - c \quad \text{[option (c)]}$$

24. (a,c) $\because A > B$

and $$3\sin x - 4\sin^3 x - k = 0$$

$$\sin 3x - k = 0$$

$$\sin 3x = k$$

$\therefore$ $$\sin 3A = k \text{ and } \sin 3B = k$$

Hence, $$\sin 3A = \sin 3B$$

$$\sin 3A = \sin(\pi - 3B)$$

$$3A = \pi - 3B \quad (\because A > B)$$

$$A + B = \frac{\pi}{3}$$

$\therefore$ $$\angle C = \pi - (A+B) = \pi - \frac{\pi}{3} = \frac{2\pi}{3}$$

25. (d) We have,

$$\cos\alpha\cos 2\alpha\cos 2^2\alpha \ldots \cos 2^{n-1}\alpha = \frac{\sin 2^n\alpha}{2^n \sin\alpha}$$

Now, $\cos\frac{2\pi}{7}\cos\frac{4\pi}{7}\cos\frac{6\pi}{7} = \cos\frac{2\pi}{7}\cos 2^2\frac{\pi}{7}\cos\left(\pi - \frac{\pi}{7}\right)$

$$= -\cos\frac{\pi}{7}\cos\frac{2\pi}{7}\cos 2^2\frac{\pi}{7}$$

$$= -\frac{\sin 2^3\frac{\pi}{7}}{2^3\sin\frac{\pi}{7}}$$

$$= -\frac{\sin\frac{8\pi}{7}}{8\sin\frac{\pi}{7}} = -\frac{\sin\left(\pi + \frac{\pi}{7}\right)}{8\sin\frac{\pi}{7}}$$

$$= \frac{\sin\frac{\pi}{7}}{8\sin\frac{\pi}{7}} = \frac{1}{8}$$

26. (a) $\because \alpha = \frac{\pi}{13}$, then

$$\prod_{r=1}^{6}\cos r\alpha = \cos\frac{\pi}{13}\cos\frac{2\pi}{13}\cos\frac{3\pi}{13}\cos\frac{4\pi}{13}\cos\frac{5\pi}{13}\cos\frac{6\pi}{13}$$

$$= \left(\cos\frac{\pi}{13}\cos\frac{2\pi}{13}\cos\frac{4\pi}{13}\cos\frac{5\pi}{13}\right)\cdot\left(\cos\frac{3\pi}{13}\cos\frac{6\pi}{13}\right)$$

$$= -\left(\cos\frac{\pi}{13}\cos\frac{2\pi}{13}\cos\frac{4\pi}{13}\cos\frac{8\pi}{13}\right)\cdot\left(\cos\frac{3\pi}{13}\cos\frac{6\pi}{13}\right)$$

$$= \frac{-\sin\frac{16\pi}{13}}{2^4\sin\frac{\pi}{13}}\cdot\frac{\sin\frac{12\pi}{13}}{2^2\sin\frac{3\pi}{13}} \quad \text{(using above information)}$$

$$= -\frac{\sin\left(\pi + \frac{3\pi}{13}\right)}{2^4\sin\frac{\pi}{13}}\cdot\frac{\sin\left(\pi - \frac{\pi}{13}\right)}{2^2\sin\frac{3\pi}{13}}$$

$$= -\frac{-\sin\frac{3\pi}{13}}{16\sin\frac{\pi}{13}}\cdot\frac{\sin\frac{\pi}{13}}{4\sin\frac{3\pi}{13}} = \frac{1}{64}$$

27. (d) We have,

$$\sin\frac{\pi}{14}\sin\frac{3\pi}{14}\sin\frac{5\pi}{14}\sin\frac{7\pi}{14}\sin\frac{9\pi}{14}\sin\frac{11\pi}{14}\sin\frac{13\pi}{14}$$

$$= \sin\frac{\pi}{14}\sin\frac{3\pi}{14}\sin\frac{5\pi}{14}\times(1)$$

$$\sin\left(\pi - \frac{5\pi}{14}\right)\sin\left(\pi - \frac{3\pi}{14}\right)\sin\left(\pi - \frac{\pi}{14}\right)$$

$$= \left(\sin\frac{\pi}{14}\sin\frac{3\pi}{14}\sin\frac{5\pi}{14}\right)^2$$

$$= \left[\cos\left(\frac{\pi}{2} + \frac{\pi}{14}\right)\cos\left(\frac{\pi}{2} - \frac{3\pi}{14}\right)\cos\left(\frac{\pi}{2} - \frac{5\pi}{14}\right)\right]^2$$

$$= \left(\cos\frac{2\pi}{14}\cos\frac{4\pi}{14}\cos\frac{8\pi}{14}\right)^2$$

$$= \left(\cos\frac{\pi}{7}\cos\frac{2\pi}{7}\cos\frac{4\pi}{7}\right)^2$$

$$= \left(\frac{\sin\frac{8\pi}{7}}{2^3\sin\frac{\pi}{7}}\right)^2$$

$$\left(\because \cos A\cos 2A\cos 2^2A \ldots \cos 2^{n-1}A = \frac{\sin 2^n A}{2^n \sin A}\right)$$

$$= \left(\frac{\sin\left(\pi + \frac{\pi}{7}\right)}{8\sin\frac{\pi}{7}}\right)^2$$

$$= \left(\frac{\sin\frac{\pi}{7}}{8\sin\frac{\pi}{7}}\right)^2 = \frac{1}{64}$$

28. (d) We have,

$$128\sqrt{3}\sin\frac{\pi}{48}\cos\frac{\pi}{48}\cos\frac{\pi}{24}\cos\frac{\pi}{12}\cos\frac{\pi}{6}$$

$$= 128\sqrt{3}\sin\frac{\pi}{48}\cdot\frac{\sin 2\cdot\frac{\pi}{6}}{2^4\sin\frac{\pi}{48}}$$

$$= \frac{128\sqrt{3}}{16}\cdot\sin\frac{\pi}{3} = 8\sqrt{3}\times\frac{\sqrt{3}}{2} = 12$$

DPP-11 Trigonometric Ratios of Multiple Angles

1. We have, $(\cos A + \cos B)^2 + (\sin A - \sin B)^2$

$$= \cos^2 A + \cos^2 B + 2\cos A \cos B + \sin^2 A + \sin^2 B - 2\sin A \sin B$$
$$= (\sin^2 A + \cos^2 A) + (\sin^2 B + \cos^2 B) + 2(\cos A \cos B - \sin A \sin B)$$
$$= 1 + 1 + 2\cos(A+B) = 2[1 + \cos(A+B)]$$
$$= 2 \times 2\cos^2\left(\frac{A+B}{2}\right) = 4\cos^2\left(\frac{A+B}{2}\right)$$

$\because \quad 4\cos^2\left(\frac{A+B}{2}\right) = k^2\cos^2\left(\frac{A+B}{2}\right)$

$\therefore \quad k^2 = 4 \Rightarrow k = \pm 2$

2. $\because \tan(45° + A) + \tan(45° - A) = \mu \sec 2A$

$$\Rightarrow \frac{1+\tan A}{1-\tan A} + \frac{1-\tan A}{1+\tan A} = \mu \sec 2A$$
$$\Rightarrow \frac{(1+\tan A)^2 + (1-\tan A)^2}{(1+\tan A)(1-\tan A)} = \mu \sec 2A$$
$$\Rightarrow \frac{2(1+\tan^2 A)}{1-\tan^2 A} = \mu \sec 2A$$
$$\Rightarrow \frac{2}{\left(\frac{1-\tan^2 A}{1+\tan^2 A}\right)} = \mu \sec 2A$$
$$\Rightarrow \frac{2}{\cos 2A} = \mu \sec 2A \quad \left(\because \cos 2\theta = \frac{1-\tan^2\theta}{1+\tan^2\theta}\right)$$
$$2\sec 2A = \mu \sec 2A$$
$\therefore \quad \mu = 2$ **Hence proved.**

3. $$\text{LHS} = \frac{\tan^3\theta}{1+\tan^2\theta} + \frac{\cot^3\theta}{1+\cot^2\theta} = \frac{\tan^3\theta}{\sec^2\theta} + \frac{\cot^3\theta}{\operatorname{cosec}^2\theta}$$
$$= \frac{\sin^3\theta}{\cos^3\theta}\cdot\frac{\cos^2\theta}{1} + \frac{\cos^3\theta}{\sin^3\theta}\cdot\frac{\sin^2\theta}{1}$$
$$= \frac{\sin^3\theta}{\cos\theta} + \frac{\cos^3\theta}{\sin\theta} = \frac{\sin^4\theta + \cos^4\theta}{\sin\theta\cos\theta}$$
$$= \frac{(\sin^2\theta + \cos^2\theta)^2 - 2\sin^2\theta\cos^2\theta}{\sin\theta\cos\theta}$$
$$= \frac{2[1 - 2\sin^2\theta\cos^2\theta]}{2\sin\theta\cos\theta} = \frac{2 - (2\sin\theta\cos\theta)^2}{2\sin\theta\cos\theta}$$
$$= \frac{2 - (\sin 2\theta)^2}{\sin 2\theta} = \frac{2-k^2}{k} = \text{RHS} \quad (\because \sin 2\theta = k)$$
Hence proved.

4. $\because \quad (\sin A + \cos A)^2 = \sin^2 A + \cos^2 A + 2\sin A\cos A$
$$= 1 + \sin 2A$$
$\therefore \quad 1 - m\sin nA = 1 + \sin 2A$

$\Rightarrow \quad m = -1$ and $n = 2$

Now, $\quad n - m = 2 + 1 = 3$

5. We have, $\tan 3A = \tan(2A + A)$
$$= \frac{\tan 2A + \tan A}{1 - \tan 2A\cdot\tan A} = \frac{\frac{2\tan A}{1-\tan^2 A} + \tan A}{1 - \frac{2\tan A}{1-\tan^2 A}\cdot\tan A}$$
$$= \frac{2\tan A + \tan A - \tan^3 A}{1 - \tan^2 A - 2\tan^2 A}$$
$$\Rightarrow \tan 3A = \frac{3\tan A - \tan^3 A}{1 - 3\tan^2 A}$$

But $\quad \tan 3A = \frac{3\tan A + k\tan^3 A}{1 - 3\tan^2 A}$

$\therefore \quad k = -1$

6. $\because \cot 75° = \cot(45° + 30°)$
$$= \frac{\cot 45°\cot 30° - 1}{\cot 45° + \cot 30°} = \frac{\sqrt{3}-1}{1+\sqrt{3}} = \frac{(\sqrt{3}-1)(\sqrt{3}-1)}{3-1}$$
$$= \frac{3 + 1 - 2\sqrt{3}}{2} = 2 - \sqrt{3}$$

But $\quad \cot 75° = 2 - \sqrt{\lambda}$

$\therefore \quad \lambda = 3$

7. True, using $\quad 2\sin^2\theta = 1 - \cos 2\theta$,

Put $\theta = \left(22\frac{1}{2}\right)^\circ$
$$\Rightarrow 2\sin^2\left(22\frac{1}{2}\right)^\circ = 1 - \cos 45°$$
$$\Rightarrow \sin^2\left(22\frac{1}{2}\right)^\circ = \frac{\sqrt{2}-1}{2\sqrt{2}} = \frac{2-\sqrt{2}}{4}$$
$$\therefore \sin\left(22\frac{1}{2}\right)^\circ = \frac{\sqrt{2-\sqrt{2}}}{2}$$

8. True, $\quad \tan A + 2\cot 2A$
$$\Rightarrow \tan A + \frac{2(1-\tan^2 A)}{2\tan A}$$
$$\Rightarrow \frac{\tan^2 A + 1 - \tan^2 A}{\tan A} = \frac{1}{\tan A} = \cot A$$

9. False, as $\tan\theta$ does not exists for $(2n+1)\frac{\pi}{2}$

$\therefore \tan A \pm \tan B = \frac{\sin(A \pm B)}{\cos A\cos B}$, is not true, for $A = n\pi + \frac{\pi}{2}$

10. (a) $\because (1 + \sec 2\theta)(1 + \sec 4\theta)(1 + \sec 8\theta)\ldots(1 + \sec 2^n\theta)$
$$= \left(\frac{1+\cos 2\theta}{\cos 2\theta}\right)\left(\frac{1+\cos 4\theta}{\cos 4\theta}\right)\left(\frac{1+\cos 8\theta}{\cos 8\theta}\right)\ldots\left(\frac{1+\cos 2^n\theta}{\cos 2^n\theta}\right)$$
$$= \frac{2\cos^2\theta\cdot 2\cos^2 2\theta\cdot 2\cos^2 4\theta\ldots 2\cos^2 2^{n-1}\theta}{\cos 2\theta\cdot\cos 4\theta\cdot\cos 8\theta\ldots\cos 2^n\theta}$$
$$= \frac{2^n\cos^2\theta\cdot\cos 2\theta\cdot\cos 4\theta\cdot\cos 8\theta\ldots\cos 2^{n-1}\theta}{\cos 2^n\theta}$$
$$= \frac{2^n\cos\theta}{\cos 2^n\theta}(\cos\theta\cos 2\theta\cos 4\theta\ldots\cos 2^{n-1}\theta)$$
$$= \frac{2^n\cos\theta}{\cos 2^n\theta}\cdot\frac{\sin 2^n\theta}{2^n\sin\theta} = \tan 2^n\theta\cot\theta$$

11. (c) $\left(1 + \cos\frac{\pi}{8}\right)\left(1 + \cos\frac{3\pi}{8}\right)\left(1 + \cos\frac{5\pi}{8}\right)\left(1 + \cos\frac{7\pi}{8}\right)$
$$= \left(1 + \cos\frac{\pi}{8}\right)\left(1 + \cos\frac{3\pi}{8}\right)\left[1 + \cos\left(\pi - \frac{3\pi}{8}\right)\right]\left[1 + \cos\left(\pi - \frac{\pi}{8}\right)\right]$$
$$= \left(1 + \cos\frac{\pi}{8}\right)\left(1 + \cos\frac{3\pi}{8}\right)\left(1 - \cos\frac{3\pi}{8}\right)\left(1 - \cos\frac{\pi}{8}\right)$$
$$= \left(1 - \cos^2\frac{\pi}{8}\right)\left(1 - \cos^2\frac{3\pi}{8}\right) = \sin^2\frac{\pi}{8}\sin^2\frac{3\pi}{8}$$
$$= \frac{1}{4}\cdot\left(2\sin\frac{3\pi}{8}\sin\frac{\pi}{8}\right)^2 = \frac{1}{4}\left[\cos\left(\frac{3\pi}{8} - \frac{\pi}{8}\right) - \cos\left(\frac{3\pi}{8} + \frac{\pi}{8}\right)\right]^2$$
$$= \frac{1}{4}\left[\cos\frac{\pi}{4} - \cos\frac{\pi}{2}\right]^2 = \frac{1}{4}\times\left[\frac{1}{\sqrt{2}} - 0\right]^2 = \frac{1}{8}$$

12. (b) $\because \tan(\theta + 50°)\tan\theta\tan(\theta - 50°) = \tan(\theta + 100°)$

$$\Rightarrow \frac{\tan(\theta + 100°)}{\tan(\theta - 50°)} = \tan(\theta + 50°)\tan\theta$$

$$\Rightarrow \frac{\sin(\theta + 100°)}{\cos(\theta + 100°)} \cdot \frac{\cos(\theta - 50°)}{\sin(\theta - 50°)} = \frac{\sin(\theta + 50°)}{\cos(\theta + 50°)} \cdot \frac{\sin\theta}{\cos\theta}$$

On applying componendo and dividendo, we get

$$\Rightarrow \frac{\sin(\theta + 100°)\cos(\theta - 50°) + \cos(\theta + 100°)\sin(\theta - 50°)}{\sin(\theta + 100°)\cos(\theta - 50°) - \cos(\theta + 100°)\sin(\theta - 50°)}$$

$$= \frac{\sin\theta\sin(\theta + 50°) + \cos(\theta + 50°)\cos\theta}{\sin\theta\sin(\theta + 50°) - \cos(\theta + 50°)\cos\theta}$$

$$\Rightarrow \frac{\sin(\theta + 100° + \theta - 50°)}{\sin(\theta + 100° - \theta + 50°)} = \frac{\cos(\theta + 50° - \theta)}{-\cos(\theta + 50° + \theta)}$$

$$\Rightarrow \frac{\sin(2\theta + 50°)}{\sin 150°} = -\frac{\cos 50°}{\cos(2\theta + 50°)}$$

$$\Rightarrow \sin(2\theta + 50°)\cos(2\theta + 50°) = -\sin 150°\cos 50°$$

$$\Rightarrow 2\sin(2\theta + 50°)\cos(2\theta + 50°) = -2\sin 150°\cos 50°$$

$$\Rightarrow \sin(4\theta + 100°) = -2\sin(180° - 30°)\cos 50°$$

$$= -2\sin 30°\cos 50°$$

$$= -2 \times \frac{1}{2}\cos 50°$$

$$\Rightarrow \sin(4\theta + 100°) = -\cos 50° = \sin(270° - 50°)$$

$$\sin(4\theta + 100°) = \sin 220°$$

$$\therefore \quad 4\theta + 100° = 220°$$

$$4\theta = 120°$$

$$\Rightarrow \theta = 30°$$

13. (c) Given, $\tan A + 2\tan 2A + 4\tan 4A + 8\cot 8A = k\cot A$

$$\Rightarrow \tan A + 2\tan 2A + 4(\tan 4A + 2\cot 8A) = k\cot A$$

$$\Rightarrow \tan A + 2\tan 2A + 4\cot 4A = k\cot A$$

$(\because \tan\theta + 2\cot 2\theta = \cot\theta)$

$$\Rightarrow \tan A + 2(\tan 2A + 2\cot 4A) = k\cot A$$

$$\tan A + 2\cot 2A = k\cot A$$

$$\Rightarrow \cot A = k\cot A$$

$$\therefore \quad k = 1$$

14. (b) $\because \cos\theta + \cos(120° + \theta) + \cos(240° + \theta)$

$$= \cos\theta + 2\cos(180° + \theta)\cos 60°$$

$$= \cos\theta + 2 \times (-\cos\theta) \cdot \frac{1}{2}$$

$$= \cos\theta - \cos\theta = 0$$

We know that, if $A + B + C = 0$, then

$$A^3 + B^3 + C^3 = 3ABC$$

Now, $\cos^3\theta + \cos^3(120° + \theta) + \cos^3(240° + \theta) = p\cos q\theta$

$$\Rightarrow 3\cos\theta\cos(120° + \theta)\cos(240° + \theta) = p\cos q\theta$$

$$\Rightarrow 3\cos\theta\cos[180° - (60° - \theta)]\cos[180° + (60° + \theta)] = p\cos q\theta$$

$$\Rightarrow 3\cos\theta\cos(60° - \theta)\cos(60° + \theta) = p\cos q\theta$$

$$\Rightarrow 3 \cdot \frac{1}{4} \cdot \cos 3\theta = p\cos q\theta$$

$$\left[\because \cos(60° - A)\cos A\cos(60° + A) = \frac{1}{4}\cos 3A\right]$$

$$\Rightarrow \frac{3}{4}\cos 3\theta = p\cos q\theta$$

$\therefore$ $p = \frac{3}{4}$ and $q = 3$, then

$$8p - q = 8 \times \frac{3}{4} - 3 = 3$$

$$\therefore \quad 8p - q = 3$$

15. (b) $\because \tan\theta + \tan\left(\frac{\pi}{3} + \theta\right) - \tan\left(\frac{\pi}{3} - \theta\right) = k\tan 3\theta$

$$\Rightarrow \tan\theta + \frac{\sqrt{3} + \tan\theta}{1 - \sqrt{3}\tan\theta} - \frac{\sqrt{3} - \tan\theta}{1 + \sqrt{3}\tan\theta} = k\tan 3\theta$$

$$\Rightarrow \tan\theta + \frac{\begin{bmatrix}\sqrt{3} + 3\tan\theta + \tan\theta + \sqrt{3}\tan^2\theta \\ -(\sqrt{3} - \tan\theta - 3\tan\theta + \sqrt{3}\tan^2\theta)\end{bmatrix}}{(1 - \sqrt{3}\tan\theta)(1 + \sqrt{3}\tan\theta)} = k\tan 3\theta$$

$$\Rightarrow \tan\theta + \frac{\begin{bmatrix}\sqrt{3} + 4\tan\theta + \sqrt{3}\tan^2\theta - \sqrt{3} \\ + 4\tan\theta - \sqrt{3}\tan^2\theta\end{bmatrix}}{1 - 3\tan^2\theta} = k\tan 3\theta$$

$$\Rightarrow \tan\theta + \frac{8\tan\theta}{1 - 3\tan^2\theta} = k\tan 3\theta$$

$$\Rightarrow \frac{\tan\theta - 3\tan^3\theta + 8\tan\theta}{1 - 3\tan^2\theta} = k\tan 3\theta$$

$$\Rightarrow \frac{9\tan\theta - 3\tan^3\theta}{1 - 3\tan^2\theta} = k\tan 3\theta$$

$$\Rightarrow 3\left(\frac{3\tan\theta - \tan^3\theta}{1 - 3\tan^2\theta}\right) = k\tan 3\theta$$

$$\Rightarrow 3\tan 3\theta = k\tan 3\theta$$

$$\therefore \quad k = 3$$

16. (b) $\because m^2\cos\frac{2\pi}{15}\cos\frac{4\pi}{15}\cos\frac{8\pi}{15}\cos\frac{14\pi}{15} = n^2$

$$\Rightarrow \cos\frac{2\pi}{15}\cos\frac{4\pi}{15}\cos\frac{8\pi}{15}\cos\left(\pi - \frac{\pi}{15}\right) = \frac{n^2}{m^2}$$

$$\Rightarrow -\cos\frac{\pi}{15}\cos\frac{2\pi}{15}\cos\frac{4\pi}{15}\cos\frac{8\pi}{15} = \frac{n^2}{m^2}$$

$$\Rightarrow -\frac{\sin\frac{16\pi}{15}}{2^4\sin\frac{\pi}{15}} = \frac{n^2}{m^2}$$

$$\Rightarrow -\frac{\sin\left(\pi + \frac{\pi}{15}\right)}{2^4\sin\frac{\pi}{15}} = \frac{n^2}{m^2}$$

$$\Rightarrow \frac{\sin\frac{\pi}{15}}{2^4\sin\frac{\pi}{15}} = \frac{n^2}{m^2} \Rightarrow \frac{1}{16} = \frac{n^2}{m^2}$$

$$\Rightarrow \frac{m^2}{n^2} = 16 \Rightarrow \frac{m^2}{n^2} - 1 = 16 - 1$$

$$\therefore \quad \frac{m^2 - n^2}{n^2} = 15$$

17. (b) $\because \cos\theta = \frac{1}{2}\left(a + \frac{1}{a}\right)$ ...(i)

On cubing both sides, we get

$$8\cos^3\theta = a^3 + \frac{1}{a^3} + 3\left(a + \frac{1}{a}\right)$$

$$8\cos^3\theta = a^3 + \frac{1}{a^3} + 3 \times 2\cos\theta \quad \text{[from Eq. (i)]}$$

$$a^3 + \frac{1}{a^3} = 8\cos^3\theta - 6\cos\theta$$

$$a^3 + \frac{1}{a^3} = 2(4\cos^3\theta - 3\cos\theta)$$

$$\frac{1}{2}\left(a^3 + \frac{1}{a^3}\right) = \cos 3\theta \quad \text{...(ii)}$$

$$\lambda\left(a^3 + \frac{1}{a^3}\right) = \cos 3\theta \quad \text{(given) ...(iii)}$$

From Eqs. (ii) and (iii), we get

$$\lambda = \frac{1}{2}$$

18. (a) $\because (2^n + 1)\,\theta = \pi$

$$2^n\theta = \pi - \theta \qquad \text{...(i)}$$

Now, $2^n \cos\theta \cos 2\theta \cos 2^2\theta \ldots \cos 2^{n-1}\theta$

$$= 2^n \cdot \frac{\sin 2^n\theta}{2^n \sin\theta} = \frac{\sin 2^n\theta}{\sin\theta} = \frac{\sin(\pi - \theta)}{\sin\theta} \qquad \text{[from Eq. (i)]}$$

$$= \frac{\sin\theta}{\sin\theta} = 1$$

19. (a) We have, $\dfrac{\cos 5\theta + \cos 4\theta}{-1 + 2\cos 3\theta}$

$$= \frac{2\cos\frac{9\theta}{2}\cos\frac{\theta}{2}}{-1 + 2\cos 3\theta} = \frac{2\cos\frac{9\theta}{2}\cos\frac{3\theta}{2}\cos\frac{\theta}{2}}{-\cos\frac{3\theta}{2} + 2\cos 3\theta \cdot \cos\frac{3\theta}{2}}$$

$$\left(\text{multiplying numerator and denominator by } \cos\frac{3\theta}{2}\right)$$

$$= \frac{2\cos\frac{9\theta}{2}\cos\frac{3\theta}{2}\cos\frac{\theta}{2}}{-\cos\frac{3\theta}{2} + \cos\frac{9\theta}{2} + \cos\frac{3\theta}{2}}$$

$$= \frac{2\cos\frac{9\theta}{2}\cos\frac{3\theta}{2}\cos\frac{\theta}{2}}{\cos\frac{9\theta}{2}} = 2\cos\frac{3\theta}{2}\cos\frac{\theta}{2}$$

$$= \cos\left(\frac{3\theta}{2} + \frac{\theta}{2}\right) + \cos\left(\frac{3\theta}{2} - \frac{\theta}{2}\right)$$

$$= \cos 2\theta + \cos\theta$$

20. (a,c) We have, $\dfrac{\sec 8\theta - 1}{\sec 4\theta - 1} = \dfrac{1 - \cos 8\theta}{\cos 8\theta} \times \dfrac{\cos 4\theta}{1 - \cos 4\theta}$

$$= \frac{2\sin^2 4\theta \cos 4\theta}{\cos 8\theta \cdot 2\sin^2 2\theta}$$

$$= \frac{2\sin 4\theta \cos 4\theta}{\cos 8\theta} \times \frac{\sin 4\theta}{2\sin^2 2\theta}$$

$$= \frac{\sin 8\theta}{\cos 8\theta} \cdot \frac{2\sin 2\theta \cos 2\theta}{2\sin^2 2\theta}$$

$$= \tan 8\theta \cot 2\theta = \frac{\tan 8\theta}{\tan 2\theta}$$

Now, $\dfrac{\sec 8\theta - 1}{\sec 4\theta - 1} = \dfrac{\tan p\theta}{\tan q\theta}$ (given)

$\Rightarrow \dfrac{\tan 8\theta}{\tan 2\theta} = \dfrac{\tan p\theta}{\tan q\theta}$

$\therefore$ $p = 8$ and $q = 2$

$\Rightarrow$ $4q = 8$

$\therefore$ $p = 4q$ i.e. $2p = 8q$ [option (c)]

$p : q = 8 : 2 = 4 : 1$ [option (a)]

21. (a,b,c) We have, $\dfrac{\tan 3\theta}{\tan\theta} = \dfrac{k}{1}$

$\Rightarrow \dfrac{\tan 3\theta - \tan\theta}{\tan\theta} = \dfrac{k - 1}{1}$ (applying dividendo)

$\Rightarrow \dfrac{\sin(3\theta - \theta)}{\dfrac{\sin\theta}{\cos\theta} \times \cos 3\theta \cos\theta} = k - 1$

$$\left(\because \tan A - \tan B = \frac{\sin(A - B)}{\cos A \cos B}\right)$$

$\Rightarrow \dfrac{\sin 2\theta}{\sin\theta \cos 3\theta} = k - 1$

$\Rightarrow \dfrac{2\sin\theta\cos\theta}{\sin\theta\cos 3\theta} = k - 1$

$\therefore \dfrac{\cos\theta}{\cos 3\theta} = \dfrac{k - 1}{2}$ ...(i)

Now, again $\dfrac{\tan 3\theta}{\tan\theta} = k$

$\Rightarrow \dfrac{\sin 3\theta}{\cos 3\theta} \times \dfrac{\cos\theta}{\sin\theta} = k$

$\Rightarrow \dfrac{\sin 3\theta}{\sin\theta} \times \dfrac{\cos\theta}{\cos 3\theta} = k$

$\Rightarrow \dfrac{\sin 3\theta}{\sin\theta} = \dfrac{2k}{k - 1}$ [from Eq. (i)][option (a)]

Now, $\dfrac{3\sin\theta - 4\sin^3\theta}{\sin\theta} = \dfrac{2k}{k - 1}$

$\Rightarrow 3 - 4\sin^2\theta = \dfrac{2k}{k - 1}$ $(\because \sin\theta \neq 0, 1)$

$\Rightarrow 4\sin^2\theta = \dfrac{k - 3}{k - 1}$ ...(ii)

We know that, $0 < \sin^2\theta < 1$ $(\because \sin\theta \neq 0, 1)$

$0 < 4\sin^2\theta < 4$

$\therefore 0 < \dfrac{k - 3}{k - 1} < 4$ [from Eq. (ii)]

If $\dfrac{k - 3}{k - 1} > 0$

$\Rightarrow (k - 3)(k - 1) > 0$

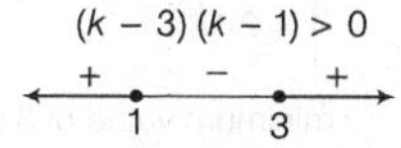

$\therefore k \in (-\infty, 1) \cup (3, \infty)$ ...(iii)

and $\dfrac{k - 3}{k - 1} < 4$

$\Rightarrow \dfrac{k - 3}{k - 1} - 4 < 0$

$\Rightarrow \dfrac{3k - 1}{k - 1} > 0$

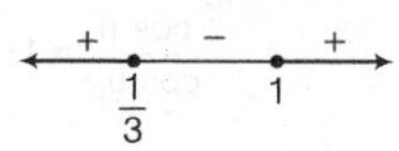

$\therefore k \in \left(-\infty, \dfrac{1}{3}\right) \cup (1, \infty)$ ...(iv)

From Eqs. (iii) and (iv), we get

$k < \dfrac{1}{3}$ and $k > 3$

22. (c,d) We have,

$$\sum_{m=1}^{6} \operatorname{cosec}\left(\theta + \frac{(m-1)\pi}{4}\right)\operatorname{cosec}\left(\theta + \frac{m\pi}{4}\right) = 4\sqrt{2}$$

$$\Rightarrow \sum_{m=1}^{6} \frac{1}{\sin\left(\theta + \frac{(m-1)\pi}{4}\right)\sin\left(\theta + \frac{m\pi}{4}\right)} = 4\sqrt{2}$$

$$\Rightarrow \sum_{m=1}^{6} \frac{\sin\left[\left(\theta + \frac{m\pi}{4}\right) - \left(\theta + \frac{(m-1)\pi}{4}\right)\right]}{\sin\left(\theta + \frac{(m-1)\pi}{4}\right)\sin\left(\theta + \frac{m\pi}{4}\right)} \cdot \frac{1}{\sin\frac{\pi}{4}} = 4\sqrt{2}$$

$$\Rightarrow \sum_{m=1}^{6}\left[\cot\left(\theta + \frac{(m-1)\pi}{4}\right) - \cot\left(\theta + \frac{m\pi}{4}\right)\right]\sqrt{2} = 4\sqrt{2}$$

$$\Rightarrow \left[\cot\theta - \cot\left(\theta + \frac{\pi}{4}\right)\right] + \left[\cot\left(\theta + \frac{\pi}{4}\right) - \cot\left(\theta + \frac{2\pi}{4}\right)\right]$$

$$+ \ldots + \left[\cot\left(\theta + \frac{5\pi}{4}\right) - \cot\left(\theta + \frac{6\pi}{4}\right)\right] = 4$$

$$\Rightarrow \cot\theta - \cot\left(\theta + \frac{3\pi}{2}\right) = 4$$

$$\Rightarrow \cot\theta + \tan\theta = 4$$

$$\Rightarrow \quad \frac{1}{\tan\theta} + \tan\theta = 4$$

$$\Rightarrow \quad \tan^2\theta - 4\tan\theta + 1 = 0$$

$$\Rightarrow \quad (\tan\theta - 2)^2 - 3 = 0$$

$$\Rightarrow \quad \tan\theta = 2 - \sqrt{3}, 2 + \sqrt{3}$$

$$\Rightarrow \quad \tan\theta = \tan\frac{\pi}{12}, \tan\frac{5\pi}{12} \qquad \left(\because 0 < \theta < \frac{\pi}{2}\right)$$

$$\therefore \quad \theta = \frac{\pi}{12}, \frac{5\pi}{12}$$

23. (2) $F(\theta) = \dfrac{1}{\sin^2\theta + 3\sin\theta\cos\theta + 5\cos^2\theta}$

$$= \frac{1}{(\sin^2\theta + \cos^2\theta) + \frac{3}{2}\sin 2\theta + 4\cos^2\theta}$$

$$= \frac{1}{1 + \frac{3}{2}\sin 2\theta + 4\cdot\frac{1+\cos 2\theta}{2}}$$

$$= \frac{2}{2 + 3\sin 2\theta + 4 + 4\cos 2\theta}$$

$$= \frac{2}{3\sin 2\theta + 4\cos 2\theta + 6}$$

$$\therefore \quad F(\theta)_{max} = \frac{2}{6-5} = \frac{2}{1} = 2$$

($\because$ minimum value of $3\sin 2\theta + 4\cos 2\theta$ is -5)

24. (1) $\because \dfrac{\cos\theta_1}{\cos\theta_2} + \dfrac{\sin\theta_1}{\sin\theta_2} = 1$

and $\quad \dfrac{\cos\theta_0}{\cos\theta_2} + \dfrac{\sin\theta_0}{\sin\theta_2} = 1$

$\therefore$ θ_1 and θ_0 be the roots of

$$\frac{\cos\theta}{\cos\theta_2} + \frac{\sin\theta}{\sin\theta_2} = 1 \qquad \text{...(i)}$$

$$\Rightarrow \quad \frac{\cos\theta}{\cos\theta_2} = 1 - \frac{\sin\theta}{\sin\theta_2}$$

On squaring both sides, we get

$$\Rightarrow \quad \frac{\cos^2\theta}{\cos^2\theta_2} = 1 + \frac{\sin^2\theta}{\sin^2\theta_2} - \frac{2\sin\theta}{\sin\theta_2}$$

$$\Rightarrow \quad \frac{1}{\cos^2\theta_2} - \frac{\sin^2\theta}{\cos^2\theta_2} = 1 + \frac{\sin^2\theta}{\sin^2\theta_2} - \frac{2\sin\theta}{\sin\theta_2}$$

$$\Rightarrow \left(\frac{1}{\cos^2\theta_2} + \frac{1}{\sin^2\theta_2}\right)\sin^2\theta - \left(\frac{2}{\sin\theta_2}\right)\sin\theta + 1 - \frac{1}{\cos^2\theta_2} = 0$$

$$\Rightarrow \quad \left(\frac{1}{\sin^2\theta_2\cos^2\theta_2}\right)\sin^2\theta - \left(\frac{2}{\sin\theta_2}\right)\sin\theta - \frac{\sin^2\theta_2}{\cos^2\theta_2} = 0$$

$\because \theta_1$ and θ_0 be the roots of above equation, then product of roots

$$\sin\theta_1\sin\theta_0 = -\frac{\dfrac{\sin^2\theta_2}{\cos^2\theta_2}}{\dfrac{1}{\sin^2\theta_2\cos^2\theta_2}} = -\frac{\sin^2\theta_2}{\cos^2\theta_2} \times \frac{\sin^2\theta_2\cos^2\theta_2}{1}$$

$$= -\sin^2\theta_2\sin^2\theta_2$$

$$\therefore \quad \frac{\sin\theta_1\sin\theta_0}{\sin^2\theta_2} = -\sin^2\theta_2 \qquad \text{...(ii)}$$

Similarly, again taking Eq. (i),

$$\frac{\cos\theta}{\cos\theta_2} + \frac{\sin\theta}{\sin\theta_2} = 1$$

$$\Rightarrow \quad \frac{\sin\theta}{\sin\theta_2} = 1 - \frac{\cos\theta}{\cos\theta_2}$$

$$\therefore \quad \frac{\cos\theta_1\cos\theta_0}{\cos^2\theta_2} = -\cos^2\theta_2 \qquad \text{...(iii)}$$

On adding Eqs. (ii) and (iii), we get

$$\frac{\cos\theta_1\cos\theta_0}{\cos^2\theta_2} + \frac{\sin\theta_1\sin\theta_0}{\sin^2\theta_2} = -(\sin^2\theta_2 + \cos^2\theta_2)$$

$$= -1$$

$$\therefore \quad \left|\frac{\cos\theta_1\cdot\cos\theta_0}{\cos^2\theta_2} + \frac{\sin\theta_1\cdot\sin\theta_0}{\sin^2\theta_2}\right| = 1$$

DPP-12 T-Ratios of Sub-Multiple Angles

1. True, we have, $\tan(60° - A)\tan A\tan(60° + A)$

$$= \frac{\sin(60° - A)\sin A\sin(60° + A)}{\cos(60° - A)\cos A\cos(60° + A)}$$

$$= \frac{\sin A(\sin^2 60° - \sin^2 A)}{\cos A(\cos^2 60° - \sin^2 A)}$$

$$\left[\begin{aligned}&\because \sin^2 A - \sin^2 B = \sin(A+B)\sin(A-B)\\&\cos^2 A - \sin^2 B = \cos(A+B)\cos(A-B)\end{aligned}\right]$$

$$= \frac{\sin A\left(\frac{3}{4} - \sin^2 A\right)}{\cos A\left(\frac{1}{4} - \sin^2 A\right)} = \frac{\frac{1}{4}(3\sin A - 4\sin^3 A)}{\frac{1}{4}\cos A(-3 + 4\cos^2 A)}$$

$$= \frac{3\sin A - 4\sin^3 A}{4\cos^3 A - 3\cos A} = \frac{\sin 3A}{\cos 3A} = \tan 3A$$

$$\left[\begin{aligned}&\because \sin 3A = 3\sin A - 4\sin^3 A\\&\cos 3A = 4\cos^3 A - 3\cos A\end{aligned}\right]$$

2. False, because $-1 \le \sin\theta \le 1$

3. True, $\because \cos 36° = \frac{1}{4}(\sqrt{5} + 1)$

$$\Rightarrow \quad \cos(90° - 54°) = \frac{1}{4}(\sqrt{5} + 1)$$

$$\therefore \quad \sin 54° = \frac{\sqrt{5} + 1}{4}$$

4. Let $A = 18°$

$$\Rightarrow \quad 5A = 90°$$

$$\Rightarrow \quad 3A = 90° - 2A$$

$$\Rightarrow \quad \cos 3A = \cos(90° - 2A)$$

$$\Rightarrow \quad 4\cos^3 A - 3\cos A = \sin 2A$$

$$\Rightarrow \quad \cos A(4\cos^2 A - 3) = 2\sin A\cos A$$

$$\Rightarrow \quad 4 - 4\sin^2 A - 3 = 2\sin A \qquad (\because \cos A \ne 0)$$

$$\Rightarrow \quad 4\sin^2 A + 2\sin A - 1 = 0$$

$$\Rightarrow \quad \sin A = \frac{-2 \pm \sqrt{4 - 4 \times 4 \times (-1)}}{2 \times 4}$$

$$\Rightarrow \quad \sin A = \frac{-2 \pm 2\sqrt{5}}{8}$$

$$= -\left(\frac{\sqrt{5} + 1}{4}\right) \text{ or } \frac{\sqrt{5} - 1}{4}$$

$$\Rightarrow \quad \sin A = \frac{\sqrt{5} - 1}{4}$$

$$\left[-\left(\frac{\sqrt{5} + 1}{4}\right) \text{is not possible because } \sin\theta > 0, \theta \in \left[0, \frac{\pi}{2}\right]\right]$$

$$\therefore \quad \sin 18° = \frac{\sqrt{5} - 1}{4}$$

Hence proved.

5. LHS $= \dfrac{1}{\tan 3A - \tan A} - \dfrac{1}{\cot 3A - \cot A}$

$= \dfrac{1}{\tan 3A - \tan A} - \dfrac{\tan 3A \tan A}{\tan A - \tan 3A}$

$= \dfrac{1 + \tan 3A \tan A}{\tan 3A - \tan A} = \dfrac{1}{\dfrac{\tan 3A - \tan A}{1 + \tan 3A \tan A}}$

$= \dfrac{1}{\tan (3A - A)} = \dfrac{1}{\tan 2A} = \cot 2A =$ RHS

Hence proved.

6. LHS $= (\sin 55° - \sin 19°) + (\sin 53° - \sin 17°)$

$= 2 \cos \left(\dfrac{55° + 19°}{2}\right) \sin \left(\dfrac{55° - 19°}{2}\right) + 2 \cos \left(\dfrac{53° + 17°}{2}\right) \sin \left(\dfrac{53° - 17°}{2}\right)$

$= 2 \cos 37° \sin 18° + 2 \cos 35° \sin 18°$

$= 2 \sin 18° (\cos 37° + \cos 35°) = 2 \sin 18° \cdot 2 \cos 36° \cos 1°$

$= 4 \cdot \dfrac{\sqrt{5} - 1}{4} \cdot \dfrac{\sqrt{5} + 1}{4} \cdot \cos 1°$

$= 4 \cdot \dfrac{(5 - 1)}{4 \cdot 4} \cdot \cos 1° = \cos 1° =$ RHS **Hence proved.**

7. (a) $\because \sin 12° \sin 48° \sin 54° = \dfrac{1}{2} (2 \sin 48° \sin 12°) \sin 54°$

$= \dfrac{1}{2} [\cos (48° - 12°) - \cos (48° + 12°)] \sin (90° - 36°)$

$= \dfrac{1}{2} (\cos 36° - \cos 60°) \cos 36°$

$= \dfrac{1}{2}\left(\dfrac{\sqrt{5} + 1}{4} - \dfrac{1}{2}\right) \dfrac{\sqrt{5} + 1}{4} = \dfrac{1}{2} \cdot \dfrac{\sqrt{5} - 1}{4} \cdot \dfrac{\sqrt{5} + 1}{4}$

$= \dfrac{1}{2} \cdot \dfrac{4}{4 \times 4} = \dfrac{1}{8}$

8. (b) $\sin^2 24° - \sin^2 6° = \sin (24° + 6°) \sin (24° - 6°)$

$= \sin 30° \sin 18°$

$= \dfrac{1}{2} \sin 18° = k \sin 18°$ (given)

$\therefore \quad k = \dfrac{1}{2}$

9. (c) $\because \cos A + \cos B + \cos C = 0$, then

$\cos^3 A + \cos^3 B + \cos^3 C = 3 \cos A \cos B \cos C$ ...(i)

(using if $a + b + c = 0$, then $a^3 + b^3 + c^3 = 3abc$)

Now, $\lambda \cos A \cos B \cos C = \cos 3A + \cos 3B + \cos 3C$

$\lambda \cos A \cos B \cos C = 4 \cos^3 A - 3 \cos A + 4 \cos^3 B - 3 \cos B + 4 \cos^3 C - 3 \cos C$

$= 4 (\cos^3 A + \cos^3 B + \cos^3 C) - 3(\cos A + \cos B + \cos C)$

$= 4 \cdot 3 \cos A \cos B \cos C - 0$ [from Eq. (i)]

$\Rightarrow \quad \lambda \cos A \cos B \cos C = 12 \cos A \cos B \cos C$

$\therefore \quad \lambda = 12$

10. (b) We have, $\tan 6° \tan 42° \tan 66° \tan 78°$

$= \dfrac{(2 \sin 66° \sin 6°)(2 \sin 78° \sin 42°)}{(2 \cos 66° \cos 6°)(2 \cos 78° \cos 42°)}$

$= \dfrac{(\cos 60° - \cos 72°)(\cos 36° - \cos 120°)}{(\cos 72° + \cos 60°)(\cos 120° + \cos 36°)}$

$= \dfrac{\left(\dfrac{1}{2} - \dfrac{\sqrt{5} - 1}{4}\right)\left(\dfrac{\sqrt{5} + 1}{4} + \dfrac{1}{2}\right)}{\left(\dfrac{\sqrt{5} - 1}{4} + \dfrac{1}{2}\right)\left(-\dfrac{1}{2} + \dfrac{\sqrt{5} + 1}{4}\right)} = \dfrac{(3 - \sqrt{5})(3 + \sqrt{5})}{(\sqrt{5} + 1)(\sqrt{5} - 1)} = \dfrac{9 - 5}{5 - 1}$

$= 1$

11. (c) $\because \sin A + \sin (120° + A) + \sin (240° + A)$

$= \sin A + 2 \sin \left(\dfrac{360° + 2A}{2}\right) \cos \left(-\dfrac{120°}{2}\right)$

$= \sin A + 2 \sin (180° + A) \cos 60°$

$= \sin A - 2 \sin A \cdot \dfrac{1}{2} = 0$

We know that, if $a + b + c = 0$, then

$a^3 + b^3 + c^3 = 3abc$

$\therefore \sin^3 A + \sin^3(120° + A) + \sin^3(240° + A)$

$= 3 \sin A \sin (120° + A) \sin (240° + A)$

$= 3 \sin A \sin [180° - (60° - A)] \sin [180° + (60° + A)]$

$= -3 \sin A \sin (60° - A) \sin (60° + A)$

$= -3 \cdot \dfrac{1}{4} \cdot \sin 3A$

[using $\sin (60° - A) \sin A \sin (60° + A) = \dfrac{1}{4} \sin 3A$]

But $-\dfrac{p}{q} \sin 3A = -\dfrac{3}{4} \sin 3A$

$\therefore \quad p : q = 3 : 4$

12. (d) We have, $\tan 9° - \tan 27° - \tan 63° + \tan 81°$

$= \tan 9° + \tan (90° - 9°) - \tan 27° - \tan (90° - 27°)$

$= (\tan 9° + \cot 9°) - (\tan 27° + \cot 27°)$

$= \dfrac{1}{\sin 9° \cos 9°} - \dfrac{1}{\sin 27° \cos 27°}$

$= \dfrac{2}{\sin 18°} - \dfrac{2}{\sin 54°} = \dfrac{2}{\dfrac{\sqrt{5} - 1}{4}} - \dfrac{2}{\dfrac{\sqrt{5} + 1}{4}}$

$= 8\left(\dfrac{1}{\sqrt{5} - 1} - \dfrac{1}{\sqrt{5} + 1}\right) = 8\left(\dfrac{\sqrt{5} + 1 - \sqrt{5} + 1}{5 - 1}\right)$

$= \dfrac{16}{4} = 4$

13. (c) Option (a) $\sin 15° = \dfrac{\sqrt{3} - 1}{2\sqrt{2}}$ (irrational)

(b) $\cos 15° = \dfrac{\sqrt{3} + 1}{2\sqrt{2}}$ (irrational)

(c) $\sin 15° \cos 15° = \dfrac{1}{2} (2 \sin 15° \cos 15°) = \dfrac{1}{2} \sin 30° = \dfrac{1}{2} \times \dfrac{1}{2}$

$= \dfrac{1}{4}$ (rational)

(d) $\sin 15° \cos 75°$ is also irrational. (irrational)

14. (c) $\cot \alpha + \cot (60° + \alpha) - \cot (60° - \alpha)$

$= \cot \alpha + \dfrac{\sin (60° - \alpha - 60° - \alpha)}{\sin (60° + \alpha) \sin (60° - \alpha)}$

$\left[\because \cot A - \cot B = \dfrac{\sin (B - A)}{\sin A \sin B}\right]$

$= \dfrac{\cos \alpha}{\sin \alpha} - \dfrac{\sin 2\alpha}{\sin^2 60° - \sin^2 \alpha}$ [$\because \sin (-\theta) = -\sin \theta$]

$= \dfrac{\cos \alpha}{\sin \alpha} - \dfrac{2 \sin \alpha \cos \alpha}{3 - 4 \sin^2 \alpha} \times 4$

$= \dfrac{3 \cos \alpha - 4 \sin^2 \alpha \cos \alpha - 8 \sin^2 \alpha \cos \alpha}{\sin \alpha (3 - 4 \sin^2 \alpha)}$

$= \dfrac{3 \cos \alpha - 12 \sin^2 \alpha \cos \alpha}{3 \sin \alpha - 4 \sin^3 \alpha} = \dfrac{3 [\cos \alpha - 4(1 - \cos^2 \alpha) \cos \alpha]}{\sin 3\alpha}$

$= \dfrac{3 [\cos \alpha - 4 \cos \alpha + 4 \cos^3 \alpha]}{\sin 3\alpha} = \dfrac{3 [4 \cos^3 \alpha - 3 \cos \alpha]}{\sin 3\alpha}$

$= \dfrac{3 \cos 3\alpha}{\sin 3\alpha} = 3 \cot 3\alpha$

15. (b) We know that, $\tan 60° = \tan(40° + 20°)$

$$\sqrt{3} = \frac{\tan 40° + \tan 20°}{1 - \tan 40° \tan 20°}$$

$$\sqrt{3} - \sqrt{3}\tan 40° \tan 20° = \tan 40° + \tan 20°$$

$$(\tan 40° + \tan 20° + \sqrt{3}\tan 40° \tan 20°) = \sqrt{3}$$

$\therefore$ $(\tan 40° + \tan 20° + \sqrt{3}\tan 40° \tan 20°)^2 = 3$

16. (a,b,c,d) We know that,

$$\cot\frac{\theta}{2} = \frac{1 + \cos\theta}{\sin\theta}$$

Put $\theta = 15°$, then we get

$$\cot\left(\frac{15}{2}\right)^\circ = \frac{1 + \cos 15°}{\sin 15°} = \frac{1 + \frac{\sqrt{3}+1}{2\sqrt{2}}}{\frac{\sqrt{3}-1}{2\sqrt{2}}}$$

$$\left[\because \cos 15° = \frac{\sqrt{3}+1}{2\sqrt{2}} \text{ and } \sin 15° = \frac{\sqrt{3}-1}{2\sqrt{2}}\right]$$

$$= \frac{2\sqrt{2} + \sqrt{3} + 1}{\sqrt{3} - 1} \times \frac{\sqrt{3}+1}{\sqrt{3}+1}$$

$$= \frac{2\sqrt{6} + 3 + \sqrt{3} + 2\sqrt{2} + \sqrt{3} + 1}{3 - 1}$$

$$= \frac{2\sqrt{6} + 2\sqrt{3} + 2\sqrt{2} + 4}{2}$$

$$= \sqrt{6} + \sqrt{3} + \sqrt{2} + 2$$

Now, $\sqrt{n_1} + \sqrt{n_2} + \sqrt{n_3} + \sqrt{n_4} = \sqrt{2} + \sqrt{3} + \sqrt{4} + \sqrt{6}$

$\therefore$ $n_1 = 2, n_2 = 3, n_3 = 4, n_4 = 6$

(a) $n_1 + n_4 = 2 + 6 = 8$

(b) $2n_4 - n_2 - n_3 = 12 - 3 - 4 = 5$

(c) $n_1 + n_2 + n_3 = 2 + 3 + 4 = 9$

(d) $n_1 + n_2 + n_3 + n_4 = 2 + 3 + 4 + 6 = 15$

17. (a,c) $\because$ $16\sin^2 27° = 8\cdot 2\sin^2 27° = 8(1 - \cos 54°)$

$$= 8\left(1 - \frac{\sqrt{10 - 2\sqrt{5}}}{4}\right)$$

$$= 8 - 2\sqrt{10 - 2\sqrt{5}}$$

$\therefore$ $4\sin 27° = \sqrt{8 - 2\sqrt{10 - 2\sqrt{5}}}$

Now, $[\sqrt{(5+\sqrt{5})} - \sqrt{(3-\sqrt{5})}]^2$

$$= 5 + \sqrt{5} + 3 - \sqrt{5} - 2\sqrt{(5+\sqrt{5})(3-\sqrt{5})}$$

$$= 8 - 2\sqrt{15 + 3\sqrt{5} - 5\sqrt{5} - 5}$$

$$= 8 - 2\sqrt{10 - 2\sqrt{5}}$$

$\therefore$ $\sqrt{(5+\sqrt{5})} - \sqrt{(3-\sqrt{5})} = \sqrt{8 - 2\sqrt{10 - 2\sqrt{5}}}$

18. (a,b,c) We have, $\dfrac{\cos 9° + \sin 9°}{\cos 9° - \sin 9°}$

On dividing numerator and denominator by cos 9°, we get

$$= \frac{1 + \tan 9°}{1 - \tan 9°} = \tan(45° + 9°) = \tan 54° \quad \text{[option (a)]}$$

$$= \tan(90° - 36°) = \cot 36° \quad \text{[option (b)]}$$

and $\tan 54° = \tan(180° - 126°) = -\tan 126°$ [option (c)]

19. (a,c,d) $\because$ $x = \sin 130° \cos 80°$

$x > 0$ $(\because \sin 130° > 0 \text{ and } \cos 80° > 0)$

$y = \sin 80° \cos 130°$

$y < 0$ $(\because \sin 80° > 0 \text{ and } \cos 130° < 0)$

and $z = 1 + xy$

$z < 1$ $(\because x > 0 \text{ and } y < 0)$

$\therefore$ $x > 0, y < 0 \text{ and } 0 < z < 1$

20. (a) If we take (p), i.e. $\left(\dfrac{13\pi}{48}, \dfrac{14\pi}{48}\right)$

$$\cos 2\alpha < 0 \text{ and } \sin 3\alpha > 0$$

Thus, $\dfrac{\sin 3\alpha}{\cos 2\alpha} < 0$

$\therefore$ $\dfrac{\sin 3\alpha}{\cos 2\alpha}$ is negative in $\left(\dfrac{13\pi}{48}, \dfrac{14\pi}{48}\right)$.

Thus, **(ii) → (p)**

Again, if we take (r), i.e. $\left(\dfrac{18\pi}{48}, \dfrac{23\pi}{48}\right)$, then

$$\sin 3\alpha < 0 \text{ and } \cos 2\alpha < 0$$

Thus, $\dfrac{\sin 3\alpha}{\cos 2\alpha} > 0$

$\Rightarrow$ $\dfrac{\sin 3\alpha}{\cos 2\alpha}$ is positive in $\left(\dfrac{18\pi}{48}, \dfrac{23\pi}{48}\right)$.

Thus, **(i) → (r)**

21. (8) We have,

$$\frac{1}{\lambda} = \sin\frac{\pi}{18}\sin\frac{5\pi}{18}\sin\frac{7\pi}{18}$$

$$= \sin 10° \sin 50° \sin 70°$$

$$= \sin 10° \sin(60° - 10°)\sin(60° + 10°)$$

$$= \frac{1}{4}\cdot\sin(3 \times 10°)$$

$$\left[\because \sin(60° - A)\sin A \sin(60° + A) = \frac{1}{4}\sin 3A\right]$$

$$= \frac{1}{4}\cdot\sin 30° = \frac{1}{4} \times \frac{1}{2}$$

$\Rightarrow$ $\dfrac{1}{\lambda} = \dfrac{1}{8} \Rightarrow \lambda = 8$

22. (6) $\because$ $\sin^3 x \sin 3x = \dfrac{1}{4}(4\sin^3 x)\sin 3x$

$$= \frac{1}{4}(3\sin x - \sin 3x)\sin 3x$$

$$= \frac{1}{4}\cdot\frac{1}{2}(2 \times 3\sin 3x \cdot \sin x - 2\sin^2 3x)$$

$$= \frac{1}{8}[3(\cos 2x - \cos 4x) - (1 - \cos 6x)]$$

$$= \frac{1}{8}[\cos 6x - 3\cos 4x + 3\cos 2x - 1]$$

and $\sum_{m=0}^{n} C_m \cos nx = \sin^3 x \sin 3x$ is an identity in x

$$= \frac{1}{8}[\cos 6x - 3\cos 4x + 3\cos 2x - 1]$$

On comparing both sides, we get

$$n = 6$$

DPP-13 **Sum of Three or More Angles**

1. We have,

$A + B + C = 180°$

$$\begin{aligned}\text{LHS} &= \sin 2A + \sin 2B + \sin 2C\\ &= 2\sin(A+B)\cos(A-B) + 2\sin C\cos C\\ &= 2\sin(180° - C)\cos(A-B) + 2\sin C\cos C\\ &= 2\sin C\cos(A-B) + 2\sin C\cos C\\ &= 2\sin C[\cos(A-B) + \cos C]\\ &= 2\sin C[\cos(A-B) + \cos(180° - \overline{A+B})]\\ &= 2\sin C[\cos(A-B) - \cos(A+B)]\\ &= 2\sin C\cdot 2\sin A\sin B\\ &= 4\sin A\sin B\sin C = \text{RHS}\end{aligned}$$

Hence proved.

2. $\because A + B + C = 180°$

and $\cos\theta(\sin B + \sin C) = \sin A$

$$\cos\theta = \frac{\sin A}{\sin B + \sin C}$$

$$\frac{1}{\cos\theta} = \frac{\sin B + \sin C}{\sin A}$$

On applying componendo and dividendo, we get

$$\frac{1-\cos\theta}{1+\cos\theta} = \frac{\sin B + \sin C - \sin A}{\sin B + \sin C + \sin A}$$

$$\tan^2\frac{\theta}{2} = \frac{4\sin\frac{B}{2}\sin\frac{C}{2}\cos\frac{A}{2}}{4\cos\frac{A}{2}\cos\frac{B}{2}\cos\frac{C}{2}}$$

$\therefore \quad \tan^2\frac{\theta}{2} = \tan\frac{B}{2}\tan\frac{C}{2}$ **Hence proved.**

3.
$$\begin{aligned}\text{LHS} &= \sin 2A + \sin 2B - \sin 2C\\ &= 2\sin(A+B)\cos(A-B) - 2\sin C\cos C\\ &= 2\sin(180° - C)\cos(A-B) - 2\sin C\cos C\\ &= 2\sin C\cos(A-B) - 2\sin C\cos C\\ &= 2\sin C[\cos(A-B) - \cos(180° - \overline{A+B})]\\ &= 2\sin C[\cos(A-B) + \cos(A+B)]\\ &= 2\sin C\cdot 2\cos A\cos B\\ &= 4\cos A\cos B\sin C = \text{RHS}\end{aligned}$$

Hence proved.

4. (b) $\text{LHS} = \dfrac{\sin A + \sin B + \sin C}{\sin A + \sin B - \sin C}$

$$= \frac{2\sin\left(\frac{A+B}{2}\right)\cos\left(\frac{A-B}{2}\right) + 2\sin\frac{C}{2}\cos\frac{C}{2}}{2\sin\left(\frac{A+B}{2}\right)\cos\left(\frac{A-B}{2}\right) - 2\sin\frac{C}{2}\cos\frac{C}{2}}$$

$$= \frac{2\sin\left(90° - \frac{C}{2}\right)\cos\left(\frac{A-B}{2}\right) + 2\sin\frac{C}{2}\cos\frac{C}{2}}{2\sin\left(90° - \frac{C}{2}\right)\cos\left(\frac{A-B}{2}\right) - 2\sin\frac{C}{2}\cos\frac{C}{2}}$$

$$= \frac{2\cos\frac{C}{2}\cos\left(\frac{A-B}{2}\right) + 2\sin\frac{C}{2}\cos\frac{C}{2}}{2\cos\frac{C}{2}\cos\left(\frac{A-B}{2}\right) - 2\sin\frac{C}{2}\cos\frac{C}{2}}$$

$$= \frac{2\cos\frac{C}{2}\left[\cos\left(\frac{A-B}{2}\right) + \sin\frac{C}{2}\right]}{2\cos\frac{C}{2}\left[\cos\left(\frac{A-B}{2}\right) - \sin\frac{C}{2}\right]}$$

$$= \frac{\cos\left(\frac{A-B}{2}\right) + \cos\left(\frac{A+B}{2}\right)}{\cos\left(\frac{A-B}{2}\right) - \cos\left(\frac{A+B}{2}\right)}$$

$$= \frac{2\cos\frac{A}{2}\cdot\cos\frac{B}{2}}{2\sin\frac{A}{2}\cdot\sin\frac{B}{2}} = \cot\frac{A}{2}\cot\frac{B}{2} \quad \ldots\text{(i)}$$

$$\text{RHS} = 2\lambda\cot\frac{A}{2}\cot\frac{B}{2} \quad \ldots\text{(ii)}$$

Then, from Eqs. (i) and (ii), we get

$$2\lambda = 1 \Rightarrow \lambda = \frac{1}{2}$$

5. (b) We have, $\dfrac{\cos A}{\sin B\sin C} + \dfrac{\cos B}{\sin C\sin A} + \dfrac{\cos C}{\sin A\sin B}$

$$= \frac{\sin A\cos A + \sin B\cos B + \sin C\cos C}{\sin A\sin B\sin C}$$

$$= \frac{1}{2}\left[\frac{\sin 2A + \sin 2B + \sin 2C}{\sin A\sin B\sin C}\right]$$

$$= \frac{1}{2}\cdot\frac{4\sin A\sin B\sin C}{\sin A\sin B\sin C} \quad \text{[from Q.1]}$$

$$= 2$$

6. (a) We have, $\dfrac{1-\cos A + \cos B + \cos C}{1 - \cos C + \cos A + \cos B}$

$$= \frac{2\sin^2\frac{A}{2} + 2\cos\left(\frac{B+C}{2}\right)\cos\left(\frac{B-C}{2}\right)}{2\sin^2\frac{C}{2} + 2\cos\left(\frac{A+B}{2}\right)\cos\left(\frac{A-B}{2}\right)}$$

$$= \frac{2\sin^2\frac{A}{2} + 2\cos\left(90° - \frac{A}{2}\right)\cos\left(\frac{B-C}{2}\right)}{2\sin^2\frac{C}{2} + 2\cos\left(90° - \frac{C}{2}\right)\cos\left(\frac{A-B}{2}\right)}$$

$$\left[\because A + B + C = 180° \therefore \frac{A}{2} + \frac{B}{2} + \frac{C}{2} = 90°\right]$$

$$= \frac{2\sin^2\frac{A}{2} + 2\sin\frac{A}{2}\cos\left(\frac{B-C}{2}\right)}{2\sin^2\frac{C}{2} + 2\sin\frac{C}{2}\cos\left(\frac{A-B}{2}\right)}$$

$$= \frac{2\sin\frac{A}{2}\left[\sin\frac{A}{2} + \cos\left(\frac{B-C}{2}\right)\right]}{2\sin\frac{C}{2}\left[\sin\frac{C}{2} + \cos\left(\frac{A-B}{2}\right)\right]}$$

$$= \frac{\sin\frac{A}{2}\left[\cos\left(\frac{B+C}{2}\right) + \cos\left(\frac{B-C}{2}\right)\right]}{\sin\frac{C}{2}\left[\cos\left(\frac{A+B}{2}\right) + \cos\left(\frac{A-B}{2}\right)\right]}$$

$$= \frac{\sin\frac{A}{2}\cdot 2\cos\frac{B}{2}\cdot\cos\frac{C}{2}}{\sin\frac{C}{2}\cdot 2\cos\frac{A}{2}\cdot\cos\frac{B}{2}} = \frac{\sin\frac{A}{2}}{\cos\frac{A}{2}}\cdot\frac{\cos\frac{C}{2}}{\sin\frac{C}{2}}$$

$$= \frac{\tan\frac{A}{2}}{\tan\frac{C}{2}} = \tan\frac{A}{2}\cot\frac{C}{2}$$

7. (b) We know that,

$$\cos A + \cos B + \cos C = 1 + 4 \sin \frac{A}{2} \sin \frac{B}{2} \sin \frac{C}{2} \quad \text{...(i)}$$

and $\sin A + \sin B + \sin C = 4 \cos \frac{A}{2} \cos \frac{B}{2} \cos \frac{C}{2}$ ...(ii)

$\because \quad A + B + C = \pi$

Now given,

$$\tan\left(\frac{B+C-A}{4}\right) \tan\left(\frac{C+A-B}{4}\right) \tan\left(\frac{A+B-C}{4}\right) = 1$$

$$\Rightarrow \tan\left(\frac{\pi-2A}{4}\right) \tan\left(\frac{\pi-2B}{4}\right) \tan\left(\frac{\pi-2C}{4}\right) = 1$$

$$\Rightarrow \sin^2\left(\frac{\pi-2A}{4}\right) \sin^2\left(\frac{\pi-2B}{4}\right) \sin^2\left(\frac{\pi-2C}{4}\right)$$

$$= \cos^2\left(\frac{\pi-2A}{4}\right) \cos^2\left(\frac{\pi-2B}{4}\right) \cos^2\left(\frac{\pi-2C}{4}\right)$$

$$\Rightarrow \left[1 - \cos\left(\frac{\pi}{2} - A\right)\right]\left[1 - \cos\left(\frac{\pi}{2} - B\right)\right]\left[1 - \cos\left(\frac{\pi}{2} - C\right)\right]$$

$$= \left[1 + \cos\left(\frac{\pi}{2} - A\right)\right]\left[1 + \cos\left(\frac{\pi}{2} - B\right)\right]\left[1 + \cos\left(\frac{\pi}{2} - C\right)\right]$$

$$\Rightarrow (1 - \sin A)(1 - \sin B)(1 - \sin C)$$

$$= (1 + \sin A)(1 + \sin B)(1 + \sin C)$$

$$\Rightarrow (\sin A + \sin B + \sin C) + \sin A \sin B \sin C = 0$$

$$\Rightarrow 4 \cos \frac{A}{2} \cos \frac{B}{2} \cos \frac{C}{2} + \left(2 \sin \frac{A}{2} \cdot \cos \frac{A}{2}\right)$$

$$\left(2 \sin \frac{B}{2} \cos \frac{B}{2}\right)\left(2 \sin \frac{C}{2} \cos \frac{C}{2}\right) = 0 \quad \text{[from Eq. (ii)]}$$

On dividing both sides by $4 \cos \frac{A}{2} \cos \frac{B}{2} \cos \frac{C}{2}$

$$\Rightarrow 1 + 2 \sin \frac{A}{2} \sin \frac{B}{2} \sin \frac{C}{2} = 0$$

$$\Rightarrow 1 + 1 + 4 \sin \frac{A}{2} \sin \frac{B}{2} \sin \frac{C}{2} = 0$$

$$\Rightarrow 1 + \cos A + \cos B + \cos C = 0 \quad \text{[from Eq. (i)]}$$

$$\therefore \cos A + \cos B + \cos C = -1$$

8. (c) We have, in ΔABC,

$$A + B + C = \pi$$

$$\Rightarrow \frac{A}{2} + \frac{B}{2} + \frac{C}{2} = \frac{\pi}{2}$$

Then, $\sin \frac{A}{2} \sin \frac{B}{2} \sin \frac{C}{2} = \frac{1}{2}\left[2 \sin \frac{A}{2} \sin \frac{B}{2}\right] \sin \frac{C}{2}$

$$= \frac{1}{2}\left[\cos\left(\frac{A-B}{2}\right) - \cos\left(\frac{A+B}{2}\right)\right] \sin \frac{C}{2}$$

$$= \frac{1}{2}\left[\cos\left(\frac{A-B}{2}\right) - \sin \frac{C}{2}\right] \sin \frac{C}{2}$$

$$\leq \frac{1}{2}\left(1 - \sin \frac{C}{2}\right) \sin \frac{C}{2}$$

$$= -\frac{1}{2}\left[\sin^2 \frac{C}{2} - \sin \frac{C}{2} + \left(\frac{1}{2}\right)^2 - \left(\frac{1}{2}\right)^2\right]$$

$$= -\frac{1}{2}\left[\left(\sin \frac{C}{2} - \frac{1}{2}\right)^2 - \frac{1}{4}\right]$$

$$= \frac{1}{2}\left[\frac{1}{4} - \left(\sin \frac{C}{2} - \frac{1}{2}\right)^2\right]$$

$$\leq \frac{1}{2} \times \frac{1}{4} = \frac{1}{8}$$

$$\therefore \sin \frac{A}{2} \sin \frac{B}{2} \sin \frac{C}{2} \leq \frac{1}{8}$$

9. (c) $\because \quad A + B + C = \pi$

$$\therefore \frac{A}{2} + \frac{B}{2} + \frac{C}{2} = \frac{\pi}{2}$$

$$\Rightarrow \tan\left(\frac{A}{2} + \frac{B}{2} + \frac{C}{2}\right) = \infty$$

$$\Rightarrow \frac{S_1 - S_3}{1 - S_2} = \frac{1}{0} \Rightarrow 1 - S_2 = 0$$

$$\Rightarrow 1 - \left(\tan \frac{A}{2} \tan \frac{B}{2} + \tan \frac{B}{2} \tan \frac{C}{2} + \tan \frac{C}{2} \tan \frac{A}{2}\right) = 0$$

$$\Rightarrow \tan \frac{A}{2} \tan \frac{B}{2} + \tan \frac{B}{2} \tan \frac{C}{2} + \tan \frac{C}{2} \tan \frac{A}{2} = 1$$

On dividing both sides by $\tan \frac{A}{2} \tan \frac{B}{2} \tan \frac{C}{2}$, we get

$$\cot \frac{A}{2} + \cot \frac{B}{2} + \cot \frac{C}{2} = \cot \frac{A}{2} \cot \frac{B}{2} \cot \frac{C}{2}$$

But $\cot \frac{A}{2} + \cot \frac{B}{2} + \cot \frac{C}{2} = \lambda \cot \frac{A}{2} \cot \frac{B}{2} \cot \frac{C}{2}$

$$\therefore \lambda = 1$$

10. (d) Let $\alpha = A$, $2\alpha = B$, $4\alpha = C$, then

$$A + B + C = 7\alpha = 2\pi$$

Then, $\tan A \tan B + \tan B \tan C + \tan C \tan A$

$$= \frac{\Sigma \sin A \sin B \sin C}{\cos A \cos B \cos C} \quad \text{...(i)}$$

and we know that,

$$\cos(A + B + C) = \cos A \cos B \cos C - \Sigma \sin A \sin B \cos C$$

$$\frac{\cos 2\pi}{\cos A \cos B \cos C} = 1 - \frac{\Sigma \sin A \sin B \cos C}{\cos A \cos B \cos C}$$

$$\Rightarrow \frac{\Sigma \sin A \sin B \cos C}{\cos A \cos B \cos C} = 1 - \frac{1}{\cos A \cos B \cos C}$$

$\Rightarrow \tan A \tan B + \tan B \tan C + \tan C \tan A$

$$= 1 - \frac{1}{\cos \frac{2\pi}{7} \cos \frac{4\pi}{7} \cos \frac{8\pi}{7}} \quad \text{[from Eq. (i)]}$$

$\Rightarrow \tan \alpha \tan 2\alpha + \tan 2\alpha \tan 4\alpha + \tan 4\alpha \tan \alpha$

$$= 1 - \frac{1}{\dfrac{\sin \frac{16\pi}{7}}{2^3 \sin \frac{2\pi}{7}}}$$

$$= 1 - \frac{8 \sin \frac{2\pi}{7}}{\sin\left(2\pi + \frac{2\pi}{7}\right)}$$

$$= 1 - \frac{8 \sin \frac{2\pi}{7}}{\sin \frac{2\pi}{7}}$$

$$= 1 - 8 = -7$$

11. (c) $\because \quad \tan A + \tan B + \tan C = 3\sqrt{3}$

Let $A = B = C$, then

$$\tan A + \tan A + \tan A = 3\sqrt{3}$$

$$3 \tan A = 3\sqrt{3}$$

$$\tan A = \sqrt{3} = \tan 60°$$

$$\Rightarrow A = 60°$$

$$\therefore A = B = C = 60°$$

So, ΔABC is an equilateral triangle.

12. (d) We have, $\cos\frac{2\pi}{7}+\cos\frac{4\pi}{7}+\cos\frac{6\pi}{7}$

$$=\cos\frac{2\pi}{7}+\cos\left(\frac{2\pi}{7}+\frac{2\pi}{7}\right)+\cos\left(\frac{2\pi}{7}+2\cdot\frac{2\pi}{7}\right)$$

$\therefore$ $\alpha=\frac{2\pi}{7},\ \beta=\frac{2\pi}{7},\ n=3$, then

$$=\frac{\sin 3\cdot\left(\frac{2\pi}{7}\right)\cdot\frac{1}{2}}{\sin\left(\frac{2\pi}{7}\times\frac{1}{2}\right)}\cdot\cos\left[\frac{\frac{2\pi}{7}+\frac{6\pi}{7}}{2}\right]$$

$$=\frac{\sin\frac{3\pi}{7}}{\sin\frac{\pi}{7}}\cdot\cos\frac{4\pi}{7}=\frac{\sin\frac{3\pi}{7}\cdot\cos\left(\pi-\frac{3\pi}{7}\right)}{\sin\frac{\pi}{7}}$$

$$=\frac{-2\sin\frac{3\pi}{7}\cdot\cos\frac{3\pi}{7}}{2\sin\frac{\pi}{7}}=-\frac{1}{2}\cdot\frac{\sin\frac{6\pi}{7}}{\sin\frac{\pi}{7}}$$

$$=-\frac{1}{2}\cdot\frac{\sin\left(\pi-\frac{\pi}{7}\right)}{\sin\frac{\pi}{7}}=-\frac{1}{2}\cdot\frac{\sin\frac{\pi}{7}}{\sin\frac{\pi}{7}}=-\frac{1}{2}$$

13. (d) $\because A+B+C=\frac{3\pi}{2}$, then

$$\begin{aligned}\cos 2A+\cos 2B+\cos 2C &= 2\cos(A+B)\cos(A-B)+\cos 2C\\ &= 2\cos\left[\frac{3\pi}{2}-C\right]\cos(A-B)+1-2\sin^2 C\\ &= -2\sin C\cos(A-B)-2\sin^2 C+1\\ &= 1-2\sin C\,[\cos(A-B)+\sin C]\\ &= 1-2\sin C\,[\cos(A-B)-\cos(A+B)]\\ &= 1-2\sin C\,[2\sin A\sin B]\\ &= 1-4\sin A\sin B\sin C\end{aligned}$$

14. (a) We have, $\alpha+\beta+\gamma=2\pi$

$$\Rightarrow\quad \frac{\alpha}{2}+\frac{\beta}{2}+\frac{\gamma}{2}=\pi$$

$$\Rightarrow\quad \frac{\alpha}{2}+\frac{\beta}{2}=\pi-\frac{\gamma}{2}$$

$$\Rightarrow\quad \tan\left(\frac{\alpha}{2}+\frac{\beta}{2}\right)=\tan\left(\pi-\frac{\gamma}{2}\right)$$

$$\Rightarrow\quad \frac{\tan\frac{\alpha}{2}+\tan\frac{\beta}{2}}{1-\tan\frac{\alpha}{2}\cdot\tan\frac{\beta}{2}}=-\tan\frac{\gamma}{2}$$

$$\Rightarrow\quad \tan\frac{\alpha}{2}+\tan\frac{\beta}{2}=-\tan\frac{\gamma}{2}+\tan\frac{\alpha}{2}\tan\frac{\beta}{2}\tan\frac{\gamma}{2}$$

$$\therefore\quad \tan\frac{\alpha}{2}+\tan\frac{\beta}{2}+\tan\frac{\gamma}{2}=\tan\frac{\alpha}{2}\tan\frac{\beta}{2}\tan\frac{\gamma}{2}$$

15. (a,c,d) We have,

$$\cos 3A+\cos 3B+\cos 3C=1$$

$$\Rightarrow\quad \cos 3A+\cos 3B+\cos 3C-1=0$$

$$\Rightarrow\quad (\cos 3A+\cos 3B)+(\cos 3C+\cos 3\pi)=0$$

$$\Rightarrow\quad 2\cos\left(\frac{3A+3B}{2}\right)\cdot\cos\left(\frac{3A-3B}{2}\right)+2\cos\left(\frac{3\pi+3C}{2}\right)\cdot\cos\left(\frac{3\pi-3C}{2}\right)=0$$

$$\Rightarrow\quad 2\cos\left(\frac{3\pi-3C}{2}\right)\cdot\cos\left(\frac{3A-3B}{2}\right)+2\cos\left(\frac{3\pi+3C}{2}\right)\cdot\cos\left(\frac{3\pi-3C}{2}\right)=0$$

$$\Rightarrow\quad 2\cos\left(\frac{3\pi-3C}{2}\right)\left[\cos\left(\frac{3A-3B}{2}\right)+\cos\left(\frac{3\pi+3C}{2}\right)\right]=0$$

$$\Rightarrow 2\cos\left(\frac{3\pi-3C}{2}\right)$$

$$\left[2\cos\left(\frac{3A-3B+3\pi+3C}{4}\right)\cdot\cos\left(\frac{3A-3B-3\pi-3C}{4}\right)\right]=0$$

$$\Rightarrow\quad 4\cos\left(\frac{3\pi-3C}{2}\right)\cos\left(\frac{3\pi-3B}{2}\right)\cos\left(\frac{3\pi-3A}{2}\right)=0$$

$$\Rightarrow\quad 4\cos\left(\frac{3\pi}{2}-\frac{3C}{2}\right)\cos\left(\frac{3\pi}{2}-\frac{3B}{2}\right)\cos\left(\frac{3\pi}{2}-\frac{3A}{2}\right)=0$$

$$\sin\frac{3A}{2}\sin\frac{3B}{2}\sin\frac{3C}{2}=0$$

$\therefore$ Either $\frac{3A}{2}=\pi$ or $\frac{3B}{2}=\pi$ or $\frac{3C}{2}=\pi$

$$A=\frac{2\pi}{3}\text{ or }B=\frac{2\pi}{3}\text{ or }C=\frac{2\pi}{3}$$

Alternate Method

$\because$

$$A+B+C=\pi$$

$$3A+3B=3\pi-3C$$

$$\frac{3A+3B}{2}=\frac{3\pi}{2}-\frac{3C}{2}$$

Now, we have

$$\cos 3A+\cos 3B+\cos 3C-1=0$$

$$\Rightarrow\quad 2\cos\left(\frac{3A+3B}{2}\right)\cdot\cos\left(\frac{3A-3B}{2}\right)-2\sin^2\frac{3C}{2}=0$$

$$\Rightarrow\quad 2\cos\left(\frac{3\pi}{2}-\frac{3C}{2}\right)\cdot\cos\left(\frac{3A-3B}{2}\right)-2\sin^2\frac{3C}{2}=0$$

$$\Rightarrow\quad -2\sin\frac{3C}{2}\cos\left(\frac{3A-3B}{2}\right)-2\sin^2\frac{3C}{2}=0$$

$$\Rightarrow\quad 2\sin\frac{3C}{2}\left[\cos\left(\frac{3A}{2}-\frac{3B}{2}\right)+\sin\frac{3C}{2}\right]=0$$

$$\Rightarrow\quad 2\sin\frac{3C}{2}\left[\cos\left(\frac{3A}{2}-\frac{3B}{2}\right)-\cos\left(\frac{3A}{2}+\frac{3B}{2}\right)\right]=0$$

$$\sin\frac{3C}{2}\cdot\sin\frac{3A}{2}\cdot\sin\frac{3B}{2}=0$$

$\therefore$ Either $\sin\frac{3C}{2}=\sin\pi$ or $\sin\frac{3A}{2}=\sin\pi$ or $\sin\frac{3B}{2}=\sin\pi$

$$\Rightarrow\quad \frac{3C}{2}=\pi\text{ or }\frac{3A}{2}=\pi\text{ or }\frac{3B}{2}=\pi$$

$$\Rightarrow\quad C=\frac{2\pi}{3}\text{ or }A=\frac{2\pi}{3}\text{ or }B=\frac{2\pi}{3}$$

16. (a,b,c) We have

$A+B+C=180°$ and A, B, C are in AP, then

$$2B=A+C$$

$$\therefore\quad 2B+B=180°$$

$$\angle B=60° \qquad \text{[option (b)]}$$

and $\sin(2A+B)=\sin(C-A)=-\sin(B+2C)=\frac{1}{2}$

$$\sin(2A+60°)=\sin(C-A)=-\sin(60°+2C)=\frac{1}{2}$$

If $\sin(2A+60°)=\frac{1}{2}=\sin 30°$ or $\sin 150°$ (but 30° is not possible)

$$\Rightarrow\quad 2A+60°=150°$$

$$\therefore\quad \angle A=45° \qquad \text{[option (a)]}$$

Now, $\angle C=180°-(A+B)$

$$=180°-(45°+60°)$$

$$\angle C=75° \qquad \text{[option (c)]}$$

17. (a,b) We know that,

$$\cos A + \cos B + \cos C = 1 + 4\sin\frac{A}{2}\cdot\sin\frac{B}{2}\cdot\sin\frac{C}{2} > 1$$

[option (a)]

and now, $\cos A + \cos B + \cos C$

$$= 2\cos\left(\frac{A+B}{2}\right)\cdot\cos\left(\frac{A-B}{2}\right) + 1 - 2\sin^2\frac{C}{2}$$

$$= 2\sin\frac{C}{2}\cdot\cos\left(\frac{A-B}{2}\right) + 1 - 2\sin^2\frac{C}{2}$$

$$\le 2\sin\frac{C}{2} + 1 - 2\sin^2\frac{C}{2} \quad \left[\because \cos\left(\frac{A-B}{2}\right) \le 1\right]$$

$$= -2\left[\sin^2\frac{C}{2} - \sin\frac{C}{2} - \frac{1}{2}\right]$$

$$= -2\left[\sin^2\frac{C}{2} - \sin\frac{C}{2} + \left(\frac{1}{2}\right)^2 - \frac{1}{4} - \frac{1}{2}\right]$$

$$= -2\left[\left(\sin\frac{C}{2} - \frac{1}{2}\right)^2 - \frac{3}{4}\right]$$

$$= \frac{3}{2} - \left(\sin\frac{C}{2} - \frac{1}{2}\right)^2$$

$$\therefore \cos A + \cos B + \cos C \le \frac{3}{2}$$

[option (b)]

18. (b) We have,

$$\sin\frac{\pi}{7} + \sin\frac{2\pi}{7} + \sin\frac{3\pi}{7}$$

$$= \sin\frac{\pi}{7} + \sin\left(\frac{\pi}{7} + \frac{\pi}{7}\right) + \sin\left(\frac{\pi}{7} + 2\cdot\frac{\pi}{7}\right)$$

$$\therefore \quad \alpha = \frac{\pi}{7}, \beta = \frac{\pi}{7}, n = 3, \text{ then}$$

$$= \frac{\sin\left(\frac{3\pi}{7}\cdot\frac{1}{2}\right)}{\sin\frac{\pi}{14}}\cdot\sin\left(\frac{\frac{\pi}{7} + \frac{3\pi}{7}}{2}\right) \quad \text{(given)}$$

$$= \frac{\sin\frac{3\pi}{14}}{\sin\frac{\pi}{14}}\cdot\sin\frac{4\pi}{14} = \frac{\sin\frac{3\pi}{14}}{\sin\frac{\pi}{14}}\cdot\sin\left(\frac{\pi}{2} - \frac{3\pi}{14}\right)$$

$$= \frac{1}{2}\cdot\frac{2\sin\frac{3\pi}{14}\cdot\cos\frac{3\pi}{14}}{\sin\frac{\pi}{14}}$$

$$= \frac{1}{2}\cdot\frac{\sin\frac{6\pi}{14}}{\sin\frac{\pi}{14}} = \frac{1}{2}\cdot\frac{\sin\left(\frac{\pi}{2} - \frac{\pi}{14}\right)}{\sin\frac{\pi}{14}}$$

$$= \frac{1}{2}\cdot\frac{\cos\frac{\pi}{14}}{\sin\frac{\pi}{14}} = \frac{1}{2}\cot\frac{\pi}{14}$$

19. (b) We have, $\sin\frac{\pi}{n} + \sin\frac{3\pi}{n} + \sin\frac{5\pi}{n} + \cdots n$ terms

$$\alpha = \frac{\pi}{n}, \beta = \frac{2\pi}{n}, n = n, \text{ then}$$

$$= \frac{\sin\left(n\cdot\frac{2\pi}{n}\cdot\frac{1}{2}\right)}{\sin\left(\frac{2\pi}{n}\cdot\frac{1}{2}\right)}\cdot\sin\left[\frac{\frac{\pi}{n} + \frac{(2n-1)\pi}{n}}{2}\right] \quad \text{(given)}$$

$$= \frac{\sin\pi}{\sin\frac{\pi}{n}}\cdot\sin\left[\frac{\pi + (2n-1)\pi}{2n}\right] = 0 \quad (\because \sin\pi = 0)$$

20. (c) We have, $\cos\frac{\pi}{11} + \cos\frac{3\pi}{11} + \cos\frac{5\pi}{11} + \cos\frac{7\pi}{11} + \cos\frac{9\pi}{11}$

$$\because \quad \alpha = \frac{\pi}{11}, \beta = \frac{2\pi}{11}, n = 5, \text{ then}$$

$$= \frac{\sin\left(5\cdot\frac{2\pi}{11}\cdot\frac{1}{2}\right)}{\sin\left(\frac{2\pi}{11}\cdot\frac{1}{2}\right)}\cdot\cos\left[\frac{\frac{\pi}{11} + \frac{9\pi}{11}}{2}\right]$$

$$= \frac{\sin\frac{5\pi}{11}}{\sin\frac{\pi}{11}}\cdot\cos\frac{5\pi}{11} = \frac{1}{2}\cdot\frac{\sin\frac{10\pi}{11}}{\sin\frac{\pi}{11}}$$

$$= \frac{1}{2}\cdot\frac{\sin\left(\pi - \frac{\pi}{11}\right)}{\sin\frac{\pi}{11}} = \frac{1}{2}\cdot\frac{\sin\frac{\pi}{11}}{\sin\frac{\pi}{11}} = \frac{1}{2}$$

21. (c) We have, $\sum_{r=0}^{n}\sin^2\left(\frac{r\pi}{n}\right) = \frac{1}{2}\sum_{r=0}^{n}\left[1 - \cos\frac{2r\pi}{n}\right]$

$$= \frac{1}{2}\left[(n+1) - 1 - \cos\frac{2\pi}{n} - \cos\frac{4\pi}{n} - \cos\frac{6\pi}{n} - \ldots - \cos\frac{2n\pi}{n}\right]$$

$$= \frac{1}{2}\left[(n-1) - \frac{1}{2}\left\{\cos\frac{2\pi}{n} + \cos\frac{4\pi}{n} + \cos\frac{6\pi}{n} + \ldots + \cos\frac{2(n-1)\pi}{n}\right\}\right]$$

$$= \frac{n-1}{2} - \frac{1}{2}\cdot\frac{\sin\left\{(n-1)\cdot\frac{2\pi}{n}\cdot\frac{1}{2}\right\}}{\sin\left(\frac{\pi}{n}\right)}\cos\left\{\frac{\frac{2\pi}{n} + \frac{2(n-1)\pi}{n}}{2}\right\}$$

$$= \frac{n-1}{2} - \frac{1}{2}\cdot\frac{\sin\left(\pi - \frac{\pi}{n}\right)}{\sin\frac{\pi}{n}}\cos\left(\frac{\pi}{n} + \pi - \frac{\pi}{n}\right)$$

$$= \frac{n-1}{2} - \frac{1}{2}\cdot\frac{\sin\frac{\pi}{n}}{\sin\frac{\pi}{n}}\cos\pi$$

$$= \frac{n-1}{2} + \frac{1}{2} = \frac{n}{2}$$

Revisal Problems (JEE Main)

1. (a) $\sin\alpha\cos^3\alpha > \sin^3\alpha\cos\alpha$

$$\sin\alpha\cos^3\alpha - \sin^3\alpha\cos\alpha > 0$$
$$\sin\alpha\cos\alpha(\cos^2\alpha - \sin^2\alpha) > 0$$
$$2\sin\alpha\cos\alpha\cdot\cos 2\alpha > 0$$
$$2\sin 2\alpha\cos 2\alpha > 0$$
$$\sin 4\alpha > 0$$
$$\because \quad 0 < \alpha < \pi$$
$$\therefore \quad 0 < 4\alpha < 4\pi$$

Thus, $\sin 4\alpha > 0$, if $0 < 4\alpha < \pi$ and $2\pi < 4\alpha < 3\pi$

i.e. $\quad 0 < \alpha < \frac{\pi}{4}$ and $\frac{\pi}{2} < \alpha < \frac{3\pi}{4}$

2. (a) We have, $\sin\left(x + \frac{\pi}{6}\right) + \cos\left(x + \frac{\pi}{6}\right)$

$$= \sqrt{2}\left[\frac{1}{\sqrt{2}}\sin\left(x + \frac{\pi}{6}\right) + \frac{1}{\sqrt{2}}\cos\left(x + \frac{\pi}{6}\right)\right]$$
$$= \sqrt{2}\left[\sin\left(x + \frac{\pi}{6}\right)\cos\frac{\pi}{4} + \cos\left(x + \frac{\pi}{6}\right)\sin\frac{\pi}{4}\right]$$
$$= \sqrt{2}\sin\left(x + \frac{\pi}{6} + \frac{\pi}{4}\right) = \sqrt{2}\sin\left(x + \frac{5\pi}{12}\right)$$
$$\because \quad \sin\theta \le 1, \text{ when } \theta = \frac{\pi}{2}$$

Now, $\sin\left(x + \frac{5\pi}{12}\right) \le 1$, when $x + \frac{5\pi}{12} = \frac{\pi}{2} \Rightarrow x = \frac{\pi}{2} - \frac{5\pi}{12}$

$$\therefore \quad x = \frac{\pi}{12} \in \left(0, \frac{\pi}{2}\right)$$

Hence, maximum value of $\sin\left(x + \frac{\pi}{6}\right) + \cos\left(x + \frac{\pi}{6}\right)$ is at $x = \frac{\pi}{12}$.

3. (c) We have,

$$\sin\frac{\pi}{2n} + \cos\frac{\pi}{2n} = \sqrt{2}\left[\frac{1}{\sqrt{2}}\sin\frac{\pi}{2n} + \frac{1}{\sqrt{2}}\cos\frac{\pi}{2n}\right]$$
$$\frac{\sqrt{n}}{2} = \sqrt{2}\sin\left(\frac{\pi}{4} + \frac{\pi}{2n}\right) \quad \text{(given)}$$
$$\therefore \quad \frac{\sqrt{n}}{2\sqrt{2}} = \sin\left(\frac{\pi}{4} + \frac{\pi}{2n}\right) > \sin\frac{\pi}{4}, \text{ when } n > 1$$
$$\Rightarrow \quad \frac{\sqrt{n}}{2\sqrt{2}} > \frac{1}{\sqrt{2}}$$
$$\Rightarrow \quad n > 4 \quad \text{...(i)}$$

and we know that, $\sin\left(\frac{\pi}{4} + \frac{\pi}{2n}\right) < 1, \forall n > 2$

Then, $\quad \frac{\sqrt{n}}{2\sqrt{2}} < 1 \Rightarrow n < 8 \quad$ ...(ii)

From Eqs. (i) and (ii), we get

$$4 < n < 8$$

$\therefore$ $n = 5$ or 6, then by putting these values, we find that $n = 6$ satisfies the given relation.

4. (a) Let $\quad p = 2 - \cos x + \sin^2 x = 2 - \cos x + 1 - \cos^2 x$

$$= -[\cos^2 x + \cos x - 3]$$
$$= -\left[\cos^2 x + \cos x + \left(\frac{1}{2}\right)^2 - \left(\frac{1}{2}\right)^2 - 3\right]$$
$$= -\left[\left(\cos x + \frac{1}{2}\right)^2 - \frac{13}{4}\right]$$
$$= \left[\frac{13}{4} - \left(\cos x + \frac{1}{2}\right)^2\right]$$

$\because p$ is maximum, when $\left(\cos x + \frac{1}{2}\right)^2$ is minimum and minimum value of $\left(\cos x + \frac{1}{2}\right)^2$ is 0, then

$$p_{max} = \frac{13}{4} - 0 = \frac{13}{4}$$

p is minimum, when $\left(\cos x + \frac{1}{2}\right)^2$ is maximum but $\left(\cos x + \frac{1}{2}\right)^2$ is maximum, when $\cos x = 1$ $\quad$ ($\because$ maximum value of $\cos x = 1$)

$$\therefore \quad p_{min} = \frac{13}{4} - \left(1 + \frac{1}{2}\right)^2 = \frac{13}{4} - \frac{9}{4} = 1$$

Now, ratio of the greatest value of $2 - \cos x + \sin^2 x$ to its least value

$$= \frac{13}{4} : 1 = 13 : 4$$

5. (a) We have, $\quad x^2 - px + q = 0 \quad$ ...(i)

$\because \sin x$ and $\cos x$ are roots of Eq. (i), then

$$\sin x + \cos x = p$$
$$\sin x\cos x = q$$

We know that, $A^3 + B^3 = (A + B)^3 - 3AB(A + B)$

Now, $\sin^3 x + \cos^3 x$

$$= (\sin x + \cos x)^3 - 3\sin x\cos x(\sin x + \cos x)$$
$$= p^3 - 3pq$$

6. (b) We have, $x + y = \frac{\pi}{6}$ and $\tan x\cdot\tan y = a$

Now, $\quad \tan(x + y) = \tan\frac{\pi}{6}$

$$\Rightarrow \quad \frac{\tan x + \tan y}{1 - \tan x\cdot\tan y} = \frac{1}{\sqrt{3}}$$
$$\Rightarrow \quad \frac{\tan x + \tan y}{1 - a} = \frac{1}{\sqrt{3}} \quad \text{(given)}$$
$$\Rightarrow \quad \tan x + \tan y = \frac{1}{\sqrt{3}}(1 - a)$$

If $\tan x$ and $\tan y$ are roots, then quadratic equation is

$$x^2 - (\tan x + \tan y)x + \tan x\tan y = 0$$
$$\Rightarrow \quad x^2 - \frac{1}{\sqrt{3}}(1 - a)x + a = 0$$
$$\therefore \quad \sqrt{3}x^2 - (1 - a)x + \sqrt{3}a = 0$$

7. (b) $\because \cot(\theta - \alpha), 3\cot\theta, \cot(\theta + \alpha)$ are in AP.

Then, $\quad 6\cot\theta = \cot(\theta - \alpha) + \cot(\theta + \alpha)$

$$\frac{6\cos\theta}{\sin\theta} = \frac{\sin(\theta - \alpha + \theta + \alpha)}{\sin(\theta - \alpha)\sin(\theta + \alpha)}$$
$$\left[\because \cot A + \cot B = \frac{\sin(A + B)}{\sin A\sin B}\right]$$
$$\Rightarrow \quad 6\cos\theta(\sin^2\theta - \sin^2\alpha) = \sin 2\theta\sin\theta$$
$$[\because \sin(A + B)\sin(A - B) = \sin^2 A - \sin^2 B]$$
$$\Rightarrow \quad 6\cos\theta(\sin^2\theta - \sin^2\alpha) = 2\sin^2\theta\cos\theta$$
$$\Rightarrow \quad 6\sin^2\theta - 6\sin^2\alpha = 2\sin^2\theta$$
$$\Rightarrow \quad 4\sin^2\theta = 6\sin^2\alpha$$
$$\Rightarrow \quad \frac{4\sin^2\theta}{6\sin^2\alpha} = 1$$
$$\Rightarrow \quad \frac{2\sin^2\theta}{3\sin^2\alpha} = 1$$
$$\therefore \quad \frac{304\sin^2\theta}{3\sin^2\alpha} = 152$$

Revisal Problems (JEE Advanced)

1. (a) $\because a\cos^2 3\alpha + b\cos^4\alpha = a(4\cos^3\alpha - 3\cos\alpha)^2 + b\cos^4\alpha$

$= a(16\cos^6\alpha + 9\cos^2\alpha - 24\cos^4\alpha) + b\cos^4\alpha$

$= 16a\cos^6\alpha + 9a\cos^2\alpha - 24a\cos^4\alpha + b\cos^4\alpha$

$= 16a\cos^6\alpha + (b - 24a)\cos^4\alpha + 9a\cos^2\alpha$

$(\because a\cos^2 3\alpha + b\cos^4\alpha = 9\cos^2\alpha + 16\cos^6\alpha \text{ given})$

$\Rightarrow 9\cos^2\alpha + 16\cos^6\alpha$

$= 16a\cos^6\alpha + (b - 24a)\cos^4\alpha + 9a\cos^2\alpha$

Now, on comparing both sides

$16a = 16 \Rightarrow a = 1$

and $b - 24a = 0$

$b - 24 \times 1 = 0 \Rightarrow b = 24$

$\therefore$ $a = 1$ and $b = 24$

2. (b) $\because 0 < \theta < \pi$, then $\sqrt{2(1+\cos\theta)} = \sqrt{2\cdot 2\cos^2\frac{\theta}{2}} = 2\cos\frac{\theta}{2}$

$\Rightarrow \sqrt{2 + 2\cos\frac{\theta}{2}} = \sqrt{2\left(1 + \cos\frac{\theta}{2}\right)} = \sqrt{2\cdot\cos^2\frac{\theta}{2^2}} = 2\cos\frac{\theta}{2^2}$

...... and so on.

Using symmetry

$\therefore$

$p = \sqrt{2 + 2 + \sqrt{2 + \cdots + \sqrt{2(1+\cos\theta)}}\ldots n \text{ times}} = 2\cos\frac{\theta}{2^n}$

and $q = \lim_{n\to\infty} p = 2\lim_{n\to\infty}\cos\frac{\theta}{2^n} = 2\cdot\cos 0 = 2$

$\therefore \frac{3 + 2q}{(2 + 3q)^2} = \frac{3 + 2\times 2}{(2 + 3\times 2)^2} = \frac{7}{64}$

3. (a) $\because \cot\frac{\alpha}{2}, \cot\frac{\beta}{2}$ are roots of $15x^2 - 26x + 8 = 0$

Then, $\cot\frac{\alpha}{2}\cot\frac{\beta}{2} = \frac{8}{15}$

$\tan\frac{\alpha}{2}\tan\frac{\beta}{2} = \frac{15}{8}$ ...(i)

and $\cot\frac{\alpha}{2} + \cot\frac{\beta}{2} = \frac{26}{15}$

$\Rightarrow \frac{\tan\frac{\alpha}{2} + \tan\frac{\beta}{2}}{\tan\frac{\alpha}{2}\cdot\tan\frac{\beta}{2}} = \frac{26}{15}$

$\Rightarrow \frac{\tan\frac{\alpha}{2} + \tan\frac{\beta}{2}}{\frac{15}{8}} = \frac{26}{15}$

$\therefore \tan\frac{\alpha}{2} + \tan\frac{\beta}{2} = \frac{26}{15}\times\frac{15}{8}$

$\tan\frac{\alpha}{2} + \tan\frac{\beta}{2} = \frac{26}{8}$ ...(ii)

We know that,

$\tan\left(\frac{\alpha}{2} + \frac{\beta}{2}\right) = \frac{\tan\frac{\alpha}{2} + \tan\frac{\beta}{2}}{1 - \tan\frac{\alpha}{2}\cdot\tan\frac{\beta}{2}} = \frac{\frac{26}{8}}{1 - \frac{15}{8}} = \frac{\frac{26}{8}}{\frac{-7}{8}}$

[from Eqs. (i) and (ii)]

$\tan\left(\frac{\alpha+\beta}{2}\right) = -\frac{26}{7}$ ...(iii)

and $\cos(\alpha+\beta) = \frac{1 - \tan^2\left(\frac{\alpha+\beta}{2}\right)}{1 + \tan^2\left(\frac{\alpha+\beta}{2}\right)}$

$= \frac{1 - \left(-\frac{26}{7}\right)^2}{1 + \left(-\frac{26}{7}\right)^2} = \frac{1 - \frac{676}{49}}{1 + \frac{676}{49}} = -\frac{627}{725}$

4. (c) We have, $\tan A, \tan B, \tan C$ are roots of $x^3 - \alpha x^2 - \beta = 0$,

then $\Sigma\tan A = \tan A + \tan B + \tan C = \alpha$,

$\Sigma\tan A\tan B = \tan A\tan B + \tan B\tan C + \tan C\tan A = 0$

and $\tan A\tan B\tan C = \beta$

Now, $(1 + \tan^2 A)(1 + \tan^2 B)(1 + \tan^2 C)$

$= 1 + \Sigma\tan^2 A + \Sigma\tan^2 A\tan^2 B + \tan^2 A\tan^2 B\tan^2 C$

$= 1 + (\Sigma\tan A)^2 - 2\Sigma\tan A\tan B + (\Sigma\tan A\tan B)^2$

$- 2\tan A\tan B\tan C\ \Sigma\tan A + (\tan A\tan B\tan C)^2$

$= 1 + \alpha^2 - 0 + 0 - 2\beta\cdot\alpha + \beta^2$

$= 1 + \alpha^2 - 2\alpha\beta + \beta^2 = 1 + (\alpha - \beta)^2$

5. (d) Given, $\cos\theta = \frac{a\cos\phi + b}{a + b\cos\phi}$ ...(i)

and $\tan\frac{\theta}{2} = \sqrt{\frac{m}{n}}\tan\frac{\phi}{2}$ ...(ii)

We know that, $\tan^2\frac{\theta}{2} = \frac{1 - \cos\theta}{1 + \cos\theta}$

$= \frac{1 - \frac{a\cos\phi + b}{a + b\cos\phi}}{1 + \frac{a\cos\phi + b}{a + b\cos\phi}}$ [from Eq. (i)]

$= \frac{a + b\cos\phi - a\cos\phi - b}{a + b\cos\phi + a\cos\phi + b}$

$= \frac{(a - b)(1 - \cos\phi)}{(a + b)(1 + \cos\phi)}$

$\Rightarrow \tan^2\frac{\theta}{2} = \frac{a - b}{a + b}\tan^2\frac{\phi}{2}$

$\tan\frac{\theta}{2} = \sqrt{\frac{a - b}{a + b}}\tan\frac{\phi}{2}$

But $\sqrt{\frac{m}{n}}\tan\frac{\phi}{2} = \sqrt{\frac{a - b}{a + b}}\tan\frac{\phi}{2}$

$\therefore \frac{m}{n} = \frac{a - b}{a + b}$

On applying componendo and dividendo, we get

$\frac{a - b + a + b}{a - b - a - b} = \frac{m + n}{m - n}$

$\Rightarrow \frac{2a}{-2b} = \frac{m + n}{m - n} \Rightarrow \frac{a}{b} = \frac{m + n}{n - m}$

$\therefore a : b = m + n : n - m$

6. (a) Given, $\lambda = \lim_{x\to 3}\frac{3x^2 - 9F(x)}{x - 3}$

$= \lim_{x\to 3}\frac{6x - 9F'(x)}{1}$ (applying L'Hospital's rule)

$= 6\times 3 - 9\times F'(3)$

$= 18 - 9\times 1$ $[\because F'(3) = 1]$

$\Rightarrow \lambda = 9$

and $\mu = \operatorname{cosec} 10^\circ - \sqrt{3}\sec 10^\circ = \frac{1}{\sin 10^\circ} - \frac{\sqrt{3}}{\cos 10^\circ}$

$= \frac{\cos 10^\circ - \sqrt{3}\sin 10^\circ}{\sin 10^\circ\cos 10^\circ} = 4\cdot\frac{\frac{1}{2}\cos 10^\circ - \frac{\sqrt{3}}{2}\sin 10^\circ}{2\sin 10^\circ\cos 10^\circ}$

$= 4\cdot\frac{\sin 30^\circ\cos 10^\circ - \cos 30^\circ\sin 10^\circ}{\sin 20^\circ} = 4\cdot\frac{\sin(30^\circ - 10^\circ)}{\sin 20^\circ}$

$\mu = 4$

$\therefore \lambda - \mu = 9 - 4 = 5$

7. (c) We have,
$$\frac{(\sin A+\sin B)(\sin B+\sin C)(\sin C+\sin A)}{\sin A\sin B\sin C}\ge p \quad \ldots(i)$$
We know that, AM≥ GM, then
$$\sin A+\sin B\ge 2\sqrt{\sin A\sin B}$$
$$\sin B+\sin C\ge 2\sqrt{\sin B\sin C}$$
and $$\sin C+\sin A\ge 2\sqrt{\sin C\sin A}$$
Now,
$$(\sin A+\sin B)(\sin B+\sin C)(\sin C+\sin A)\ge 8\sin A\sin B\sin C$$
$$\Rightarrow \frac{(\sin A+\sin B)(\sin B+\sin C)(\sin C+\sin A)}{\sin A\sin B\sin C}\ge 8 \quad \ldots(ii)$$
($\because$ A, B, C are positive acute angles)

From Eqs. (i) and (ii), $p=8$

8. (a,b,c,d) $\because F_n(\theta)=\tan\frac{\theta}{2}(1+\sec\theta)(1+\sec 2\theta)(1+\sec 4\theta)\cdots(1+\sec 2^n\theta)$
$$=\tan\frac{\theta}{2}\left(\frac{1+\cos\theta}{\cos\theta}\right)\left(\frac{1+\cos 2\theta}{\cos 2\theta}\right)\left(\frac{1+\cos 4\theta}{\cos 4\theta}\right)\cdots\left(\frac{1+\cos 2^n\theta}{\cos 2^n\theta}\right)$$
$$=\tan\frac{\theta}{2}\cdot\frac{2\cos^2\frac{\theta}{2}\cdot 2\cos^2\theta\, 2\cos^2 2\theta\cdots 2\cos^2 2^{n-1}\theta}{\cos\theta\cos 2\theta\cos 4\theta\cdots\cos 2^n\theta}$$
$$=\frac{\sin\frac{\theta}{2}}{\cos\frac{\theta}{2}}\cdot\frac{2^n\cos^2\frac{\theta}{2}\cos^2\theta\cos^2 2\theta\cdots\cos^2 2^{n-1}\theta}{\cos\theta\cos 2\theta\cos 4\theta\cdots\cos 2^n\theta}$$
$$=\frac{2^n\sin\frac{\theta}{2}\cos\frac{\theta}{2}\cdot\cos\theta\cos 2\theta\cdots\cos 2^{n-1}\theta}{\cos 2^n\theta}$$
$$=\frac{\sin 2^n\theta}{\cos 2^n\theta}=\tan 2^n\theta \quad \text{(using formula)}$$
$$F_n(\theta)=\tan 2^n\theta$$
Now,
$$F_3\left(\frac{\pi}{32}\right)=\tan\left(2^3\cdot\frac{\pi}{32}\right)=\tan\frac{\pi}{4}=1$$
$$F_4\left(\frac{\pi}{64}\right)=\tan\left(2^4\cdot\frac{\pi}{64}\right)=\tan\frac{\pi}{4}=1$$
$$F_5\left(\frac{\pi}{128}\right)=\tan\left(2^5\cdot\frac{\pi}{128}\right)=\tan\frac{\pi}{4}=1$$
$$F_6\left(\frac{\pi}{256}\right)=\tan\left(2^6\cdot\frac{\pi}{256}\right)=\tan\frac{\pi}{4}=1$$

9. (b,c) $\because \cos^2\theta+\alpha,\ \sin^2\theta+\alpha$ are roots of $x^2+2bx+b=0$,

then
$$(\cos^2\theta+\alpha)+(\sin^2\theta+\alpha)=-2b$$
$$(\cos^2\theta+\alpha)(\sin^2\theta+\alpha)=b$$
We know that, $(a-b)^2=(a+b)^2-4ab$, then
$$(\cos^2\theta+\alpha-\sin^2\theta-\alpha)^2$$
$$=[(\cos^2\theta+\alpha)+(\sin^2\theta+\alpha)]^2-4(\cos^2\theta+\alpha)(\sin^2\theta+\alpha)$$
$$=(-2b)^2-4\times b=4b^2-4b$$
$$\therefore (\cos^2\theta-\sin^2\theta)^2=4b^2-4b \quad \ldots(i)$$
Similarly, $\cos^2\theta+\beta,\ \sin^2\theta+\beta$ are roots of $x^2+4x+2=0$.
$$\therefore (\cos^2\theta-\sin^2\theta)^2=(-4)^2-4\times 2$$
$$=16-8=8$$
Using Eq. (i), we get
$$4b^2-4b=8$$
$$b^2-b-2=0$$
$$b^2-2b+b-2=0$$
$$b(b-2)+1(b-2)=0$$
$$(b-2)(b+1)=0$$
$$\therefore b=-1,2$$

10. (a,b,c) Given, $\cos x=\tan y$, $\cos y=\tan z$, $\cos z=\tan x$

Then, by symmetry $\cos x=\tan x$
$$\cos x=\frac{\sin x}{\cos x}$$
$$\sin x=\cos^2 x$$
$$\sin x=1-\sin^2 x$$
$$\sin^2 x+\sin x-1=0$$
$$\therefore \sin x=\frac{-1\pm\sqrt{1-4\times 1\times(-1)}}{2}$$
$$\sin x=\frac{-1\pm\sqrt{5}}{2}$$
$$\because \sin x=\frac{-1-\sqrt{5}}{2} \quad \text{(impossible)}$$
Thus, $$\sin x=\frac{\sqrt{5}-1}{2} \quad \text{[option (b)]}$$
$$=2\cdot\frac{\sqrt{5}-1}{4}$$
$$\Rightarrow \sin x=2\cdot\sin 18° \quad \text{[option (a)]}$$
$$\left(\because \sin 18°=\frac{\sqrt{5}-1}{4}\right)$$
$$=2\sin(90°-72°)$$
$$=2\cos 72° \quad \text{[option (c)]}$$

11. (a,b) Given, $\sin n\theta=\sum_{r=0}^{n} b_r\sin^r\theta$
$$\sin n\theta=b_0+b_1\sin\theta+b_2\sin^2\theta+b_3\sin^3\theta+\ldots+b_n\sin^n\theta \quad \ldots(i)$$
Put $\theta=0$, we get
$$b_0=0$$
On differentiating Eq. (i) w.r.t. θ, we have
$$n\cos n\theta=0+b_1\cos\theta+2b_2\sin\theta\cos\theta+3b_3\sin^2\theta\cos\theta+\ldots+nb_n\sin^{n-1}\theta\cos\theta$$
Again, put $\theta=0$, we get
$$n=b_1+0+0\ldots$$
$$\therefore b_1=n$$
Thus, $b_0=0$ and $b_1=n$

12. (b,d) We have
$$a=\sin\frac{\pi}{18}\cdot\sin\frac{5\pi}{18}\cdot\sin\frac{7\pi}{18}$$
$$=\sin 10°\cdot\sin 50°\cdot\sin 70°$$
$$=\sin 10°\sin(60°-10°)\sin(60°+10°)$$
$$=\frac{1}{4}\cdot\sin 3\times 10°$$
$$\left[\because \sin\theta\cdot\sin(60°-\theta)\cdot\sin(60°+\theta)=\frac{1}{4}\sin 3\theta\right]$$
$$=\frac{1}{4}\cdot\sin 30°=\frac{1}{4}\times\frac{1}{2}=\frac{1}{8}$$
$$\Rightarrow a=\frac{1}{8} \quad \text{[option (d)]}\ldots(i)$$
Now, $$y=2[x]+2 \quad \ldots(ii)$$
$$y=3[x-2]=3[x]-[6]$$
$$y=3[x]-6 \quad \ldots(iii)$$
From Eqs. (ii) and (iii), we get
$$3[x]-6=2[x]+2$$
$$\Rightarrow 3[x]-2[x]=8$$
$$[x]=8$$
Then, $[x]=8$ put in Eq. (i), $a=\frac{1}{[x]}$ [option (b)]

13. (a,b,c) We have,

$$\sin^4 \alpha \cos^2 \alpha = \sum_{k=0}^{3} C_k \cos 2k\alpha$$

$\therefore \quad \sin^4 \alpha \cos^2 \alpha = C_0 + C_1 \cos 2\alpha + C_2 \cos 4\alpha + C_3 \cos 6\alpha$

Put $\alpha = 0$, then

$$C_0 + C_1 + C_2 + C_3 = 0 \qquad \text{[option (c)]...(i)}$$

Put $\alpha = \frac{\pi}{2}$, then

$$C_0 - C_1 + C_2 - C_3 = 0 \qquad \text{...(ii)}$$

On solving Eqs. (i) and (ii), we get

$$C_1 + C_3 = 0 \text{ and } C_0 + C_2 = 0$$

14. (b,c,d) We have,

$$F(x) = ax + b, a < 0$$
$$F'(x) = a < 0$$

$\therefore$ $F(x)$ is decreasing function.

It is given that $F(x)$ is an onto function, then

$$F(-1) = 2 \text{ and } F(1) = 0$$

$\Rightarrow \quad -a + b = 2 \qquad$ ...(i)

and $\quad a + b = 0 \qquad$ ...(ii)

On solving Eqs. (i) and (ii), we get

$$a = -1 \text{ and } b = 1$$

Thus, $\quad F(x) = -x + 1$

Now, $F(0) = 1, F\left(\frac{1}{4}\right) = \frac{3}{4}, F(-2) = 3, F\left(\frac{1}{3}\right) = \frac{2}{3}, F(-1) = 2$

Also, we have

$$A = \cos^2 \theta + \sin^4 \theta$$
$$= \frac{1 + \cos 2\theta}{2} + \left(\frac{1 - \cos 2\theta}{2}\right)^2$$
$$= \frac{1 + \cos 2\theta}{2} + \frac{1}{4} + \frac{1}{4}\cos^2 2\theta - \frac{1}{2}\cos 2\theta$$
$$= \frac{3}{4} + \frac{1}{4}\cos^2 2\theta = \frac{3}{4} + \frac{1}{4}\cdot\left(\frac{1 + \cos 4\theta}{2}\right)$$
$$= \frac{3}{4} + \frac{1}{8} + \frac{1}{8}\cos 4\theta$$
$$A = \frac{7}{8} + \frac{1}{8}\cos 4\theta$$

We know that,

$$-1 \le \cos 4\theta \le 1$$
$$-\frac{1}{8} \le \frac{1}{8}\cos 4\theta \le \frac{1}{8}$$
$$\frac{3}{4} \le \frac{7}{8} + \frac{1}{8}\cos 4\theta \le 1$$
$$\frac{3}{4} \le A \le 1$$

$\therefore \quad F\left(\frac{1}{4}\right) \le A \le F(0)$

15. (a,b,c,d) $\because \quad 3 \sin \beta = \sin (2\alpha + \beta) \qquad$ ...(i)

$$2 \sin \beta = \sin (2\alpha + \beta) - \sin \beta$$
$$= 2 \cos\left(\frac{2\alpha + \beta + \beta}{2}\right)\cdot \sin\left(\frac{2\alpha + \beta - \beta}{2}\right)$$

$\Rightarrow \quad 2 \sin \beta = 2 \cos (\alpha + \beta)\cdot \sin \alpha \qquad$ [option (b)]

$\Rightarrow \quad \sin \beta = \cos (\alpha + \beta) \sin \alpha \qquad$ ...(ii)

Now, $\quad \sin \beta = \sin [(\alpha + \beta) - \alpha]$

$$= \sin (\alpha + \beta)\cdot \cos \alpha - \cos (\alpha + \beta)\cdot \sin \alpha$$

$\Rightarrow \quad \sin \beta = \sin (\alpha + \beta)\cdot \cos \alpha - \sin \beta \qquad$ [from Eq. (ii)]

$\Rightarrow \quad 2 \sin \beta = \sin (\alpha + \beta)\cdot \cos \alpha \qquad$ [option (c)]

Now, from option (a)

LHS $= [\cot \alpha + \cot (\alpha + \beta)] [\cot \beta - 3 \cot (2\alpha + \beta)]$

$$= \frac{\sin (2\alpha + \beta)}{\sin \alpha \cdot \sin (\alpha + \beta)} \cdot \left[\frac{\cos \beta}{\sin \beta} - \frac{3 \cos (2\alpha + \beta)}{\sin (2\alpha + \beta)}\right]$$
$$= \frac{3 \sin \beta}{\sin \alpha \cdot \sin (\alpha + \beta)} \cdot \left[\frac{\cos \beta}{\sin \beta} - \frac{3 \cos (2\alpha + \beta)}{3 \sin \beta}\right] \qquad \text{[from Eq. (i)]}$$
$$= \frac{3 \sin \beta}{\sin \alpha \cdot \sin (\alpha + \beta)} \cdot \left[\frac{\cos \beta - \cos (2\alpha + \beta)}{\sin \beta}\right]$$
$$= \frac{3}{\sin \alpha \cdot \sin (\alpha + \beta)} \cdot 2 \sin\left(\frac{\beta + 2\alpha + \beta}{2}\right) \cdot \sin\left(\frac{2\alpha + \beta - \beta}{2}\right)$$
$$= \frac{3}{\sin \alpha \cdot \sin (\alpha + \beta)} \cdot 2 \sin (\alpha + \beta) \sin \alpha = 6 = \text{RHS}$$

$\therefore$ Option (a) is correct.

Now, from option (d)

$$\tan (\alpha + \beta) = 2 \tan \alpha$$

$\Rightarrow \quad \dfrac{\sin (\alpha + \beta)}{\cos (\alpha + \beta)} = \dfrac{2 \sin \alpha}{\cos \alpha}$

$\Rightarrow \quad \sin (\alpha + \beta) \cos \alpha = 2 \sin \alpha \cdot \cos (\alpha + \beta)$

$\Rightarrow \quad 2 \sin \beta = 2 \sin \beta \qquad$ [using option (b) and (c)]

$\therefore$ Option (d) is correct.

16. (b,c,d) $\because x = \sec \theta - \tan \theta$

$\Rightarrow \quad x = \dfrac{1}{\cos \theta} - \dfrac{\sin \theta}{\cos \theta},$

$\Rightarrow \quad x = \dfrac{1 - \sin \theta}{\cos \theta}$

and $\quad y = \operatorname{cosec} \theta + \cot \theta$

$$y = \frac{1 + \cos \theta}{\sin \theta}$$

Now, $\quad x - y = \dfrac{1 - \sin \theta}{\cos \theta} - \dfrac{1 + \cos \theta}{\sin \theta}$

$$= \frac{\sin \theta - \sin^2 \theta - \cos \theta - \cos^2 \theta}{\sin \theta \cdot \cos \theta}$$

$\Rightarrow \quad x - y = \dfrac{\sin \theta - \cos \theta - 1}{\sin \theta \cdot \cos \theta} \qquad$ ...(i)

and $\quad xy + 1 = \dfrac{1 - \sin \theta}{\cos \theta} \cdot \dfrac{1 + \cos \theta}{\sin \theta} + 1$

$$= \frac{1 - \sin \theta + \cos \theta - \sin \theta \cdot \cos \theta}{\sin \theta \cdot \cos \theta} + 1$$
$$= \frac{1 - \sin \theta + \cos \theta - \sin \theta \cdot \cos \theta + \sin \theta \cdot \cos \theta}{\sin \theta \cdot \cos \theta}$$

$$xy + 1 = \frac{1 - \sin \theta + \cos \theta}{\sin \theta \cdot \cos \theta} \qquad \text{...(ii)}$$

On adding Eqs. (i) and (ii), we get

$$xy + x - y + 1 = \frac{\sin \theta - \cos \theta - 1}{\sin \theta \cdot \cos \theta} + \frac{1 - \sin \theta + \cos \theta}{\sin \theta \cdot \cos \theta}$$
$$= \frac{\sin \theta - \cos \theta - 1 + 1 - \sin \theta + \cos \theta}{\sin \theta \cdot \cos \theta}$$

$\Rightarrow \quad xy + x - y + 1 = 0 \qquad$ [option (d)]...(iii)

Now, $\quad x(y + 1) = y - 1$

$$x = \frac{y - 1}{y + 1} \qquad \text{[option (b)]}$$

Again, from Eq. (iii),

$$xy + x - y + 1 = 0$$
$$y(-x + 1) = 1 + x$$
$$y = \frac{1 + x}{1 - x} \qquad \text{[option (c)]}$$

17. (8) $\because A + B + C = \pi$

and $\lambda \sin\frac{A}{2}\sin\frac{B}{2}\sin\frac{C}{2} = \frac{\sin 2A + \sin 2B + \sin 2C}{\sin A + \sin B + \sin C}$

$$= \frac{2\sin(A+B)\cos(A-B) + 2\sin C\cos C}{2\sin\left(\frac{A+B}{2}\right)\cos\left(\frac{A-B}{2}\right) + 2\sin\frac{C}{2}\cos\frac{C}{2}}$$

$$= \frac{2\sin C\cdot\cos(A-B) + 2\sin C\cos C}{2\cos\frac{C}{2}\cos\left(\frac{A-B}{2}\right) + 2\sin\frac{C}{2}\cos\frac{C}{2}}$$

$$= \frac{2\sin C\,[\cos(A-B) + \cos C]}{2\cos\frac{C}{2}\left[\cos\left(\frac{A-B}{2}\right) + \sin\frac{C}{2}\right]}$$

$$= \frac{\sin C\,[\cos(A-B) - \cos(A+B)]}{\cos\frac{C}{2}\left[\cos\left(\frac{A-B}{2}\right) + \cos\left(\frac{A+B}{2}\right)\right]} = \frac{\sin C\cdot 2\sin A\sin B}{\cos\frac{C}{2}\cdot 2\cos\frac{A}{2}\cos\frac{B}{2}}$$

$$= \frac{\sin A\sin B\sin C}{\cos\frac{A}{2}\cos\frac{B}{2}\cos\frac{C}{2}} = \frac{2\sin\frac{A}{2}\cos\frac{A}{2}\cdot 2\sin\frac{B}{2}\cos\frac{B}{2}\cdot 2\sin\frac{C}{2}\cos\frac{C}{2}}{\cos\frac{A}{2}\cos\frac{B}{2}\cos\frac{C}{2}}$$

$$= 8\sin\frac{A}{2}\sin\frac{B}{2}\sin\frac{C}{2}$$

$\therefore \qquad \lambda = 8$

18. (3) We have, $\lambda = \tan\frac{3\pi}{11} + 4\sin\frac{2\pi}{11} = \frac{\sin\frac{3\pi}{11}}{\cos\frac{3\pi}{11}} + 4\sin\frac{2\pi}{11}$

$\Rightarrow \qquad \lambda\cdot\cos\frac{3\pi}{11} = \sin\frac{3\pi}{11} + 4\sin\frac{2\pi}{11}\cdot\cos 2\frac{3\pi}{11}$

On squaring both sides, we get

$$\lambda^2\cos^2\frac{3\lambda}{11} = \sin^2\frac{3\pi}{11} + 16\sin^2\frac{2\pi}{11}\cdot\cos^2\frac{3\pi}{11} + 8\sin\frac{3\pi}{11}\cdot\cos\frac{3\pi}{11}\cdot\sin\frac{2\pi}{11}$$

$$\Rightarrow \lambda^2\left(2\cos^2\frac{3\pi}{11}\right) = 2\sin^2\frac{3\pi}{11} + 8\left(2\sin\frac{2\pi}{11}\cdot\cos\frac{3\pi}{11}\right)^2 + 8\sin\frac{6\pi}{11}\cdot\sin\frac{2\pi}{11}$$

$$= 1 - \cos\frac{6\pi}{11} + 8\left(\sin\frac{5\pi}{11} - \sin\frac{\pi}{11}\right)^2 + 4\left(\cos\frac{4\pi}{11} - \cos\frac{8\pi}{11}\right)$$

$$= 1 - \cos\frac{6\pi}{11} + 8\left(\sin^2\frac{5\pi}{11} + \sin^2\frac{\pi}{11} - 2\sin\frac{5\pi}{11}\sin\frac{\pi}{11}\right) + 4\cos\frac{4\pi}{11} - 4\cos\frac{8\pi}{11}$$

$$= 1 - \cos\frac{6\pi}{11} + 4\cdot 2\sin^2\frac{5\pi}{11} + 4\cdot 2\sin^2\frac{\pi}{11} - 8\left(\cos\frac{4\pi}{11} - \cos\frac{6\pi}{11}\right) + 4\cos\frac{4\pi}{11} - 4\cos\frac{8\pi}{11}$$

$$= 1 - \cos\frac{6\pi}{11} + 4\left(1 - \cos\frac{10\pi}{11}\right) + 4\left(1 - \cos\frac{2\pi}{11}\right) - 8\cos\frac{4\pi}{11} + 8\cos\frac{6\pi}{11} + 4\cos\frac{4\pi}{11} - 4\cos\frac{8\pi}{11}$$

$$= 9 - 4\cos\frac{2\pi}{11} - 4\cos\frac{4\pi}{11} + 7\cos\frac{6\pi}{11} - 4\cos\frac{8\pi}{11} - 4\cos\frac{10\pi}{11}$$

$$= 9 + 11\cos\frac{6\pi}{11} - 4\left(\cos\frac{2\pi}{11} + \cos\frac{4\pi}{11} + \cos\frac{6\pi}{11} + \cos\frac{8\pi}{11} + \cos\frac{10\pi}{11}\right)$$

$$= 9 + 11\cos\frac{6\pi}{11} - 4\cdot\frac{\sin\frac{5\pi}{11}}{\sin\frac{\pi}{11}}\cdot\cos\left(\frac{\frac{2\pi}{11} + \frac{10\pi}{11}}{2}\right)$$

[using formula $\cos\alpha + \cos(\alpha+\beta) + \cos(\alpha + 2\beta)\cdots\cos(\alpha + (n-1)\beta)$]

$$= 9 + 11\cos\frac{6\pi}{11} - \frac{4\sin\frac{5\pi}{11}}{\sin\frac{\pi}{11}}\times\cos\frac{6\pi}{11}$$

$$= 9 + 11\cos\frac{6\pi}{11} + \frac{2\cdot 2\sin\frac{5\pi}{11}\cdot\cos\frac{5\pi}{11}}{\sin\frac{\pi}{11}}$$

$$\left[\because \cos\frac{6\pi}{11} = \cos\left(\pi - \frac{5\pi}{11}\right) = -\cos\frac{5\pi}{11}\right]$$

$$= 9 + 11\cos\frac{6\pi}{11} + \frac{2\cdot\sin\frac{10\pi}{11}}{\sin\frac{\pi}{11}} = 9 + 11\cos\frac{6\pi}{11} + 2$$

$$= 11\left(1 + \cos\frac{6\pi}{11}\right)$$

$\Rightarrow \qquad \lambda^2\cdot 2\cos^2\frac{3\pi}{11} = 11\cdot 2\cos^2\frac{3\pi}{11}$

$\therefore \qquad \lambda^2 = 11 \Rightarrow [\lambda^2] = 3$

19. (3) $\because 4\cos 36° = 4\cdot\frac{\sqrt{5}+1}{4} = 1 + \sqrt{5}$

and we know that, $\cot\frac{\theta}{2} = \frac{1 + \cos\theta}{\sin\theta}$

Put $\theta = 15°$, we get

$$\cot 7\frac{1}{2}° = \frac{1 + \cos 15°}{\sin 15°} = \frac{1 + \frac{\sqrt{3}+1}{2\sqrt{2}}}{\frac{\sqrt{3}-1}{2\sqrt{2}}} = \frac{2\sqrt{2} + \sqrt{3} + 1}{\sqrt{3} - 1}\times\frac{\sqrt{3}+1}{\sqrt{3}+1}$$

$$= \frac{2\sqrt{6} + 2\sqrt{2} + 3 + \sqrt{3} + \sqrt{3} + 1}{2}$$

$$= \sqrt{6} + \sqrt{2} + \sqrt{3} + 2 = \sqrt{2} + \sqrt{3} + \sqrt{4} + \sqrt{6}$$

Now, $4\cos 36° + \cot 7\frac{1}{2}° = \sqrt{1} + \sqrt{2} + \sqrt{3} + \sqrt{4} + \sqrt{5} + \sqrt{6}$

$\Rightarrow \quad \sqrt{n_1} + \sqrt{n_2} + \sqrt{n_3} + \sqrt{n_4} + \sqrt{n_5} + \sqrt{n_6} = \sqrt{1} + \sqrt{2} + \sqrt{3} + \sqrt{4} + \sqrt{5} + \sqrt{6}$

$\Rightarrow \quad n_1 = 1, n_2 = 2, n_3 = 3, n_4 = 4, n_5 = 5, n_6 = 6$

$\therefore \qquad \frac{\sum_{i=1}^{6} n_i}{7} = 3$

20. (7) $\because \tan^2\left(\frac{\pi}{16}\right) + \tan^2\left(\frac{2\pi}{16}\right) + \tan^2\left(\frac{3\pi}{16}\right) + \cdots + \tan^2\left(\frac{7\pi}{16}\right) = \lambda$

$$\Rightarrow \lambda = \left[\tan^2\left(\frac{\pi}{16}\right) + \tan^2\left(\frac{7\pi}{16}\right) + \left\{\tan^2\left(\frac{2\pi}{16}\right) + \tan^2\left(\frac{6\pi}{16}\right)\right\} + \left\{\tan^2\left(\frac{3\pi}{16}\right) + \tan^2\left(\frac{5\pi}{16}\right)\right\} + \tan^2\left(\frac{4\pi}{16}\right)\right]$$

$$= \left[\tan^2\left(\frac{\pi}{16}\right) + \tan^2\left(\frac{\pi}{2} - \frac{\pi}{16}\right)\right] + \left[\tan^2\left(\frac{2\pi}{16}\right) + \tan^2\left(\frac{\pi}{2} - \frac{2\pi}{16}\right)\right] + \left[\tan^2\left(\frac{3\pi}{16}\right) + \tan^2\left(\frac{\pi}{2} - \frac{3\pi}{16}\right)\right] + (1)^2$$

$$= \left[\tan^2\left(\frac{\pi}{16}\right) + \cot^2\left(\frac{\pi}{16}\right)\right] + \left[\tan^2\left(\frac{2\pi}{16}\right) + \cot^2\left(\frac{2\pi}{16}\right)\right] + \left[\tan^2\left(\frac{3\pi}{16}\right) + \cot^2\left(\frac{3\pi}{16}\right)\right] + 1$$

$$= \left[\tan\left(\frac{\pi}{16}\right) + \cot\left(\frac{\pi}{16}\right)\right]^2 - 2 + \left[\tan\left(\frac{2\pi}{16}\right) + \cot\left(\frac{2\pi}{16}\right)\right]^2 - 2 + \left[\tan\left(\frac{3\pi}{16}\right) + \cot\left(\frac{3\pi}{16}\right)\right]^2 - 2 + 1$$

$$= \frac{1}{\sin^2\left(\frac{\pi}{16}\right)\cos^2\left(\frac{\pi}{16}\right)} + \frac{1}{\sin^2\left(\frac{2\pi}{16}\right)\cos^2\left(\frac{2\pi}{16}\right)} + \frac{1}{\sin^2\left(\frac{3\pi}{16}\right)\cos^2\left(\frac{3\pi}{16}\right)} - 5$$

$$= \frac{4}{\sin^2\left(\frac{\pi}{8}\right)} + \frac{4}{\sin^2\left(\frac{\pi}{4}\right)} + \frac{4}{\sin^2\left(\frac{3\pi}{8}\right)} - 5$$

$$= \frac{4}{\sin^2\left(\frac{\pi}{8}\right)} + \frac{4}{\sin^2\left(\frac{\pi}{2} - \frac{\pi}{8}\right)} + \frac{4}{\left(\frac{1}{\sqrt{2}}\right)^2} - 5$$

$$= 4\left[\frac{1}{\sin^2\left(\frac{\pi}{8}\right)} + \frac{1}{\cos^2\left(\frac{\pi}{8}\right)}\right] + 8 - 5 = 4 \cdot \frac{1}{\sin^2\frac{\pi}{8}\cos^2\frac{\pi}{8}} + 8 - 5$$

$$= \frac{16}{\sin^2\left(\frac{\pi}{4}\right)} + 3 = \frac{16}{\left(\frac{1}{\sqrt{2}}\right)^2} + 3$$

$= 16 \times 2 + 3 = 35$

$\therefore \quad \lambda = 35$

Now, $\quad x^y + y^x = 35$

$\therefore$ Either $\quad x = 1, y = 34$

or $x = 34, y = 1$, then

$x + y = 35$ in both cases

$$\therefore \quad \left(\frac{x+y}{5}\right) = \frac{35}{5} = 7$$

21. (1) We have, $\quad \cos\alpha = \tan 45°$

$\cos\alpha = 1$

$\therefore \quad \alpha = 0°$

and $\quad 4x^2 - 16x + 15 < 0$

$4x^2 - 10x - 6x + 15 < 0$

$2x(2x - 5) - 3(2x - 5) < 0$

$(2x - 5)(2x - 3) < 0$

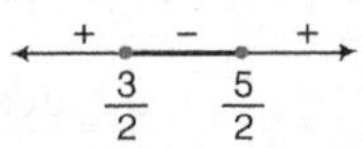

$\Rightarrow \quad \frac{3}{2} < x < \frac{5}{2}$

$\therefore \quad x = 2 \quad (\because x \in I)$

$\therefore \quad \tan\beta = 2 \quad$ (given)

Now, $\quad \frac{\sin(\alpha + \beta)}{\sin(\alpha - \beta)} = \frac{\sin(0 + \beta)}{\sin(0 - \beta)} = \frac{\sin\beta}{-\sin\beta}$

($\because \sin\beta \neq 0$ because $\tan\beta = 2$)

$\therefore \quad \frac{\sin(\alpha + \beta)}{\sin(\alpha - \beta)} = -1$

$\Rightarrow \quad -\lambda = -1 \Rightarrow \lambda = 1$

22. (1) We have, $\tan^2\theta + \sec\theta = \lambda$

$\sec^2\theta - 1 + \sec\theta - \lambda = 0$

$\sec^2\theta + \sec\theta - (\lambda + 1) = 0$

Now, $\quad \sec\theta = \frac{-1 \pm \sqrt{1 + 4(\lambda + 1)}}{2}$

$= \frac{-1 \pm \sqrt{1 + 4\lambda + 4}}{2}$

$\sec\theta = \frac{-1 \pm \sqrt{4\lambda + 5}}{2} \quad$ ...(i)

But we know that

$\sec\theta \le -1$ and $\sec\theta \ge 1$

$\Rightarrow \quad \frac{-1 - \sqrt{4\lambda + 5}}{2} \le -1$

$\Rightarrow \quad -1 - \sqrt{4\lambda + 5} \le -2$

$\Rightarrow \quad \sqrt{4\lambda + 5} \ge 1 \quad$ ...(ii)

and $\quad \frac{-1 + \sqrt{4\lambda + 5}}{2} \ge 1$

$\Rightarrow \quad -1 + \sqrt{4\lambda + 5} \ge 2$

$\Rightarrow \quad \sqrt{4\lambda + 5} \ge 3 \quad$ ...(iii)

From Eqs. (ii) and (iii), we get

$\sqrt{4\lambda + 5} \ge 3 \quad \Rightarrow \quad 4\lambda + 5 \ge 9$

$4\lambda \ge 4$

$\lambda \ge 1$ and $\lambda \in R$

So, $\quad \lambda \in [1, \infty)$ but $\lambda \in [a, \infty)$

$\therefore \quad a = 1$

23. (3) $\because \quad A + B + C = \pi \quad$ ...(i)

$\therefore \quad \frac{A}{2} + \frac{B}{2} + \frac{C}{2} = \frac{\pi}{2}$

We have,

$$\frac{\cos\left(\frac{A-B}{2}\right)}{\cos\left(\frac{A+B}{2}\right)} = \frac{2\sin\left(\frac{A+B}{2}\right)\cdot\cos\left(\frac{A-B}{2}\right)}{2\sin\left(\frac{A+B}{2}\right)\cdot\cos\left(\frac{A+B}{2}\right)}$$

$$= \frac{\sin\left(\frac{A+B}{2} + \frac{A-B}{2}\right) + \sin\left(\frac{A+B}{2} - \frac{A-B}{2}\right)}{\sin(A+B)}$$

$$= \frac{\sin A + \sin B}{\sin(A+B)} = \frac{\sin A + \sin B}{\sin(\pi - C)}$$

$$= \frac{\sin A + \sin B}{\sin C} = \frac{\sin A}{\sin C} + \frac{\sin B}{\sin C}$$

$$\therefore \quad \frac{\cos\left(\frac{A-B}{2}\right)}{\cos\left(\frac{A+B}{2}\right)} = \frac{\sin A}{\sin C} + \frac{\sin B}{\sin C} \quad \text{...(ii)}$$

$$\text{Similarly,} \quad \frac{\cos\left(\frac{B-C}{2}\right)}{\cos\left(\frac{B+C}{2}\right)} = \frac{\sin B}{\sin A} + \frac{\sin C}{\sin A} \quad \text{...(iii)}$$

$$\frac{\cos\left(\frac{C-A}{2}\right)}{\cos\left(\frac{C+A}{2}\right)} = \frac{\sin C}{\sin B} + \frac{\sin A}{\sin B} \quad \text{...(iv)}$$

On adding Eqs. (ii), (iii) and (iv), we get

$$\frac{\cos\left(\frac{A-B}{2}\right)}{\cos\left(\frac{A+B}{2}\right)} + \frac{\cos\left(\frac{B-C}{2}\right)}{\cos\left(\frac{B+C}{2}\right)} + \frac{\cos\left(\frac{C-A}{2}\right)}{\cos\left(\frac{C+A}{2}\right)}$$

$$= \frac{\sin A}{\sin C} + \frac{\sin B}{\sin C} + \frac{\sin B}{\sin A} + \frac{\sin C}{\sin A} + \frac{\sin C}{\sin B} + \frac{\sin A}{\sin B}$$

$$\ge 6 \times 6\sqrt{\frac{\sin A}{\sin C}\cdot\frac{\sin B}{\sin C}\cdot\frac{\sin B}{\sin A}\cdot\frac{\sin C}{\sin A}\cdot\frac{\sin C}{\sin B}\cdot\frac{\sin A}{\sin B}}$$

($\because$ AM $\ge$ GM)

$= 6 \times 1 = 6$

But $\quad 2k = 6$

$\therefore \quad k = 3$

24. (0) $\because$ α is the common root of equations $x^2 - ax + 12 = 0$, $x^2 - bx + 15 = 0$ and $x^2 - (a + b)x + 36 = 0$, then

$\alpha^2 - a\alpha + 12 = 0 \quad$ ...(i)

$\alpha^2 - b\alpha + 15 = 0 \quad$ ...(ii)

$\alpha^2 - (a + b)\alpha + 36 = 0 \quad$ ...(iii)

On solving Eqs. (i), (ii) and (iii), we get

$\alpha = \pm 3$

$\alpha = 3 \quad (\because \alpha > 0)$

Thus, $\cos x + \cos 2x + \cos 3x = \alpha = 3$

which is possible only when

$\cos x = 1, \cos 2x = 1$ and $\cos 3x = 1$

Then, $\quad \sin x = 0, \sin 2x = 0$ and $\sin 3x = 0$

$\therefore \quad \sin x + \sin 2x + \sin 3x = 0$

25. (5) We have,

$$\frac{\lambda}{2} = 1 + 2(1-\cos\theta) + 3(1-\cos\theta)^2 + 4(1-\cos\theta)^3 + \dots\infty \quad \dots(i)$$

We know that, $1 + 2x + 3x^2 + 4x^3 + \dots\infty = (1-x)^{-2}$

Put $x = 1 - \cos\theta$, we get

$$1 + 2(1-\cos\theta) + 3(1-\cos\theta)^2 + 4(1-\cos\theta)^3 + \dots\infty$$

$$= (1 - 1 + \cos\theta)^{-2} = \sec^2\theta = 1 + \tan^2\theta$$

$$= 1 + \frac{3}{2} = \frac{5}{2} \qquad \left(\because \tan\theta = \sqrt{\frac{3}{2}}\right)$$

Now, $\frac{\lambda}{2} = \frac{5}{2}$ [from Eq. (i)]

$\therefore \quad \lambda = 5$

26. (2) We have, $\alpha, \beta, \gamma, \delta$ are smallest positive angles satisfying the conditions

$\alpha < \beta < \gamma < \delta$ and $\sin\alpha = \sin\beta = \sin\gamma = \sin\delta = k$ (given)

Thus, we can take

$$\left.\begin{aligned}\beta &= \pi - \alpha,\\ \gamma &= 2\pi + \alpha\\ \delta &= 3\pi - \alpha\end{aligned}\right\} \quad \dots(i)$$

Also, $4\sin\frac{\alpha}{2} + 3\sin\frac{\beta}{2} + 2\sin\frac{\gamma}{2} + \sin\frac{\delta}{2} = p\sqrt{1+k}$

$$\Rightarrow 4\sin\frac{\alpha}{2} + 3\sin\left(\frac{\pi-\alpha}{2}\right) + 2\sin\left(\frac{2\pi+\alpha}{2}\right) + \sin\left(\frac{3\pi-\alpha}{2}\right) = p\sqrt{1+k} \quad \text{[from Eq. (i)]}$$

$$\Rightarrow 4\sin\frac{\alpha}{2} + 3\cos\frac{\alpha}{2} - 2\sin\frac{\alpha}{2} - \cos\frac{\alpha}{2} = p\sqrt{1+k}$$

$$\Rightarrow 2\sin\frac{\alpha}{2} + 2\cos\frac{\alpha}{2} = p\sqrt{1+k}$$

$$\Rightarrow 2\sqrt{\left(\sin\frac{\alpha}{2} + \cos\frac{\alpha}{2}\right)^2} = p\sqrt{1+k}$$

$$\Rightarrow 2\sqrt{1+\sin\alpha} = p\sqrt{1+k}$$

$$\Rightarrow 2\sqrt{1+k} = p\sqrt{1+k}$$

$$\therefore p = 2$$

27. (a) $\because \sin^2 A + \sin^2 B + \sin^2 C = \frac{1}{2}[2\sin^2 A + 2\sin^2 B] + \sin^2 C$

$$= \frac{1}{2}[1 - \cos 2A + 1 - \cos 2B] + \sin^2 C$$

$$= \frac{1}{2}[2 - (\cos 2A + \cos 2B)] + \sin^2 C$$

$$= 1 - \cos(A+B)\cos(A-B) + \sin^2 C$$

$$= 1 + \cos C\cos(A-B) - \cos^2 C + 1$$

$$= 2 - \cos^2 C + \cos C\cos(A-B)$$

$$\therefore \sin^2 A + \sin^2 B + \sin^2 C \le 2 - \cos^2 C + \cos C \quad \dots(i)$$

$$= 2 - \left(\cos^2 C - \cos C + \frac{1}{4} - \frac{1}{4}\right) = 2 - \left(\cos C - \frac{1}{2}\right)^2 + \frac{1}{4}$$

$$= \frac{9}{4} - \left(\cos C - \frac{1}{2}\right)^2 \le \frac{9}{4}$$

$$\therefore \sin^2 A + \sin^2 B + \sin^2 C \le \frac{9}{4}$$

Thus, **(ii) → (r)**

We know that, AM ≥ GM, then

$$\frac{\sin^2 A + \sin^2 B + \sin^2 C}{3} \ge \sqrt[3]{\sin^2 A \sin^2 B \sin^2 C}$$

$$\Rightarrow \frac{\frac{9}{4}}{3} \ge (\sin A \sin B \sin C)^{2/3} \quad \text{[using condition (ii)]}$$

$$\Rightarrow \sin A \sin B \sin C \le \left(\frac{3}{4}\right)^{\frac{3}{2}}$$

$$\Rightarrow \sin A \sin B \sin C \le \frac{3\sqrt{3}}{8}$$

Thus, **(i) → (q)**

Also, $\sin A \sin B \sin C > 0$

Thus, **(iii) → (s)**

From Eq. (i), we get

$$\sin^2 A + \sin^2 B + \sin^2 C \le 2 - \cos^2 C + \cos C$$

$$\Rightarrow \sin^2 A + \sin^2 B + \sin^2 C \le 1 + \sin^2 C + \cos C$$

$$\Rightarrow \sin^2 A + \sin^2 B \le 1 + \cos C$$

Thus, **(iv) → (p)**

28. (a) We have, $\prod_{i=1}^{n} \cos\alpha_i = 1$

$$\prod_{i=1}^{n}\cos\alpha_i = \prod_{i=1}^{n}\sin\alpha_i$$

$$\prod_{i=1}^{n}\cos^2\alpha_i = \prod_{i=1}^{n}\left(\frac{\sin 2\alpha_i}{2}\right) \quad \dots(i)$$

$\because \quad 0 \le \alpha_1, \alpha_2, \alpha_3, \cdots, \alpha_n \le \frac{\pi}{2}$ (given)

$$0 \le \alpha_i \le \frac{\pi}{2}$$

$$0 \le 2\alpha_i \le \pi$$

We know that, maximum value of $\sin 2\alpha_i$ is 1, $\forall i$, then

$$\prod_{i=1}^{n}\cos^2\alpha_i \le \frac{1}{2^n}$$

$$\prod_{i=1}^{n}\cos^2\alpha_i \le \frac{1}{2^{n/2}}$$

29. (a) $\because \cos\alpha + \cos\left(\alpha + \frac{2\pi}{3}\right) + \cos\left(\alpha + \frac{4\pi}{3}\right)$

$$= \cos\alpha + 2\cos(\pi + \alpha)\cdot\cos\frac{\pi}{3}$$

$$= \cos\alpha - 2\cos\alpha\cdot\frac{1}{2} = 0$$

$\because$ We know that, $a^3 + b^3 + c^3 = 3abc$

if $a + b + c = 0$

$$\therefore \cos^3\alpha + \cos^3\left(\alpha + \frac{2\pi}{3}\right) + \cos^3\left(\alpha + \frac{4\pi}{3}\right)$$

$$= 3\cos\alpha\cos\left(\alpha + \frac{2\pi}{3}\right)\cos\left(\alpha + \frac{4\pi}{3}\right)$$

30. (b) $\because$ Given, $\cos\frac{\pi}{7}, \cos\frac{3\pi}{7}, \cos\frac{5\pi}{7}$ are roots of the equation

$$8x^3 - 4x^2 - 4x + 1 = 0 \quad \dots(i)$$

On replacing x by $\frac{1}{x}$, we get

$$x^3 - 4x^2 - 4x + 8 = 0 \quad \dots(ii)$$

$\therefore \quad \sec\frac{\pi}{7}, \sec\frac{3\pi}{7}, \sec\frac{5\pi}{7}$ are roots of Eq. (ii).

Now, from Eq. (i), we have

$$8x^3 - 4x^2 - 4x + 1 = 8\left(x - \cos\frac{\pi}{7}\right)\left(x - \cos\frac{3\pi}{7}\right)\left(x - \cos\frac{5\pi}{7}\right) \quad \dots(iii)$$

Put $x = 1$, then

$$8 - 4 - 4 + 1 = 8\left(1 - \cos\frac{\pi}{7}\right)\left(1 - \cos\frac{3\pi}{7}\right)\left(1 - \cos\frac{5\pi}{7}\right)$$

$$\Rightarrow \frac{1}{8} = 2\sin^2\frac{\pi}{14}\cdot 2\sin^2\frac{3\pi}{14}\cdot 2\sin^2\frac{5\pi}{14}$$

$$\Rightarrow \sin^2\frac{\pi}{14}\sin^2\frac{3\pi}{14}\sin^2\frac{5\pi}{14} = \frac{1}{64}$$

$$\therefore \sin\frac{\pi}{14}\sin\frac{3\pi}{14}\sin\frac{5\pi}{14} = \frac{1}{8}$$

31. (d) Put $x = -1$, from Q. 30 in Eq. (iii), we get

$$-8-4+4+1 = 8\left(-1-\cos\frac{\pi}{7}\right)\left(-1-\cos\frac{3\pi}{7}\right)\left(-1-\cos\frac{5\pi}{7}\right)$$

$$\Rightarrow \quad -7 = -8\left(1+\cos\frac{\pi}{7}\right)\left(1+\cos\frac{3\pi}{7}\right)\left(1+\cos\frac{5\pi}{7}\right)$$

$$\Rightarrow \quad \frac{7}{8} = 2\cos^2\frac{\pi}{14}\cdot 2\cos^2\frac{3\pi}{14}\cdot 2\cos^2\frac{5\pi}{14}$$

$$\Rightarrow \quad \cos^2\frac{\pi}{14}\cos^2\frac{3\pi}{14}\cos^2\frac{5\pi}{14} = \frac{7}{64}$$

$$\therefore \quad \cos\frac{\pi}{14}\cos\frac{3\pi}{14}\cos\frac{5\pi}{14} = \frac{\sqrt{7}}{8}$$

32. (c) From Q. 30 in Eq. (ii), $x^3 - 4x^2 - 4x + 8 = 0$

$$x^3 - 4x = 4x^2 - 8$$

$$\left[\begin{array}{l}\text{Let } y = \tan^2\theta = \sec^2\theta - 1 = x^2 - 1 \\ \therefore x^2 = y + 1, \text{ where } \theta = \frac{\pi}{7}, \frac{3\pi}{7}, \frac{5\pi}{7}\end{array}\right]$$

$$\Rightarrow \quad x(x^2-4) = 4(x^2-2)$$

On squaring both sides, we get

$$x^2(x^2-4)^2 = 16(x^2-2)^2$$

$$(y+1)(y+1-4)^2 = 16(y+1-2)^2$$

$$(y+1)(y-3)^2 = 16(y-1)^2$$

$$y^3 - 21y^2 + 35y - 7 = 0$$

Put $y = x$, then $x^3 - 21x^2 + 35x - 7 = 0$

$\therefore \quad \tan^2\frac{\pi}{7}, \tan^2\frac{3\pi}{7}, \tan^2\frac{5\pi}{7}$ are the roots.

33. (b) From Q. 32, we have

$\tan^2\frac{\pi}{7}, \tan^2\frac{3\pi}{7}, \tan^2\frac{5\pi}{7}$ are roots of

$$x^3 - 21x^2 + 35x - 7 = 0 \quad \text{...(i)}$$

Then, $\quad \tan^2\frac{\pi}{7} + \tan^2\frac{3\pi}{7} + \tan^2\frac{5\pi}{7} = 21$

$$\therefore \quad \sum_{r=1}^{3}\tan^2\left(\frac{2r-1}{7}\right) = 21$$

On replacing x by $\frac{1}{x}$ in Eq. (i), we get

$$7x^3 - 35x^2 + 21x - 1 = 0$$

$\therefore \quad \cot^2\frac{\pi}{7}, \cot^2\frac{3\pi}{7}, \cot^2\frac{5\pi}{7}$ are its roots.

Then, $\quad \cot^2\frac{\pi}{7} + \cot^2\frac{3\pi}{7} + \cot^2\frac{5\pi}{7} = \frac{35}{7}$

$$\sum_{r=1}^{3}\cot^2\left(\frac{2r-1}{7}\right) = 5$$

Now, $\quad \sum_{r=1}^{3}\tan^2\left(\frac{2r-1}{7}\right)\cdot\sum_{r=1}^{3}\cot^2\left(\frac{2r-1}{7}\right) = 21 \times 5 = 105$

JEE Main & AIEEE Archive

1. (d) We have, $f_k(x) = \frac{1}{k}(\sin^k x + \cos^k x)$, where $x \in R$ and $k \geq 1$

Now, $f_4(x) - f_6(x)$

$$= \frac{1}{4}(\sin^4 x + \cos^4 x) - \frac{1}{6}(\sin^6 x + \cos^6 x)$$

$$= \frac{1}{4}(1 - 2\sin^2 x\cos^2 x) - \frac{1}{6}(1 - 3\sin^2 x\cos^2 x)$$

$$= \frac{1}{4} - \frac{1}{6} = \frac{1}{12}$$

2. (b) Given expression is

$$\frac{\tan A}{1-\cot A} + \frac{\cot A}{1-\tan A}$$

$$= \frac{\sin A}{\cos A} \times \frac{\sin A}{\sin A - \cos A} + \frac{\cos A}{\sin A} \times \frac{\cos A}{\cos A - \sin A}$$

$$= \frac{1}{\sin A - \cos A}\left\{\frac{\sin^3 A - \cos^3 A}{\cos A \sin A}\right\}$$

$$= \frac{\sin^2 A + \sin A\cos A + \cos^2 A}{\sin A\cos A}$$

$$= \frac{1 + \sin A\cos A}{\sin A\cos A}$$

$$= 1 + \sec A\operatorname{cosec} A$$

3. (d) Given,

$$A = \sin^2 x + \cos^4 x$$

$$\Rightarrow \quad A = 1 - \cos^2 x + \cos^4 x$$

$$= \cos^4 x - \cos^2 x + \frac{1}{4} + \frac{3}{4}$$

$$= \left(\cos^2 x - \frac{1}{2}\right)^2 + \frac{3}{4} \quad \text{...(i)}$$

where, $0 \leq \left(\cos^2 x - \frac{1}{2}\right)^2 \leq \frac{1}{4} \quad \text{...(ii)}$

$$\therefore \quad \frac{3}{4} \leq A \leq 1$$

4. (b) $\because \quad \cos(\alpha+\beta) = \frac{4}{5} \Rightarrow \alpha+\beta \in$ Ist quadrant

and $\quad \sin(\alpha-\beta) = \frac{5}{13} \Rightarrow \alpha-\beta \in$ Ist quadrant

Now, $\quad 2\alpha = (\alpha+\beta) + (\alpha-\beta)$

$$\therefore \quad \tan 2\alpha = \frac{\tan(\alpha+\beta) + \tan(\alpha-\beta)}{1 - \tan(\alpha+\beta)\tan(\alpha-\beta)}$$

$$= \frac{\frac{3}{4} + \frac{5}{12}}{1 - \frac{3}{4}\cdot\frac{5}{12}} = \frac{56}{33}$$

5. (c) $\because \cos(\beta-\gamma) + \cos(\gamma-\alpha) + \cos(\alpha-\beta) = -\frac{3}{2}$

$\Rightarrow 2[\cos(\beta-\gamma) + \cos(\gamma-\alpha) + \cos(\alpha-\beta)] + 3 = 0$

$\Rightarrow 2[\cos(\beta-\gamma) + \cos(\gamma-\alpha) + \cos(\alpha-\beta)]$
$+ \sin^2\alpha + \cos^2\alpha + \sin^2\beta + \cos^2\beta + \sin^2\gamma + \cos^2\gamma = 0$

$\Rightarrow (\sin\alpha + \sin\beta + \sin\gamma)^2 + (\cos\alpha + \cos\beta + \cos\gamma)^2 = 0$

It is possible when, $\sin\alpha + \sin\beta + \sin\gamma = 0$

and $\quad \cos\alpha + \cos\beta + \cos\gamma = 0$

Hence, both Statements A and B are true.

6. (b) Area $= \frac{1}{2} \times$ Base $\times$ Altitude

$$= \frac{1}{2} \times (2x\cos\theta) \times (x\sin\theta) = \frac{1}{2}x^2\sin 2\theta$$

(since, maximum value of $\sin 2\theta$ is 1)

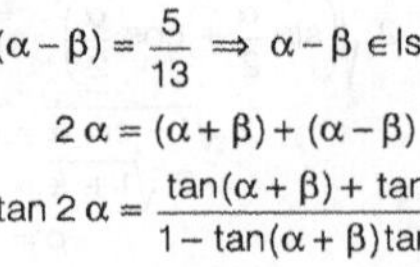

$\therefore \quad$ Maximum area $= \frac{1}{2}x^2$

7. (b) Given, $\cos x + \sin x = \frac{1}{2}$

$\therefore \quad \frac{1-\tan^2 \frac{x}{2}}{1+\tan^2 \frac{x}{2}} + \frac{2\tan \frac{x}{2}}{1+\tan^2 \frac{x}{2}} = \frac{1}{2}$

Let $\tan \frac{x}{2} = t$

$\Rightarrow \quad \frac{1-t^2}{1+t^2} + \frac{2t}{1+t^2} = \frac{1}{2}$

$\Rightarrow \quad 2(1-t^2+2t) = 1+t^2$

$\Rightarrow \quad 3t^2 - 4t - 1 = 0$

$\Rightarrow \quad t = \frac{2 \pm \sqrt{7}}{3}$

As, $\quad 0 < x < \pi$

$\Rightarrow \quad 0 < \frac{x}{2} < \frac{\pi}{2}$

So, $\tan \frac{x}{2}$ is positive.

$\therefore \quad t = \tan \frac{x}{2} = \frac{2+\sqrt{7}}{3}$

Now, $\quad \tan x = \frac{2\tan \frac{x}{2}}{1-\tan^2 \frac{x}{2}} = \frac{2t}{1-t^2}$

$\Rightarrow \quad \tan x = \frac{2\left(\frac{2+\sqrt{7}}{3}\right)}{1-\left(\frac{2+\sqrt{7}}{3}\right)^2}$

$\Rightarrow \quad \tan x = \frac{-3(2+\sqrt{7})}{1+2\sqrt{7}} \times \frac{1-2\sqrt{7}}{1-2\sqrt{7}}$

$\Rightarrow \quad \tan x = -\left(\frac{4+\sqrt{7}}{3}\right)$

8. (c) Since, $\tan\left(\frac{P}{2}\right)$ and $\tan\left(\frac{Q}{2}\right)$ are the roots of equation

$$ax^2 + bx + c = 0.$$

$\therefore \quad \tan\left(\frac{P}{2}\right) + \tan\left(\frac{Q}{2}\right) = -\frac{b}{a}$

and $\quad \tan\left(\frac{P}{2}\right)\tan\left(\frac{Q}{2}\right) = \frac{c}{a} \quad \ldots(i)$

Also, $\quad \frac{P}{2} + \frac{Q}{2} + \frac{R}{2} = \frac{\pi}{2} \quad (\because P+Q+R=\pi)$

$\Rightarrow \quad \frac{P+Q}{2} = \frac{\pi}{2} - \frac{R}{2}$

$\Rightarrow \quad \frac{P+Q}{2} = \frac{\pi}{4} \quad \left(\because \angle R = \frac{\pi}{2}\right)$

$\Rightarrow \quad \tan\left(\frac{P}{2}+\frac{Q}{2}\right) = 1$

$\Rightarrow \quad \frac{\tan\left(\frac{P}{2}\right)+\tan\left(\frac{Q}{2}\right)}{1-\tan\left(\frac{P}{2}\right)\tan\left(\frac{Q}{2}\right)} = 1$

$\Rightarrow \quad \frac{-\frac{b}{a}}{1-\frac{c}{a}} = 1 \quad$ [from Eq. (i)]

$\Rightarrow \quad -\frac{b}{a} = 1 - \frac{c}{a}$

$\Rightarrow \quad -b = a - c$

$\therefore \quad c = a + b$

Alternate Method

Since, $\quad \angle R = \frac{\pi}{2} \Rightarrow \angle P + \angle Q = \frac{\pi}{2}$

$\Rightarrow \quad \frac{\angle P}{2} = \frac{\pi}{4} - \frac{\angle Q}{2}$

$\therefore \quad \tan\left(\frac{P}{2}\right) = \tan\left(\frac{\pi}{4} - \frac{Q}{2}\right) = \frac{\tan\left(\frac{\pi}{4}\right) - \tan\left(\frac{Q}{2}\right)}{1+\tan\left(\frac{\pi}{4}\right)\tan\left(\frac{Q}{2}\right)}$

$\Rightarrow \quad \tan\left(\frac{P}{2}\right) + \tan\left(\frac{P}{2}\right)\tan\left(\frac{Q}{2}\right) = 1 - \tan\left(\frac{Q}{2}\right)$

$\Rightarrow \quad \tan\left(\frac{P}{2}\right) + \tan\left(\frac{Q}{2}\right) = 1 - \tan\left(\frac{P}{2}\right)\tan\left(\frac{Q}{2}\right)$

$\Rightarrow \quad -\frac{b}{a} = 1 - \frac{c}{a} \Rightarrow -b = a - c$

$\therefore \quad c = a + b$

9. (a) Given that, $\quad \sin\alpha + \sin\beta = -\frac{21}{65} \quad \ldots(i)$

and $\quad \cos\alpha + \cos\beta = -\frac{27}{65} \quad \ldots(ii)$

On squaring and adding Eqs. (i) and (ii), we get

$\sin^2\alpha + \sin^2\beta + 2\sin\alpha\sin\beta + \cos^2\alpha + \cos^2\beta + 2\cos\alpha\cos\beta$

$$= \left(-\frac{21}{65}\right)^2 + \left(-\frac{27}{65}\right)^2$$

$\Rightarrow \quad 2 + 2(\cos\alpha\cos\beta + \sin\alpha\sin\beta)$

$$= \frac{441}{4225} + \frac{729}{4225}$$

$\Rightarrow \quad 2[1+\cos(\alpha-\beta)] = \frac{1170}{4225}$

$\Rightarrow \quad \cos^2\left(\frac{\alpha-\beta}{2}\right) = \frac{1170}{4 \times 4225}$

$\Rightarrow \quad \cos^2\left(\frac{\alpha-\beta}{2}\right) = \frac{9}{130}$

$\therefore \quad \cos\left(\frac{\alpha-\beta}{2}\right) = -\frac{3}{\sqrt{130}}$

$$\left(\because \pi < \alpha - \beta < 3\pi \Rightarrow \frac{\pi}{2} < \frac{\alpha-\beta}{2} < \frac{3\pi}{2}\right)$$

10. (c) $\because \quad \sin^2\theta \le 1$

$\Rightarrow \quad \frac{4xy}{(x+y)^2} \le 1 \quad \left(\because \sin^2\theta = \frac{4xy}{(x+y)^2}\right)$

$\Rightarrow \quad x^2 + y^2 + 2xy - 4xy \ge 0$

$\Rightarrow \quad (x-y)^2 \ge 0$

which is true for all real values of x and y provided $x + y \ne 0$, otherwise $\frac{4xy}{(x+y)^2}$ will be meaningless.

11. (c) We know that,

$$\frac{1-\tan^2 15^\circ}{1+\tan^2 15^\circ} = \cos 30^\circ = \frac{\sqrt{3}}{2}$$

12. (a) $\because \quad \sin(\alpha+\beta) = 1$

$\Rightarrow \quad \alpha + \beta = \frac{\pi}{2} \quad \ldots(i)$

and $\quad \sin(\alpha-\beta) = \frac{1}{2}$

$\Rightarrow \quad \alpha - \beta = \frac{\pi}{6} \quad \ldots(ii)$

On solving Eqs. (i) and (ii), we get

$$\alpha = \frac{\pi}{3} \text{ and } \beta = \frac{\pi}{6}$$

$\therefore \tan(\alpha + 2\beta)\tan(2\alpha + \beta)$

$$= \tan\left(\frac{2\pi}{3}\right)\tan\left(\frac{5\pi}{6}\right) = \tan\left(\pi - \frac{\pi}{3}\right)\tan\left(\pi - \frac{\pi}{6}\right)$$

$$= \left(-\cot\frac{\pi}{3}\right)\left(-\cot\frac{\pi}{6}\right) = \frac{1}{\sqrt{3}} \times \sqrt{3} = 1$$

13. (d) Given that, $y = \sin^2\theta + \operatorname{cosec}^2\theta$

$\therefore \quad y = (\sin\theta - \operatorname{cosec}\theta)^2 + 2$

$\Rightarrow \quad y \geq 2, \quad \theta \neq 0$

14. (c) Since, $|c| > \sqrt{a^2 + b^2} \quad \Rightarrow \quad c < -\sqrt{a^2 + b^2}$

and $\quad c > \sqrt{a^2 + b^2}$

But $\quad -\sqrt{a^2 + b^2} \leq a\sin x + b\cos x \leq \sqrt{a^2 + b^2}$ ...(i)

and $\quad a\sin x + b\cos x = c$ ...(ii)

From Eqs. (i) and (ii), we see that there is no solution exists.

15. (b) Since, α is a root of

$$25\cos^2\theta + 5\cos\theta - 12 = 0$$

$\therefore \quad 25\cos^2\alpha + 5\cos\alpha - 12 = 0$

$\Rightarrow \quad (5\cos\alpha - 3)(5\cos\alpha + 4) = 0$

$\Rightarrow \quad \cos\alpha = -\frac{4}{5}$ and $\frac{3}{5}$

But $\frac{\pi}{2} < \alpha < \pi$, i.e. in second quadrant.

$\therefore \quad \cos\alpha = -\frac{4}{5}$

$\Rightarrow \quad \sin\alpha = \frac{3}{5}$

Now, $\quad \sin 2\alpha = 2\sin\alpha\cos\alpha$

$$= 2 \times \frac{3}{5} \times \left(-\frac{4}{5}\right) = -\frac{24}{25}$$

JEE Advanced & IIT JEE Archive

1. (a,b) Given, $f(\cos 4\theta) = \dfrac{2}{2 - \sec^2\theta}$ ...(i)

At, $\quad \cos 4\theta = \frac{1}{3} \Rightarrow 2\cos^2 2\theta - 1 = \frac{1}{3}$

$\Rightarrow \quad \cos^2 2\theta = \frac{2}{3} \Rightarrow \cos 2\theta = \pm\sqrt{\frac{2}{3}}$ ...(ii)

$\therefore \quad f(\cos 4\theta) = \dfrac{2 \cdot \cos^2\theta}{2\cos^2\theta - 1} = \dfrac{1 + \cos 2\theta}{\cos 2\theta}$

$\Rightarrow \quad f\left(\frac{1}{3}\right) = 1 \pm \sqrt{\frac{3}{2}}$ [from Eq. (ii)]

2. (3) Given equations can be written as

$$x\sin 3\theta - \frac{\cos 3\theta}{y} - \frac{\cos 3\theta}{z} = 0 \quad \text{...(i)}$$

$$x\sin 3\theta - \frac{2\cos 3\theta}{y} - \frac{2\sin 3\theta}{z} = 0 \quad \text{...(ii)}$$

and $x\sin 3\theta - \frac{2}{y}\cos 3\theta - \frac{1}{z}(\cos 3\theta + \sin 3\theta) = 0$ ...(iii)

Eqs. (ii) and (iii), implies

$$2\sin 3\theta = \cos 3\theta + \sin 3\theta \Rightarrow \sin 3\theta = \cos 3\theta$$

$\therefore \quad \tan 3\theta = 1$

$\Rightarrow 3\theta = \frac{\pi}{4}, \frac{5\pi}{4}, \frac{9\pi}{4}$ or $\theta = \frac{\pi}{12}, \frac{5\pi}{12}, \frac{9\pi}{12}$.

3. (a,b) Given, $\dfrac{\sin^4 x}{2} + \dfrac{\cos^4 x}{3} = \dfrac{1}{5}$

$\Rightarrow \quad \dfrac{\sin^4 x}{2} + \dfrac{(1 - \sin^2 x)^2}{3} = \dfrac{1}{5}$

$\Rightarrow \quad \dfrac{\sin^4 x}{2} + \dfrac{1 + \sin^4 x - 2\sin^2 x}{3} = \dfrac{1}{5}$

$\Rightarrow \quad 5\sin^4 x - 4\sin^2 x + 2 = \frac{6}{5}$

$\Rightarrow \quad 25\sin^4 x - 20\sin^2 x + 4 = 0$

$\Rightarrow \quad (5\sin^2 x - 2)^2 = 0$

$\Rightarrow \quad \sin^2 x = \frac{2}{5}, \cos^2 x = \frac{3}{5}, \tan^2 x = \frac{2}{3}$

$\therefore \quad \dfrac{\sin^8 x}{8} + \dfrac{\cos^8 x}{27} = \dfrac{1}{125}$

4. (c,d) For $0 < \theta < \frac{\pi}{2}$

$$\sum_{m=1}^{6} \operatorname{cosec}\left(\theta + \frac{(m-1)\pi}{4}\right)\operatorname{cosec}\left(\theta + \frac{m\pi}{4}\right) = 4\sqrt{2}$$

$$\Rightarrow \sum_{m=1}^{6} \frac{1}{\sin\left(\theta + \frac{(m-1)\pi}{4}\right)\sin\left(\theta + \frac{m\pi}{4}\right)} = 4\sqrt{2}$$

$$\Rightarrow \sum_{m=1}^{6} \frac{\sin\left[\theta + \frac{m\pi}{4} - \left(\theta + \frac{(m-1)\pi}{4}\right)\right]}{\sin\frac{\pi}{4}\left\{\sin\left(\theta + \frac{(m-1)\pi}{4}\right)\sin\left(\theta + \frac{m\pi}{4}\right)\right\}} = 4\sqrt{2}$$

$$\Rightarrow \sum_{m=1}^{6} \frac{\cot\left(\theta + \frac{(m-1)\pi}{4}\right) - \cot\left(\theta + \frac{m\pi}{4}\right)}{1/\sqrt{2}} = 4\sqrt{2}$$

$$\Rightarrow \sum_{m=1}^{6} \left[\cot\left(\theta + \frac{(m-1)\pi}{4}\right) - \cot\left(\theta + \frac{m\pi}{4}\right)\right] = 4$$

$$\Rightarrow \cot(\theta) - \cot\left(\theta + \frac{\pi}{4}\right) + \cot\left(\theta + \frac{\pi}{4}\right) - \cot\left(\theta + \frac{2\pi}{4}\right)$$

$$+ \ldots + \cot\left(\theta + \frac{5\pi}{4}\right) - \cot\left(\theta + \frac{6\pi}{4}\right) = 4$$

$\Rightarrow \quad \cot\theta - \cot\left(\frac{3\pi}{2} + \theta\right) = 4$

$\Rightarrow \quad \cot\theta + \tan\theta = 4$

$\Rightarrow \quad \tan^2\theta - 4\tan\theta + 1 = 0$

$\Rightarrow \quad (\tan\theta - 2)^2 - 3 = 0$

$\Rightarrow \quad (\tan\theta - 2 + \sqrt{3})(\tan\theta - 2 - \sqrt{3}) = 0$

$\Rightarrow \quad \tan\theta = 2 - \sqrt{3}$ or $\tan\theta = 2 + \sqrt{3}$

$\therefore \quad \theta = \frac{\pi}{12}; \ \theta = \frac{5\pi}{12} \quad \left[\because \theta \in \left(0, \frac{\pi}{2}\right)\right]$

5. (b) As, when $\theta \in \left(0, \frac{\pi}{4}\right)$, $\tan\theta < \cot\theta$

Since, $\quad \tan\theta < 1$ and $\cot\theta > 1$

$\therefore \quad (\tan\theta)^{\cot\theta} < 1$ and $(\cot\theta)^{\tan\theta} > 1$

$\therefore \; t_4 > t_1$ which only holds in (b).

6. (d) Since, $\quad \cos(\alpha - \beta) = 1$

$\Rightarrow \quad \alpha - \beta = n\pi,$

But $\quad -2\pi < \alpha - \beta < 2\pi \quad$ [as, $\alpha, \beta \in (-\pi, \pi)$]

$\therefore \quad \alpha - \beta = 0$ ...(i)

Given, $\quad \cos(\alpha + \beta) = \frac{1}{e}$

$\Rightarrow \cos 2\alpha = \frac{1}{e} < 1$, which is true for four values of α.

As, $\quad -2\pi < 2\alpha < 2\pi$

2. Height and Distance

DPP-1 Height and Distance

1. Let h be the height of the tower.

In ΔABD,

$$\tan 30° = \frac{h}{x + 60}$$

$$\Rightarrow \quad \frac{1}{\sqrt{3}} = \frac{h}{x + 60}$$

$$\Rightarrow \quad x = \sqrt{3}h - 60 \quad \ldots \text{(i)}$$

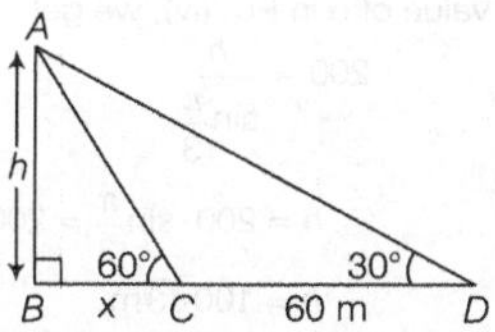

Now, in ΔABC,

$$\tan 60° = \frac{h}{x}$$

$$\Rightarrow \quad \sqrt{3} = \frac{h}{x}$$

$$\Rightarrow \quad h = \sqrt{3}x = \sqrt{3}(\sqrt{3}\,h - 60) \quad \text{[from Eq. (i)]}$$

$$\Rightarrow \quad h = 3h - 60\sqrt{3}$$

$$\Rightarrow \quad 2h = 60\sqrt{3}$$

$$\Rightarrow \quad h = 30\sqrt{3}$$

$\therefore$ Height of tower $= 30\sqrt{3}$ m

2. Let h be the height of pole.

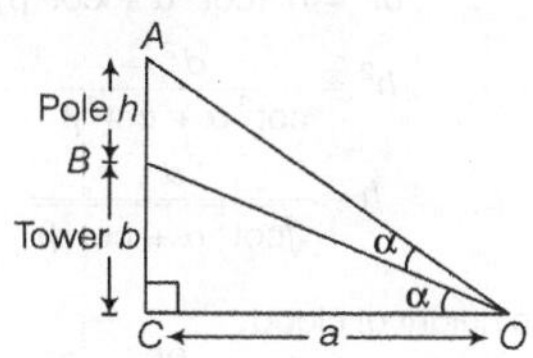

In ΔBCO,

$$\tan\alpha = \frac{b}{a} \quad \ldots \text{(i)}$$

and in ΔACO,

$$\tan 2\alpha = \frac{h + b}{a}$$

$$\Rightarrow \quad \frac{2\tan\alpha}{1 - \tan^2\alpha} = \frac{h + b}{a}$$

$$\Rightarrow \quad \frac{2 \cdot \frac{b}{a}}{1 - \frac{b^2}{a^2}} = \frac{h + b}{a} \quad \text{[from Eq. (i)]}$$

$$\Rightarrow \quad \frac{2ab}{a^2 - b^2} = \frac{h + b}{a}$$

$$\Rightarrow \quad h + b = \frac{2a^2b}{a^2 - b^2}$$

$$\Rightarrow \quad h = \frac{2a^2b - a^2b + b^3}{a^2 - b^2}$$

$$\Rightarrow \quad h = \frac{a^2b + b^3}{a^2 - b^2}$$

$$\therefore \quad h = \frac{b(a^2 + b^2)}{a^2 - b^2}$$

3. Let h be the height of the tower and A, B, C are three collinear points on a road making angles 30°, 45°, 60° respectively.

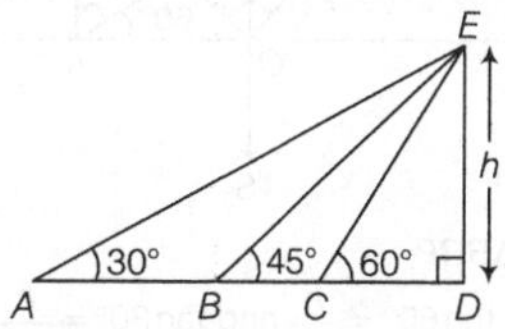

In ΔCDE,

$$\tan 60° = \frac{h}{CD}$$

$$\Rightarrow \quad CD = \frac{h}{\sqrt{3}} \quad \ldots \text{(i)}$$

In ΔBDE,

$$\tan 45° = \frac{h}{BD}$$

$$\Rightarrow \quad BD = h \quad \ldots \text{(ii)}$$

Now,

$$BC = BD - CD$$

$$BC = h - \frac{h}{\sqrt{3}} = \frac{h(\sqrt{3} - 1)}{\sqrt{3}}$$

[from Eqs. (i) and (ii)]

In ΔADE,

$$\tan 30° = \frac{h}{AD}$$

$$\Rightarrow \quad AD = \sqrt{3}h \quad \ldots \text{(iii)}$$

Now,

$$AB = AD - BD = \sqrt{3}h - h$$

[from Eqs. (ii) and (iii)]

$$= (\sqrt{3} - 1)h$$

So,

$$\frac{AB}{BC} = \frac{(\sqrt{3} - 1)h}{\frac{h(\sqrt{3} - 1)}{\sqrt{3}}} = \frac{\sqrt{3}}{1}$$

$$\therefore \quad AB : BC = \sqrt{3} : 1$$

Hence proved.

4. (a) Let h be the height of a tower.

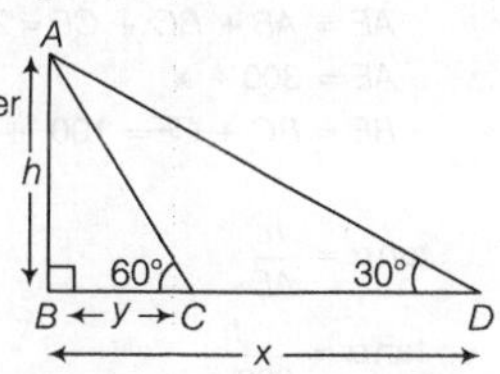

In ΔABC,

$$\tan 60° = \frac{h}{y}$$

$$\Rightarrow \quad y\sqrt{3} = h = \frac{45\sqrt{3}}{2} \quad \text{(given)}$$

$$\Rightarrow \quad y = \frac{45}{2}$$

In ΔABD,

$$\tan 30° = \frac{h}{x}$$

$$\Rightarrow \quad \frac{1}{\sqrt{3}} = \frac{h}{x}$$

$$\Rightarrow \quad x = \sqrt{3}h = \sqrt{3} \times \frac{45\sqrt{3}}{2}$$

$$\Rightarrow \quad x = \frac{45 \cdot 3}{2}$$

Now,

$$x - y = \frac{45 \cdot 3}{2} - \frac{45}{2} = 45 \text{ m}$$

5. (a) Let height of the temple, $OP = h$

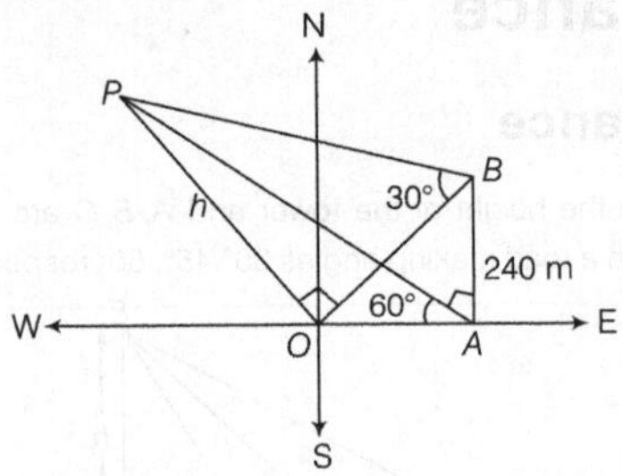

In ΔAOP and ΔBOP,

$$\tan 60° = \frac{h}{OA} \text{ and } \tan 30° = \frac{h}{OB}$$

$$\Rightarrow \quad OA = \frac{h}{\sqrt{3}} \text{ and } OB = h\sqrt{3}$$

In ΔOAB,

$$OB^2 = OA^2 + AB^2$$

$$(h\sqrt{3})^2 = \left(\frac{h}{\sqrt{3}}\right)^2 + (240)^2$$

$$3h^2 - \frac{h^2}{3} = (240)^2$$

$$\frac{8h^2}{3} = (240)^2$$

$$h = \frac{\sqrt{3}}{2\sqrt{2}} \times 240$$

$$= \frac{\sqrt{6} \times 24}{4} = 60\sqrt{6} \text{ m}$$

6. (d) Let h be the height of balloon and $CE = x$ m.

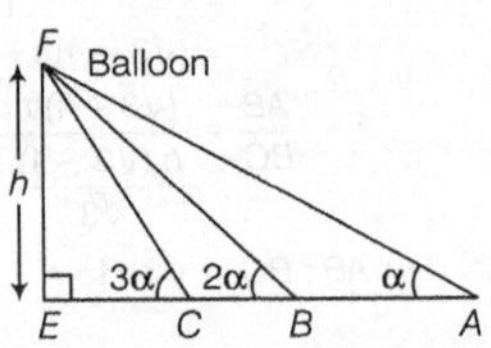

Given, $\quad AB = 200$ m, $BC = 100$ m

Now, $\quad AE = AB + BC + CE = 200 + 100 + x$

$$AE = 300 + x$$

and $\quad BE = BC + CE = 100 + x$

In ΔAEF,

$$\tan\alpha = \frac{h}{AE}$$

$$\tan\alpha = \frac{h}{300 + x}$$

$$300 + x = h\cot\alpha \quad \ldots(i)$$

In ΔBEF, $\quad \tan 2\alpha = \dfrac{h}{100 + x}$

$$\Rightarrow \quad 100 + x = h\cot 2\alpha \quad \ldots(ii)$$

and in ΔCEF $\quad \tan 3\alpha = \dfrac{h}{x}$

$$\Rightarrow \quad x = h\cot 3\alpha \quad \ldots(iii)$$

Subtracting Eq. (ii) from Eq. (i) and Eq. (iii) from Eq. (ii), we get

$$200 = h(\cot\alpha - \cot 2\alpha) \text{ and } 100 = h(\cot 2\alpha - \cot 3\alpha)$$

$$200 = \frac{h\sin(2\alpha - \alpha)}{\sin 2\alpha \sin\alpha} \text{ and } 100 = \frac{h\sin(3\alpha - 2\alpha)}{\sin 2\alpha \sin 3\alpha}$$

$$200 = \frac{h\sin\alpha}{\sin 2\alpha \sin\alpha} = \frac{h}{\sin 2\alpha} \quad \ldots(iv)$$

and $\quad 100 = \dfrac{h\sin\alpha}{\sin 3\alpha \sin 2\alpha} \quad \ldots(v)$

On dividing Eq. (iv) by Eq. (v), we get

$$\frac{h}{\sin 2\alpha} \times \frac{\sin 3\alpha \sin 2\alpha}{h\sin\alpha} = \frac{200}{100}$$

$$\frac{\sin 3\alpha}{\sin\alpha} = 2$$

$$\Rightarrow \quad 3\sin\alpha - 4\sin^3\alpha = 2\sin\alpha$$

$$\Rightarrow \quad \sin\alpha - 4\sin^3\alpha = 0$$

$$\Rightarrow \quad \sin\alpha(1 - 4\sin^2\alpha) = 0$$

$$\Rightarrow \quad 1 - 4\sin^2\alpha = 0 \quad (\because \sin\alpha \neq 0)$$

$$\Rightarrow \quad \sin^2\alpha = \left(\frac{1}{2}\right)^2 = \sin^2\frac{\pi}{6} \Rightarrow \alpha = \frac{\pi}{6}$$

On putting the value of α in Eq. (iv), we get

$$200 = \frac{h}{\sin\frac{\pi}{3}}$$

$$\Rightarrow \quad h = 200 \cdot \sin\frac{\pi}{3} = 200 \cdot \frac{\sqrt{3}}{2}$$

$$\therefore \quad h = 100\sqrt{3} \text{ m}$$

7. (c) $\because \quad OB = h\cot\beta$ and $OA = h\cot\alpha$

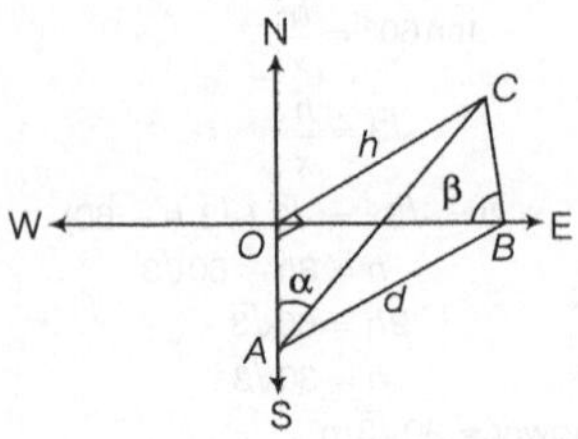

Then, $\quad OB^2 + OA^2 = h^2\cot^2\beta + h^2\cot^2\alpha$

$$\Rightarrow \quad d^2 = h^2(\cot^2\alpha + \cot^2\beta)$$

$$\Rightarrow \quad h^2 = \frac{d^2}{\cot^2\alpha + \cot^2\beta}$$

$$\Rightarrow \quad h = \frac{d}{\sqrt{\cot^2\alpha + \cot^2\beta}}$$

8. (a) Let h be the height of cloud.

In ΔABF, $\quad \cot 15° = \dfrac{BF}{AB}$

$$\Rightarrow \quad BF = (h - 2500)\cot 15° \quad \ldots(i)$$

In ΔBDF, $\quad \cot 45° = \dfrac{BF}{BD}$

$$\Rightarrow \quad BF = (h + 2500)\cot 45°$$

$$\Rightarrow \quad BF = (h + 2500) \quad \ldots(ii)$$

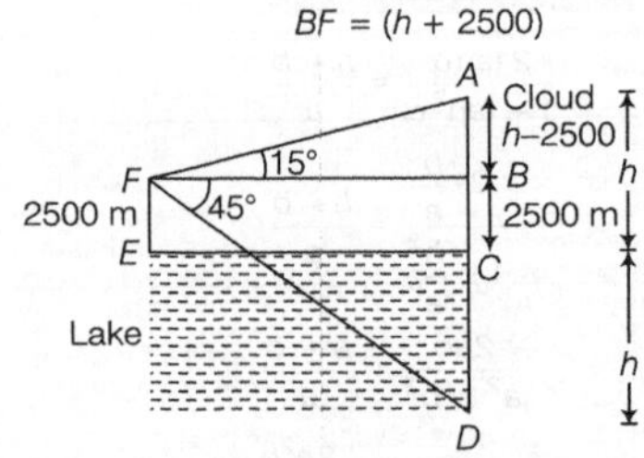

From Eqs. (i) and (ii), we get

$$(h - 2500)\cot 15° = h + 2500$$

$$\Rightarrow \quad (h - 2500)(\sqrt{3} + 2) = h + 2500 \quad [\because \cot 15° = (2 + \sqrt{3})]$$

$$\Rightarrow \quad (\sqrt{3} + 2)h - 2500(\sqrt{3} + 2) = h + 2500$$

$$\Rightarrow \quad (\sqrt{3} + 2 - 1)h = 2500 + 2500(\sqrt{3} + 2)$$

$$\Rightarrow \quad (\sqrt{3} + 1)h = 2500(\sqrt{3} + 3)$$

$$\Rightarrow \quad (\sqrt{3} + 1)h = 2500\sqrt{3}(\sqrt{3} + 1)$$

$$\Rightarrow \quad h = 2500\sqrt{3}$$

$\therefore$ Height of cloud $= 2500\sqrt{3}$ m

9. (a) Let x be the breadth of river and h be the height of a tree.

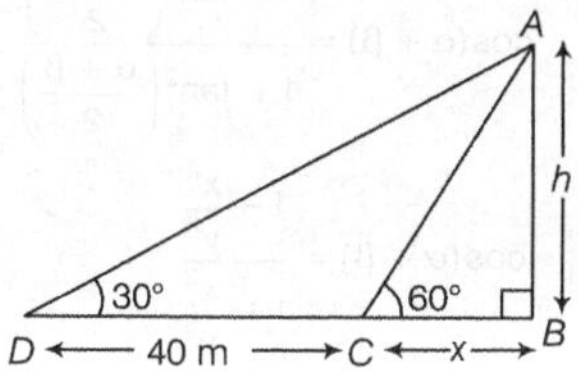

In ΔABC,

$$\cot 60^\circ = \frac{BC}{AB}$$

$$\Rightarrow \quad \frac{1}{\sqrt{3}} = \frac{x}{h}$$

$$\Rightarrow \quad h = \sqrt{3}x \quad \ldots\text{(i)}$$

In ΔABD, $\cot 30^\circ = \dfrac{x+40}{h}$

$$\Rightarrow \quad x + 40 = h \cdot \sqrt{3}$$

$$\Rightarrow \quad x + 40 = \sqrt{3}x \cdot \sqrt{3} \quad \text{[from Eq. (i)]}$$

$$\Rightarrow \quad x + 40 = 3x$$

$$\Rightarrow \quad 2x = 40$$

$$\Rightarrow \quad x = 20$$

$\therefore$ Breadth of the river = 20 m

10. (c) Let h be the height of the tower.

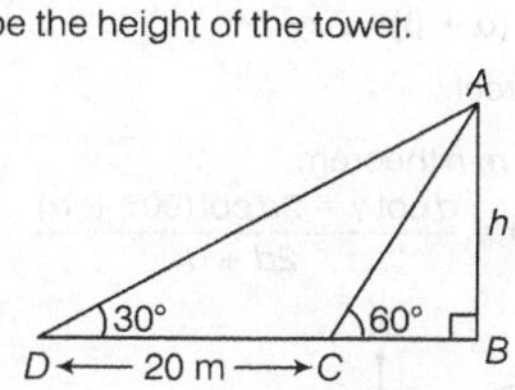

In ΔABC,

$$\cot 60^\circ = \frac{BC}{AB}$$

$$\Rightarrow \quad \frac{1}{\sqrt{3}} = \frac{BC}{h}$$

$$\Rightarrow \quad BC = \frac{h}{\sqrt{3}} \quad \ldots\text{(i)}$$

In ΔABD, $\cot 30^\circ = \dfrac{BD}{AB}$

$$\Rightarrow \quad \sqrt{3} = \frac{BC + CD}{h}$$

$$\Rightarrow \quad \sqrt{3}h = BC + 20$$

$$\Rightarrow \quad \sqrt{3}h = \frac{h}{\sqrt{3}} + 20 \quad \text{[from Eq. (i)]}$$

$$\Rightarrow \quad \sqrt{3}h - \frac{h}{\sqrt{3}} = 20$$

$$\Rightarrow \quad h = \frac{20\sqrt{3}}{2} = 10\sqrt{3}$$

$\therefore$ Height of the tower = $10\sqrt{3}$ m

11. (b) Let x be the distance of boat from the foot of the light house.

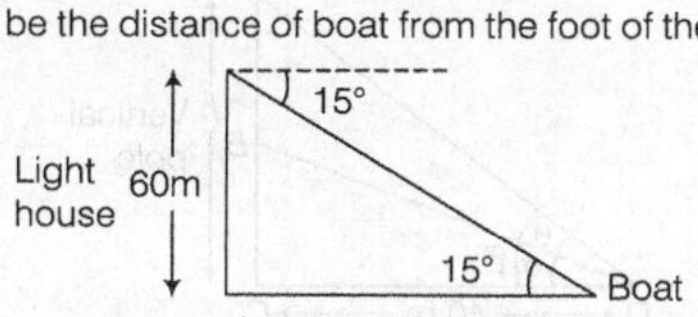

$$\therefore \quad \cot 15^\circ = \frac{x}{60}$$

$$x = 60\left(\frac{\sqrt{3}+1}{\sqrt{3}-1}\right) \text{m} \quad \left(\because \cot 15^\circ = \frac{\sqrt{3}+1}{\sqrt{3}-1}\right)$$

12. (b) Let height of plane to the ground

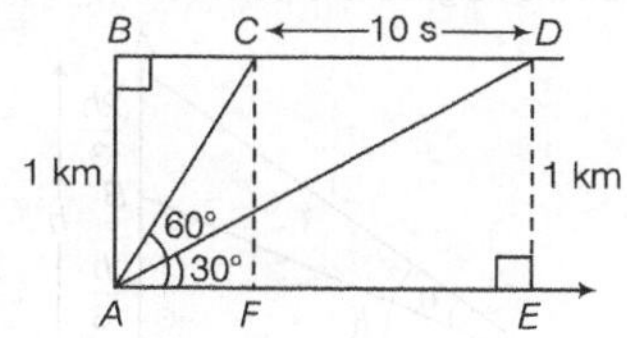

$$AB = CF = DE = 1\text{ km}$$

and $AF = BC$, $AE = BD$

In ΔAFC, $\cot 60^\circ = \dfrac{AF}{CF}$

$$\Rightarrow \quad \frac{1}{\sqrt{3}} = \frac{BC}{1} \quad (\because AF = BC)$$

$$\Rightarrow \quad BC = \frac{1}{\sqrt{3}}$$

and in ΔADE, $\cot 30^\circ = \dfrac{AE}{DE}$

$$\Rightarrow \quad \sqrt{3} = \frac{BD}{1} \quad (\because AE = BD)$$

$$\Rightarrow \quad BD = \sqrt{3}$$

Now, $CD = BD - BC = \sqrt{3} - \dfrac{1}{\sqrt{3}}$

$$\Rightarrow \quad CD = \frac{2}{\sqrt{3}} \text{ km}$$

$$\therefore \quad \text{Time} = 10\text{s} = \frac{10}{3600} = \frac{1}{360}\text{h}$$

$$\therefore \quad \text{Speed} = \frac{\text{Distance}}{\text{Time}} = \frac{CD}{\frac{1}{360}} = \frac{2}{\sqrt{3}} \times 360$$

$$= \frac{2\sqrt{3} \times 360}{3} = 240\sqrt{3} \text{ km/h}$$

13. (a) Let h be the height of pole.

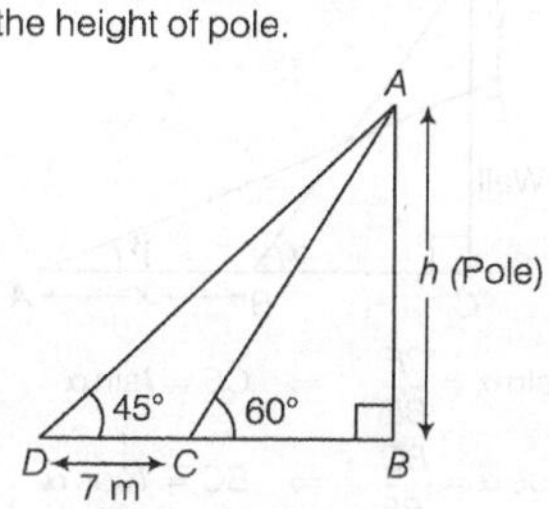

In ΔABC, $\cot 60^\circ = \dfrac{BC}{AB} = \dfrac{BC}{h}$

$$\Rightarrow \quad BC = \frac{h}{\sqrt{3}} \quad \ldots\text{(i)}$$

In ΔABD, $\tan 45^\circ = \dfrac{AB}{BD}$

$$\Rightarrow \quad 1 = \frac{AB}{BC + CD}$$

$$\Rightarrow \quad h = BC + CD$$

$$\Rightarrow \quad h = \frac{h}{\sqrt{3}} + 7 \quad [\because CD = 7\text{ m and from Eq. (i)}]$$

$$\Rightarrow \quad \left(\frac{\sqrt{3}-1}{\sqrt{3}}\right)h = 7 \Rightarrow h = \frac{7\sqrt{3}}{\sqrt{3}-1}$$

$$\Rightarrow \quad h = \frac{7\sqrt{3}}{\sqrt{3}-1} \times \frac{\sqrt{3}+1}{\sqrt{3}+1}$$

$$\Rightarrow \quad h = \frac{7\sqrt{3}(\sqrt{3}+1)}{2} \text{ m}$$

$\therefore$ Height of the vertical pole = $\dfrac{7\sqrt{3}}{2}(\sqrt{3}+1)$ m

14. (a,b) Let h be the height of the pole.

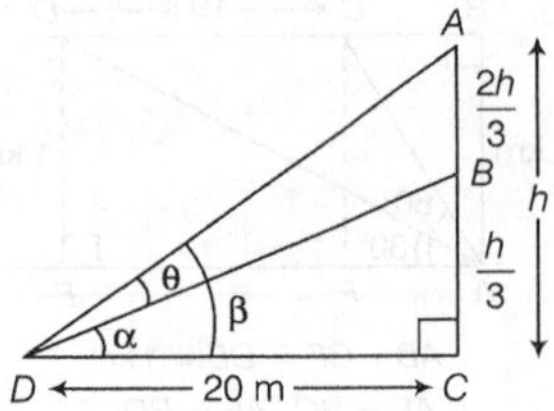

In ΔBCD, $\tan\alpha = \dfrac{\frac{h}{3}}{20}$

$\Rightarrow$ $\tan\alpha = \dfrac{h}{60}$ ...(i)

In ΔACD, $\tan\beta = \dfrac{h}{20}$ ...(ii)

We have, $\tan(\beta - \alpha) = \tan\theta$ $\left(\tan\theta = \dfrac{1}{2}, \text{ given}\right)$

$\Rightarrow$ $\dfrac{\tan\beta - \tan\alpha}{1 + \tan\alpha \cdot \tan\beta} = \dfrac{1}{2}$

$\Rightarrow$ $\dfrac{\frac{h}{20} - \frac{h}{60}}{1 + \frac{h}{20} \cdot \frac{h}{60}} = \dfrac{1}{2}$ $\Rightarrow$ $\dfrac{60h - 20h}{1200 + h^2} = \dfrac{1}{2}$

$\Rightarrow$ $h^2 - 80h + 1200 = 0$

$\Rightarrow$ $(h - 20)(h - 60) = 0$

$\Rightarrow$ $h = 20$ or 60

$\therefore$ Height of pole = 20 m or 60 m

15. (a,b,c) Let l be the length of the ladder, then $AD = BE = l$

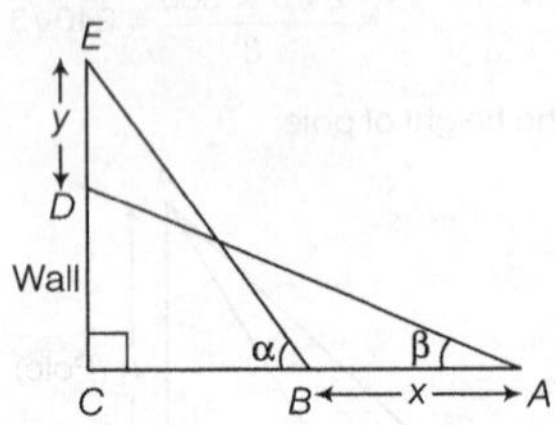

In ΔBCE, $\sin\alpha = \dfrac{CE}{BE}$ $\Rightarrow$ $CE = l\sin\alpha$

and $\cos\alpha = \dfrac{BC}{BE}$ $\Rightarrow$ $BC = l\cos\alpha$

In ΔACD, $\sin\beta = \dfrac{CD}{AD}$ $\Rightarrow$ $CD = l\sin\beta$

and $\cos\beta = \dfrac{AC}{AD}$ $\Rightarrow$ $AC = l\cos\beta$

$\because$ $x = AC - BC = l\cos\beta - l\cos\alpha$

$\Rightarrow$ $x = l(\cos\beta - \cos\alpha)$ ...(i)

and $y = CE - CD$ $\Rightarrow$ $y = l\sin\alpha - l\sin\beta$

$\Rightarrow$ $y = l(\sin\alpha - \sin\beta)$ ...(ii)

From Eqs. (i) and (ii), we get

$$\frac{x}{y} = \frac{\cos\beta - \cos\alpha}{\sin\alpha - \sin\beta} = \frac{2\sin\left(\frac{\alpha+\beta}{2}\right)\sin\left(\frac{\alpha-\beta}{2}\right)}{2\cos\left(\frac{\alpha+\beta}{2}\right)\sin\left(\frac{\alpha-\beta}{2}\right)}$$

$\Rightarrow$ $\dfrac{x}{y} = \tan\left(\dfrac{\alpha+\beta}{2}\right)$ ...(iii)

$\Rightarrow$ $x = y\tan\left(\dfrac{\alpha+\beta}{2}\right)$

$\therefore$ Option (a) is correct.

$\because$ $\cos(\alpha+\beta) = \dfrac{1 - \tan^2\left(\frac{\alpha+\beta}{2}\right)}{1 + \tan^2\left(\frac{\alpha+\beta}{2}\right)}$

$\Rightarrow$ $\cos(\alpha+\beta) = \dfrac{1 - \frac{x^2}{y^2}}{1 + \frac{x^2}{y^2}}$ [from Eq. (iii)]

$\Rightarrow$ $\cos(\alpha+\beta) = \dfrac{y^2 - x^2}{x^2 + y^2}$

$\Rightarrow$ $y^2 - x^2 = (x^2 + y^2)\cos(\alpha+\beta)$

$\therefore$ Option (b) is correct.

$\because$ $\sin(\alpha+\beta) = \dfrac{2\tan\left(\frac{\alpha+\beta}{2}\right)}{1 + \tan^2\left(\frac{\alpha+\beta}{2}\right)}$

$\Rightarrow$ $\sin(\alpha+\beta) = \dfrac{2 \cdot \frac{x}{y}}{1 + \frac{x^2}{y^2}}$ [from Eq. (iii)]

$\Rightarrow$ $\sin(\alpha+\beta) = \dfrac{2xy}{x^2 + y^2}$

$\Rightarrow$ $(x^2 + y^2)\sin(\alpha+\beta) = 2xy$

$\therefore$ Option (c) is correct.

16. (c) In ΔABD, using m-n theorem,

$$\cot\beta = \frac{d\cot\gamma - 2d\cot(90° + \alpha)}{2d + d}$$

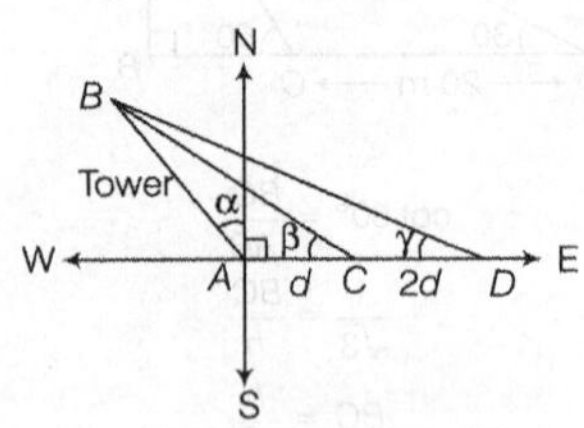

$\Rightarrow$ $\cot\beta = \dfrac{\cot\gamma + 2\tan\alpha}{3}$

$\Rightarrow$ $3\cot\beta = \cot\gamma + 2\tan\alpha$

$\therefore$ $2\tan\alpha = 3\cot\beta - \cot\gamma$

17. (4) Let h be the height of vertical pole, then

$$AB = \frac{3}{4}h,\ BC = \frac{1}{4}h$$

In ΔBCD, $\tan\alpha = \dfrac{BC}{CD} = \dfrac{\frac{1}{4}h}{40}$

$\Rightarrow$ $\tan\alpha = \dfrac{h}{160}$ ...(i)

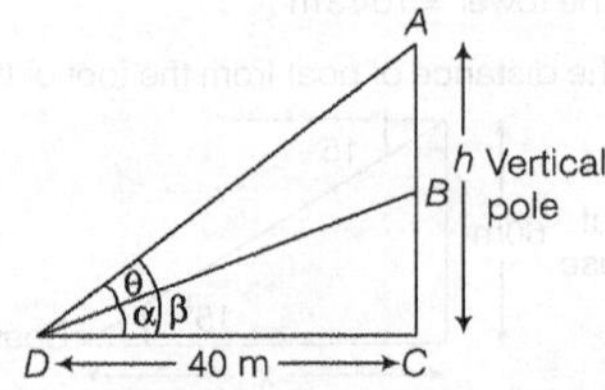

In ΔACD, $\tan\beta = \dfrac{AC}{CD}$

$\Rightarrow$ $\tan\beta = \dfrac{h}{40}$ ...(ii)

We have,

$$\tan\theta = \tan(\beta - \alpha)$$

$\Rightarrow \quad \frac{3}{5} = \frac{\tan\beta - \tan\alpha}{1 + \tan\alpha \cdot \tan\beta}$ [Given, $\theta = \tan^{-1}(3/5)$]

$\Rightarrow \quad \frac{3}{5} = \frac{\frac{h}{40} - \frac{h}{160}}{1 + \frac{h}{160} \cdot \frac{h}{40}}$

$\Rightarrow \quad \frac{3}{5} = \frac{\frac{3h}{160}}{\frac{h^2 + 6400}{160 \times 40}}$

$\Rightarrow \quad \frac{5}{3} = \frac{3h}{160} \times \frac{160 \times 40}{h^2 + 6400}$

$\Rightarrow \quad \frac{3}{5} = \frac{120h}{h^2 + 6400}$

$\Rightarrow \quad 3h^2 + 19200 - 600h = 0$

$\Rightarrow \quad h^2 - 200h + 6400 = 0$

$\Rightarrow \quad (h - 40)(h - 160) = 0$

$\Rightarrow \quad h = 40$ or 160

$\therefore$ Possible height = 40 m

Thus, $\quad h \times 10 = 4 \times 10$

$\Rightarrow \quad h = 4$

18. (3) In ΔABC,

$$\cot 60° = \frac{BC}{100}$$

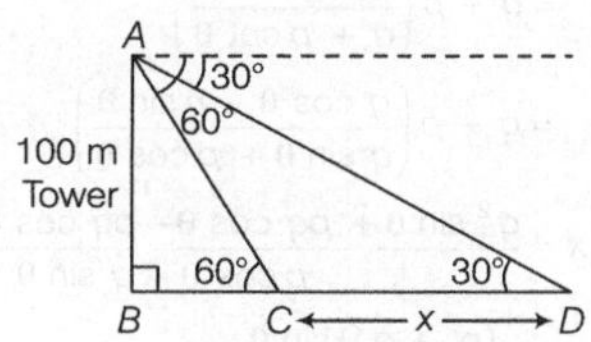

$\Rightarrow \quad BC = 100 \cot 60°$

$\Rightarrow \quad BC = \frac{100}{\sqrt{3}}$ ...(i)

In ΔABD, $\cot 30° = \frac{BD}{100}$

$\Rightarrow \quad BD = 100 \cot 30°$

$\Rightarrow \quad BD = 100\sqrt{3}$ ...(ii)

We have, $\quad x = \frac{200}{\sqrt{\lambda}}$ (given)

$\Rightarrow \quad BD - BC = \frac{200}{\sqrt{\lambda}}$

$\Rightarrow \quad 100\sqrt{3} - \frac{100}{\sqrt{3}} = \frac{200}{\sqrt{\lambda}}$ [from Eqs. (i) and (ii)]

$\Rightarrow \quad \frac{200}{\sqrt{3}} = \frac{200}{\sqrt{\lambda}}$

On comparing, $\quad \lambda = 3$

19. (4) Let x be the distance between two chimney and y be the distance between their tops.

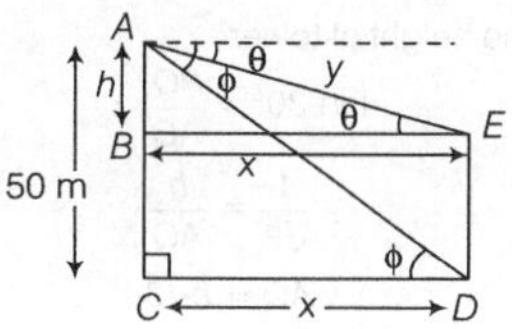

In ΔACD,

$$\tan\phi = \frac{150}{x}$$

$\Rightarrow \quad x = \frac{150}{\tan\phi}$

$\Rightarrow \quad x = \frac{150}{\frac{5}{2}} = 150 \times \frac{2}{5}$ $\left(\because \tan\phi = \frac{5}{2}\right)$

$\therefore \quad x = 60$

In ΔABE, $\quad \tan\theta = \frac{AB}{x}$

$\Rightarrow \quad \frac{4}{3} = \frac{h}{60}$ $\left(\because \tan\theta = \frac{4}{3} \text{ and } x = 60\right)$

$\Rightarrow \quad h = 60 \times \frac{4}{3}$

$\therefore \quad h = 80$

$\therefore$ In ΔABE, using Pythagoras theorem,

$$y^2 = x^2 + h^2 = 60^2 + 80^2$$

$\Rightarrow \quad y = \sqrt{10000} = 100$

$\therefore$ Distance between their tops = 100 m = 25λ $\Rightarrow \lambda = 4$

20. (4) Given, $\cot\alpha = \frac{3}{5}$ and $\cot\beta = \frac{2}{5}$

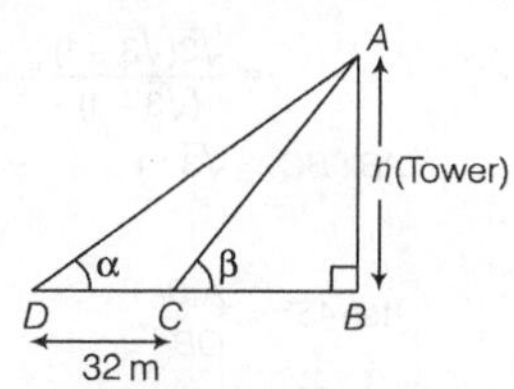

$\frac{BD}{h} = \frac{3}{5}$ and $\frac{BC}{h} = \frac{2}{5}$ ...(i)

Now, $\quad \frac{BC + CD}{h} = \frac{3}{5}$

$\Rightarrow \quad \frac{BC}{h} + \frac{32}{h} = \frac{3}{5}$

$\Rightarrow \quad \frac{2}{5} + \frac{32}{h} = \frac{3}{5}$ [from Eq. (i)]

$\Rightarrow \quad \frac{32}{h} = \frac{1}{5}$

$\Rightarrow \quad h = 160$

$\therefore$ Height of tower = 160 m = 40λ $\Rightarrow \lambda = 4$

JEE Main & AIEEE Archive

1. (a) Let h be the height of tower.

In ΔAPQ, $\tan 30° = \frac{PQ}{AQ}$

$$\Rightarrow \frac{1}{\sqrt{3}} = \frac{h}{AQ}$$

$$\Rightarrow AQ = h\sqrt{3}$$

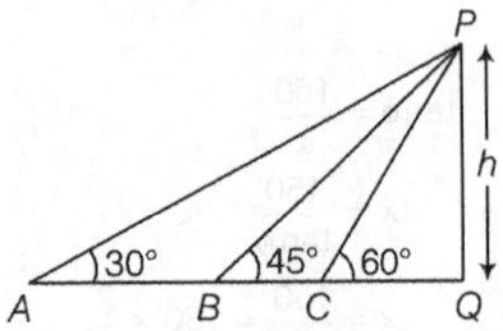

In ΔBPQ, $\tan 45° = \frac{PQ}{BQ}$

$$\Rightarrow 1 = \frac{h}{BQ}$$

$$\Rightarrow BQ = h$$

and in ΔCPQ, $\tan 60° = \frac{PQ}{CQ}$

$$\Rightarrow \sqrt{3} = \frac{h}{CQ}$$

$$\Rightarrow CQ = \frac{h}{\sqrt{3}}$$

$$\therefore \frac{AB}{BC} = \frac{AQ - BQ}{BQ - CQ}$$

$$= \frac{h\sqrt{3} - h}{h - \frac{h}{\sqrt{3}}} = \frac{h(\sqrt{3}-1)}{h\left(1 - \frac{1}{\sqrt{3}}\right)}$$

$$= \frac{\sqrt{3}(\sqrt{3}-1)}{(\sqrt{3}-1)} = \frac{\sqrt{3}}{1}$$

$$\therefore AB : BC = \sqrt{3} : 1$$

2. (d) In ΔOA_1B_1,

$$\tan 45° = \frac{A_1B_1}{OB_1}$$

$$\Rightarrow \frac{20}{OB_1} = 1$$

$$\Rightarrow OB_1 = 20 \quad \ldots(i)$$

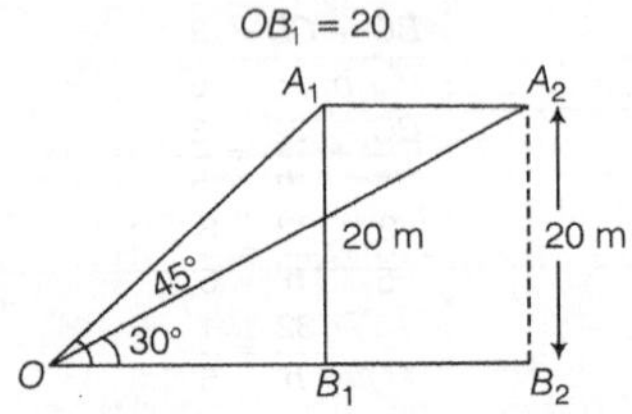

In ΔOA_2B_2,

$$\tan 30° = \frac{20}{OB_2}$$

$$\Rightarrow OB_2 = 20\sqrt{3}$$

$$\Rightarrow B_1B_2 + OB_1 = 20\sqrt{3}$$

$$\Rightarrow B_1B_2 = 20\sqrt{3} - 20 \quad \text{[from Eq. (i)]}$$

$$\Rightarrow B_1B_2 = 20(\sqrt{3}-1)\text{ m}$$

$$\therefore \text{Speed} = \frac{\text{Distance}}{\text{Time}}$$

$$= \frac{20(\sqrt{3}-1)}{1}$$

$$= 20(\sqrt{3}-1)\text{ ms}^{-1}$$

3. (a) Let $AB = x$

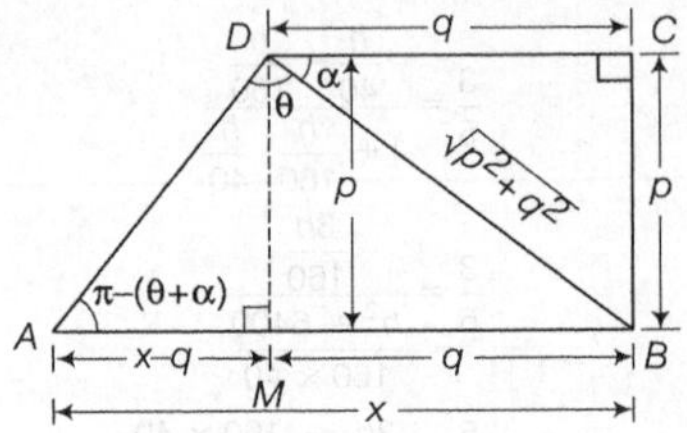

In ΔDAM,

$$\tan(\pi - \theta - \alpha) = \frac{p}{x-q}$$

$$\Rightarrow \tan(\theta + \alpha) = \frac{p}{q - x}$$

$$\Rightarrow q - x = p\cot(\theta + \alpha)$$

$$\Rightarrow x = q - p\cot(\theta + \alpha)$$

$$= q - p\left(\frac{\cot\theta\cot\alpha - 1}{\cot\alpha + \cot\theta}\right)$$

$$= q - p\left(\frac{\frac{q}{p}\cot\theta - 1}{\frac{q}{p} + \cot\theta}\right) \quad \left(\text{in } \Delta BDC, \cot\alpha = \frac{q}{p}\right)$$

$$= q - p\left(\frac{q\cot\theta - p}{q + p\cot\theta}\right)$$

$$= q - p\left(\frac{q\cos\theta - p\sin\theta}{q\sin\theta + p\cos\theta}\right)$$

$$\Rightarrow x = \frac{q^2\sin\theta + pq\cos\theta - pq\cos\theta + p^2\sin\theta}{p\cos\theta + q\sin\theta}$$

$$\Rightarrow AB = \frac{(p^2+q^2)\sin\theta}{p\cos\theta + q\sin\theta}$$

Alternate Method

Applying sine rule in ΔABD,

$$\frac{a}{\sin A} = \frac{b}{\sin B} = \frac{c}{\sin C}$$

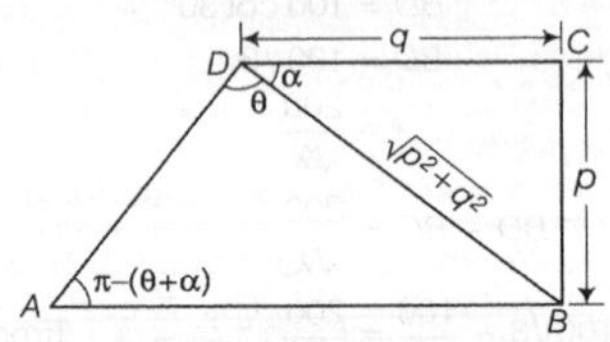

$$\frac{AB}{\sin\theta} = \frac{\sqrt{p^2+q^2}}{\sin\{\pi - (\theta + \alpha)\}}$$

$$\Rightarrow \frac{AB}{\sin\theta} = \frac{\sqrt{p^2+q^2}}{\sin(\theta+\alpha)}$$

$$\Rightarrow AB = \frac{\sqrt{p^2+q^2}\sin\theta}{\sin\theta\cos\alpha + \cos\theta\sin\alpha}$$

$$= \frac{(p^2+q^2)\sin\theta}{q\sin\theta + p\cos\theta}$$

$$\left(\because \cos\alpha = \frac{q}{\sqrt{p^2+q^2}} \text{ and } \sin\alpha = \frac{p}{\sqrt{p^2+q^2}}\right)$$

$$= \frac{(p^2+q^2)\sin\theta}{p\cos\theta + q\sin\theta}$$

4. (c) Let h be the height of the pole.

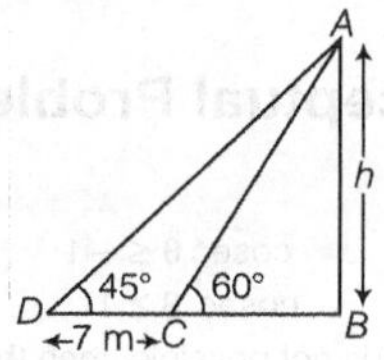

In ΔABC,

$$BC = h\cot 60°$$

and in ΔABD, $BD = h\cot 45°$

$\because$ $BD - BC = DC$

$\Rightarrow$ $h\cot 45° - h\cot 60° = 7$

$$h = \frac{7}{\cot 45° - \cot 60°} = \frac{7}{\left(1 - \frac{1}{\sqrt{3}}\right)} = \frac{7\sqrt{3}}{\sqrt{3}-1} \times \frac{\sqrt{3}+1}{\sqrt{3}+1} = \frac{7\sqrt{3}}{2}(\sqrt{3}+1)\text{ m}$$

5. (c) Let h be the height of a tower.

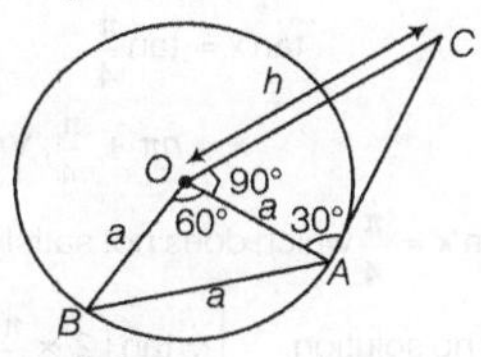

Since, $\angle AOB = 60°$

Also, $OB = OA$ = radii

$\therefore$ $\angle OBA = \angle OAB = 60°$

So, ΔOAB is an equilateral.

$\therefore$ $OA = OB = AB = a$

In ΔOAC, $\tan 30° = \frac{h}{a}$

$\Rightarrow$ $\frac{1}{\sqrt{3}} = \frac{h}{a}$

$\therefore$ $h = \frac{a}{\sqrt{3}}$

6. (a) Let $CD = h$ be the height of the tree

and $BC = x$ be the width of the river.

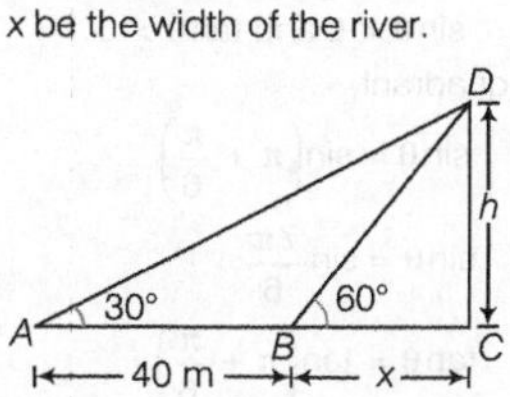

In ΔCBD,

$$\tan 60° = \frac{CD}{BC}$$

$\Rightarrow$ $\sqrt{3} = \frac{h}{x}$

$\Rightarrow$ $h = x\sqrt{3}$ ...(i)

In ΔCAD,

$$\tan 30° = \frac{CD}{AC}$$

$\Rightarrow$ $\frac{1}{\sqrt{3}} = \frac{h}{40+x}$

$\Rightarrow$ $h\sqrt{3} = 40 + x$

$\Rightarrow$ $3x = 40 + x$ [from Eq. (i)]

$\Rightarrow$ $2x = 40$

$\therefore$ $x = 20$ m

7. (b) Here, from the figure we see that,

$$\theta_2 = \tan^{-1}\frac{3}{5}$$

$\Rightarrow$ $\tan\theta_2 = \frac{3}{5}$ ...(i)

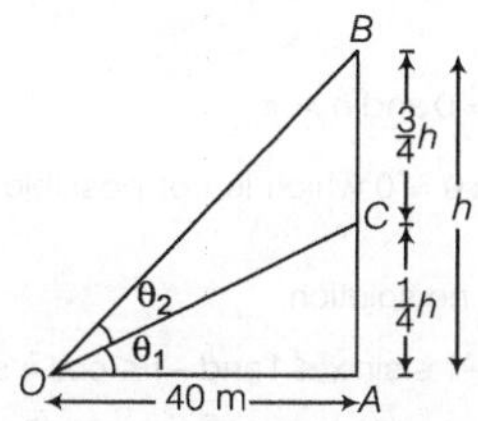

In ΔAOC,

$$\tan\theta_1 = \frac{AC}{AO}$$

$\Rightarrow$ $\tan\theta_1 = \frac{\frac{1}{4}h}{40} = \frac{h}{160}$ ...(ii)

In ΔAOB,

$$\tan(\theta_1 + \theta_2) = \frac{AB}{AO} = \frac{h}{40}$$

$\Rightarrow$ $\frac{\tan\theta_1 + \tan\theta_2}{1 - \tan\theta_1\tan\theta_2} = \frac{h}{40}$

$\Rightarrow$ $\frac{\frac{h}{160} + \frac{3}{5}}{1 - \frac{h}{160} \times \frac{3}{5}} = \frac{h}{40}$ [from Eqs. (i) and (ii)]

$\Rightarrow$ $\frac{5(h+96)}{800-3h} = \frac{h}{40}$

$\Rightarrow$ $200(h+96) = 800h - 3h^2$

$\Rightarrow$ $3h^2 - 600h + 19200 = 0$

$\Rightarrow$ $h^2 - 200h + 6400 = 0$

$\Rightarrow$ $(h-160)(h-40) = 0$

$\therefore$ $h = 160$ or $h = 40$

Hence, height of the vertical pole is 40 m.

3. Trigonometric Equations

DPP-1 For Trigonometric Equation with Conceptual Problems

1. True

2. True

3. True, we know that a trigonometric identity is satisfied for every value of unknown angle (a variable quantity), while a trigonometric equation is satisfied for some particular values of unknown angle.

So, $\sin x + \cos x = 1$ is not satisfied at $x = \frac{\pi}{4}$, then it is trigonometric equation.

4. False, we know that, $\tan\theta = \tan\alpha$ is not defined, when θ and α are odd multiple of $\frac{\pi}{2}$.

Hence, $\tan\theta = \tan\alpha \Rightarrow \theta = n\pi + \alpha, n \in I$, when θ and α are not odd multiple of $\pi/2$.

5. $\alpha \in \left[-\frac{\pi}{2}, \frac{\pi}{2}\right]$

6. $\alpha \in [0, \pi]$, $\therefore a = 0$ and $b = \pi$

7. Given that, $\sec\theta = 0$ which is not possible because $\sec\theta \le -1$ and $\sec\theta \ge 1$.

Hence, there is no solution.

8. We know that, $-1 \le \sin x \le 1$ and $-1 \le \cos x \le 1$

Then, $-\sqrt{2} \le \sin x + \cos x \le \sqrt{2}$

So, $\sin x + \cos x = 3$ cannot be possible.

Hence, number of solutions is 0.

9. (d) We have,

$$\operatorname{cosec}\theta + 2 = 0$$
$$\operatorname{cosec}\theta = -2$$
$$\sin\theta = -\frac{1}{2} = -\sin 30^\circ$$
$$\sin\theta = \sin(180^\circ + 30^\circ), \sin(360^\circ - 30^\circ) \quad (\because 0^\circ < \theta < 360^\circ)$$
$$\therefore \quad \theta = 210^\circ, 330^\circ$$

10. (b) We have, $\sin\theta = -\frac{1}{2}$

$$\sin\theta = -\sin\frac{\pi}{6}$$
$$\sin\theta = \sin\left(\pi + \frac{\pi}{6}\right)$$
$$\therefore \quad \theta = \frac{7\pi}{6}$$

11. (c) We have, $\cot x = -\sqrt{3}$

$$\cot x = -\cot\frac{\pi}{6} = \cot\left(2\pi - \frac{\pi}{6}\right)$$
$$\therefore \quad x = \frac{11\pi}{6}$$

12. (a) We have, $\sin 5\theta = 0$

$$\therefore \quad 5\theta = n\pi, n \in I$$
$$\theta = \frac{n\pi}{5}, n \in I$$

13. (a) We know that,

$$\operatorname{cosec}\theta \le -1$$
and
$$\operatorname{cosec}\theta \ge 1$$

Thus, $\operatorname{cosec}\theta = 0$ is not possible, then there is no solution.

14. (c) Given,

$$\sqrt{3}\sec 2\theta = 2$$
$$\sec 2\theta = \frac{2}{\sqrt{3}}$$
$$\cos 2\theta = \frac{\sqrt{3}}{2} = \cos\frac{\pi}{6}$$
$$\therefore \quad 2\theta = 2n\pi \pm \frac{\pi}{6}, \forall n \in I$$
$$\theta = n\pi \pm \frac{\pi}{12}, \forall n \in I$$

15. (d) We have,

$$\frac{\tan 3x - \tan 2x}{1 + \tan 3x \tan 2x} = 1$$
$$\tan(3x - 2x) = \tan\frac{\pi}{4}$$
$$\tan x = \tan\frac{\pi}{4}$$
$$\therefore \quad x = n\pi + \frac{\pi}{4}, \forall n \in I$$

But if $n = 0$, then $x = \frac{\pi}{4}$ which does not satisfy the given equation.

Hence, there is no solution. $\left[\because \tan\left(2 \times \frac{\pi}{4}\right) = \tan\frac{\pi}{2} = \infty, \therefore \frac{\infty}{\infty}\right]$

16. (a) We have,

$$\sin x + \sin y + \sin z = -3$$
$$0 \le x \le 2\pi$$
$$0 \le y \le 2\pi$$
$$0 \le z \le 2\pi$$

It is possible only, when $\sin x = \sin y = \sin z = -1$

$$\therefore \quad x = y = z = \frac{3\pi}{2} \text{ for } x, y, z \in [0, 2\pi]$$

17. (d) We have,

$$\sin\theta = -\frac{1}{2} \text{ and } \tan\theta = \frac{1}{\sqrt{3}}$$
$$\because \quad \sin\theta < 0 \text{ and } \tan\theta > 0$$

So, θ lies in III quadrant.

$$\therefore \quad \sin\theta = \sin\left(\pi + \frac{\pi}{6}\right)$$
$$\sin\theta = \sin\frac{7\pi}{6}$$
and
$$\tan\theta = \tan\left(\pi + \frac{\pi}{6}\right)$$
$$\tan\theta = \tan\frac{7\pi}{6}$$
$$\therefore \quad \theta = 2n\pi + \frac{7\pi}{6}$$

18. (a,c) We have, $2\sin^2\theta = 3\cos\theta,\ \theta \in [0, 2\pi]$

$$2(1-\cos^2\theta) = 3\cos\theta$$

$$2\cos^2\theta + 3\cos\theta - 2 = 0$$

$$(\cos\theta + 2)(2\cos\theta - 1) = 0$$

$\because$ $\cos\theta + 2 \neq 0$

Then, $2\cos\theta - 1 = 0$

$$\cos\theta = \frac{1}{2} = \cos\frac{\pi}{3}, \cos\left(2\pi - \frac{\pi}{3}\right)$$

$\therefore$ $\theta = \frac{\pi}{3}, \frac{5\pi}{3}$

19. (a,d) We have, $\cot\theta = \sin 2\theta,\ \theta \neq n\pi$

$$\frac{\cos\theta}{\sin\theta} = 2\sin\theta\cos\theta$$

$$2\sin^2\theta\cos\theta = \cos\theta$$

$$2\sin^2\theta\cos\theta - \cos\theta = 0$$

$$\cos\theta(2\sin^2\theta - 1) = 0$$

Either $\cos\theta = 0$ or $2\sin^2\theta - 1 = 0$

$\theta = (2n+1)\frac{\pi}{2}$ or $\sin^2\theta = \sin^2\frac{\pi}{4}$

$\theta = n\pi + \frac{\pi}{2}$ or $\theta = n\pi \pm \frac{\pi}{4}$

$\therefore$ $\theta = \frac{\pi}{2}, \frac{\pi}{4}$, i.e. 90° and 45°

20. (a,b) We have, $(2\cos x - 1)(3 + 2\cos x) = 0$

$\because$ $3 + 2\cos x \neq 0$

Then, $2\cos x - 1 = 0$

$$\cos x = \frac{1}{2} = \cos\frac{\pi}{3}, \cos\left(2\pi - \frac{\pi}{3}\right)$$

$x = \frac{\pi}{3}, \frac{5\pi}{3}$ $\quad(\because 0 \le x \le 2x)$

21. (3) $\because \cos A\sin\left(A - \frac{\pi}{6}\right) = \frac{1}{2}\left[2\cos A\sin\left(A - \frac{\pi}{6}\right)\right]$

$$= \frac{1}{2}\left[\sin\left(A + A - \frac{\pi}{6}\right) - \sin\left(A - A + \frac{\pi}{6}\right)\right]$$

$$= \frac{1}{2}\left[\sin\left(2A - \frac{\pi}{6}\right) - \sin\frac{\pi}{6}\right]$$

So, it is maximum, when $\sin\left(2A - \frac{\pi}{6}\right)$ is maximum.

i.e. $2A - \frac{\pi}{6} = \frac{\pi}{2}$

$$2A = \frac{\pi}{2} + \frac{\pi}{6}$$

$$A = \frac{\pi}{3}$$

but $A = \frac{\pi}{\lambda}$

Hence, $\lambda = 3$

22. (4) We have, $2\sin^2 x + 5\sin x - 3 = 0$

$$2\sin^2 x + 6\sin x - \sin x - 3 = 0$$

$$(\sin x + 3)(2\sin x - 1) = 0$$

$\because$ $\sin x + 3 \neq 0$, then $2\sin x - 1 = 0$

$$\sin x = \frac{1}{2}$$

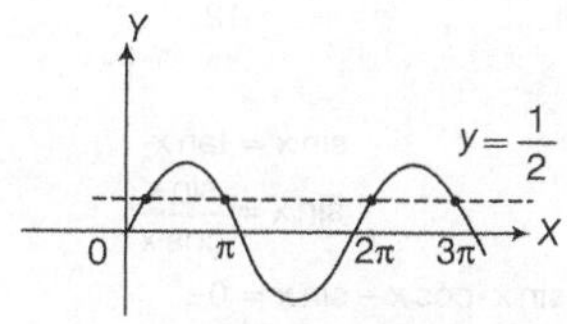

Since, $y = \frac{1}{2}$ and $y = \sin x$ has four common points.

$\therefore$ Number of solutions is 4.

DPP-2 Equations of the Form $\sin^2\theta = \sin^2\alpha, \cos^2\theta = \cos^2\alpha, \tan^2\theta = \tan^2\alpha$

1. We have,

$$3\tan\theta + \cot\theta = 5\,\mathrm{cosec}\,\theta$$

$\frac{3\sin\theta}{\cos\theta} + \frac{\cos\theta}{\sin\theta} = \frac{5}{\sin\theta}$ $\quad(\because \sin\theta$ and $\cos\theta \neq 0)$

$$3\sin^2\theta + \cos^2\theta = 5\cos\theta$$

$$3(1-\cos^2\theta) + \cos^2\theta = 5\cos\theta$$

$$2\cos^2\theta + 5\cos\theta - 3 = 0$$

$$2\cos^2\theta + 6\cos\theta - \cos\theta - 3 = 0$$

$$2\cos\theta(\cos\theta + 3) - 1(\cos\theta + 3) = 0$$

$$(\cos\theta + 3)(2\cos\theta - 1) = 0$$

$\therefore$ $2\cos\theta - 1 = 0$ $\quad(\because \cos\theta + 3 \neq 0)$

$$\cos\theta = \frac{1}{2}$$

$$\cos\theta = \cos\frac{\pi}{3}$$

$\therefore$ $\theta = 2n\pi \pm \frac{\pi}{3}, \forall n \in I$

2. We have, $7\cos^2\theta + 3\sin^2\theta = 4$

$$4\cos^2\theta + 3(\cos^2\theta + \sin^2\theta) = 4$$

$$4\cos^2\theta = 4 - 3 = 1 \Rightarrow \cos^2\theta = \frac{1}{4} = \left(\frac{1}{2}\right)^2$$

$$\cos^2\theta = \cos^2\frac{\pi}{3}$$

$\therefore$ $\theta = n\pi \pm \frac{\pi}{3}, \forall n \in I$

3. $\because \sqrt{3}\tan x + \sqrt{3}\tan 2x + \tan x\tan 2x = 1$

$$\sqrt{3}(\tan x + \tan 2x) = 1 - \tan x\tan 2x$$

$$\frac{\tan x + \tan 2x}{1 - \tan x\tan 2x} = \frac{1}{\sqrt{3}}$$

$$\tan(x + 2x) = \tan\frac{\pi}{6}$$

$\therefore$ $3x = n\pi + \frac{\pi}{6}, \forall n \in I$

$$x = \frac{n\pi}{3} + \frac{\pi}{18}, \forall n \in I$$

4. (c) We have

$$\tan\theta + \tan\left(\theta + \frac{\pi}{3}\right) + \tan\left(\theta + \frac{2\pi}{3}\right) = 3$$

$$\tan\theta + \frac{\tan\theta + \tan\frac{\pi}{3}}{1 - \tan\theta \cdot \tan\frac{\pi}{3}} + \frac{\tan\theta + \tan\frac{2\pi}{3}}{1 - \tan\theta \cdot \tan\frac{2\pi}{3}} = 3$$

$$\tan\theta + \frac{\tan\theta + \sqrt{3}}{1 - \sqrt{3}\tan\theta} + \frac{\tan\theta - \sqrt{3}}{1 + \sqrt{3}\tan\theta} = 3$$

$$\tan\theta + \frac{8\tan\theta}{1 - 3\tan^2\theta} = 3$$

$$\frac{\tan\theta - 3\tan^3\theta + 8\tan\theta}{1 - 3\tan^2\theta} = 3$$

$$\frac{3(3\tan\theta - \tan^3\theta)}{1 - 3\tan^2\theta} = 3$$

$$\tan 3\theta = 1 = \tan\frac{\pi}{4}$$

$$3\theta = n\pi + \frac{\pi}{4}$$

$\Rightarrow$ $$\theta = (4n + 1)\frac{\pi}{12}$$

But $\theta = k(4n + 1)$ (given)

$\therefore$ $$k = \frac{\pi}{12}$$

5. (a) We have,

$$\sin x = \tan x$$

$$\sin x = \frac{\sin x}{\cos x} \quad (\because \cos x \neq 0)$$

$$\sin x \cdot \cos x - \sin x = 0$$

$$\sin x(\cos x - 1) = 0$$

Either $\sin x = 0$

$$x = n\pi, \forall n \in I \quad \ldots \text{(i)}$$

or $\cos x - 1 = 0$

$$\cos x = 1$$

$\therefore$ $$x = 2n\pi, \forall n \in I \quad \ldots \text{(ii)}$$

From Eqs. (i) and (ii), we get

$$x = n\pi, \forall n \in I$$

6. (b) It is given that,

$$\sin x + \sin 3x + \sin 5x = 0$$

$$\sin 5x + \sin x + \sin 3x = 0$$

$$2\sin 3x \cdot \cos 2x + \sin 3x = 0$$

$$\sin 3x\,(2\cos 2x + 1) = 0$$

$\therefore$ Either $\sin 3x = 0$

$\Rightarrow$ $3x = n\pi$

$\Rightarrow$ $$x = \frac{n\pi}{3}, \forall n \in I \quad \ldots \text{(i)}$$

or $2\cos 2x + 1 = 0$

$\Rightarrow$ $$\cos 2x = \frac{-1}{2} = \cos\frac{2\pi}{3}$$

$\Rightarrow$ $$2x = 2n\pi \pm \frac{2\pi}{3}, \forall n \in I$$

$\Rightarrow$ $$x = n\pi \pm \frac{\pi}{3}, \forall n \in I$$

$\therefore$ $$x = (3n \pm 1)\frac{\pi}{3}, \forall n \in I \quad \ldots \text{(ii)}$$

But solutions in Eq. (ii) are included in $x = \frac{n\pi}{3}$, $n \in I$, because $(3n \pm 1)$ is also an integer, then required solution $x = \frac{n\pi}{3}, \forall n \in I$.

7. (b) We have

$$\tan x + \sec x = 2\cos x \quad (\because \cos x \neq 0)$$

$$\frac{\sin x}{\cos x} + \frac{1}{\cos x} = 2\cos x \quad \left(\because x \neq \text{odd multiple } \frac{\pi}{2}\right)$$

$$\sin x + 1 = 2\cos^2 x = 2(1 - \sin^2 x)$$

$$2\sin^2 x + \sin x - 1 = 0$$

$$(2\sin x - 1)(\sin x + 1) = 0$$

$\therefore$ Either $2\sin x = 1$ or $\sin x + 1 = 0$

$$\sin x = \frac{1}{2} \quad \text{or} \quad \sin x = -1$$

$$\sin x = \sin\left(\frac{\pi}{6}\right) \quad \text{or} \quad \sin x = \sin\left(-\frac{\pi}{2}\right)$$

$\Rightarrow$ $$x = n\pi + (-1)^n\frac{\pi}{6} \quad \text{or} \quad x = n\pi + (-1)^{n+1}\frac{\pi}{2}$$

$$\left(\text{not possible because } x \text{ is odd multiple of } \frac{\pi}{2}\right)$$

$\therefore$ $$x = n\pi + (-1)^n\frac{\pi}{6}, n \in I$$

Note

(i) If R $(\tan x, \sec x) = 0$, then $\cos x \neq 0 \Rightarrow x \neq$ odd multiple of $\frac{\pi}{2}$

(ii) If R $(\cot x, \operatorname{cosec} x) = 0$, then $\sin x \neq 0 \Rightarrow x \neq m\pi$

8. (a) We have, $4\cos x - 3\sec x = \tan x$

$$\left(\cos x \neq 0, \text{ i.e } x \text{ odd multiple of } \frac{\pi}{2}\right)$$

Then, $$4\cos x - \frac{3}{\cos x} = \frac{\sin x}{\cos x}$$

$$4\cos^2 x - 3 = \sin x$$

$$4 - 4\sin^2 x - 3 = \sin x$$

$$4\sin^2 x + \sin x - 1 = 0$$

$\therefore$ $$\sin x = \frac{-1 \pm \sqrt{17}}{8}$$

Either $$\sin x = \frac{-1 + \sqrt{17}}{8} = \sin\alpha \text{ (let)}$$

or $$\sin x = \frac{-1 - \sqrt{17}}{8} = \sin\beta \text{ (let)}$$

$\therefore$ $$x = n\pi + (-1)^n\alpha \quad \text{or} \quad x = n\pi + (-1)^n\beta$$

$\therefore$ $$\sin\alpha = \frac{-1 + \sqrt{17}}{8}, \sin\beta = \frac{-1 - \sqrt{17}}{8}$$

So, $$4(\sin\alpha + \sin\beta) = 4\left[\frac{-1 + \sqrt{17}}{8} + \frac{-1 - \sqrt{17}}{8}\right]$$

$$= 4\left[\frac{-2}{8}\right] = -1$$

9. (c) We have,

$$\sqrt{3}\cos x - \sin x = 1$$

$$\frac{\sqrt{3}}{2}\cos x - \frac{1}{2}\sin x = \frac{1}{2}$$

$$\left(\text{dividing both sides by } \sqrt{(\sqrt{3})^2 + (1)^2}, \text{ i.e. } 2\right)$$

$$\cos x \cdot \cos\frac{\pi}{6} - \sin x \cdot \sin\frac{\pi}{6} = \frac{1}{2}$$

$$\cos\left(x + \frac{\pi}{6}\right) = \cos\frac{\pi}{3}$$

$\therefore$ $$x + \frac{\pi}{6} = 2n\pi \pm \frac{\pi}{3}, \forall n \in I \quad \text{(using formula)}$$

$$x = 2n\pi \pm \frac{\pi}{3} - \frac{\pi}{6}, \forall n \in I$$

10. (b) We have

$$\sqrt{2}\sec x + \tan x = 1$$

$$\frac{\sqrt{2}}{\cos x} + \frac{\sin x}{\cos x} = 1 \qquad (\because \cos x \neq 0)$$

$$\sqrt{2} + \sin x = \cos x$$

$$\cos x - \sin x = \sqrt{2}$$

$$\frac{1}{\sqrt{2}}\cos x - \frac{1}{\sqrt{2}}\sin x = 1$$

$$\cos\left(x + \frac{\pi}{4}\right) = 1$$

$$\therefore \quad x + \frac{\pi}{4} = 2n\pi, \forall n \in I \qquad \text{(using formula)}$$

$$x = 2n\pi - \frac{\pi}{4}, \forall n \in I$$

11. (a) We have,

$$a^2 - 4a + 6 = (a^2 - 4a + 4) + 2$$

$$= (a-2)^2 + 2 \geq 2 \qquad [\because (-2)^2 \geq 0]$$

Now, $\sin\alpha + \cos\alpha = \min_{a \in R}\{1, a^2 - 4a + 6\}$

Then, $\sin\alpha + \cos\alpha = 1$

$$\frac{1}{\sqrt{2}}\sin\alpha + \frac{1}{\sqrt{2}}\cos\alpha = \frac{1}{\sqrt{2}}$$

$$\sin\left(\alpha + \frac{\pi}{4}\right) = \sin\frac{\pi}{4}$$

$$\Rightarrow \quad \alpha + \frac{\pi}{4} = n\pi + (-1)^n\frac{\pi}{4}, \forall n \in I \qquad \text{(using formula)}$$

$$\therefore \quad \alpha = n\pi + (-1)^n\frac{\pi}{4} - \frac{\pi}{4}, \forall n \in I$$

12. (c) We have,

$$\tan m\theta = \tan n\theta$$

$$m\theta = k\pi + n\theta, \forall k \in I \qquad \text{(using formula)}$$

$$m\theta - n\theta = k\pi, \forall k \in I$$

$$\theta = k \cdot \frac{\pi}{m-n}, \forall k \in I \quad (\because m \text{ and } n \text{ are constants})$$

Then, if we put $k = 1, 2, 3, \ldots$, we get

$$\theta = \frac{\pi}{m-n}, \frac{2\pi}{m-n}, \frac{3\pi}{m-n}, \ldots \text{ which is AP.}$$

Thus, common difference $= \frac{\pi}{m-n}$

13. (b) $\because \sin x - 3\sin 2x + \sin 3x = \cos x - 3\cos 2x + \cos 3x$

$$(\sin 3x + \sin x) - 3\sin 2x = (\cos 3x + \cos x) - 3\cos 2x$$

$$2\sin 2x \cdot \cos x - 3\sin 2x = 2\cos 2x \cdot \cos x - 3\cos 2x$$

$$\sin 2x(2\cos x - 3) - \cos 2x(2\cos x - 3) = 0$$

$$(\sin 2x - \cos 2x)(2\cos x - 3) = 0$$

So, $\sin 2x - \cos 2x = 0$

as $2\cos x - 3 \neq 0$

Now, $\sin 2x = \cos 2x$

$$\tan 2x = 1 = \tan\frac{\pi}{4}$$

$$2x = n\pi + \frac{\pi}{4}$$

$$\therefore \quad x = \frac{n\pi}{2} + \frac{\pi}{8}$$

14. (a,b) $\because \cos\theta + \cos 3\theta - 2\cos 2\theta = 0$

$$2\cos\left(\frac{3\theta + \theta}{2}\right) \cdot \cos\left(\frac{3\theta - \theta}{2}\right) - 2\cos 2\theta = 0$$

$$2\cos 2\theta \cdot \cos\theta - 2\cos 2\theta = 0$$

$$2\cos 2\theta(\cos\theta - 1) = 0$$

Either $2\cos 2\theta = 0$ or $\cos\theta - 1 = 0$

$\cos 2\theta = 0$ or $\cos\theta = 1$

$2\theta = (2n+1)\frac{\pi}{2}$ or $\theta = 2n\pi, \forall n \in I$

$\therefore \quad \theta = (2n+1)\frac{\pi}{4}, \forall n \in I$ or $2n\pi, \forall n \in I$

$$\therefore \quad \theta = (2n+1)\frac{\pi}{4}, 2n\pi, \forall n \in I$$

15. (a,b,c) We have,

$$\sin\theta - \cos\theta = \sqrt{2}\cos\theta$$

$$\sin\theta = (\sqrt{2} + 1)\cos\theta$$

$$\tan\theta = \sqrt{2} + 1 = \tan\left(\frac{3\pi}{8}\right)$$

$$\therefore \quad \theta = n\pi + \frac{3\pi}{8}, \forall n \in I$$

$$\therefore \quad P = \left\{\theta = n\pi + \frac{3\pi}{8}, \forall n \in I\right\} \qquad \ldots\text{(i)}$$

and $\sin\theta + \cos\theta = \sqrt{2}\sin\theta$

$$\cos\theta = \sqrt{2}\sin\theta - \sin\theta$$

$$\Rightarrow \quad \cos\theta = (\sqrt{2} - 1)\sin\theta$$

$$\tan\theta = \frac{1}{\sqrt{2} - 1}$$

$$= \sqrt{2} + 1 = \tan\left(\frac{3\pi}{8}\right)$$

$$\therefore \quad \theta = n\pi + \frac{3\pi}{8}, \forall n \in I$$

$$\therefore \quad Q = \left\{\theta = n\pi + \frac{3\pi}{8}, \forall n \in I\right\} \qquad \ldots\text{(ii)}$$

From Eqs. (i) and (ii), we get

$$P = Q$$

16. (a,b) We have,

$$\tan(x - y) = 1$$

$$= \tan\left(\frac{\pi}{4}\right), \tan\left(\pi + \frac{\pi}{4}\right)$$

$$x - y = \frac{\pi}{4}, \frac{5\pi}{4}$$

and $\sec(x + y) = \frac{2}{\sqrt{3}}$

$$\cos(x + y) = \frac{\sqrt{3}}{2}$$

$$= \cos\frac{\pi}{6}, \cos\left(2\pi - \frac{\pi}{6}\right)$$

$$x + y = \frac{\pi}{6}, \frac{11\pi}{6}$$

$\because x, y$ are positive values, then $x + y > x - y$

So, $x + y = \frac{11\pi}{6}$

and $x - y = \frac{\pi}{4}$ ...(i)

or $x + y = \frac{11\pi}{6}$

and $x - y = \frac{5\pi}{4}$ ...(ii)

Now, solving Eqs. (i) and (ii), we get

$$x = \frac{25\pi}{24} \text{ and } y = \frac{19\pi}{24}$$

or

$$x = \frac{37\pi}{24} \text{ and } y = \frac{7\pi}{24}$$

17. (a,d) $\because$ $R = \frac{1}{2}\sin^2\theta + \frac{1}{3}\cos^2\theta$

$= \frac{1}{6}[3\sin^2\theta + 2\cos^2\theta]$

$= \frac{1}{6}[\sin^2\theta + 2(\sin^2\theta + \cos^2\theta)]$

$R = \frac{1}{6}[\sin^2\theta + 2]$ ...(i)

We know that, $0 \le \sin^2\theta \le 1$

$2 \le \sin^2\theta + 2 \le 3$

$\frac{1}{3} \le \frac{\sin^2\theta + 2}{6} \le \frac{1}{2}$

$\Rightarrow$ $\frac{1}{3} \le R \le \frac{1}{2}$ [from Eq. (i)]

But $a \le R \le b$ (given)

Then, $a = \frac{1}{3}$ and $b = \frac{1}{2}$

18. (0) $\because$ $3\cos x + 4\sin x = 6$

On dividing both sides by $\sqrt{3^2 + 4^2} = 5$, then

$\frac{3}{5}\cos x + \frac{4}{5}\sin x = \frac{6}{5} > 1$

Then, it has no solution.

Note *If $a\sin x + b\cos x = c$ and $|c| > \sqrt{a^2 + b^2}$, then this equation has no solution. Number of solutions = 0.*

19. (6) $\because$ $3\sin^2 x - 7\sin x + 2 = 0$

$3\sin^2 x - 6\sin x - \sin x + 2 = 0$

$(\sin x - 2)(3\sin x - 1) = 0$

Either $\sin x - 2 = 0$

$\Rightarrow$ $\sin x = 2$ (not possible)

or $3\sin x - 1 = 0$

$\sin x = \frac{1}{3} = +\text{ve}$

$\because \sin x$ is +ve in I and II quadrants.

Now, $[0, 2\pi] \rightarrow 2$ values in I and II quadrants

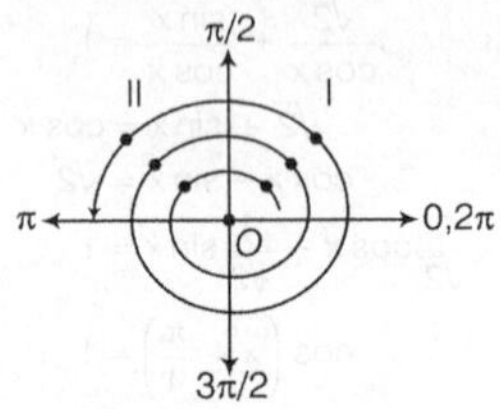

$(2\pi, 4\pi] \rightarrow 2$ values in I and II quadrants

$(4\pi, 5\pi] \rightarrow 2$ values in I and II quadrants

$\therefore$ Total number of values are 6.

20. (3) We have, $\sin 2t = \cos 3t$ and $0° < t < 90°$

$2\sin t\cos t = 4\cos^3 t - 3\cos t$

$2\sin t = 4\cos^2 t - 3$ (dividing both sides by $\cos t$)

$2\sin t = 4(1 - \sin^2 t) - 3$

$2\sin t = 4 - 4\sin^2 t - 3$

$4\sin^2 t + 2\sin t - 1 = 0$

$\therefore$ $\sin t = \frac{-2 \pm \sqrt{4 + 16}}{8}$

$= \frac{-2 \pm 2\sqrt{5}}{8} = \frac{-1 \pm \sqrt{5}}{4}$

Now, $\sin t = \frac{-1 + \sqrt{5}}{4}$

(neglecting – ve value because $0° < t < 90°$)

$\sin t = \sin 18°$ $\left(\because \sin 18° = \frac{\sqrt{5} - 1}{4}\right)$

$\therefore$ $t = 18°$

or $\frac{t}{6} = 3$

DPP-3 Homogeneous, Extreme Values and T-Inequations

1. $\because (1 + \tan\alpha)(1 + \tan\beta) = 2$

$1 + \tan\alpha + \tan\beta + \tan\alpha\tan\beta = 2$

$\tan\alpha + \tan\beta = 1 - \tan\alpha\tan\beta$

$\frac{\tan\alpha + \tan\beta}{1 - \tan\alpha\tan\beta} = 1$

$\tan(\alpha + \beta) = \tan 45°$ (using formula)

$\therefore$ $\alpha + \beta = 45°$ **Hence proved.**

2. True, we have,

$2^{\frac{1}{1 - |\cos x|}} = 4$

$\Rightarrow$ $2^{\frac{1}{1 - |\cos x|}} = 2^2$

$\Rightarrow$ $\frac{1}{1 - |\cos x|} = 2$

$\Rightarrow$ $1 - |\cos x| = \frac{1}{2}$

$\Rightarrow$ $|\cos x| = \frac{1}{2}$

$\Rightarrow$ $\cos x = \pm\frac{1}{2} = \pm\cos\frac{\pi}{3}$

$\Rightarrow$ $\cos x = \cos\frac{\pi}{3}, \cos\left(\pi - \frac{\pi}{3}\right)$

$\therefore$ $x = 2n\pi \pm \frac{\pi}{3}, 2n\pi \pm \left(\pi - \frac{\pi}{3}\right), \forall n \in I$

$= 2n\pi \pm \frac{\pi}{3}, (2n \pm 1)\pi \pm \frac{\pi}{3}, \forall n \in I$

$\therefore$ $x = n\pi \pm \frac{\pi}{3}, \forall n \in I$

3. True, we have,

$\cos 40° = t$

and $\cos\theta = 1 - 2t^2 = 1 - 2\cos^2 40°$

$= -(2\cos^2 40° - 1) = -\cos 80°$

$= \cos(180° - 80°), \cos(180° + 80°)$

$\cos\theta = \cos 100°, \cos 260°$

$\therefore$ $\theta = 100°, 260°$

4. (b) We have, $\sin^2\theta\sec\theta + \sqrt{3}\tan\theta = 0$

$\frac{\sin^2\theta}{\cos\theta} + \sqrt{3}\frac{\sin\theta}{\cos\theta} = 0$ $(\because \cos\theta \ne 0)$

$\sin^2\theta + \sqrt{3}\sin\theta = 0$

$\sin\theta(\sin\theta + \sqrt{3}) = 0$

Either $\sin\theta = 0$ or $\sin\theta + \sqrt{3} = 0$ (which is not possible)

$\theta = n\pi, \forall n \in Z$

5. (a) $\because$ $\sin 3\alpha = 4\sin\alpha \sin(x-\alpha)\sin(x+\alpha)$

$$3\sin\alpha - 4\sin^3\alpha = 4\sin\alpha(\sin^2 x - \sin^2\alpha)$$

On dividing both sides by $\sin\alpha$, we get

$$3 - 4\sin^2\alpha = 4\sin^2 x - 4\sin^2\alpha$$

$$\Rightarrow \quad 4\sin^2 x = 3$$

$$\Rightarrow \quad \sin^2 x = \frac{3}{4}$$

$$\Rightarrow \quad \sin^2 x = \left(\frac{\sqrt{3}}{2}\right)^2 = \sin^2\frac{\pi}{3}$$

$$\therefore \quad x = n\pi \pm \frac{\pi}{3}, \forall n \in I \quad \text{(using formula)}$$

6. (a) $\because \sin x + \cos x - 2\sqrt{2}\sin x \cdot \cos x = 0$ ...(i)

[from $R(\sin x + \cos x, \sin x - \cos x) = 0$]

Let $\sin x + \cos x = t$

$$1 + 2\sin x\cos x = t^2$$

$$2\sin x \cos x = t^2 - 1$$

Now, Eq. (i) becomes

$$t - \sqrt{2}(t^2 - 1) = 0$$

$$\sqrt{2}t^2 - t - \sqrt{2} = 0$$

$$\sqrt{2}t^2 - 2t + t - \sqrt{2} = 0$$

$$\sqrt{2}t(t - \sqrt{2}) + 1(t - \sqrt{2}) = 0$$

$$(t - \sqrt{2})(\sqrt{2}t + 1) = 0$$

$$t = \sqrt{2}, \frac{-1}{\sqrt{2}}$$

Now, $\sin x + \cos x = \sqrt{2}, \sin x + \cos x = \frac{-1}{\sqrt{2}}$

$$\frac{1}{\sqrt{2}}\sin x + \frac{1}{\sqrt{2}}\cos x = 1, \cos\left(x - \frac{\pi}{4}\right) = -\frac{1}{2}$$

$$\Rightarrow \quad \cos\left(\frac{-\pi}{4} + x\right) = 1$$

$$\therefore \quad x - \frac{\pi}{4} = 2n\pi$$

$$x = 2n\pi + \frac{\pi}{4}$$

7. (d) $\because \sqrt{3}\cos^2\theta - 2\sqrt{3}\sin\theta\cos\theta - 3\sin^2\theta = 0$

On dividing both sides by $\cos^2\theta$, we get

$$\sqrt{3} - 2\sqrt{3}\tan\theta - 3\tan^2\theta = 0$$

$$\sqrt{3}\tan^2\theta + 2\tan\theta - 1 = 0$$

$$\tan\theta = \frac{-2 \pm \sqrt{4 + 4\sqrt{3}}}{2\sqrt{3}}$$

$$\tan\theta = \frac{-1 \pm \sqrt{1 + \sqrt{3}}}{\sqrt{3}} \quad \left(\text{which are not equal to } \pm\frac{\pi}{6}, \frac{\pi}{3}\right)$$

8. (d) We have,

$$\sin 5\theta \cos 3\theta = \sin 9\theta \cos 7\theta$$

$$2\sin 5\theta \cos 3\theta = 2\sin 9\theta \cos 7\theta$$

$$\sin 8\theta + \sin 2\theta = \sin 16\theta + \sin 2\theta$$

$$\sin 16\theta = \sin 8\theta$$

$$\sin 16\theta - \sin 8\theta = 0$$

$$2\cos\left(\frac{16\theta + 8\theta}{2}\right) \cdot \sin\left(\frac{16\theta - 8\theta}{2}\right) = 0$$

$$\cos 12\theta \cdot \sin 4\theta = 0$$

Either $\cos 12\theta = 0$ or $\sin 4\theta = 0$

$$12\theta = (2n+1)\frac{\pi}{2}, \forall n \in I \text{ or } 4\theta = n\pi, \forall n \in I$$

$$\theta = (2n+1)\frac{\pi}{24}, \forall n \in I \text{ or } \theta = \frac{n\pi}{4}, \forall n \in I$$

If $n = 0, 1, 2, 3, 4, 5$, then θ lies in $\left[0, \frac{\pi}{2}\right]$

and number of solutions in $\left[0, \frac{\pi}{2}\right]$ is 6.

If $n = 0, 1, 2$, then θ lies in $\left[0, \frac{\pi}{2}\right]$

and number of solutions in $\left[0, \frac{\pi}{2}\right]$ is 3.

$\therefore$ Total number of solutions is 9.

9. (c) We know that,

$a\sin\theta \pm b\cos\theta = c$ is solvable, if $|c| \le \sqrt{a^2 + b^2}$.

Now, $\lambda\cos x - 3\sin x = \lambda + 1$ is solvable, if

$$|\lambda + 1| \le \sqrt{\lambda^2 + 9}$$

$$(\lambda + 1)^2 \le \lambda^2 + 9$$

$$\lambda^2 + 1 + 2\lambda \le \lambda^2 + 9$$

$$2\lambda \le 9 - 1$$

$$\Rightarrow \quad \lambda \le 4$$

$$\therefore \quad \lambda \in (-\infty, 4]$$

10. (d) We have

$$\sqrt{3}\sec x - \sqrt{3}\tan x = 3$$

$$\sec x - \tan x = \frac{1}{\sqrt{3}} \quad \text{...(i)}$$

We know that

$$\sec^2 x - \tan^2 x = 1$$

Then, $$\sec x + \tan x = \frac{1}{\sec x - \tan x} = \sqrt{3}$$

[from Eq. (i)]

$$\sec x + \tan x = \sqrt{3} \quad \text{...(ii)}$$

On adding Eqs. (i) and (ii), we get

$$2\sec x = \sqrt{3} + \frac{1}{\sqrt{3}} = \frac{4}{\sqrt{3}}$$

$\therefore \cos x = \frac{\sqrt{3}}{2}$, then

$$\cos x = \cos\frac{\pi}{6}, \cos\left(2\pi - \frac{\pi}{6}\right), \cos\left(2\pi + \frac{\pi}{6}\right)$$

$$\therefore \quad x = \frac{\pi}{6}, \frac{11\pi}{6}, \frac{13\pi}{6}$$

$\therefore$ Number of distinct solutions is 3.

11. (a) We have,

$$\cos 2x - 3\cos x + 1 = \frac{1}{(\cot 2x - \cot x)\sin(x - \pi)}$$

$$2\cos^2 x - 1 - 3\cos x + 1 = \frac{1}{\frac{\sin(x - 2x)}{\sin 2x \cdot \sin x} \cdot (-\sin x)}$$

$$\left[\text{using formula, } \cot A - \cot B = \frac{\sin(B - A)}{\sin A \cdot \sin B}\right]$$

$$2\cos^2 x - 3\cos x = \frac{2\sin x \cdot \cos x \cdot \sin x}{\sin^2 x}$$

$$2\cos^2 x - 3\cos x - 2\cos x = 0$$

$$2\cos^2 x - 5\cos x = 0$$

$$\cos x(2\cos x - 5) = 0$$

Either $2\cos x - 5 = 0$

$$\Rightarrow \quad \cos x = \frac{5}{2} \quad \text{(which is not possible)}$$

Then, $\cos x = 0$

12. (d) $\because$ $\sin\theta\tan 4\theta = \cos\theta$ and $\theta \in (0, \pi)$

$$\sin\theta \cdot \frac{\sin 4\theta}{\cos 4\theta} = \cos\theta \qquad (\because \cos 4\theta \neq 0)$$

$$\sin 4\theta \cdot \sin\theta = \cos 4\theta \cdot \cos\theta$$

$$\cos 4\theta \cdot \cos\theta - \sin 4\theta \cdot \sin\theta = 0$$

$$\cos(4\theta + \theta) = 0$$

$$\Rightarrow \quad \cos 5\theta = 0$$

$$5\theta = (2n+1)\frac{\pi}{2}, \forall n \in I$$

$$\Rightarrow \quad \theta = (2n+1)\frac{\pi}{10}, \forall n \in I$$

$$\therefore \quad \theta = \frac{\pi}{10}, \frac{3\pi}{10}, \frac{5\pi}{10}, \frac{7\pi}{10}, \frac{9\pi}{10} \qquad [\because \theta \in (0, \pi)]$$

So, number of solutions is 5.

13. (c) $\because \sin^4\theta + \cos^4\theta + 2\sin\theta \cdot \cos\theta + \alpha = 0$

$$1 - 2\sin^2\theta\cos^2\theta + \sin 2\theta + \alpha = 0$$

$$2 - (2\sin\theta \cdot \cos\theta)^2 + 2\sin 2\theta + 2\alpha = 0$$

$$\sin^2 2\theta - 2\sin 2\theta - 2(\alpha + 1) = 0$$

$$\therefore \quad \sin 2\theta = \frac{2 \pm \sqrt{4 + 8(\alpha+1)}}{2} = 1 \pm \sqrt{2\alpha + 3}$$

We know that

$$-1 \leq \sin 2\theta \leq 1$$

$$-1 \leq 1 \pm \sqrt{2\alpha+3} \leq 1$$

$$-2 \leq \pm\sqrt{2\alpha+3} \leq 0$$

$$0 \leq 2\alpha + 3 \leq 4$$

$$\frac{-3}{2} \leq \alpha \leq \frac{1}{2}$$

$$\therefore \quad \alpha \in \left[\frac{-3}{2}, \frac{1}{2}\right]$$

14. (a,b,c) We have,

$$\cos^2 x + \sin^2 y - 2\tan^2 z = 2$$

$$\cos^2 x + \sin^2 y = 2 + 2\tan^2 z$$

$$\cos^2 x + \sin^2 y = 2\sec^2 z$$

We know that

$$\cos^2 x + \sin^2 y \leq 2 \qquad \ldots(i)$$

$$(\because \sin^2 y \leq 1, \cos^2 x \leq 1)$$

and $\qquad \sec^2 z \geq 1$

$$2\sec^2 z \geq 2 \qquad \ldots(ii)$$

From Eqs. (i) and (ii), we get

$$\cos^2 x + \sin^2 y = 2\sec^2 z$$

is possible only, when

$$\cos^2 x + \sin^2 y = 2$$

and $\qquad 2\sec^2 z = 2$

Then, $\cos^2 x = 1, \sin^2 y = 1$ and $\sec^2 z = 1$

$\Rightarrow \quad \sin^2 x = 0, \cos^2 y = 0$ and $\sin^2 z = 0$

$\therefore \quad \sin x = 0, \cos y = 0$ and $\sin z = 0$

$x = n\pi, n \in I;$

$y = (2m+1)\frac{\pi}{2}, m \in I$ and $z = t\pi, t \in I$

15. (c) $\because$ $x + y = \frac{4\pi}{3}$ $\qquad \ldots(i)$

and $\qquad \sin x = 2\sin y \qquad$ (given)

$$\sin x = 2\sin\left(\frac{4\pi}{3} - x\right) \qquad \text{[from Eq. (i)]}$$

$$\sin x = 2\left[\sin\frac{4\pi}{3} \cdot \cos x - \cos\frac{4\pi}{3} \cdot \sin x\right]$$

$$\sin x = 2\left[\frac{-\sqrt{3}}{2}\cos x + \frac{1}{2}\sin x\right]$$

$$\sin x = -\sqrt{3}\cos x + \sin x$$

$$\cos x = 0$$

$$x = (2n+1)\frac{\pi}{2}, \forall n \in I$$

$$x = n\pi + \frac{\pi}{2}$$

Putting the value of x in Eq. (i), we get

$$y = \frac{4\pi}{3} - n\pi - \frac{\pi}{2}$$

$$y = \frac{5\pi}{6} - n\pi, \forall n \in I$$

$$\therefore \quad (x, y) = \left(n\pi + \frac{\pi}{2}, \frac{5\pi}{6} - n\pi\right), \forall n \in I$$

16. (a,c) $\max\limits_{\theta \in R} \{5\sin\theta + 3\sin(\theta - \alpha)\} = 7 \qquad \ldots(i)$

$$\because \quad 5\sin\theta + 3(\sin\theta \cdot \cos\alpha - \cos\theta \cdot \sin\alpha)$$

$$= 5\sin\theta + 3\cos\alpha \cdot \sin\theta - 3\sin\alpha \cdot \cos\theta$$

$$= (5 + 3\cos\alpha)\sin\theta - 3\sin\alpha\cos\theta$$

Thus, maximum value of $(5 + 3\cos\alpha)\sin\theta - 3\sin\alpha\cos\theta$

$$= \sqrt{(5 + 3\cos\alpha)^2 + (-3\sin\alpha)^2}$$

$$= \sqrt{25 + 9\cos^2\alpha + 30\cos\alpha + 9\sin^2\alpha}$$

$$= \sqrt{34 + 30\cos\alpha}$$

Now, using Eq. (i), we get

$$\sqrt{34 + 30\cos\alpha} = 7$$

$$34 + 30\cos\alpha = 49$$

$$30\cos\alpha = 15$$

$$\cos\alpha = \frac{1}{2} = \cos\frac{\pi}{3}$$

$$\alpha = 2n\pi \pm \frac{\pi}{3}$$

17. (4) We have,

$$|\sqrt{3}\cos x - \sin x| \geq 2 \text{ for } x \in [0, 4\pi] \qquad \ldots(i)$$

Now, $|\sqrt{3}\cos x - \sin x| \leq \sqrt{(\sqrt{3})^2 + (1)^2} = 2$

$$|\sqrt{3}\cos x - \sin x| \leq 2 \qquad \ldots(ii)$$

From Eqs. (i) and (ii), we get

$$|\sqrt{3}\cos x - \sin x| = 2$$

$$\Rightarrow \quad \left|\frac{\sqrt{3}}{2}\cos x - \frac{1}{2}\sin x\right| = 1$$

$$\Rightarrow \quad \left|\cos\frac{\pi}{6}\cos x - \sin\frac{\pi}{6}\sin x\right| = 1$$

$$\Rightarrow \quad \left|\cos\left(x + \frac{\pi}{6}\right)\right| = 1$$

$$\Rightarrow \quad \cos\left(x + \frac{\pi}{6}\right) = \pm 1$$

$$\therefore \quad x + \frac{\pi}{6} = 0, 2\pi, 4\pi, \ldots \text{ and } \pi, 3\pi, 5\pi, \ldots$$

$$\therefore \quad x = \frac{11\pi}{6}, \frac{23\pi}{6} \text{ and } \frac{5\pi}{6}, \frac{17\pi}{6}$$

So, number of solutions is 4.

18. (0) We know that, $\sin^4 t \le 1$ ...(i)

and $1 + \tan^4 t \ge 1$ ...(ii)

Using Eqs. (i) and (ii) equality holds only when

$$\sin^4 t = 1 \quad \text{and} \quad 1 + \tan^4 t = 1$$

$(\because \sin^4 t = 1 + \tan^4 t, \text{given})$

$$\sin^4 t = 1 \quad \text{and} \quad \tan^4 t = 0$$

which is not possible because $\sin x$ and $\tan x$ vanish simultaneously.

So, number of solutions is 0.

19. (c) We have, $\sin x = \frac{x}{10}$, so let $f(x) = \sin x$, $g(x) = \frac{x}{10}$

Now, draw the graph of $f(x)$ and $g(x)$.

$\because$ $-1 \le \sin x \le 1$

$$-1 \le \frac{x}{10} \le 1$$

$$-10 \le x \le 10$$

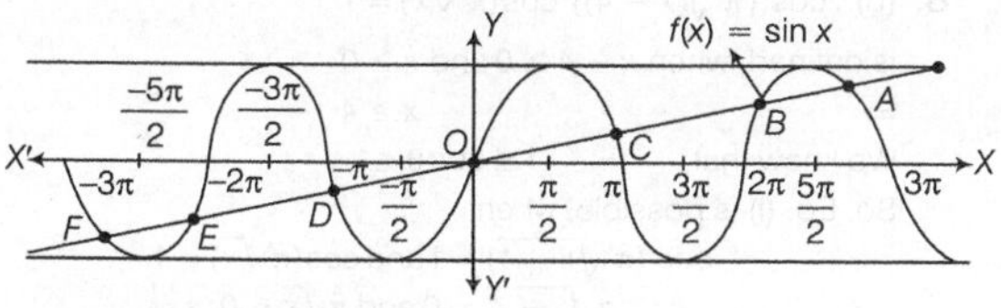

Intersection points are A, B, C, O, D, E, F.

$\therefore$ Number of intersection points = 7

i.e. Number of solutions is 7.

20. (a) Given, $\cos 2x = |\sin x|$

$\Rightarrow \quad \cos^2 2x = \sin^2 x$

$\Rightarrow \quad (1 - 2\sin^2 x)^2 = \sin^2 x$

$\Rightarrow \quad 4\sin^4 x - 5\sin^2 x + 1 = 0$

Put $\sin^2 x = t$

$\Rightarrow \quad 4t^2 - 5t + 1 = 0$

$\Rightarrow \quad t = \frac{5 \pm \sqrt{25 - 16}}{8}$

$\Rightarrow \quad t = \frac{5 \pm \sqrt{9}}{8} = \frac{5 \pm 3}{8}$

$\Rightarrow \quad t = 1, \frac{1}{4}$

When, $t = 1$

$\Rightarrow \quad \sin^2 x = 1 = \sin^2 \frac{\pi}{2}$

$\Rightarrow \quad x = n\pi \pm \frac{\pi}{2}$

$\Rightarrow \quad x = \frac{\pi}{2} \quad \left[\because x \in \left(\frac{-\pi}{2}, \pi\right)\right]$

when, $t = \frac{1}{4}$

$\Rightarrow \quad \sin^2 x = \frac{1}{4}$

$\Rightarrow \quad \sin x = \pm \frac{1}{2}$

Taking + ve sign, $\sin x = \frac{1}{2} = \sin \frac{\pi}{6}$

$$x = n\pi + (-1)^n \frac{\pi}{6} \Rightarrow x = \frac{\pi}{6} \quad \left[\because x \in \left(\frac{-\pi}{2}, \pi\right)\right]$$

Taking − ve sign,

$$\sin x = \frac{-1}{2}$$

$\Rightarrow \quad \sin x = -\sin \frac{\pi}{6} = \sin\left(\pi + \frac{\pi}{6}\right)$

$\Rightarrow \quad \sin x = \sin \frac{7\pi}{6}$

$\Rightarrow \quad x = n\pi + (-1)^n \frac{7\pi}{6}$

$\therefore \quad x = -\frac{\pi}{6} \quad \left[\because x \in \left(\frac{-\pi}{2}, \pi\right)\right]$

Hence, total number of solutions is 3.

Alternate Method

Graphically,

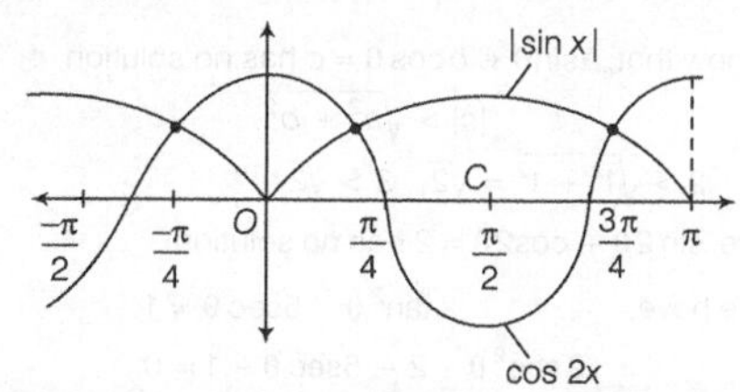

Clearly, graph of $|\sin x|$ and $\cos 2x$ intersect at three points between $\frac{-\pi}{2}$ to π.

Hence, total number of solutions is 3.

Revisal Problems (JEE Main)

1. (c) $\because \sqrt{3}\cos\theta + \sin\theta = \sqrt{2}$

and $\theta = n\pi + (-1)^n \alpha - \beta$ ...(i)

On dividing both sides by $\sqrt{(\sqrt{3})^2 + (1)^2} = 2$, we get

$$\frac{\sqrt{3}}{2}\cos\theta + \frac{1}{2}\sin\theta = \frac{1}{\sqrt{2}}$$

$$\sin\frac{\pi}{3}\cos\theta + \cos\frac{\pi}{3}\sin\theta = \frac{1}{\sqrt{2}}$$

$$\sin\left(\theta + \frac{\pi}{3}\right) = \sin\frac{\pi}{4}$$

$$\theta + \frac{\pi}{3} = n\pi + (-1)^n \frac{\pi}{4}, \forall n \in I \quad \text{(using formula)}$$

$$\theta = n\pi + (-1)^n \frac{\pi}{4} - \frac{\pi}{3}, \forall n \in I \quad ...(ii)$$

Now, on comparing Eqs. (i) and (ii), we get

$$\alpha = \frac{\pi}{4} \text{ and } \beta = \frac{\pi}{3}$$

$\therefore \quad \alpha + \beta = \frac{\pi}{4} + \frac{\pi}{3} = \frac{7\pi}{12}$

2. (d) We have, $1 + \cot\alpha = \operatorname{cosec}\alpha$

$\operatorname{cosec}\alpha - \cot\alpha = 1$ ...(i)

Then, $\operatorname{cosec}\alpha + \cot\alpha = 1$ ...(ii)

On solving Eqs. (i) and (ii), we get

$$2\operatorname{cosec}\alpha = 2 \Rightarrow \sin\alpha = 1$$

$\Rightarrow \quad \alpha = 2n\pi + \frac{\pi}{2}, \forall n \in I$

$\Rightarrow \quad \alpha = (4n + 1)\frac{\pi}{2}, \forall n \in I$

3. (a) $\because \sec x \cdot \cos 5x = -1$ and $0 < x < \frac{\pi}{4}$

$$\frac{\cos 5x}{\cos x} = -1$$

$\Rightarrow$ $\cos 5x = -\cos x$

$\cos 5x + \cos x = 0$

$\Rightarrow$ $2\cos 3x \cdot \cos 2x = 0$

Either $2\cos 3x = 0$ or $\cos 2x = 0$

$\cos 3x = 0$ or $2x = (2n+1)\frac{\pi}{2}$

$3x = (2n+1)\frac{\pi}{2}$ or $x = (2n+1)\frac{\pi}{4}, \forall n \in I$ (not possible)

$\therefore$ $x = (2n+1)\frac{\pi}{6}$

Put $n = 0$, then

$x = \frac{\pi}{6}$

$\therefore$ $x = \frac{\pi}{6}$

4. (b) $\because$ $\cos^2 \alpha + \sin\alpha + 1 = 0$

$1 - \sin^2 \alpha + \sin\alpha + 1 = 0$

$\sin^2 \alpha - \sin\alpha - 2 = 0$

$(\sin\alpha - 2)(\sin\alpha + 1) = 0$

$\therefore$ $\sin\alpha = 2$, which is not possible.

or $\sin\alpha = -1 = \sin\frac{3\pi}{2}$

$\therefore$ $\alpha = \frac{3\pi}{2} \in [0, 2\pi]$

5. (a) $\because$ $\sin 2\theta + \cos 2\theta = 2$

We know that, $a\sin\theta + b\cos\theta = c$ has no solution, if

$$|c| > \sqrt{a^2 + b^2}.$$

Here, $2 > \sqrt{1^2 + 1^2} = \sqrt{2}$, $2 > \sqrt{2}$

Hence, $\sin 2\theta + \cos 2\theta = 2$ has no solution.

6. (d) We have, $2\tan^2\theta - 5\sec\theta = 1$

$2\sec^2\theta - 2 - 5\sec\theta - 1 = 0$

$2\sec^2\theta - 5\sec\theta - 3 = 0$

$(\sec\theta - 3)\left(\sec\theta + \frac{1}{2}\right) = 0$

$\therefore$ $\sec\theta = 3$ and $\sec\theta = \frac{-1}{2}$ (neglecting)

$\because \sec\theta$ is +ve in I and IV quadrants, then

2 values in $[0, 2\pi]$

2 values in $(2\pi, 4\pi]$

2 values in $(4\pi, 6\pi]$

1 value in $\left[6\pi, 6\pi + \frac{3\pi}{2}\right]$ ($\because n$ is greatest)

We get 7 distinct values in $\left[0, \frac{15\pi}{2}\right]$

Now, comparing to $\left[0, \frac{n\pi}{2}\right]$, we get $n = 15$

7. (b) $\because$ $\sin^{100}\theta - \cos^{100}\theta = 1$

$\sin^{100}\theta = 1 + \cos^{100}\theta$ ...(i)

This equation is valid, if $\cos^{100}\theta = 0$ and $\sin^{100}\theta = 1$

($\because 0 \le \cos^2\theta \le 1, 0 \le \sin^2\theta \le 1$)

$\therefore$ $\cos\theta = 0$, then $\sin\theta = 1$

$\theta = (2n+1)\frac{\pi}{2}, \forall n \in I$

$\theta = n\pi + \frac{\pi}{2}, \forall n \in I$

8. (b) $\because \cos\{\pi\sqrt{(x-4)}\}\cos(\pi\sqrt{x}) = 1$ ... (i)

is defined, when $x - 4 \ge 0$ and $x \ge 0$

$\Rightarrow$ $x \ge 4$ ...(ii)

We know that, $-1 \le \cos\theta \le 1$

So, Eq. (i) is possible, when

$\cos(\pi\sqrt{x-4}) = 1$ and $\cos(\pi\sqrt{x}) = 1$

$\therefore$ $\pi\sqrt{x-4} = 0$ and $\pi\sqrt{x} = 0$

$x = 4$ and $x = 0$

But $x \ge 4$

$\therefore$ Number of solutions = 1 as $x = 4$.

9. (b) $\because \alpha^2 - 4\alpha + 7 = \alpha^2 - 4\alpha + 4 + 3$

$= (\alpha - 2)^2 + 3 \ge 3$

and $2 < e < 3$

and $\pi = 3.14$

$4 < e^2 < 9$

Now, $\sqrt{3}\sin x - \cos x = \min\limits_{\alpha \in R}\{2, e^2, \pi, \alpha^2 - 4\alpha + 7\}$

$\sqrt{3}\sin x - \cos x = 2$

$\frac{\sqrt{3}}{2}\sin x - \frac{1}{2}\cos x = 1$

$\sin x \cdot \cos\frac{\pi}{6} - \cos x \cdot \sin\frac{\pi}{6} = 1$

$\sin\left(x - \frac{\pi}{6}\right) = 1$

$\therefore$ $x - \frac{\pi}{6} = 2n\pi + \frac{\pi}{2}, \forall n \in I$

$x = 2n\pi + \frac{\pi}{2} + \frac{\pi}{6}, \forall n \in I$

$x = 2n\pi + \frac{2\pi}{3}, \forall n \in I$

Revisal Problems (JEE Advanced)

1. (b) Given equation $(\cos\alpha - 1)t^2 + (\cos\alpha)t + \sin\alpha = 0$ has real roots, then

$D \ge 0$

$\cos^2\alpha - 4(\cos\alpha - 1)\sin\alpha \ge 0$

$\cos^2\alpha - 4\sin\alpha\cos\alpha + 4\sin\alpha \ge 0$

$(\cos^2\alpha - 4\sin\alpha\cos\alpha + 4\sin^2\alpha) + 4\sin\alpha - 4\sin^2\alpha \ge 0$

$(\cos\alpha - 2\sin\alpha)^2 + 4\sin\alpha(1 - \sin\alpha) \ge 0$ ...(i)

We know that, $(\cos\alpha - 2\sin\alpha)^2 \ge 0$ and $1 - \sin\alpha \ge 0$, then inequality Eq. (i) possible only when,

$4\sin\alpha \ge 0$

$\sin\alpha \ge 0$

($\because \sin\alpha$ is + ve in I and II quadrants)

$\therefore$ $\alpha \in [0, \pi]$

2. (a) We have, $\cos 4x + 6 = 7\cos 2x$

$$2\cos^2 2x - 1 + 6 - 7\cos 2x = 0$$

$$2\cos^2 2x - 7\cos 2x + 5 = 0$$

$$(2\cos x - 5)(\cos 2x - 1) = 0$$

Thus, $\cos 2x = 1$

and $\cos 2x = \frac{5}{2}$

(which is not possible, $-1 \le \cos\theta \le 1$)

$\therefore$ $\cos 2x = 1$

$2x = 2n\pi, \forall n \in I$ (using formula)

$x = n\pi, \forall n \in I$

If we take $n = 100$, then $x = 100\pi = 314° < 315°$

and if we take $n = 101$, then

$$x = 101\pi = 317.14° > 317°$$

So, $x = n\pi \notin [315°, 317°], \forall x \in I$

Hence, number of solutions is 0.

3. (a) $\because \cot\left[\frac{\pi}{3}\cos 2\pi x\right] = \sqrt{3}$

$$\frac{\pi}{3}\cos 2\pi x = \cot^{-1}(\sqrt{3})$$

$$\frac{\pi}{3}\cos 2\pi x = \frac{\pi}{6} \Rightarrow \cos 2\pi x = \frac{1}{2}$$

$$\cos 2\pi x = \cos\frac{\pi}{3}$$

$$2\pi x = 2n\pi \pm \frac{\pi}{3}, \forall n \in I$$

$$x = n \pm \frac{1}{6}, \forall n \in I$$

4. (d) We have, $\sin\theta + \sin 2\theta + \sin 3\theta + \sin 4\theta = 4$

It is possible only when $\sin\theta = \sin 2\theta = \sin 3\theta = \sin 4\theta = 1$

But, it is not possible simultaneously for a particular value of x.

($\because 1 \le \sin\theta \le 1$)

$\therefore$ Number of solutions is 0.

5. (a) We have, $F(t) = \sin t + \cos t$, then

$$F\left(\frac{\pi}{10}\right) = \sin 18° + \cos 18°$$

$$= \sqrt{2}\left(\frac{1}{\sqrt{2}}\sin 18° + \frac{1}{\sqrt{2}}\cos 18°\right)$$

$$= \sqrt{2}\sin(45° + 18°) = \sqrt{2}\sin 63°$$

We know that

$$\sin 45° < \sin 63° < \sin 90°$$

$$\frac{1}{\sqrt{2}} < \sin 63° < 1$$

$$1 < \sqrt{2}\sin 63° < \sqrt{2}$$

Thus, $\sqrt{2}\sin 63° = 1$ ($\because t < x < t + 1 \Rightarrow [x] = t$)

Now, $F(t) = \left[F\left(\frac{\pi}{10}\right)\right]$ (given)

$$\sin t + \cos t = 1$$

$$\frac{1}{\sqrt{2}}\sin t + \frac{1}{\sqrt{2}}\cos t = \frac{1}{\sqrt{2}}$$

$$\cos\left(t - \frac{\pi}{4}\right) = \cos\frac{\pi}{4}$$

$t - \frac{\pi}{4} = 2n\pi \pm \frac{\pi}{4}, \forall n \in I$ (using formula)

$\therefore$ $t = 2n\pi + \frac{\pi}{4} + \frac{\pi}{4}, 2n\pi - \frac{\pi}{4} + \frac{\pi}{4}, \forall n \in I$

$$t = 2n\pi + \frac{\pi}{2}, 2n\pi, \forall n \in I$$

6. (a) $\because \log_{\cos\alpha}\sin\alpha + \log_{\sin\alpha}\cos\alpha = 2$ is valid, if $\sin\alpha > 0, \cos\alpha > 0$ and $\sin\alpha \ne 1$ and $\cos\alpha \ne 1$.

Now, $\log_{\cos\alpha}\sin\alpha + \frac{1}{\log_{\cos\alpha}\sin\alpha} - 2 = 0$ ...(i)

$$(\log_{\cos\alpha}\sin\alpha - 1)^2 = 0$$

$$\log_{\cos\alpha}\sin\alpha - 1 = 0$$

$$\log_{\cos\alpha}\sin\alpha = 1$$

$\sin\alpha = \cos\alpha$ ($\because \log_a b = m \Rightarrow b = a^m$)

$$\tan\alpha = 1 = \tan\frac{\pi}{4}$$

$\therefore$ $\alpha = 2n\pi + \frac{\pi}{4}, \forall n \in I$ [from Eq. (i)]

7. (c) We have, $\cos^4 x - (p + 2)\cos^2 x - (p + 3) = 0$

Then, $\cos^2 x = \frac{(p + 2) \pm \sqrt{(p + 2)^2 + 4(p + 3)}}{2}$

$$= \frac{(p + 2) \pm \sqrt{p^2 + 8p + 16}}{2}$$

$$= \frac{(p + 2) \pm (p + 4)}{2} = (p + 3), -1$$

Thus, $\cos^2 x = -1$ cannot possible, then $\cos^2 x = p + 3$

We know that, $0 \le \cos^2 x \le 1$

$$0 \le p + 3 \le 1$$

$$-3 \le p \le -2$$

8. (a) We have, $\cos\theta + \sin\theta = \sqrt{x + \frac{1}{x}}$ ...(i)

We know that, $\cos\theta + \sin\theta \le \sqrt{2}$ ...(ii)

and AM $\ge$ GM

Now, $\frac{x + \frac{1}{x}}{2} \ge \sqrt{x \cdot \frac{1}{x}} \Rightarrow x + \frac{1}{x} \ge 2$

$$\sqrt{x + \frac{1}{x}} \ge \sqrt{2}$$

$\cos\theta + \sin\theta \ge \sqrt{2}$ ...(iii)

From Eqs. (ii) and (iii), we get

$$\sin\theta + \cos\theta = \sqrt{2}$$

$$\frac{1}{\sqrt{2}}\sin\theta + \frac{1}{\sqrt{2}}\cos\theta = 1 \Rightarrow \cos\left(\theta - \frac{\pi}{4}\right) = 1$$

$\therefore$ $\theta - \frac{\pi}{4} = 2n\pi, \forall n \in I$

$$\theta = 2n\pi + \frac{\pi}{4}, \forall n \in I$$

$\therefore$ $\theta = \frac{\pi}{4} \in [0, \pi]$ and $x + \frac{1}{x} = 2$

which is possible, when $x = 1$.

9. (b) Let $A = 5\pi\sin\theta$ and $B = 5\pi\cos\theta$

Then, $\cot A - \tan B = 0$

$$\frac{\cos A}{\sin A} - \frac{\sin B}{\cos B} = 0$$

$\Rightarrow$ $\cos A \cdot \cos B - \sin A \cdot \sin B = 0$

$\therefore$ $\cos(A + B) = 0$

$\Rightarrow$ $A + B = (2n + 1)\frac{\pi}{2}, \forall n \in I$ ...(i)

Now, $5\pi\sin\theta + 5\pi\cos\theta = (2n + 1)\frac{\pi}{2}, \forall n \in I$

$$\sin\theta + \cos\theta = \frac{2n + 1}{10}$$

$$\sin\theta \cdot \frac{1}{\sqrt{2}} + \cos\theta \cdot \frac{1}{\sqrt{2}} = \frac{2n + 1}{10\sqrt{2}}$$

$$\sin\left(\theta + \frac{\pi}{4}\right) = \frac{2n+1}{10\sqrt{2}} \quad \ldots(ii)$$

$$\because \quad -1 \le \sin\left(\theta + \frac{\pi}{4}\right) \le 1$$

$$-1 \le \frac{2n+1}{10\sqrt{2}} \le 1$$

$$\frac{-10\sqrt{2}-1}{2} \le n \le \frac{10\sqrt{2}-1}{2}$$

$$-7.5 \le n \le 6.5 \qquad [\because \sqrt{2} = 1.4 \text{ (let)}]$$

$\therefore \quad n = -7, -6, -5, -4, -3, -2, -1, 0, 1, 2, 3, 4, 5, 6$

Hence, number of solutions is 14.

10. (a,b,c) We have

$$\alpha \in [0, \pi] \text{ and } 81^{\sin^2\alpha} + 81^{\cos^2\alpha} = 30$$

$$\Rightarrow \quad 81^{\sin^2\alpha} + \frac{81}{81^{\sin^2\alpha}} = 30$$

Let $\quad (81)^{\sin^2\alpha} = t$

Then, $\quad t + \frac{81}{t} = 30$

$$t^2 - 30t + 81 = 0$$

$$(t-27)(t-3) = 0$$

$$t = 27 \text{ and } t = 3$$

If $t = 27$

$$(81)^{\sin^2\alpha} = 27$$

$$3^{4\sin^2\alpha} = 3^3$$

$$4\sin^2\alpha = 3$$

$$\sin^2\alpha = \frac{3}{4}$$

$$\sin\alpha = \frac{\sqrt{3}}{2} \quad \text{(neglecting – ve value)}$$

$$\sin\alpha = \sin\frac{\pi}{3}, \sin\left(\pi - \frac{\pi}{3}\right)$$

$$\Rightarrow \quad \alpha = \frac{\pi}{3}, \frac{2\pi}{3}$$

and if $t = 3$ $\quad 81^{\sin^2\alpha} = 3$

$$3^{4\sin^2\alpha} = 3^1$$

$$4\sin^2\alpha = 1$$

$$\sin\alpha = \frac{1}{2}$$

$$\sin\alpha = \sin\frac{\pi}{6}, \sin\left(\pi - \frac{\pi}{6}\right)$$

$$\Rightarrow \quad \alpha = \frac{\pi}{6}, \frac{5\pi}{6}$$

$$\therefore \quad \alpha = \frac{\pi}{6}, \frac{5\pi}{6}, \frac{\pi}{3}, \frac{2\pi}{3}$$

11. (a,c,d) $\because \cos(\sin\theta) = \frac{1}{\sqrt{2}} = \cos\left(\pm\frac{\pi}{4}\right)$

$$\sin\theta = \pm\frac{\pi}{4}$$

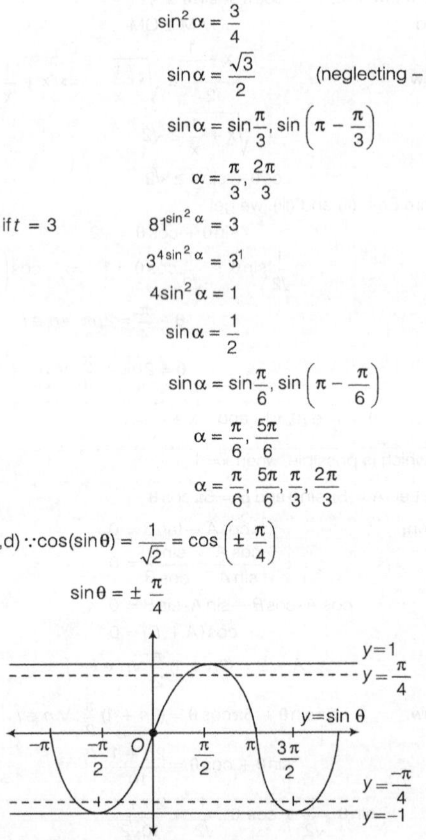

which satisfies the options (a), (c) and (d).

12. (a,c) We have,

$$\cos 2x + \sin 7x + 2 = 0$$

$$\cos 2x + \sin 7x = -2$$

It holds only, if $\cos 2x = -1$ and $\sin 7x = -1$

$$2x = 2n\pi + \pi, \forall n \in I$$

$$x = n\pi + \frac{\pi}{2}, \forall n \in I$$

and $\quad \sin 7x = -1 = \sin\frac{3\pi}{2}$

$$7x = 2n\pi + \frac{3\pi}{2}, \forall n \in I$$

$$x = \frac{2n\pi}{7} + \frac{3\pi}{14}, \forall n \in I$$

13. (a) We know that,

$$[x + I] = [x] + I, \text{ if } I \text{ is an integer,}$$

then $\quad y = \frac{1}{3}[\sin\theta + [\sin\theta + [\sin\theta]]]$

$$= \frac{1}{3}[\sin\theta + [\sin\theta] + [\sin\theta]]$$

$$= \frac{1}{3}([\sin\theta] + [\sin\theta] + [\sin\theta])$$

$$y = [\sin\theta] \quad \ldots(i)$$

and $\quad [y + [y]] = 2\cos\theta$

$$[y] + [y] = 2\cos\theta$$

$$[y] = \cos\theta \quad \ldots(ii)$$

$\because -1 \le \sin\theta \le 1$, then three cases arise:

Case I If $-1 \le \sin\theta < 0$, then $[\sin\theta] = -1$

$y = -1$ put in Eq. (ii), then

$$\cos\theta = [-1] = -1$$

$$\Rightarrow \quad \sin\theta = 0$$

But $\quad -1 \le \sin\theta < 0$

Hence, this case is not possible.

Case II If $0 \le \sin\theta < 1$, then $[\sin\theta] = 0$

and $\quad y = 0 \quad$ [from Eq. (i)]

Put in Eq. (ii), $\quad \cos\theta = 0$

$$\Rightarrow \quad \sin\theta = 1$$

But we have $0 \le \sin\theta < 1$, so this case is not possible.

Case III If $\sin\theta = 1$, then $[\sin\theta] = 1$

$y = 1$ put in Eq. (ii), $\quad \cos\theta = 1$

$$\Rightarrow \quad \sin\theta = 0$$

But we have $\sin\theta = 1$, so this case is not possible.

$\therefore$ Number of solutions is 0.

14. (a) $\because [\sin x] + [\sqrt{2}\cos x] = -3, x \in [0, 2\pi]$

It is possible only when

$$[\sin x] = -1$$

and $\quad [\sqrt{2}\cos x] = -2$

If $\quad [\sin x] = -1$

$$\Rightarrow \quad -1 \le \sin x < 0$$

$$\therefore \quad x \in (\pi, 2\pi) \quad \ldots(i)$$

If $\quad [\sqrt{2}\cos x] = -2$

$$\Rightarrow \quad -2 \le \sqrt{2}\cos x < -1$$

$$\Rightarrow \quad -\sqrt{2} \le \cos x < -\frac{1}{\sqrt{2}}$$

$$\Rightarrow \quad -1 \le \cos x < -\frac{1}{\sqrt{2}}$$

$$\therefore \quad x \in \left[\pi, \frac{5\pi}{4}\right) \quad \ldots(ii)$$

From Eqs. (i) and (ii), we get

$$x \in \left(\pi, \frac{5\pi}{4}\right)$$

15. (a,b,d) $\because \sin\theta + \cos\theta + \tan\theta + \cot\theta + \sec\theta + \text{cosec}\,\theta = 7$

$$(\sin\theta + \cos\theta) + \frac{\sin^2\theta + \cos^2\theta}{\sin\theta\cos\theta} + \frac{\sin\theta + \cos\theta}{\sin\theta\cos\theta} = 7$$

$(\because \sin\theta = \cos\theta \neq 0)$

$$(\sin\theta + \cos\theta) + \frac{2}{\sin 2\theta} + \frac{2(\sin\theta + \cos\theta)}{\sin 2\theta} = 7$$

$$(\sin\theta + \cos\theta)\sin 2\theta + 2 + 2(\sin\theta + \cos\theta) = 7 \cdot \sin 2\theta$$

Let $\sin\theta + \cos\theta = t$, then $\sin 2\theta = t^2 - 1$

Now, $t(t^2 - 1) + 2 + 2t = 7(t^2 - 1)$

$$t^3 - t + 2 + 2t - 7t^2 + 7 = 0$$

$$t^3 - 7t^2 + t + 9 = 0$$

$$(t + 1)(t^2 - 8t + 9) = 0$$

$$t^2 - 8t + 9 = 0 \quad (\because t + 1 \neq 0)$$

$$t = \frac{8 \pm \sqrt{28}}{2} = 4 \pm \sqrt{7}$$

$\therefore$ $\sin\theta + \cos\theta = 4 \pm \sqrt{7}$

On squaring both sides, we get

$$1 + \sin 2\theta = 16 + 7 \pm 8\sqrt{7}$$

$$\sin 2\theta = 22 \pm 8\sqrt{7}$$

$\Rightarrow$ $\sin 2\theta = 22 - 8\sqrt{7}$

(neglecting $22 + 8\sqrt{7}$ because $-1 \leq \sin 2\theta \leq 1$)

but $a - b\sqrt{c} = 22 - 8\sqrt{7}$

On comparing, $a = 22, b = 8$ and $c = 7$

$\therefore$ $a - b + 2c = 22 - 8 + 14 = 28$

16. (a,b,c,d) We have,

$$\sin x = [1 + \sin x] + [1 - \cos x]$$

$$\sin x = 1 + [\sin x] + 1 + [-\cos x]$$

Now, at $x = -\frac{\pi}{2}$, $-1 = 1 - 1 + 1 + 0$

$-1 = 1$ (which is not possible)

At $x = \frac{\pi}{2}$, $1 = 1 + 1 + 1 + 0$

$1 = 3$ (not possible)

At $x = \pi$, $0 = 1$ (not possible)

If $x \in \left(\frac{-\pi}{2}, 0\right)$, $\sin x < 0$

$\sin x = 2 + (-1) + 0 = 1$ (not possible)

If $x \in \left(0, \frac{\pi}{2}\right)$, $0 < \sin x < 1$

$\sin x = 2 + 0 - 1 = 1$ (not possible)

If $x \in \left(\frac{\pi}{2}, \pi\right)$, $\sin x = 2 + 0 + 0 = 2$ (not possible)

If $x \in \left(\pi, \frac{3\pi}{2}\right]$, $\sin x < 0$

$\sin x = 2 - 1 + 0 = 1$ (not possible)

17. (b,c) We have, $(1 - \tan\theta)(1 + \tan\theta)\sec^2\theta + 2^{\tan^2\theta} = 0$

$$(1 - \tan^2\theta)(1 + \tan^2\theta) + 2^{\tan^2\theta} = 0$$

$$(1 - \tan^4\theta) + 2^{\tan^2\theta} = 0$$

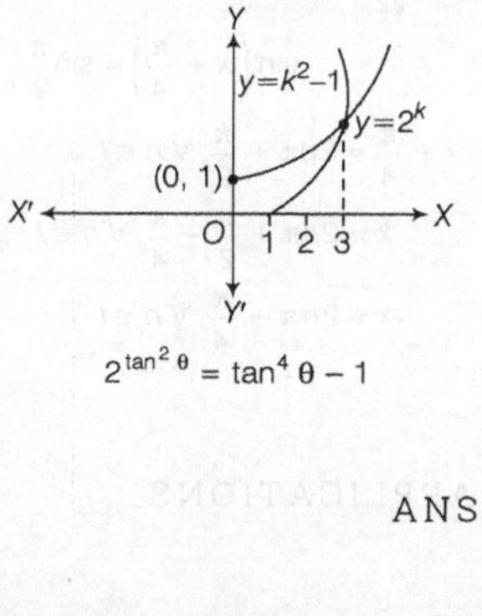

$$2^{\tan^2\theta} = \tan^4\theta - 1$$

Let $\tan^2\theta = k, k \in [0, \infty)$

$\therefore$ $y = 2^k = k^2 - 1$

$\Rightarrow$ $k = 3$

Now, $\tan^2\theta = 3 = \tan^2\frac{\pi}{3}$

$\Rightarrow$ $\theta = \pm\frac{\pi}{3}$

18. (a,b,c) We have, $\sqrt{\cos 2x} + \sqrt{1 + \sin 2x} = 2\sqrt{\sin x + \cos x}$

$$\sqrt{\cos^2 x - \sin^2 x} + \sqrt{(\cos x + \sin x)^2} = 2\sqrt{\sin x + \cos x}$$

$$\sqrt{\cos x + \sin x}\cdot\sqrt{\cos x - \sin x} + (\cos x + \sin x) - 2\sqrt{\sin x + \cos x} = 0$$

$$\sqrt{\cos x + \sin x}\,[\sqrt{\cos x - \sin x} + \sqrt{\cos x + \sin x} - 2] = 0$$

Either $\sqrt{\cos x + \sin x} = 0$

$\Rightarrow$ $\cos x + \sin x = 0$ [option (a)]

$\Rightarrow$ $\tan x = -1 = \tan\left(-\frac{\pi}{4}\right)$

$\therefore$ $x = n\pi - \frac{\pi}{4}, n \in I$ [option (c)]

or $\sqrt{\cos x - \sin x} + \sqrt{\cos x + \sin x} - 2 = 0$

$$\sqrt{\cos x - \sin x} + \sqrt{\cos x + \sin x} = 2$$

On squaring both sides, we get

$$\cos x - \sin x + \cos x + \sin x + 2\sqrt{\cos^2 x - \sin^2 x} = 4$$

$\Rightarrow$ $\cos x + \sqrt{\cos^2 x - \sin^2 x} = 2$

$\Rightarrow$ $2 - \cos x = \sqrt{\cos^2 x - \sin^2 x}$

$\Rightarrow$ $\cos^2 x + 4 - 4\cos x = \cos^2 x - \sin^2 x$

$\Rightarrow$ $-\sin^2 x + 4\cos x - 4 = 0$

$\Rightarrow$ $-1 + \cos^2 x + 4\cos x - 4 = 0$

$\Rightarrow$ $\cos^2 x + 4\cos x - 5 = 0$

$\Rightarrow$ $(\cos x + 5)(\cos x - 1) = 0$

Either $\cos x = -5$ (not possible)

or $\cos x - 1 = 0$

$\Rightarrow$ $\cos x = 1$

$\therefore$ $x = 2n\pi, n \in I$ [option (b)]

19. (a)

(i) We have,

$$\sin x = \frac{-1}{2} = -\sin\frac{\pi}{6}$$

$$\sin x = \sin\left(\pi + \frac{\pi}{6}\right), \sin\left(2\pi - \frac{\pi}{6}\right) \quad (\because x \in [0, 2\pi])$$

$$x = \frac{7\pi}{6}, \frac{11\pi}{6}$$

$\therefore$ $\alpha = \frac{7\pi}{6}, \beta = \frac{11\pi}{6}$

and $\cos x = \frac{-\sqrt{3}}{2} = -\cos\frac{\pi}{6}$

$$\cos x = \cos\left(\pi - \frac{\pi}{6}\right), \cos\left(\pi + \frac{\pi}{6}\right) \quad (\because x \in [0, 2\pi])$$

$$x = \frac{5\pi}{6}, \frac{7\pi}{6}$$

$\therefore$ $\alpha = \frac{7\pi}{6}, \gamma = \frac{5\pi}{6}$

Now, $\alpha + \beta = 3\pi, \beta - \gamma = \pi$

Thus, **(i) → (q,s)**

(ii) We have, $\cot x = -\sqrt{3} = -\cot\frac{\pi}{6}$

$$\cot x = \cot\left(\pi - \frac{\pi}{6}\right), \cot\left(2\pi - \frac{\pi}{6}\right)$$

$$x = \frac{5\pi}{6}, \frac{11\pi}{6} \Rightarrow \beta = \frac{5\pi}{6}, \alpha = \frac{11\pi}{6}$$

and cosec $x = -2$

$$\sin x = -\frac{1}{2} = -\sin\frac{\pi}{6}$$

$$\sin x = \sin\left(\pi + \frac{\pi}{6}\right), \left(2\pi - \frac{\pi}{6}\right)$$

$$x = \frac{7\pi}{6}, \frac{11\pi}{6}$$

$$\Rightarrow \quad \alpha = \frac{11\pi}{6}, \gamma = \frac{7\pi}{6}$$

$$\alpha - \beta = \pi, \beta + \gamma = 2\pi$$

Thus, **(ii) → (p,t)**

(iii) We have

$$\sin x = \frac{-1}{2}$$

$$\therefore \quad \alpha = \frac{7\pi}{6}, \beta = \frac{11\pi}{6} \quad \text{[by part (i)]}$$

and $\quad \tan x = \frac{1}{\sqrt{3}} = \tan\frac{\pi}{6}, \tan\left(\pi + \frac{\pi}{6}\right)$

$$x = \frac{\pi}{6}, \frac{7\pi}{6} \Rightarrow \alpha = \frac{7\pi}{6}, \gamma = \frac{\pi}{6}$$

Now, $\quad \alpha + \beta = 3\pi, \beta + \gamma = 2\pi, \alpha - \gamma = \pi$

Thus, **(iii) → (r,s,t)**

20. (d) $\because \quad 1 - \sin 2x = \cos x - \sin x \quad \ldots(i)$

Let $\cos x - \sin x = t$, then $1 - \sin 2x = t^2$, put in Eq. (i), we get

$$t^2 = t \Rightarrow t^2 - t = 0$$

$$t(t-1) = 0$$

$$\therefore \quad t = 0, \quad t = 1$$

$$\cos x - \sin x = 0, \quad \cos x - \sin x = 1$$

$$\tan x = 1; \quad \frac{1}{\sqrt{2}}\cos x - \frac{1}{\sqrt{2}}\sin x = \frac{1}{\sqrt{2}}$$

$$x = n\pi + \frac{\pi}{4}, \forall n \in I; \quad \cos\left(x + \frac{\pi}{4}\right) = \cos\frac{\pi}{4}$$

$$x + \frac{\pi}{4} = 2n\pi \pm \frac{\pi}{4}$$

$$x = 2n\pi + \frac{\pi}{4} - \frac{\pi}{4}, 2n\pi - \frac{\pi}{4} - \frac{\pi}{4}$$

$$x = 2n\pi, 2n\pi - \frac{\pi}{2}, \forall n \in I$$

21. (a) $\because \sin x + \cos x = 1 + \sin x \cos x \quad \ldots(i)$

Let $\quad \sin x + \cos x = t$

$$\Rightarrow \quad \sin x \cdot \cos x = \frac{t^2 - 1}{2}, \text{ put in Eq.(i)}$$

$$t = 1 + \frac{t^2 - 1}{2}$$

$$\Rightarrow \quad 2t = 2 + t^2 - 1$$

$$\Rightarrow \quad t^2 - 2t + 1 = 0$$

$$\Rightarrow \quad (t-1)^2 = 0 \Rightarrow t = 1$$

$$\therefore \quad \sin x + \cos x = 1$$

$$\frac{1}{\sqrt{2}}\sin x + \frac{1}{\sqrt{2}}\cos x = \frac{1}{\sqrt{2}}$$

$$\cos\left(x - \frac{\pi}{4}\right) = \cos\frac{\pi}{4}$$

$$x - \frac{\pi}{4} = 2n\pi \pm \frac{\pi}{4} \quad \text{(using formula)}$$

$$x = 2n\pi + \frac{\pi}{4} + \frac{\pi}{4}, 2n\pi - \frac{\pi}{4} + \frac{\pi}{4}, \forall n \in I$$

$$\therefore \quad x = 2n\pi + \frac{\pi}{2}, 2n\pi, \forall n \in I$$

22. (c) We have, $\sin^4 x + \cos^4 x = \sin x \cos x$

$$1 - 2(\sin x \cos x)^2 = \sin x \cos x$$

$$1 - \frac{1}{2}\sin^2 2x = \frac{1}{2}\sin 2x$$

$$2 - \sin^2 2x = \sin 2x$$

$$\sin^2 2x + \sin 2x - 2 = 0$$

$$(\sin 2x + 2)(\sin 2x - 1) = 0$$

$$\therefore \quad \sin 2x = 1, \quad \sin 2x = -2 \quad \text{(not possible)}$$

$$2x = 2n\pi + \frac{\pi}{2}, \forall n \in I \quad \text{(using formula)}$$

$$x = n\pi + \frac{\pi}{4}, \forall n \in I$$

$$x = (4n+1)\frac{\pi}{4}, \forall n \in I$$

23. (b) $\because 1 + \sin^3 x + \cos^3 x = \frac{3}{2}\sin 2x$

$$1 + (\sin x + \cos x)(1 - \sin x \cdot \cos x) = \frac{3}{2} \cdot 2\sin x \cos x$$

Let $\quad \sin x + \cos x = t \Rightarrow \sin x \cdot \cos x = \frac{t^2 - 1}{2}$

$$1 + t\left(1 - \frac{t^2 - 1}{2}\right) = 3 \cdot \frac{t^2 - 1}{2}$$

$$t^3 + 3t^2 - 3t - 5 = 0$$

$$(t+1)(t^2 + 2t - 5) = 0$$

$$\therefore \quad t = -1, \quad -1 \pm \sqrt{6}$$

$\sin x + \cos x = -1, \quad \sin x + \cos x = -1 \pm \sqrt{6}$ which is not possible.

$$\frac{1}{\sqrt{2}}\sin x + \frac{1}{\sqrt{2}}\cos x = -\frac{1}{\sqrt{2}} \quad (\because -\sqrt{2} \le \sin x + \cos x \le \sqrt{2})$$

$$\cos\left(x - \frac{\pi}{4}\right) = \cos\frac{3\pi}{4}$$

$$x - \frac{\pi}{4} = 2n\pi \pm \frac{3\pi}{4}, \forall n \in I \quad \text{(using formula)}$$

$$x = 2n\pi \pm \frac{3\pi}{4} + \frac{\pi}{4}, \forall n \in I$$

$$x = 2n\pi + \frac{3\pi}{4} + \frac{\pi}{4}, 2n\pi - \frac{3\pi}{4} + \frac{\pi}{4}$$

$$x = (2n+1)\pi, \quad 2n\pi - \frac{\pi}{2}$$

$$\therefore \quad x = (2n+1)\pi, \left(2n\pi - \frac{\pi}{2}\right), \forall n \in I$$

24. (a) Given, $\sin x + \cos x - 2\sqrt{2}\sin x \cos x = 0$

Let $\sin x + \cos x = t \Rightarrow 2\sin x \cdot \cos x = t^2 - 1$

$$t - \sqrt{2}(t^2 - 1) = 0$$

$$t - \sqrt{2}t^2 + \sqrt{2} = 0 \Rightarrow \sqrt{2}t^2 - t - \sqrt{2} = 0$$

$$\Rightarrow \quad (t - \sqrt{2})(\sqrt{2}t + 1) = 0$$

$$t = \sqrt{2}, \frac{-1}{\sqrt{2}}$$

$$\therefore \quad \sin x + \cos x = \sqrt{2} \quad \text{and} \quad \sin x + \cos x = \frac{-1}{\sqrt{2}}$$

$$\frac{1}{\sqrt{2}}\sin x + \frac{1}{\sqrt{2}}\cos x = 1$$

$$\sin\left(x + \frac{\pi}{4}\right) = \sin\frac{\pi}{2}$$

$$x + \frac{\pi}{4} = 2n\pi + \frac{\pi}{2}, \forall n \in I$$

$$x = 2n\pi + \frac{\pi}{2} - \frac{\pi}{4}, \forall n \in I$$

$$x = 2n\pi + \frac{\pi}{4}, \forall n \in I$$

and if $\sin x + \cos x = -\frac{1}{\sqrt{2}}$

$$\sin\left(x + \frac{\pi}{4}\right) = -\frac{1}{2} = \sin\left(-\frac{\pi}{6}\right)$$

$$x + \frac{\pi}{4} = n\pi + (-1)^n\left(-\frac{\pi}{6}\right)$$

$$x = n\pi + (-1)^n\left(-\frac{\pi}{6}\right) - \frac{\pi}{4}, \forall n \in I$$

25. (a) We know that, AM $\geq$ GM

$$\frac{2^{\sin\theta} + 2^{\cos\theta}}{2} \geq \sqrt{2^{\sin\theta} \cdot 2^{\cos\theta}}$$

$$2^{\sin\theta} + 2^{\cos\theta} \geq 2 \cdot 2^{\frac{1}{2}(\sin\theta + \cos\theta)} = 2 \cdot 2^{\left(\frac{1}{\sqrt{2}}\sin\theta + \frac{1}{\sqrt{2}}\cos\theta\right)\cdot\frac{1}{\sqrt{2}}}$$

$$= 2 \cdot 2^{\frac{1}{\sqrt{2}}\sin\left(\theta + \frac{\pi}{4}\right)}$$

$$2^{\sin\theta} + 2^{\cos\theta} \geq 2^{1 - \frac{1}{\sqrt{2}}} \quad \left[\because \sin\left(\theta + \frac{\pi}{2}\right) \geq -1\right]$$

Now, equality holds, if $2^{\sin\theta} = 2^{\cos\theta}$

$$\Rightarrow \quad \sin\theta = \cos\theta \quad \left[\begin{array}{l}\because \text{AM of } a, b = \text{GM of } a, b \\ \text{If } a = b\end{array}\right]$$

$$\tan\theta = 1 = \tan\frac{\pi}{4}$$

$$\theta = n\pi + \frac{\pi}{4}, \forall n \in I \quad \text{(using formula)}$$

26. (b) We have, $\tan\theta + \tan 2\theta + \tan 3\theta = \tan\theta \tan 2\theta \tan 3\theta$

Statement I $\tan(\theta + 2\theta + 3\theta) = 0$

$$\left(\begin{array}{l}\because \text{if } \tan A + \tan B + \tan C = \tan A \tan B \tan C, \\ \text{then } \tan(A + B + C) = 0\end{array}\right)$$

$$\tan 6\theta = 0$$

$$6\theta = n\pi, \forall n \in I \quad \text{(using formula)}$$

$$\theta = \frac{n\pi}{6}, \forall n \in I$$

$\because \quad \theta \in (0, \pi)$, then $\theta = \frac{\pi}{6}, \frac{2\pi}{6}, \frac{3\pi}{6}, \frac{4\pi}{6}, \frac{5\pi}{6}$

$$\theta = \frac{\pi}{6}, \frac{\pi}{3}, \frac{\pi}{2}, \frac{2\pi}{3}, \frac{5\pi}{6}$$

$\because \tan 3\theta$ is not defined, when $\theta = \frac{\pi}{6}$

$\tan\theta$ is not defined, when $\theta = \frac{\pi}{2}$

$\tan\theta$ is not defined, when $\theta = \frac{5\pi}{6}$

So, neglecting all above these values

$$\therefore \quad \theta = \frac{\pi}{3}, \frac{2\pi}{3}$$

$\therefore$ Number of solutions is 2.

Statement II $\tan 6\theta$ is not defined at $\theta = (2n + 1)\frac{\pi}{12}$.

27. (2) We have, $\sum_{r=1}^{5} \cos rx = 5$ in $[0, 2\pi]$

$$\cos x + \cos 2x + \cos 3x + \cos 4x + \cos 5x = 5$$

It is possible only when

$\cos x = \cos 2x = \cos 3x = \cos 4x = \cos 5x = 1$ $(\because 1 \leq \cos\theta \leq 1)$

and it is satisfied by $x = 0, 2\pi$

$\therefore$ Number of solutions is 2.

28. (5) We have, $\cos 4t + 6 = 7\cos 2t$

$$2\cos^2 2t - 1 + 6 - 7\cos 2t = 0$$

$$2\cos^2 2t - 7\cos 2t + 5 = 0$$

$$(\cos 2t - 1)(2\cos 2t - 5) = 0$$

Either $\quad \cos 2t = \frac{5}{2}$ (neglecting)

or $\quad \cos 2t - 1 = 0$

$$\cos 2t = 1$$

$$\therefore \quad 2t = 2n\pi, \forall n \in I$$

$$t = n\pi, \forall n \in I$$

Now, over the interval $[0, 314]$,

$$t = \pi, 2\pi, 3\pi, \ldots, 99\pi$$

Sum of the roots $= \pi\,[1 + 2 + 3 + \cdots + 99]$

$$= \pi \cdot \frac{1}{2}\,99 \cdot (99 + 1) = 990\pi \times 5$$

Now, $\quad 990\pi\lambda = 990\pi \times 5$

$$\therefore \quad \lambda = 5$$

29. (4) Draw the curves $p|\cos\theta|$ and $q|\sin\theta|$ and find the number of intersection points.

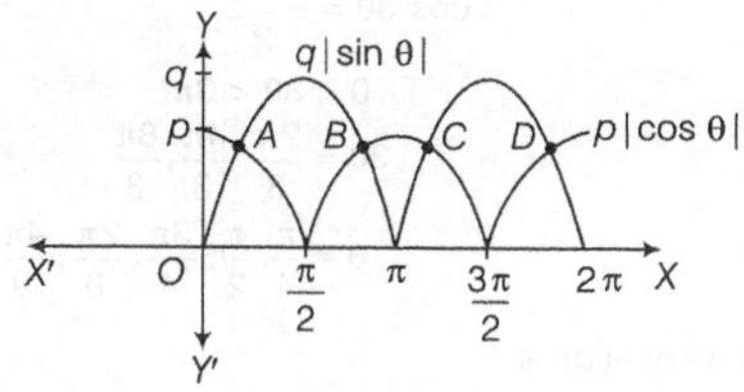

Hence, intersection points are A, B, C, D.

$\therefore$ Number of solutions is 4.

30. (3) $\because \tan^2 x - \sec^{10} x = -1$ and $x \in (0, 10)$

$$1 + \tan^2 x - \sec^{10} x = 0$$

$$\sec^2 x - \sec^{10} x = 0$$

$$\sec^2 x\,(1 - \sec^8 x) = 0$$

$\therefore$ Either $\sec^2 x = 0$ or $\sec^8 x = 1$

$\sec x = 0$ or $\sec x = \pm 1$

(not possible) or $\cos x = \pm 1$

If $\cos x = 1$, then $x = 0, 2\pi$, but $x \in (0,10)$

$$\therefore \quad x = 2\pi$$

If $\quad \cos x = -1$, then $x = \pi, 3\pi$, as $x \in (0, 10)$

So, number of solutions is 3.

31. (2) Given, $4\operatorname{cosec}^2[\pi(p + t)] + p^2 - 4p = 0$

$$4\operatorname{cosec}^2[\pi(p + t)] = 4p - p^2$$

$$\sin^2[\pi(p + t)] = \frac{4}{4p - p^2}$$

We know that, $\quad 0 \leq \sin^2[\pi(p + t)] \leq 1$

$$0 \leq \frac{4}{4p - p^2} \leq 1$$

If $\frac{4}{4p - p^2} \geq 0 \Rightarrow 4p - p^2 > 0$

$$\Rightarrow \quad p(-4 + p) < 0$$

$$\therefore \quad 0 < p < 4 \quad \ldots\text{(i)}$$

If $\frac{4}{4p - p^2} \leq 1 \Rightarrow \frac{4 - 4p + p^2}{4p - p^2} \leq 0$

$$\Rightarrow \quad \frac{(p - 2)^2}{p(4 - p)} \leq 0$$

$$\therefore \quad p \in (-\infty, 0) \cup \{2\} \cup (4, \infty) \quad \ldots\text{(ii)}$$

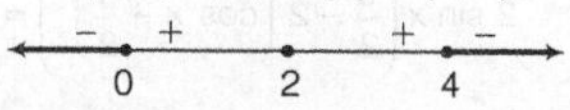

Now, on taking common from Eqs. (i) and (ii), we get

$$p = 2$$

JEE Archive

1. (a) We have,

$$\sin\theta + \sin 4\theta + \sin 7\theta = 0$$

$\Rightarrow \quad \sin 4\theta + (\sin\theta + \sin 7\theta) = 0$

$\Rightarrow \quad \sin 4\theta + 2\sin 4\theta \cdot \cos 3\theta = 0$

$\Rightarrow \quad \sin 4\theta\,(1 + 2\cos 3\theta) = 0$

$\Rightarrow \quad \sin 4\theta = 0$

$\cos 3\theta = -\frac{1}{2}$

Now, $\quad \sin 4\theta = 0$

Given, $\quad 0 < \theta < \pi$

$\Rightarrow \quad 0 < 4\theta < 4\pi$

$\Rightarrow \quad 4\theta = \pi, 2\pi, 3\pi$

and $\quad \cos 3\theta = -\frac{1}{2}$

$\Rightarrow \quad 0 < 3\theta < 3\pi$

$\Rightarrow \quad 3\theta = \frac{2\pi}{3}, \frac{4\pi}{3}, \frac{8\pi}{3}$

$\therefore \quad \theta = \frac{\pi}{4}, \frac{\pi}{2}, \frac{3\pi}{4}, \frac{2\pi}{9}, \frac{4\pi}{9}, \frac{8\pi}{9}$

2. (d) Given equation is

$$2\sin^2 x + 5\sin x - 3 = 0$$

$\Rightarrow \quad (2\sin x - 1)(\sin x + 3) = 0$

$\Rightarrow \quad \sin x = \frac{1}{2} \qquad (\because \sin x \neq -3)$

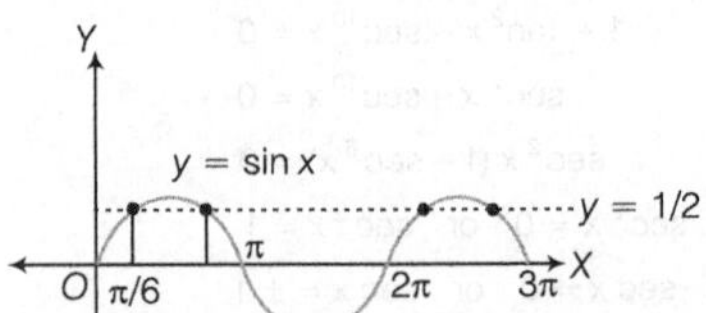

It is clear from figure that the curve intersect the line at four points in the given interval.
Hence, number of solutions is 4.

3. (c) Since, $\quad |c| > \sqrt{a^2 + b^2} \qquad \ldots(\text{i})$

$\Rightarrow \quad c < -\sqrt{a^2 + b^2}$

and $\quad c > \sqrt{a^2 + b^2}$

But $\quad -\sqrt{a^2 + b^2} \leq a\sin x + b\cos x \leq \sqrt{a^2 + b^2} \qquad \ldots(\text{ii})$

$\therefore \quad a\sin x + b\cos x = c$

We see that no solution exists.

4. (d) **Plan** For solving this type of questions, obtain the LHS and RHS in equation and examine, both are equal or not for a given interval.

Given, trigonometrical equation $(\sin x - \sin 3x) + 2\sin 2x = 3$

$\Rightarrow \quad -2\cos 2x \sin x + 4\sin x\cos x = 3$

$$\left[\begin{array}{l}\because \sin C - \sin D = 2\cos\left(\frac{C+D}{2}\right)\sin\left(\frac{C-D}{2}\right) \\ \text{and } \sin 2\theta = 2\sin\theta\cos\theta\end{array}\right]$$

$\Rightarrow \quad 2\sin x\,(2\cos x - \cos 2x) = 3$

$\Rightarrow \quad 2\sin x\,(2\cos x - 2\cos^2 x + 1) = 3$

$\Rightarrow \quad 2\sin x\left(\frac{3}{2} - 2\left(\cos x - \frac{1}{2}\right)^2\right) = 3$

$\Rightarrow \quad 3\sin x - 3 = 4\left(\cos x - \frac{1}{2}\right)^2 \sin x$

As, $x \in (0, \pi)$

$\text{LHS} \leq 0$ and $\text{RHS} \geq 0$

For solution to exist,

$\text{LHS} = \text{RHS} = 0$

Now, $\quad \text{LHS} = 0$

$\Rightarrow \quad 3\sin x - 3 = 0$

$\Rightarrow \quad \sin x = 1$

$\Rightarrow \quad x = \frac{\pi}{2}$

For $x = \frac{\pi}{2}$, $\quad \text{RHS} = 4\left(\cos\frac{\pi}{2} - \frac{1}{2}\right)^2 \sin\frac{\pi}{2}$

$= 4\left(\frac{1}{4}\right)(1) = 1 \neq 0$

$\therefore$ No solution of the equation exists.

5. (a,c,d) **Concept Involved** It is based on range of $\sin x$, i.e. $[-1, 1]$ and the interval for $a < x < b$.

Description of Situation As $\theta, \phi \in [0, 2\pi]$ and

$\tan(2\pi - \theta) > 0$

$-1 < \sin\theta < -\frac{\sqrt{3}}{2}$

$\tan(2\pi - \theta) > 0$

$\Rightarrow \quad -\tan\theta > 0$

$\therefore \quad \theta \in$ II or IV quadrant.

Also, $\quad -1 < \sin\theta < -\frac{\sqrt{3}}{2}$

$\Rightarrow \quad \frac{4\pi}{3} < \theta < \frac{5\pi}{3}$ but $\theta \in$ II or IV quadrant

$\therefore \quad \frac{3\pi}{2} < \theta < \frac{5\pi}{3} \qquad \ldots(\text{i})$

Here, $\quad 2\cos\theta(1 - \sin\phi) = \sin^2\theta\left(\tan\frac{\theta}{2} + \cot\frac{\theta}{2}\right)\cos\phi - 1$

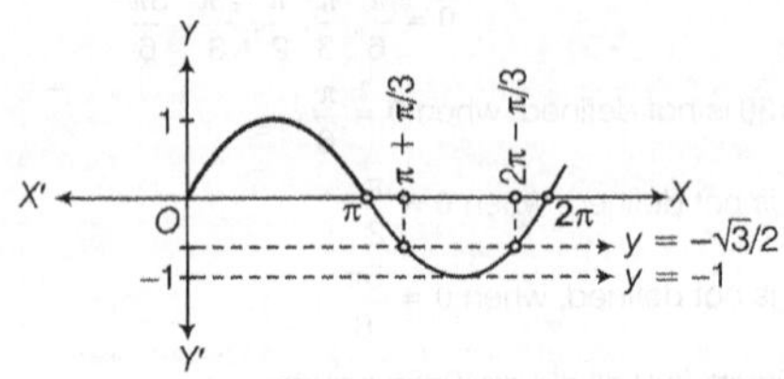

$\Rightarrow \quad 2\cos\theta - 2\cos\theta\sin\phi = \sin^2\theta\left(\frac{\sin^2\frac{\theta}{2} + \cos^2\frac{\theta}{2}}{\sin\frac{\theta}{2}\cos\frac{\theta}{2}}\right)\cos\phi - 1$

$\Rightarrow \quad 2\cos\theta - 2\cos\theta\sin\phi = 2\sin^2\theta\left(\frac{1}{\sin\theta}\right)\cos\phi - 1$

$\Rightarrow \quad 2\cos\theta + 1 = 2\sin\phi\cos\theta + 2\sin\theta\cos\phi$

$\Rightarrow \quad 2\cos\theta + 1 = 2\sin(\theta + \phi) \qquad \ldots(\text{ii})$

From Eq. (i), $\quad \frac{3\pi}{2} < \theta < \frac{5\pi}{3}$

$\Rightarrow \quad 2\cos\theta + 1 \in (1, 2)$

$\therefore \quad 1 < 2\sin(\theta + \phi) < 2$

$\Rightarrow \quad \frac{1}{2} < \sin(\theta + \phi) < 1 \qquad \ldots(\text{iii})$

$$\Rightarrow \quad \frac{\pi}{6} < \theta + \phi < \frac{5\pi}{6} \text{ or } \frac{13\pi}{6} < \theta + \phi < \frac{17\pi}{6}$$

$$\therefore \quad \frac{\pi}{6} - \theta < \phi < \frac{5\pi}{6} - \theta \text{ or } \frac{13\pi}{6} - \theta < \phi < \left(\frac{17\pi}{6}\right) - \theta$$

$$\Rightarrow \quad \phi \in \left(-\frac{3\pi}{2}, -\frac{2\pi}{3}\right) \text{ or } \left(\frac{2\pi}{3}, \frac{7\pi}{6}\right)$$

as $\quad \theta \in \left(\frac{3\pi}{2}, \frac{5\pi}{3}\right)$

6. (d) Given, $\quad P = \{\theta : \sin\theta - \cos\theta = \sqrt{2}\cos\theta\}$

$\Rightarrow \cos\theta(\sqrt{2} + 1) = \sin\theta$

$\Rightarrow \tan\theta = \sqrt{2} + 1 \quad$ ...(i)

and $\quad Q = \{\theta : \sin\theta + \cos\theta = \sqrt{2}\sin\theta\}$

$\Rightarrow \sin\theta(\sqrt{2} - 1) = \cos\theta$

$$\tan\theta = \frac{1}{\sqrt{2} - 1} \times \frac{\sqrt{2} + 1}{\sqrt{2} + 1} = (\sqrt{2} + 1) \quad \text{...(ii)}$$

From Eqs. (i) and (ii), we get

$\therefore \quad P = Q$

7. (7) Given, $n > 3 \in I$

and $$\frac{1}{\sin\left(\frac{\pi}{n}\right)} = \frac{1}{\sin\left(\frac{2\pi}{n}\right)} + \frac{1}{\sin\left(\frac{3\pi}{n}\right)}$$

$$\Rightarrow \quad \frac{1}{\sin\left(\frac{\pi}{n}\right)} - \frac{1}{\sin\left(\frac{3\pi}{n}\right)} = \frac{1}{\sin\left(\frac{2\pi}{n}\right)}$$

$$\Rightarrow \quad \frac{\sin\left(\frac{3\pi}{n}\right) - \sin\left(\frac{\pi}{n}\right)}{\sin\left(\frac{\pi}{n}\right) \cdot \sin\left(\frac{3\pi}{n}\right)} = \frac{1}{\sin\left(\frac{2\pi}{n}\right)}$$

$$\Rightarrow \quad 2\cos\left(\frac{2\pi}{n}\right) \cdot \sin\left(\frac{\pi}{n}\right) = \frac{\sin\left(\frac{\pi}{n}\right) \cdot \sin\left(\frac{3\pi}{n}\right)}{\sin\left(\frac{2\pi}{n}\right)}$$

$$\Rightarrow \quad 2\sin\left(\frac{2\pi}{n}\right) \cdot \cos\left(\frac{2\pi}{n}\right) = \sin\left(\frac{3\pi}{n}\right)$$

$$\Rightarrow \quad \sin\left(\frac{4\pi}{n}\right) = \sin\left(\frac{3\pi}{n}\right)$$

$$\Rightarrow \quad \frac{4\pi}{n} = \pi - \frac{3\pi}{n}$$

$$\Rightarrow \quad \frac{7\pi}{n} = \pi$$

$\therefore \quad n = 7$

8. (3) Given, $\quad \tan\theta = \cot 5\theta$

$$\Rightarrow \quad \tan\theta = \tan\left(\frac{\pi}{2} - 5\theta\right)$$

$$\Rightarrow \quad \frac{\pi}{2} - 5\theta = n\pi + \theta$$

$$\Rightarrow \quad 6\theta = \frac{\pi}{2} - n\pi$$

$$\Rightarrow \quad \theta = \frac{\pi}{12} - \frac{n\pi}{6}$$

Also, $$\cos 4\theta = \sin 2\theta = \cos\left(\frac{\pi}{2} - 2\theta\right)$$

$$\Rightarrow \quad 4\theta = 2n\pi \pm \left(\frac{\pi}{2} - 2\theta\right)$$

On taking + ve sign,

$$6\theta = 2n\pi + \frac{\pi}{2}$$

$$\Rightarrow \quad \theta = \frac{n\pi}{3} + \frac{\pi}{12}$$

On taking –ve sign,

$$2\theta = 2n\pi - \frac{\pi}{2}$$

$$\Rightarrow \quad \theta = n\pi - \frac{\pi}{4}$$

Above values of θ suggests that there are only 3 common solutions.

$\therefore$ There are 3 values of θ.

9. (c) Given, $\quad 2\sin^2\theta - \cos 2\theta = 0$

$$\Rightarrow \quad \sin^2\theta = \frac{1}{4}$$

Also, $\quad 2\cos^2\theta = 3\sin\theta$

$$\Rightarrow \quad \sin\theta = \frac{1}{2}$$

$\therefore$ Two solutions exist in the interval $[0, 2\pi]$.

10. (b) Given, $\quad 2\sin^2\theta - 5\sin\theta + 2 > 0$

$\Rightarrow \quad (2\sin\theta - 1)(\sin\theta - 2) > 0$

[where, $(\sin\theta - 2) < 0, \forall\, \theta \in R$]

$\therefore \quad (2\sin\theta - 1) < 0$

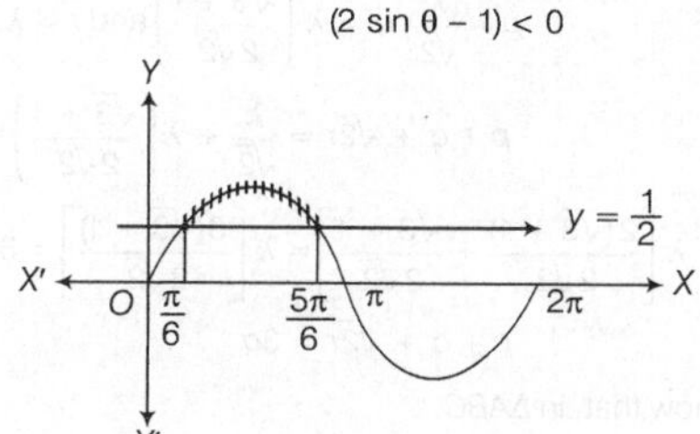

$$\Rightarrow \quad \sin\theta < \frac{1}{2}$$

$\therefore$ From the graph, $\theta \in \left[0, \frac{\pi}{6}\right) \cup \left(\frac{5\pi}{6}, 2\pi\right]$

4. Solutions of Triangles

DPP-1 Sine Rule and Its Applications

1. We have, in ΔABC, $a = 3$, $b = 8$ and $\sin A = \frac{5}{13}$

Using sine rule in ΔABC,

$$\frac{\sin A}{a} = \frac{\sin B}{b} = \frac{\sin C}{c}$$

If we take $\frac{\sin A}{a} = \frac{\sin B}{b}$

$$\frac{\frac{5}{13}}{3} = \frac{\sin B}{8}$$

$$\sin B = 8 \times \frac{5}{13} \times \frac{1}{3} = \frac{40}{39} > 1$$

which is not possible because $-1 \le \sin\theta \le 1$.

$\therefore$ Number of triangles is 0.

2. $\because$ In ΔPQR, $P : Q : R = 3 : 5 : 4$

Let $P = 3x$, $Q = 5x$, $R = 4x$

Then, $P + Q + R = 180°$

$$3x + 5x + 4x = 180°$$

$$x = 15°$$

$\therefore$ $P = 45°, Q = 75°, R = 60°$

Then, $\sin P = \frac{1}{\sqrt{2}}, \sin Q = \frac{\sqrt{3}+1}{2\sqrt{2}}, \sin R = \frac{\sqrt{3}}{2}$

Now, using sine rule, we get

$$\frac{p}{\sin P} = \frac{q}{\sin Q} = \frac{r}{\sin R} = \lambda \text{ (let)}$$

$\therefore$ $p = \lambda \sin P, q = \lambda \sin Q$ and $r = \lambda \sin R$

$$p = \frac{\lambda}{\sqrt{2}}, q = \lambda\left(\frac{\sqrt{3}+1}{2\sqrt{2}}\right) \text{ and } r = \lambda\frac{\sqrt{3}}{2}$$

Now, $p + q + \sqrt{2}r = \frac{\lambda}{\sqrt{2}} + \lambda\left(\frac{\sqrt{3}+1}{2\sqrt{2}}\right) + \frac{\lambda\sqrt{3}}{2} \times \sqrt{2}$

$$\lambda\left[\frac{2(\sqrt{3}+1)}{2\sqrt{2}} + \frac{\sqrt{3}+1}{2\sqrt{2}}\right] = \lambda\left[\frac{3(\sqrt{3}+1)}{2\sqrt{2}}\right] = 3\left(\frac{\sqrt{3}+1}{2\sqrt{2}}\cdot\lambda\right)$$

$\therefore$ $p + q + \sqrt{2}r = 3q$

3. We know that, in ΔABC,

$$\frac{a}{\sin A} = \frac{b}{\sin B} = \frac{c}{\sin C} = \lambda \text{ (let)}$$

$\therefore$ $a = \lambda \sin A, b = \lambda \sin B, c = \lambda \sin C$

$\therefore$ LHS $= a^3\cos(B - C) + b^3\cos(C - A) + c^3\cos(A - B)$

$= \lambda^3\sin^3 A\cos(B - C) + \lambda^3\sin^3 B\cos(C - A) + \lambda^3\sin^3 C\cos(A - B)$

$= \frac{\lambda^3}{2}[\sin^2 A\cdot\{2\sin A\cos(B - C)\} + \sin^2 B\cdot\{2\sin B\cos(C - A)\} + \sin^2 C\cdot\{2\sin C\cos(A - B)\}]$

$= \frac{\lambda^3}{2}[\sin^2 A\cdot 2\sin(B + C)\cos(B - C) + \sin^2 B\cdot 2\sin(C + A)\cos(C - A) + \sin^2 C\cdot 2\sin(A + B)\cos(A + B)]$

$= \frac{\lambda^3}{2}[\sin^2 A(\sin 2B + \sin 2C) + \sin^2 B(\sin 2C + \sin 2A) + \sin^2 C(\sin 2A + \sin 2B)]$

$= \lambda^3[\sin^2 A\sin B\cos B + \sin^2 A\sin C\cos C + \sin^2 B\sin C\cos C + \sin^2 B\sin A\cos A + \sin^2 C\sin A\cos A + \sin^2 C\sin B\cos B]$

$= \lambda^3[\sin A\sin B(\sin A\cos B + \cos A\sin B) + \sin B\sin C(\sin B\cos C + \cos B\sin C) + \sin C\sin A(\sin C\cos A + \cos C\sin A)]$

$= \lambda^3[\sin A\sin B\sin(A + B) + \sin B\sin C\sin(B + C) + \sin C\sin A\sin(C + A)]$

$= \lambda^3[\sin A\sin B\sin C + \sin A\sin B\sin C + \sin A\sin B\sin C]$ $(\because A + B + C = \pi)$

$= 3(\lambda\sin A)(\lambda\sin B)(\lambda\sin C)$

$= 3abc =$ RHS **Hence proved.**

4. In ΔABC,

$$A + B + C = 180°$$

and $\frac{\sin A}{a} = \frac{\sin B}{b} = \frac{\sin C}{c} = \lambda$ (let)

Then, LHS $= \frac{\sin B}{\sin(A + B)}$

$$= \frac{\sin B}{\sin(180° - C)} = \frac{\sin B}{\sin C}$$

$$= \frac{\lambda b}{\lambda c} = \frac{b}{c} = \text{RHS}$$ **Hence proved.**

5. (a) $\because$ $p\sin(Q - R) = \lambda\sin P\sin(Q - R)$

$\left(\text{using sine rule, } \frac{p}{\sin P} = \frac{q}{\sin Q} = \frac{r}{\sin R} = \lambda\right)$

$= \lambda\sin(180° - \overline{Q + R})\sin(Q - R)$

$= \lambda\sin(Q + R)\sin(Q - R)$

$= \lambda(\sin^2 Q - \sin^2 R)$

Similarly, $q\sin(R - P) = \lambda(\sin^2 R - \sin^2 P)$

and $r\sin(P - Q) = \lambda(\sin^2 P - \sin^2 Q)$

Now, $p\sin(Q - R) + q\sin(R - P) + r\sin(P - Q)$

$= \lambda(\sin^2 Q - \sin^2 R + \sin^2 R - \sin^2 P + \sin^2 P - \sin^2 Q)$

$= 0$

6. (c) We know that, in ΔABC,

$$\frac{\sin A}{a} = \frac{\sin B}{b}$$

$$\frac{\sin 3B}{a} = \frac{\sin B}{\frac{\sqrt{3}-1}{2}\cdot a}$$

$\Rightarrow$ $\frac{\sin 3B}{\sin B} = \frac{2}{\sqrt{3}-1}$

$\Rightarrow$ $\frac{3\sin B - 4\sin^3 B}{\sin B} = \frac{2(\sqrt{3}+1)}{3-1}$

$\Rightarrow$ $3 - 4\sin^2 B = \sqrt{3} + 1$

Now, $\sin B = \sqrt{\frac{2-\sqrt{3}}{4}} = \sqrt{\frac{4-2\sqrt{3}}{8}}$

$$= \sqrt{\frac{(\sqrt{3})^2 - 2\sqrt{3} + 1}{8}} = \frac{\sqrt{3}-1}{2\sqrt{2}}$$

$\therefore$ $\sin B = \sin 15°$

$\Rightarrow$ $B = 15°$

and $A = 3B = 3 \times 15° = 45°$

$\therefore$ $C = 180° - (A + B)$

$= 180° - (45° + 15°)$

$= 120°$

7. (b) In ΔPQR,

$$\frac{\sin P}{p} = \frac{\sin Q}{q} = \frac{\sin R}{r} = \lambda \quad \text{(using sine rule)}$$

Given, $\sin P \sin Q = \frac{pq}{r^2}$

$\Rightarrow \sin P \sin Q = \frac{\sin P \sin Q}{\sin^2 R}$

$\Rightarrow \sin^2 R = 1$

$\sin R = 1 = \sin\frac{\pi}{2}$

$\therefore R = \frac{\pi}{2} = 90°$

8. (c) We know that, in ΔABC,

$$\frac{a}{\sin A} = \frac{b}{\sin B} = \frac{c}{\sin C} = \lambda \text{ (let)}$$

Given, $\frac{\cos A}{a} = \frac{\cos B}{b} = \frac{\cos C}{c}, a = 2$

$\Rightarrow \frac{\cos A}{\lambda \sin A} = \frac{\cos B}{\lambda \sin B} = \frac{\cos C}{\lambda \sin C}$

$\Rightarrow \cot A = \cot B = \cot C$

$\Rightarrow A = B = C$

Thus, ΔABC is an equilateral triangle, then

Area of $\Delta ABC = \frac{\sqrt{3}}{4}\cdot(\text{side})^2$

$= \frac{\sqrt{3}}{4} \times (2)^2 = \sqrt{3}$ sq units

9. (a) In ΔABC, $\frac{\sin A}{a} = \frac{\sin B}{b} = \frac{\sin C}{c} = \lambda$ (let)

Now, $a^2(\cos^2 B - \cos^2 C) + b^2(\cos^2 C - \cos^2 A) + c^2(\cos^2 A - \cos^2 B)$

$= a^2(\sin^2 C - \sin^2 B) + b^2(\sin^2 A - \sin^2 C) + c^2(\sin^2 B - \sin^2 A)$

$= a^2(\lambda^2c^2 - \lambda^2b^2) + b^2(\lambda^2a^2 - \lambda^2c^2) + c^2(\lambda^2b^2 - \lambda^2a^2)$

$= \lambda^2(a^2c^2 - a^2b^2 + a^2b^2 - b^2c^2 + b^2c^2 - a^2c^2)$

$= 0$

10. (a) In ΔABC, we have

$AB = 2BC \Rightarrow c = 2a$

Now, $\frac{\tan\frac{B}{2}}{\cot\left(\frac{C-A}{2}\right)} = \frac{\sin\frac{B}{2}}{\cos\frac{B}{2}} \times \frac{\sin\left(\frac{C-A}{2}\right)}{\cos\left(\frac{C-A}{2}\right)}$

$= \frac{\sin\left(90° - \frac{C+A}{2}\right)\sin\left(\frac{C-A}{2}\right)}{\cos\left(90° - \frac{C+A}{2}\right)\cos\left(\frac{C-A}{2}\right)}$

$= \frac{2\cos\left(\frac{C+A}{2}\right)\sin\left(\frac{C-A}{2}\right)}{2\sin\left(\frac{C+A}{2}\right)\cos\left(\frac{C-A}{2}\right)}$

$= \frac{\sin C - \sin A}{\sin C + \sin A}$

$= \frac{\lambda c - \lambda a}{\lambda c + \lambda a}$ (using sine rule)

$= \frac{c-a}{c+a} = \frac{2a-a}{2a+a}$ ($\because c = 2a$, given)

$= \frac{a}{3a} = \frac{1}{3} = 1:3$

11. (a) Let in ΔABC,

$A : B : C = 4 : 1 : 1$

and $A = 4x, B = x, C = x$

We know that, $A + B + C = 180°$

$4x + x + x = 180°$

$x = 30°$

$\therefore A = 120°, B = 30°, C = 30°$

$\sin A = \frac{\sqrt{3}}{2}, \sin B = \frac{1}{2}, \sin C = \frac{1}{2}$

Now, using sine rule, we get

$a = \lambda \sin A, b = \lambda \sin B, c = \lambda \sin C$

$a = \frac{\lambda\sqrt{3}}{2}, b = \frac{\lambda}{2}, c = \frac{\lambda}{2}$

Perimeter of $\Delta ABC = a + b + c = \frac{\lambda(\sqrt{3}+2)}{2}$

and largest side $= \frac{\lambda\sqrt{3}}{2}$

$\therefore$ Required ratio $= \frac{\lambda\sqrt{3}}{2} : \frac{\lambda(\sqrt{3}+2)}{2}$

$= \sqrt{3} : (2 + \sqrt{3})$

12. (c) In ΔABC, $\frac{a}{\sin A} = \frac{b}{\sin B} = \frac{c}{\sin C} = \lambda$ (let)

We have, $\frac{a^2 - b^2}{a^2 + b^2} = \frac{\sin(A-B)}{\sin(A+B)}$

$\frac{\lambda^2\sin^2 A - \lambda^2\sin^2 B}{\lambda^2\sin^2 A + \lambda^2\sin^2 B} = \frac{\sin(A-B)}{\sin(A+B)}$

$\sin^2(A+B)\sin(A-B) = \sin(A-B)(\sin^2 A + \sin^2 B)$

$\sin^2(\pi - C)\sin(A-B) - \sin(A-B)(\sin^2 A + \sin^2 B) = 0$

$\sin(A-B)[\sin^2 C - (\sin^2 A + \sin^2 B)] = 0$

Either $\sin(A-B) = 0$

$\Rightarrow A - B = 0$

$\Rightarrow A = B$

or $\sin^2 C - (\sin^2 A + \sin^2 B) = 0$

$\sin^2 C = \sin^2 A + \sin^2 B$

$c^2 = a^2 + b^2$

$\therefore \Delta ABC$ is a right angled or isosceles.

13. (a) In ΔABC, using sine rule,

$\frac{\sin B}{b} = \frac{\sin C}{c}$

$\sin C = \frac{c \sin B}{b} > 1$ ($\because c\sin B > b$, given)

which is not possible because $-1 \le \sin\theta \le 1$.

14. (b) We have, $\sin P, \sin Q, \sin R$ are in AP, then p, q, r are also in AP.

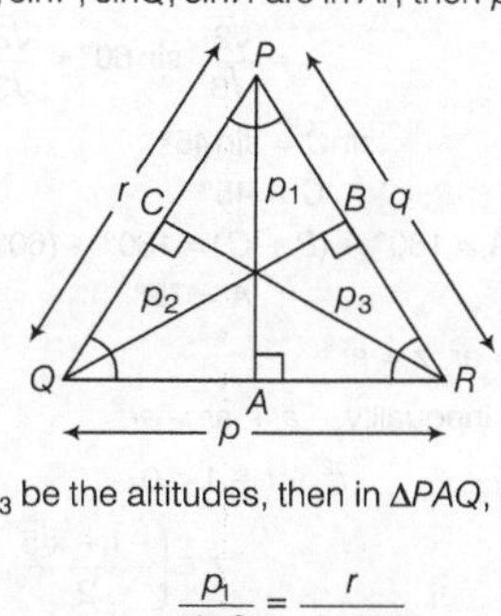

Let p_1, p_2, p_3 be the altitudes, then in ΔPAQ, using sine rule, we get

$\frac{p_1}{\sin Q} = \frac{r}{\sin 90°}$

$\Rightarrow p_1 = r \sin Q$

$p_1 = \lambda qr$ ...(i)

Similarly, in ΔBQR and ΔPCR,

$$p_2 = \lambda pr \quad \ldots(\text{ii})$$
$$p_3 = \lambda pq \quad \ldots(\text{iii})$$

$\because$ p, q, r are in AP, then $\frac{1}{p}, \frac{1}{q}, \frac{1}{r}$ are in HP.

$\therefore$ $\frac{\lambda pqr}{p}, \frac{\lambda pqr}{q}, \frac{\lambda pqr}{r}$ are in HP, then $\lambda qr, \lambda pr, \lambda pq$ are in HP.

So, p_1, p_2, p_3 are in HP.

15. (a,b,c,d) We have, $A = 15°$, $b = 10\sqrt{2}$ and for unique triangle $b \le a$.

$$\because \quad \frac{\sin A}{a} = \frac{\sin B}{b}$$

$$\Rightarrow \quad \sin B = \frac{b}{a}\sin A = \frac{10\sqrt{2}\cdot\sin 15°}{a}$$

$$= \frac{10\sqrt{2}\left(\frac{\sqrt{3}-1}{2\sqrt{2}}\right)}{a} = \frac{5(\sqrt{3}-1)}{a}$$

Thus, for triangle $\sin B \le 1$

$$\frac{5(\sqrt{3}-1)}{a} \le 1$$

$$\Rightarrow \quad a \ge 5(\sqrt{3}-1)\text{ cm}$$

16. (a,c,d) $\because A : B : C = 2 : 3 : 7$

Let $A = 2\alpha, B = 3\alpha, C = 7\alpha$

We know that, $A + B + C = 180°$

$$2\alpha + 3\alpha + 7\alpha = 180°$$
$$\alpha = 15°$$

$\therefore$ $A = 30°, B = 45°, C = 105°$

Now, using sine rule

$$a : b : c = \sin A : \sin B : \sin C$$
$$= \sin 30° : \sin 45° : \sin 105°$$
$$= \frac{1}{2} : \frac{1}{\sqrt{2}} : \frac{\sqrt{3}+1}{2\sqrt{2}}$$

$$a : b : c = \frac{1}{\sqrt{2}} : 1 : \frac{\sqrt{3}+1}{2} \quad \text{[option (d)]}$$
$$= \sqrt{2} : 2 : \sqrt{3}+1 \quad \text{[option (a)]}$$
$$= 1 : \sqrt{2} : \frac{\sqrt{2}}{\sqrt{3}-1} \quad \text{[option (c)]}$$

17. (a,b,c) We have, $A + B + C = 180° \quad \ldots(\text{i})$

and $2B = A + C$ (given)

Thus, $B = 60°$ [option (c)]

We know that, $\frac{a}{\sin A} = \frac{b}{\sin B} = \frac{c}{\sin C}$

Now, $\sin C = \frac{c}{b}\cdot\sin B$

$$= \frac{\sqrt{2}}{\sqrt{3}}\cdot\sin 60° = \frac{\sqrt{2}}{\sqrt{3}}\cdot\frac{\sqrt{3}}{2} = \frac{1}{\sqrt{2}}$$

$$\sin C = \sin 45°$$
$$C = 45° \quad \text{[option (b)]}$$

From Eq. (i), $A = 180° - (B + C) = 180° - (60° + 45°)$

$$A = 75° \quad \text{[option (a)]}$$

18. (a) Let $x = a$, $y = ar$, $z = ar^2$

Using triangle inequality, $a + ar > ar^2$

$$\Rightarrow \quad r^2 - r - 1 < 0$$

$$\Rightarrow \quad r \in \left(\frac{-1+\sqrt{5}}{2}, \frac{1+\sqrt{5}}{2}\right) \quad \ldots(\text{i})$$

$$a + ar^2 > ar$$

$$\Rightarrow \quad r^2 - r + 1 > 0$$

$$\Rightarrow \quad r \in R \quad \ldots(\text{ii})$$

$$ar + ar^2 > a$$

$$\Rightarrow \quad r^2 + r - 1 > 0$$

$$\Rightarrow \quad r \in \left(-\infty, \frac{-1-\sqrt{5}}{2}\right) \cup \left(\frac{-1+\sqrt{5}}{2}, \infty\right) \quad \ldots(\text{iii})$$

From Eqs. (i), (ii) and (iii), we get

$$\Rightarrow \quad r \in \left(\frac{-1+\sqrt{5}}{2}, \frac{1+\sqrt{5}}{2}\right)$$

19. (d) $\because \frac{\sin Y}{\sin X} = \frac{y}{x} = r$ and it can take only one integral value, i.e. 1.

(using sine rule)

20. (d) $\frac{\sin Z}{\sin Y} = \frac{z}{y} = r$

$\Rightarrow$ Maximum value of $\frac{\sin Z}{\sin Y} = \max(r) =$ not defined.

21. (4) Let h be the height of pole and $AD = x$ m

$\because$ In ΔBAD, $\angle ABD = 180° - (90° + 60°) = 30°$

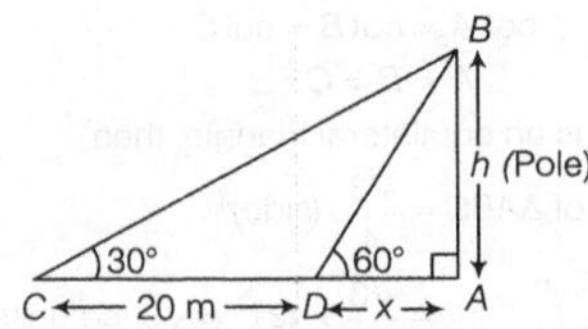

In ΔABC, $\angle ABC = 180° - (90° + 30°) = 60°$

$\therefore$ $\angle CBD = \angle ABC - \angle ABD$

$$= 60° - 30° = 30° = \angle BCD$$

Then, $BD = CD = 20$ m

Now, in ΔABD using sine rule, we get

$$\frac{x}{\sin 30°} = \frac{h}{\sin 60°} = \frac{BD}{\sin 90°}$$

$$\frac{x}{\frac{1}{2}} = \frac{h}{\frac{\sqrt{3}}{2}} = \frac{20}{1}$$

If $\frac{x}{1/2} = 20$, then $x = 20 \times \frac{1}{2} = 10$ m

If $\frac{h}{\sqrt{3}/2} = 20$, then $h = 20 \times \frac{\sqrt{3}}{2} = 10\sqrt{3}$ m

$\therefore$ Breadth of river $= 10$ m $= \mu$

and height of pole $= 10\sqrt{3}$m

$$10\sqrt{\lambda} = 10\sqrt{3}$$

$$\Rightarrow \quad \lambda = 3$$

So, $\mu - 2\lambda = 10 - 2 \times 3 = 4$

22. (2) Let Q be an object and its height be h.

$\because$ $\angle PCQ = 3\theta$

$\angle QCB = 180° - \angle PCQ = 180° - 3\theta$

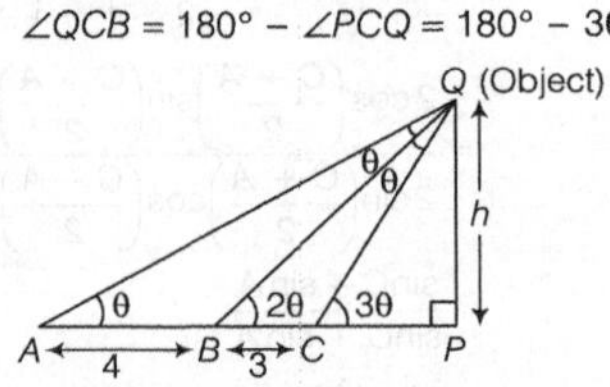

Now, in ΔBQC, using sine rule

$$\frac{BQ}{\sin(180° - 3\theta)} = \frac{BC}{\sin\theta}$$

$$\frac{BQ}{BC} = \frac{\sin 3\theta}{\sin\theta}$$

$$\Rightarrow \quad \frac{3\sin\theta - 4\sin^3\theta}{\sin\theta} = \frac{4}{3} \quad (\because BQ = AB = 4)$$

$$\Rightarrow \quad 3 - 4\sin^2\theta = \frac{4}{3} \quad (\because \sin\theta \neq 0)$$

$$\Rightarrow \quad \sin^2\theta = \frac{5}{12}$$

$$\Rightarrow \quad \sin\theta = \sqrt{\frac{5}{12}} \quad \ldots(i)$$

and $$\cos\theta = \sqrt{\frac{7}{12}} \quad \ldots(ii)$$

Now, in ΔBPQ, $$\sin 2\theta = \frac{h}{BQ}$$

$$\Rightarrow \quad h = 4\sin 2\theta$$

$$h = 8\sin\theta\cdot\cos\theta$$

$$= 8\cdot\sqrt{\frac{5}{12}}\cdot\sqrt{\frac{7}{12}} \quad \text{[from Eqs. (i) and (ii)]}$$

$$= 8\times\frac{\sqrt{35}}{12} = \frac{2}{3}\sqrt{35} \text{ unit}$$

Now, $$\frac{\lambda}{3}\sqrt{35} = \frac{2}{3}\sqrt{35} \Rightarrow \lambda = 2$$

23. (0) $\because (b-c)\cot\frac{A}{2} = \lambda(\sin B - \sin C)\cot\frac{A}{2}$

$$= \lambda\left[2\cos\left(\frac{B+C}{2}\right)\cdot\sin\left(\frac{B-C}{2}\right)\right]\cdot\frac{\cos\frac{A}{2}}{\sin\frac{A}{2}}$$

$$= \lambda\left[2\sin\frac{A}{2}\sin\left(\frac{B-C}{2}\right)\right]\frac{\cos\frac{A}{2}}{\sin\frac{A}{2}}$$

$$= \lambda\left[2\sin\left(\frac{B-C}{2}\right)\sin\left(\frac{B+C}{2}\right)\right] \quad \left(\begin{array}{l}\because A+B+C = 180^\circ\\ \therefore \frac{A}{2}+\frac{B}{2}+\frac{C}{2} = 90^\circ\end{array}\right)$$

$$= \lambda(\cos C - \cos B)$$

Similarly, $$(c-a)\cot\frac{B}{2} = \lambda(\cos A - \cos C)$$

$$(a-b)\cot\frac{C}{2} = \lambda(\cos B - \cos A)$$

$$\therefore \quad (b-c)\cot\frac{A}{2} + (c-a)\cot\frac{B}{2} + (a-b)\cot\frac{C}{2}$$

$$= \lambda(\cos C - \cos B + \cos A - \cos C + \cos B - \cos A)$$

$$= 0$$

DPP-2 Cosine Rule and Projection Formulae

1. $\because (b-c)^2\cos^2\frac{A}{2} + (b+c)^2\sin^2\frac{A}{2}$

$$= (b^2+c^2-2bc)\cos^2\frac{A}{2} + (b^2+c^2+2bc)\sin^2\frac{A}{2}$$

$$= (b^2+c^2)\left(\sin^2\frac{A}{2}+\cos^2\frac{A}{2}\right) - 2bc\left(\cos^2\frac{A}{2}-\sin^2\frac{A}{2}\right)$$

$$= b^2+c^2-2bc\cos A$$

$$= b^2+c^2-2bc\cdot\frac{b^2+c^2-a^2}{2bc} \quad \text{(using cosine rule)}$$

$$= b^2+c^2-b^2-c^2+a^2 = a^2$$

2. LHS $= a(\cos B + \cos C - 1) + b(\cos C + \cos A - 1) + c(\cos A + \cos B - 1)$

$$= (a\cos B + b\cos A) + (a\cos C + c\cos A) - a + (b\cos C + c\cos B) - b - c$$

$$= c + b - a + a - b - c = 0 = \text{RHS}$$

(using projection formula)

Hence proved.

3. Let x be the distance between pole A and B.

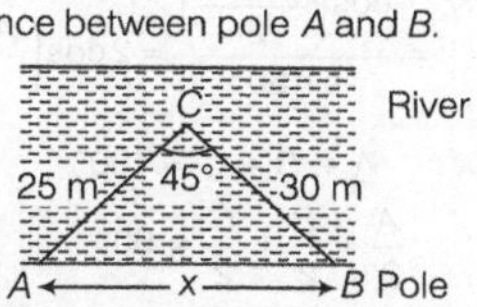

In ΔABC, using cosine rule,

$$\cos 45^\circ = \frac{(25)^2 + (30)^2 - x^2}{2\times 25\times 30}$$

$$\frac{1}{\sqrt{2}} = \frac{625 + 900 - x^2}{1500}$$

$$\frac{\sqrt{2}}{2}\times 1500 = 1525 - x^2$$

$$1.44\times 750 = 1525 - x^2$$

$$x^2 = 1525 - 1080 = 445$$

$$x = \sqrt{445} = 21.09$$

$\therefore$ The distance between two poles is 21.09 m.

4. (d) $\because \dfrac{\cos A}{b\cos C + c\cos B} + \dfrac{\cos B}{c\cos A + a\cos C} + \dfrac{\cos C}{a\cos B + b\cos A}$

$$= \frac{\cos A}{a} + \frac{\cos B}{b} + \frac{\cos C}{c} \quad \text{(using projection formula)}$$

$$= \frac{b^2+c^2-a^2}{2abc} + \frac{c^2+a^2-b^2}{2abc} + \frac{a^2+b^2-c^2}{2abc}$$

(using cosine rule)

$$= \frac{b^2+c^2-a^2+c^2+a^2-b^2+a^2+b^2-c^2}{2abc}$$

$$= \frac{a^2+b^2+c^2}{2abc}$$

Now, we have $$\frac{a^2+b^2+c^2}{\lambda abc} = \frac{a^2+b^2+c^2}{2abc}$$

$$\therefore \quad \lambda = 2$$

5. (b) We have, A, B, C are in AP, then

$$2B = A + C \quad \ldots(i)$$

We know that, $$A + B + C = 180^\circ$$

$$2B + B = 180^\circ$$

$$B = 60^\circ$$

$$\therefore \quad \cos B = \frac{a^2+c^2-b^2}{2ac}$$

$$\Rightarrow \quad \cos 60^\circ = \frac{a^2+c^2-b^2}{2ac}$$

$$\Rightarrow \quad \frac{1}{2}\times 2ac = a^2+c^2-b^2$$

$$b^2 = a^2+c^2-ac$$

6. (c) $\because a^2 + b^2 = 3c^2 \quad$ (given) $\ldots$(i)

$$\cot A + \cot B - \cot C = \frac{\cos A}{\sin A} + \frac{\cos B}{\sin B} - \frac{\cos C}{\sin C}$$

$$= \frac{\frac{b^2+c^2-a^2}{2bc}}{ka} + \frac{\frac{a^2+c^2-b^2}{2ac}}{kb} - \frac{\frac{a^2+b^2-c^2}{2ab}}{kc}$$

(using sine and cosine rule)

$$= \frac{b^2+c^2-a^2}{2kabc} + \frac{a^2+c^2-b^2}{2kabc} - \frac{a^2+b^2-c^2}{2kabc}$$

$$= \frac{b^2 + c^2 - a^2 + a^2 + c^2 - b^2 - a^2 - b^2 + c^2}{2kabc}$$

$$= \frac{3c^2 - a^2 - b^2}{2abck} = \frac{0}{2abck} \quad \text{[from Eq. (i)]}$$

$$= 0$$

7. (a) We have,

$$(a + b + c)(a - b + c) = 3ac$$
$$(a + c)^2 - b^2 = 3ac$$
$$(a^2 + c^2 - b^2) + 2ac = 3ac$$
$$a^2 + c^2 - b^2 = ac$$
$$\frac{a^2 + c^2 - b^2}{2ac} = \frac{1}{2}$$
$$\Rightarrow \quad \cos B = \cos 60°$$
$$\therefore \quad \angle B = 60°$$

8. (c) $\because$

$$\cos C = \frac{a^2 + b^2 - c^2}{2ab}$$
$$\cos 60° = \frac{a^2 + b^2 - c^2}{2ab}$$
$$\frac{1}{2}\cdot 2ab = a^2 + b^2 - c^2$$
$$a^2 + b^2 = c^2 + ab$$
$$a^2 + b^2 + bc + ca = c^2 + bc + ca + ab$$
$$a(a + c) + b(b + c) = c(b + c) + a(b + c)$$
$$a(a + c) + b(b + c) = (b + c)(c + a)$$

On dividing both sides by $(a + c)(b + c)$, we get

$$\frac{a}{b + c} + \frac{b}{c + a} = 1$$
$$1 + \frac{a}{b + c} + 1 + \frac{b}{c + a} = 3$$
$$\frac{a + b + c}{b + c} + \frac{a + b + c}{c + a} = 3$$
$$\therefore \quad \frac{1}{b + c} + \frac{1}{c + a} = \frac{3}{a + b + c}$$

9. (a) Ist term $= (b^2 - c^2)\cot A$

$$= k(\sin^2 B - \sin^2 C)\cot A$$
$$= k\sin(B + C)\sin(B - C)\cdot\frac{\cos A}{\sin A}$$
$$= k\sin(\pi - A)\sin(B - C)\cdot\frac{\cos(\pi - \overline{B + C})}{\sin A}$$
$$= \frac{-1}{2}k\sin A\cdot 2\sin(B - C)\cdot\frac{\cos(B + C)}{\sin A}$$
$$= \frac{-1}{2}k(\sin 2B - \sin 2C) = \frac{k}{2}(\sin 2C - \sin 2B)$$

Similarly, $(c^2 - a^2)\cot B = \frac{k}{2}(\sin 2A - \sin 2C)$

and $(a^2 - b^2)\cot C = \frac{k}{2}(\sin 2B - \sin 2A)$

$$\therefore \quad (b^2 - c^2)\cot A + (c^2 - a^2)\cot B + (a^2 - b^2)\cot C$$
$$= \frac{k}{2}(\sin 2C - \sin 2B + \sin 2A - \sin 2C + \sin 2B - \sin 2A)$$
$$= 0$$

Alternate Method

$$(b^2 - c^2)\cot A = (b^2 - c^2)\cdot\frac{\cos A}{\sin A}$$
$$= \frac{(b^2 - c^2)(b^2 + c^2 - a^2)}{2kabc}$$
$$= \frac{(b^4 - c^4) - a^2(b^2 - c^2)}{2kabc}$$

Similarly, $(c^2 - a^2)\cot B = \frac{(c^4 - a^4) - b^2(c^2 - a^2)}{2kabc}$

and $(a^2 - b^2)\cot C = \frac{(a^4 - b^4) - c^2(a^2 - b^2)}{2kabc}$

$$\therefore \quad (b^2 - c^2)\cot A + (c^2 - a^2)\cot B + (a^2 - b^2)\cot C$$
$$= \frac{1}{2kabc}[(b^4 - c^4 + c^4 - a^4 + a^4 - b^4) - \{a^2(b^2 - c^2) + b^2(c^2 - a^2) + c^2(a^2 - b^2)\}]$$
$$= 0$$

10. (c) We have,

$$\frac{2\cos A}{a} + \frac{\cos B}{b} + \frac{2\cos C}{c} = \frac{a}{bc} + \frac{b}{ca}$$
$$\frac{2(b^2 + c^2 - a^2)}{2abc} + \frac{c^2 + a^2 - b^2}{2abc} + \frac{2(a^2 + b^2 - c^2)}{2abc} = \frac{a^2 + b^2}{abc}$$
$$2b^2 + 2c^2 - 2a^2 + c^2 + a^2 - b^2 + 2a^2 + 2b^2 - 2c^2 = 2a^2 + 2b^2$$
$$b^2 + c^2 = a^2$$

Hence, the value of angle A is 90°.

11. (a) We know that,

$$A + B + C = 180°$$
$$2B + B = 180° \quad (\because 2B = A + C, \text{ given})$$
$$B = 60°$$

Then, $\cos B = \frac{c^2 + a^2 - b^2}{2ca}$ (using cosine rule)

$$\cos 60° = \frac{c^2 + a^2 - b^2}{2ca}$$
$$\frac{1}{2} = \frac{c^2 + a^2 - b^2}{2ca}$$
$$b^2 = c^2 + a^2 - ca \quad \ldots(i)$$

Now, $\frac{a + c}{\sqrt{a^2 + c^2 - ac}} = \frac{a + c}{b}$ [from Eq. (i)]

$$= \frac{\sin A + \sin C}{\sin B} \quad \text{(using sine rule)}$$
$$= \frac{2\sin\left(\frac{A + C}{2}\right)\cos\left(\frac{A - C}{2}\right)}{2\sin\frac{B}{2}\cos\frac{B}{2}}$$
$$= \frac{\cos\frac{B}{2}\cos\left(\frac{A - C}{2}\right)}{\sin\frac{B}{2}\cos\frac{B}{2}} \quad \left(\because A + B + C = 180°,\ \therefore \frac{A}{2} + \frac{B}{2} + \frac{C}{2} = 90°\right)$$
$$= \frac{\cos\left(\frac{A - C}{2}\right)}{\sin 30°} = 2\cos\left(\frac{A - C}{2}\right) \quad (\because B = 60°)$$

12. (a) $\because$ In ΔABC, $A + B + C = 180°$

$$\frac{A}{2} + \frac{B}{2} + \frac{C}{2} = 90°$$

Then, $2ab\sin\left(\frac{A - C + B}{2}\right) = 2ab\sin\left(\frac{180° - 2C}{2}\right)$

$$= 2ab\sin(90° - C)$$
$$= 2ab\cdot\cos C$$
$$= 2ab \times \frac{a^2 + b^2 - c^2}{2ab}$$
$$= a^2 + b^2 - c^2$$

13. (b) $\because \cos A + \cos B + \cos C = \frac{3}{2}$

If we put $A = B = C = 60°$ satisfy the given equation, i.e. $\cos 60° = \frac{1}{2}$.

Hence, ΔABC is an equilateral triangle.

14. (b) $\because$
$$AB^2 = (a-c)^2 + (b-d)^2$$
$$OA^2 = a^2 + b^2$$
$$OB^2 = c^2 + d^2$$

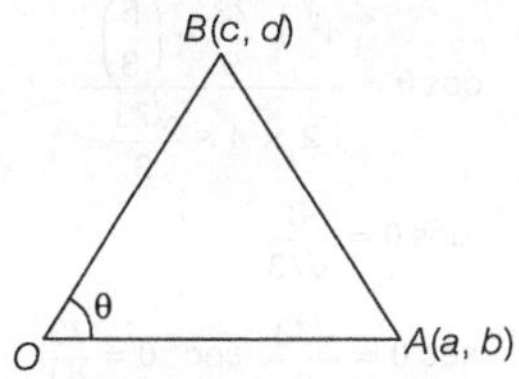

Now, in ΔAOB using cosine rule, we get

$$\cos\theta = \frac{OA^2 + OB^2 - AB^2}{2OA \cdot OB}$$
$$= \frac{a^2 + b^2 + c^2 + d^2 - (a-c)^2 - (b-d)^2}{2\sqrt{a^2+b^2}\sqrt{c^2+d^2}}$$
$$= \frac{a^2 + b^2 + c^2 + d^2 - a^2 - c^2 + 2ac - b^2 - d^2 + 2bd}{2\sqrt{(a^2+b^2)(c^2+d^2)}}$$
$$= \frac{2(ac+bd)}{2\sqrt{(a^2+b^2)(c^2+d^2)}} = \frac{ac+bd}{\sqrt{(a^2+b^2)(c^2+d^2)}}$$

15. (a,b,c) We have,

$$\frac{a+b}{11} = \frac{b+c}{13} = \frac{c+a}{12} = \lambda \text{ (let)}$$

Then, $a + b = 11\lambda$ ...(i)

$b + c = 13\lambda$ ...(ii)

$c + a = 12\lambda$ ...(iii)

On adding Eqs. (i), (ii) and (iii), we get

$$2(a+b+c) = 36\lambda$$

$a + b + c = 18\lambda$ ...(iv)

On solving Eqs. (i) and (iv), (ii) and (iv), (iii) and (iv), we get

$$c = 7\lambda,\ a = 5\lambda,\ b = 6\lambda$$

Now, $\cos A = \dfrac{b^2 + c^2 - a^2}{2bc}$

$$= \frac{36\lambda^2 + 49\lambda^2 - 25\lambda^2}{2 \times 6\lambda \times 7\lambda}$$
$$= \frac{36 + 49 - 25}{2 \times 7 \times 6} = \frac{5}{7}$$

$\therefore$ $\dfrac{\cos A}{\frac{5}{7}} = 1$ ...(v)

Similarly, $\dfrac{\cos B}{19/35} = 1$ and $\dfrac{\cos C}{1/5} = 1$

$\therefore$ $\dfrac{\cos A}{5/7} = \dfrac{\cos B}{19/35} = \dfrac{\cos C}{1/5}$

$$\frac{\cos A}{25} = \frac{\cos B}{19} = \frac{\cos C}{7}$$

But given that $\dfrac{\cos A}{p} = \dfrac{\cos B}{q} = \dfrac{\cos C}{r}$

$\therefore$ $p = 25, q = 19, r = 7$

16. (c,d) We know that,

$$\cos B = \frac{c^2 + a^2 - b^2}{2ca}$$
$$2ca\cos 60^\circ = c^2 + a^2 - b^2$$

$c^2 + a^2 - b^2 = ca$ ...(i)

$$c^2 + a^2 - 2ca = b^2 + ca - 2ca$$

$(c - a)^2 = b^2 - ca$ [option (c)]

From Eq. (i), $c^2 + a^2 - b^2 = ca$

$$c^2 + a^2 - b^2 + 2b^2 = 2b^2 + ca$$

$a^2 + b^2 + c^2 = 2b^2 + ca$ [option (d)]

17. (c,d) In ΔABC, $\tan A + \tan B + \tan C = \tan A \tan B \tan C$

So, $\tan A + \tan B + \tan C \neq 0$ $(\because A, B, C \neq 0)$

From option (b),

$$\frac{\sin A}{2} = \frac{\sin B}{3} = \frac{\sin C}{7}$$

$\dfrac{a}{2} = \dfrac{b}{3} = \dfrac{c}{7}$ (using sine rule)

$\Rightarrow$ $\dfrac{a+b}{5} = \dfrac{c}{7}$

$\Rightarrow$ $\dfrac{a+b}{c} = \dfrac{5}{7} < 1$

$\Rightarrow$ $a + b < c$ (not possible)

From option (c), $(a+b)^2 = c^2 + ab$

$$a^2 + b^2 + 2ab = c^2 + ab$$
$$a^2 + b^2 - c^2 = -ab$$
$$\frac{a^2 + b^2 - c^2}{2ab} = -\frac{1}{2} = -\cos 60^\circ$$
$$\cos C = \cos(180^\circ - 60^\circ)$$
$$C = 120^\circ$$

Now, $\sqrt{2}(\sin A + \cos A) = \sqrt{3}$

$$\sqrt{2} \cdot \sqrt{2}\left[\frac{1}{\sqrt{2}}\sin A + \frac{1}{\sqrt{2}}\cos A\right] = \sqrt{3}$$
$$\sin(A + 45^\circ) = \frac{\sqrt{3}}{2} = \sin 60^\circ$$
$$A + 45^\circ = 60^\circ$$

$A = 15^\circ$ (possible)

From option (d),

$\sin A + \sin B = \dfrac{\sqrt{3}+1}{2}$ ...(i)

and $\cos A \cos B = \dfrac{\sqrt{3}}{4} = \sin A \sin B$

Then, $\cos A \cos B - \sin A \sin B = 0$

$$\cos(A + B) = \cos\frac{\pi}{2}$$
$$A + B = \frac{\pi}{2}$$
$$B = \frac{\pi}{2} - A$$

Put in Eq. (i), we get

$$\sin A + \sin\left(\frac{\pi}{2} - A\right) = \frac{\sqrt{3}+1}{2}$$
$$\sin A + \cos A = \frac{\sqrt{3}+1}{2}$$
$$\sin\left(A + \frac{\pi}{4}\right) = \frac{\sqrt{3}+1}{2\sqrt{2}} = \sin\frac{5\pi}{12}$$
$$A + \frac{\pi}{4} = \frac{5\pi}{12}$$

$\Rightarrow$ $A = \dfrac{5\pi}{12} - \dfrac{\pi}{4} = \dfrac{2\pi}{12}$

$\Rightarrow$ $A = \dfrac{\pi}{6}$, then $B = \dfrac{\pi}{2} - \dfrac{\pi}{6}$

$$B = \frac{2\pi}{6}$$
$$B = \frac{\pi}{3}$$

$\because$ $A = \dfrac{\pi}{6}, B = \dfrac{\pi}{3}$, then $C = \dfrac{\pi}{2}$ which is possible.

18. (a,b) Given, $a = \alpha - \beta, b = \alpha + \beta, c = \sqrt{3\alpha^2 + \beta^2}$

$$\because \quad \cos C = \frac{a^2 + b^2 - c^2}{2ab}$$

$$= \frac{(\alpha - \beta)^2 + (\alpha + \beta)^2 - \left(\sqrt{3\alpha^2 + \beta^2}\right)^2}{2(\alpha - \beta)(\alpha + \beta)}$$

$$= \frac{\alpha^2 + \beta^2 - 2\alpha\beta + \alpha^2 + \beta^2 + 2\alpha\beta - 3\alpha^2 - \beta^2}{2(\alpha^2 - \beta^2)}$$

$$= \frac{-(\alpha^2 - \beta^2)}{2(\alpha^2 - \beta^2)} = -\frac{1}{2} = \cos\frac{2\pi}{3}$$

$$\therefore \quad C = \frac{2\pi}{3}$$

$$\therefore \quad A + B = \pi - C = \pi - \frac{2\pi}{3} = \frac{\pi}{3}$$

19. (5) We have, $AB = 2$, $BC = 4$, $CA = 3$

D is the mid-point of BC, then $BD = CD = 2$

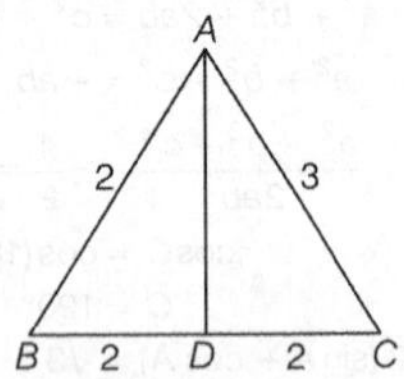

In ΔABC, using cosine rule,

$$\cos B = \frac{AB^2 + BC^2 - AC^2}{2\,AB \cdot BC}$$

$$= \frac{2^2 + 4^2 - 3^2}{2 \times 2 \times 4} = \frac{11}{16}$$

Now, in ΔABD, $\cos B = \dfrac{AB^2 + BD^2 - AD^2}{2\,AB \cdot BD}$

$$\frac{11}{16} = \frac{4 + 4 - AD^2}{2 \times 2 \times 2}$$

$$\Rightarrow \quad AD^2 = 8 - \frac{11}{2} = \frac{16 - 11}{2} = \frac{5}{2}$$

$$\therefore \quad 2\,AD^2 = 5$$

20. (6) Let $\angle CAE$ be θ.

We have, $AB = 3$, $BC = 5$, $CA = 4$

$$\because \quad 3^2 + 4^2 = 5^2$$

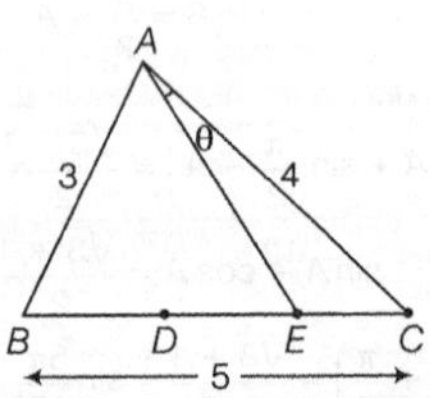

$$\Rightarrow \quad (AB)^2 + (CA)^2 = (BC)^2$$

$$\therefore \quad \angle BAC = 90^\circ \text{ and } \cos C = \frac{4}{5}$$

and $\quad BD = DE = CE = \dfrac{5}{3}$

Then, in ΔACE, $\cos C = \dfrac{AC^2 + CE^2 - AE^2}{2\,AC \cdot CE}$

$$\frac{4}{5} = \frac{4^2 + \left(\frac{5}{3}\right)^2 - AE^2}{2 \times 4 \times \frac{5}{3}}$$

$$AE^2 = \frac{73}{9}$$

Now, again in ΔACE,

$$\cos\theta = \frac{AC^2 + AE^2 - CE^2}{2\,AC \cdot AE}$$

$$\Rightarrow \quad \cos\theta = \frac{4^2 + \frac{73}{9} - \left(\frac{5}{3}\right)^2}{2 \times 4 \times \frac{\sqrt{73}}{3}}$$

$$\Rightarrow \quad \cos\theta = \frac{8}{\sqrt{73}}$$

Then, $\quad \sec\theta = \dfrac{\sqrt{73}}{8}$, $\sec^2\theta = \dfrac{73}{64}$

$$\therefore \quad 1 + \tan^2\theta = \frac{73}{64}$$

$$\tan^2\theta = \frac{73}{64} - 1 = \frac{9}{64}$$

$$\tan\theta = \frac{3}{8}$$

Thus, $\quad 16\tan(\angle CAE) = 16 \times \dfrac{3}{8} = 6$

21. (a)

(i) We have,

$$a^4 - 2(b^2 + c^2)a^2 + b^4 + b^2c^2 + c^4 = 0$$

$$\Rightarrow (b^2 + c^2 - a^2)^2 - (b^2 + c^2)^2 + b^4 + b^2c^2 + c^4 = 0$$

$$\Rightarrow \quad (b^2 + c^2 - a^2)^2 - b^2c^2 = 0$$

$$\Rightarrow \quad (b^2 + c^2 - a^2)^2 = b^2c^2$$

$$\Rightarrow \quad b^2 + c^2 - a^2 = \pm bc$$

$$\Rightarrow \quad \frac{b^2 + c^2 - a^2}{2bc} = \pm\frac{1}{2}$$

$$\Rightarrow \quad \cos A = \cos 60^\circ, \cos 120^\circ$$

$$A = 60^\circ, 120^\circ$$

Thus, **(i) → (q,s)**

(ii) We have,

$$a^4 + b^4 + c^4 = a^2b^2 + 2b^2c^2 + 2c^2a^2$$

$$a^4 + b^4 + c^4 - 2a^2c^2 - 2b^2c^2 = a^2b^2 \qquad \text{...(i)}$$

We know that,

$$\cos C = \frac{a^2 + b^2 - c^2}{2ab}$$

$$a^2 + b^2 - c^2 = 2ab\cos C$$

On squaring both sides, we get

$$a^4 + b^4 + c^4 + 2a^2b^2 - 2b^2c^2 - 2c^2a^2 = 4a^2b^2\cos^2 C$$

$$a^2b^2 + 2a^2b^2 = 4a^2b^2\cos^2 C \qquad \text{[from Eq. (i)]}$$

$$3a^2b^2 = 4a^2b^2\cos^2 C$$

$$\cos C = \pm\frac{\sqrt{3}}{2}$$

$$\cos C = \cos 30^\circ, \cos 150^\circ$$

$$C = 30^\circ, 150^\circ$$

Thus, **(ii) → (p,t)**

(iii) We have,

$$a^4 + b^4 + c^4 + 2a^2c^2 = 2a^2b^2 + 2b^2c^2$$

$$b^4 + (a^2 + c^2)^2 - 2(a^2 + c^2)b^2 = 0$$

$$[b^2 - (a^2 + c^2)]^2 = 0$$

$$b^2 - (a^2 + c^2) = 0$$

$$a^2 + c^2 = b^2$$

$$\therefore \quad \cos B = 90^\circ$$

Thus, **(iii) → (r)**

DPP-3 **Napier's Analogy, Area of Triangle** (Heron's Formula), **Half-Angle Formulae**

1. $\text{LHS} = \left(\cot\frac{B}{2} + \cot\frac{C}{2}\right)\left(a\sin^2\frac{B}{2} + b\sin^2\frac{A}{2}\right)$

$$= \left[\sqrt{\frac{s(s-b)}{(s-a)(s-c)}} + \sqrt{\frac{s(s-c)}{(s-a)(s-b)}}\right]\left[\frac{a(s-a)(s-c)}{ac} + \frac{b(s-b)(s-c)}{bc}\right]$$

$$= \sqrt{\frac{s}{s-a}}\left[\sqrt{\frac{s-b}{s-c}} + \sqrt{\frac{s-c}{s-b}}\right]\left[\frac{(s-a)(s-c)}{c} + \frac{(s-b)(s-c)}{c}\right]$$

$$= \sqrt{\frac{s}{s-a}}\left[\frac{s-b+s-c}{\sqrt{(s-c)(s-b)}}\right]\left[(s-c)\left\{\frac{(s-a)+(s-b)}{c}\right\}\right]$$

$$= \frac{\sqrt{s}}{\sqrt{(s-a)}} \times \frac{2s-b-c}{\sqrt{(s-c)(s-b)}} \times (s-c)\left\{\frac{2s-a-b}{c}\right\}$$

$$= \sqrt{\frac{s(s-c)}{(s-a)(s-b)}} \times a$$

$$= a\cot\frac{C}{2}$$

= RHS **Hence proved.**

2. $\because \cos B + \cos C = 4\sin^2\frac{A}{2}$

$$2\cos\left(\frac{B+C}{2}\right)\cos\left(\frac{B-C}{2}\right) = 4\sin^2\frac{A}{2}$$

$$\Rightarrow 2\sin\frac{A}{2}\cos\left(\frac{B-C}{2}\right) = 4\sin^2\frac{A}{2} \qquad (\because A+B+C = 180°)$$

$$\Rightarrow \cos\left(\frac{B}{2} - \frac{C}{2}\right) = 2\sin\frac{A}{2}$$

$$\Rightarrow \cos\frac{B}{2}\cos\frac{C}{2} + \sin\frac{B}{2}\sin\frac{C}{2} = 2\sin\frac{A}{2}$$

$$\Rightarrow \sqrt{\frac{s(s-b)}{ac}}\cdot\sqrt{\frac{s(s-c)}{ab}} + \sqrt{\frac{(s-a)(s-c)}{ac}}\cdot\sqrt{\frac{(s-a)(s-b)}{ab}} = 2\sqrt{\frac{(s-b)(s-c)}{bc}}$$

$$\Rightarrow \frac{s\sqrt{(s-b)(s-c)}}{a\sqrt{bc}} + \frac{(s-a)\sqrt{(s-b)(s-c)}}{a\sqrt{bc}} = \frac{2\sqrt{(s-b)(s-c)}}{\sqrt{bc}}$$

$$\Rightarrow \frac{s}{a} + \frac{s-a}{a} = 2$$

$$\Rightarrow s + s - a = 2a$$

$$\Rightarrow 2s - a = 2a$$

$$\Rightarrow a + b + c - a = 2a \qquad (\because 2s = a+b+c)$$

$$\therefore b + c = 2a$$

Hence, b, a, c are in AP. **Hence proved.**

3. $\because \tan\frac{A}{2}\tan\frac{C}{2} = \frac{1}{2}$

$$\sqrt{\frac{(s-b)(s-c)}{s(s-a)}}\cdot\sqrt{\frac{(s-a)(s-b)}{s(s-c)}} = \frac{1}{2}$$

$$\Rightarrow \frac{(s-b)}{s} = \frac{1}{2}$$

$$\Rightarrow 2s - 2b = s$$

$$\Rightarrow s - 2b = 0$$

$$\Rightarrow \frac{a+b+c}{2} - 2b = 0$$

$$\Rightarrow a + b + c - 4b = 0$$

$$\therefore a + c = 3b$$

Hence proved.

4. True,

$$\because (s-b)(s-c) = s(s-a)$$

$$\sqrt{\frac{(s-b)(s-c)}{s(s-a)}} = 1$$

$$\tan\frac{A}{2} = \tan 45° \qquad \text{(using formula)}$$

$$\Rightarrow \frac{A}{2} = 45°$$

$$\therefore \angle A = 90°$$

5. False, we know that,

$$\sin\frac{C}{2} = \sqrt{\frac{(s-a)(s-b)}{ab}}$$

We have,

$$(s-a)(s-b) = \frac{\lambda(s-a)(s-b)}{ab}$$

$$\therefore \lambda = ab$$

6. (b) $\because a, b, c$ are in AP, then $2b = a + c$ ...(i)

$$\text{Now, } \frac{\sin\frac{A}{2}\sin\frac{C}{2}}{\sin\frac{B}{2}} = \frac{\sqrt{\frac{(s-b)(s-c)}{bc}}\cdot\sqrt{\frac{(s-a)(s-b)}{ab}}}{\sqrt{\frac{(s-a)(s-c)}{ac}}}$$

$$= \frac{\frac{s-b}{b}\cdot\sqrt{\frac{(s-a)(s-c)}{ac}}}{\sqrt{\frac{(s-a)(s-c)}{ac}}}$$

$$= \frac{s-b}{b} = \frac{a+b+c-2b}{2b}$$

$$= \frac{a+c-b}{2b} = \frac{2b-b}{2b} = \frac{1}{2}$$

7. (a) $1 - \tan\frac{B}{2}\cdot\tan\frac{C}{2} = 1 - \sqrt{\frac{(s-a)(s-c)}{s(s-b)}}\cdot\sqrt{\frac{(s-a)(s-b)}{s(s-c)}}$

$$= 1 - \frac{s-a}{s} = \frac{s-s+a}{s}$$

$$= \frac{a}{s} = \frac{a}{\frac{a+b+c}{2}} = \frac{2a}{a+b+c}$$

8. (d) We have, $\sin\frac{B}{2}\sin\frac{C}{2} = \frac{b+c-a}{a}$

$$\Rightarrow 2\sin\frac{B}{2}\sin\frac{C}{2} = \frac{2(b+c-a)}{a}$$

$$\Rightarrow \cos\left(\frac{B}{2} - \frac{C}{2}\right) - \cos\left(\frac{B}{2} + \frac{C}{2}\right) = \frac{2(b+c-a)}{a}$$

$$\Rightarrow \cos\frac{B}{2}\cos\frac{C}{2} + \sin\frac{B}{2}\sin\frac{C}{2} - \cos\left(\frac{\pi}{2} - \frac{A}{2}\right) = \frac{2(b+c-a)}{a}$$

$$\Rightarrow \sqrt{\frac{s(s-b)}{ac}}\cdot\sqrt{\frac{s(s-c)}{ab}} + \sqrt{\frac{(s-a)(s-c)}{ac}}\cdot\sqrt{\frac{(s-a)(s-b)}{ab}} - \sin\frac{A}{2} = \frac{2(b+c-a)}{a}$$

$$\Rightarrow \frac{s}{a}\cdot\sqrt{\frac{(s-b)(s-c)}{bc}} + \frac{s-a}{a}\cdot\sqrt{\frac{(s-b)(s-c)}{bc}} - \sqrt{\frac{(s-b)(s-c)}{bc}} = \frac{2(b+c-a)}{a}$$

$$\Rightarrow \sqrt{\frac{(s-b)(s-c)}{bc}}\left[\frac{s}{a} + \frac{s-a}{a} - 1\right] = \frac{2(b+c-a)}{a}$$

$$\Rightarrow \sin\frac{A}{2}\left[\frac{s+s-a-a}{a}\right] = \frac{2(b+c-a)}{a}$$

$\Rightarrow \quad \sin\frac{A}{2}\left[\frac{2s-2a}{a}\right] = \frac{2(b+c-a)}{a}$

$\Rightarrow \quad \sin\frac{A}{2}\left[\frac{a+b+c-2a}{a}\right] = \frac{2(b+c-a)}{a}$

$\Rightarrow \quad \sin\frac{A}{2}\left[\frac{b+c-a}{a}\right] = \frac{2(b+c-a)}{a}$

$\therefore \quad \sin\frac{A}{2} = 2$, which is not possible.

9. (a) We know that, $\tan\left(\frac{A-B}{2}\right) = \frac{a-b}{a+b}\cot\frac{C}{2}$

(using Napier's analogy)

Put $\quad A - B = 90°, a = \sqrt{3}, b = 1$

Then, $\quad \tan 45° = \frac{\sqrt{3}-1}{\sqrt{3}+1}\cdot\cot\frac{C}{2}$

$\tan\frac{C}{2} = \frac{\sqrt{3}-1}{\sqrt{3}+1} = \tan 15° \quad \left[\because \tan 15° = \frac{\sqrt{3}-1}{\sqrt{3}+1}\right]$

$\Rightarrow \quad \frac{C}{2} = 15°$

$\therefore \quad \angle C = 30°$

10. (a) We have, $\quad a\cos^2\frac{B}{2} + b\cos^2\frac{A}{2} = \frac{3c}{2}$

$a\cdot\frac{s(s-b)}{ac} + b\cdot\frac{s(s-a)}{bc} = \frac{3c}{2}$

$\frac{s(s-b+s-a)}{c} = \frac{3c}{2}$

$2s(2s-a-b) = 3c^2$

$2s(a+b+c-a-b) = 3c^2$

$(a+b+c)\cdot c = 3c^2$

$a+b+c = 3c$

$2c = a+b$

$\therefore$ *a*, *c*, *b* are in AP.

11. (a) We have, $\sin\frac{A}{2}\sin\frac{B}{2} = \sin\frac{C}{2}$

$\sqrt{\frac{(s-b)(s-c)}{bc}}\cdot\sqrt{\frac{(s-a)(s-c)}{ac}} = \sqrt{\frac{(s-a)(s-b)}{ab}}$

$\Rightarrow \quad \frac{s-c}{c} = 1$

$\Rightarrow \quad s - c = c$

$\therefore \quad s = 2c$

12. (b) Given, $\frac{\tan\frac{B}{2} - \tan\frac{C}{2}}{\tan\frac{B}{2} + \tan\frac{C}{2}} = \frac{\sqrt{\frac{(s-a)(s-c)}{s(s-b)}} - \sqrt{\frac{(s-a)(s-b)}{s(s-c)}}}{\sqrt{\frac{(s-a)(s-c)}{s(s-b)}} + \sqrt{\frac{(s-a)(s-b)}{s(s-c)}}}$

$= \frac{\sqrt{\frac{s-a}{s}}\cdot\frac{[s-c-s+b]}{\sqrt{(s-b)(s-c)}}}{\sqrt{\frac{s-a}{s}}\cdot\frac{[s-c+s-b]}{\sqrt{(s-b)(s-c)}}}$

$= \frac{b-c}{2s-b-c} = \frac{b-c}{a+b+c-b-c}$

$= \frac{b-c}{a}$

13. (a) ***Case I*** LHS $= b\cos^2\frac{C}{2} + c\cos^2\frac{B}{2}$

$= b\cdot\frac{s(s-c)}{ab} + c\cdot\frac{s(s-b)}{ac}$

$= \frac{s}{a}(s-c+s-b)$

$= \frac{s}{a}(2s-b-c) = \frac{s}{a}\times a \quad (\because 2s = a+b+c)$

$= s =$ RHS

Case II Given, $\cot\frac{A}{2} = \frac{b+c}{a}$

$\Rightarrow \quad \frac{\cos\frac{A}{2}}{\sin\frac{A}{2}} = \frac{\sin B + \sin C}{\sin A}$

$\Rightarrow \quad \frac{\cos\frac{A}{2}}{\sin\frac{A}{2}} = \frac{2\sin\left(\frac{B+C}{2}\right)\cos\left(\frac{B-C}{2}\right)}{2\sin\frac{A}{2}\cos\frac{A}{2}}$

$\Rightarrow \quad \cos\frac{A}{2} = \frac{2\cos\frac{A}{2}\cdot\cos\left(\frac{B-C}{2}\right)}{2\cos\frac{A}{2}} \quad (\because A+B+C = 180°)$

$\Rightarrow \quad \cos\frac{A}{2} = \cos\left(\frac{B-C}{2}\right)$

$\Rightarrow \quad \frac{A}{2} = \frac{B-C}{2}$

$\Rightarrow \quad A = B - C$

$\Rightarrow \quad A + C = B \quad \ldots(i)$

$\because \quad A + B + C = 180°$

$B + B = 180° \quad$ [from Eq. (i)]

$\therefore \quad \angle B = 90° \quad$ (true)

14. (b) We have, $\cot\frac{A}{2}\cot\frac{B}{2} = c$

$\sqrt{\frac{s(s-a)}{(s-b)(s-c)}}\cdot\sqrt{\frac{s(s-b)}{(s-a)(s-c)}} = c$

$\frac{s}{s-c} = c$

$\frac{1}{s-c} = \frac{c}{s} \quad \ldots(i)$

Similarly, $\quad \cot\frac{B}{2}\cot\frac{C}{2} = a$

$\Rightarrow \quad \frac{1}{s-a} = \frac{a}{s} \quad \ldots(ii)$

and $\quad \cot\frac{C}{2}\cot\frac{A}{2} = b$

$\Rightarrow \quad \frac{1}{s-b} = \frac{b}{s} \quad \ldots(iii)$

Now, $\quad \frac{1}{s-a} + \frac{1}{s-b} + \frac{1}{s-c} = \frac{a}{s} + \frac{b}{s} + \frac{c}{s}$

[from Eqs. (i), (ii) and (iii)]

$= \frac{a+b+c}{s}$

$= \frac{2s}{s} = 2$

15. (b) We have, $\quad \tan\frac{A}{2} = \frac{5}{6}, \tan\frac{C}{2} = \frac{2}{5}$

Then, $\quad \cot\frac{A}{2} = \frac{6}{5}, \cot\frac{C}{2} = \frac{5}{2}$

$\therefore \quad \cot\frac{A}{2}\cot\frac{C}{2} = 3$

$\Rightarrow \quad \sqrt{\frac{s(s-a)}{(s-b)(s-c)}}\cdot\sqrt{\frac{s(s-c)}{(s-a)(s-b)}} = 3$

$\Rightarrow \quad \frac{s}{s-b} = 3$

$\Rightarrow \quad s = 3s - 3b$

$\Rightarrow \quad 2s - 3b = 0$

$\Rightarrow \quad a + c - 2b = 0$

$\therefore \quad 2b = a + c$

Hence, *a*, *b*, *c* are in AP.

16. (a,d) We know that, $\frac{a}{\sin A} = \frac{b}{\sin B} = \frac{c}{\sin C}$

$\Rightarrow \quad a\sin B = b\sin A$

From option (a), $b\sin A = a$

$\Rightarrow \quad a\sin B = a$

$\Rightarrow \quad \sin B = 1 = \sin\frac{\pi}{2}$

$\Rightarrow \quad B = \frac{\pi}{2}$

Thus, $A < \frac{\pi}{2}$ (correct)

From option (b), $b\sin A > a$

$a\sin B > a$

$\sin B > 1$ (not possible)

Similarly, option (c) is not possible.

From option (d),

$b\sin A < a$

$a\sin B < a$

$\sin B < 1$ (possible)

Now, if $b > a$

$\Rightarrow \quad B > A \quad \left(\because A < \frac{\pi}{2}\right)$

$\therefore$ Triangle is possible.

17. (b,c) We have, $\cos B + \cos C = 4\sin^2\frac{A}{2}$

$\Rightarrow \quad 2\cos\left(\frac{B+C}{2}\right)\cdot\cos\left(\frac{B-C}{2}\right) = 4\sin^2\frac{A}{2}$

$\Rightarrow \quad 2\sin\frac{A}{2}\cos\left(\frac{B-C}{2}\right) - 4\sin^2\frac{A}{2} = 0 \quad (\because A + B + C = 180°)$

$\Rightarrow \quad 2\sin\frac{A}{2}\left[\cos\left(\frac{B-C}{2}\right) - 2\sin\frac{A}{2}\right] = 0$

$\Rightarrow \quad \cos\left(\frac{B-C}{2}\right) - 2\cos\left(\frac{B+C}{2}\right) = 0 \quad \left(\because 2\sin\frac{A}{2} \neq 0\right)$

$\Rightarrow \cos\frac{B}{2}\cos\frac{C}{2} + \sin\frac{B}{2}\sin\frac{C}{2} - 2\cos\frac{B}{2}\cos\frac{C}{2} + 2\sin\frac{B}{2}\sin\frac{C}{2} = 0$

$\Rightarrow \quad 3\sin\frac{B}{2}\sin\frac{C}{2} - \cos\frac{B}{2}\cos\frac{C}{2} = 0$

$\Rightarrow \quad \tan\frac{B}{2}\tan\frac{C}{2} = \frac{1}{3}$

$\Rightarrow \quad \sqrt{\frac{(s-a)(s-c)}{s(s-b)}}\cdot\sqrt{\frac{(s-a)(s-b)}{s(s-c)}} = \frac{1}{3}$

$\Rightarrow \quad \frac{s-a}{s} = \frac{1}{3}$

$\Rightarrow \quad b + c = 2a$ [option (b)]

$\therefore$ Locus of A is an ellipse. [option (c)]

18. (a,b,d) We have, $\frac{b-c}{a} = \frac{\sin B - \sin C}{\sin A}$ (by sine rule)

$$= \frac{2\cos\left(\frac{B+C}{2}\right)\sin\left(\frac{B-C}{2}\right)}{2\sin\frac{A}{2}\cos\frac{A}{2}}$$

$$= \frac{2\sin\frac{A}{2}\sin\left(\frac{B-C}{2}\right)}{2\sin\frac{A}{2}\cos\frac{A}{2}}$$

$$\Rightarrow \quad \frac{b-c}{a} = \frac{\sin\left(\frac{B-C}{2}\right)}{\cos\frac{A}{2}}$$

$\therefore \quad (b-c)\cos\frac{A}{2} = a\sin\left(\frac{B-C}{2}\right)$ [option (c)]

19. (d) $\because PG = \frac{1}{3}AD = \frac{1}{3}\cdot\frac{2\Delta}{a}$

$= \frac{2}{3a}\cdot\frac{1}{2}ab\sin C = \frac{b\sin C}{3}$

Also, $PG = \frac{2}{3a}\cdot\frac{1}{2}ac\sin B = \frac{1}{3}c\sin B$

20. (b) $\because$ Area of $\Delta GPL = \frac{1}{2}(PL)(PG)$

Area of $\Delta ALD = \frac{1}{2}(DL)(AD)$

$\because \quad PL = \frac{1}{3}DL$ and $PG = \frac{1}{3}AD$

$$\therefore \quad \frac{\text{Area of } \Delta GPL}{\text{Area of } \Delta ALD} = \frac{\frac{1}{2}(PL)(PG)}{\frac{1}{2}(DL)(AD)} = \frac{\frac{1}{3}(DL)\cdot\frac{1}{3}(AD)}{DL\cdot AD} = \frac{1}{9}$$

21. (b) $\because$ Area of ΔPQR = Area of ΔPGQ + Area of ΔQGR + Area of ΔRGP

$\because \quad$ Area of $\Delta PGR = \frac{1}{2}(PG)\cdot(GQ)\cdot\sin(\angle PGQ)$

$= \frac{1}{2}\times\frac{1}{3}AD\times\frac{1}{3}BE\times\sin(\pi - C)$

$= \frac{1}{18}\times\frac{2\Delta}{a}\times\frac{2\Delta}{b}\times\sin C$

$= \frac{2}{9ab}\times\frac{1}{2}bc\sin A\times\frac{1}{2}ac\sin B\times\sin C$

$= \frac{c^2}{18}\cdot\sin A\sin B\sin C$

Similarly, area of $\Delta QGR = \frac{a^2}{18}\cdot\sin A\sin B\sin C$

and area of $\Delta RGP = \frac{b^2}{18}\cdot\sin A\sin B\sin C$

$\Rightarrow \quad$ Area of $\Delta PQR = \frac{1}{18}(a^2 + b^2 + c^2)(\sin A\sin B\sin C)$

22. (4) $\because \quad \Delta = a^2 - (b-c)^2$

$= (a + b - c)(a - b + c)$

$= (2s - c - c))(2s - b - b)$

$= 2\times 2(s-c)(s-b)$

$\Rightarrow \sqrt{s(s-a)(s-b)(s-c)} = 4(s-c)(s-b)$

(using Hero's Formula)

$\Rightarrow \quad \frac{1}{4} = \sqrt{\frac{(s-b)(s-c)}{s(s-a)}}$

$\Rightarrow \quad \tan\frac{A}{2} = \frac{1}{4}$

But $\tan\frac{A}{2} = \frac{1}{\lambda}$

$\therefore \quad \lambda = 4$

23. (5) We have, $\frac{s-a}{11} = \frac{s-b}{12} = \frac{s-c}{13}$

Now, $\frac{s-a+s-b+s-c}{11+12+13} = \frac{3s-(a+b+c)}{36} = \frac{3s-2s}{36}$

$\Rightarrow \quad \frac{s-a}{11} = \frac{s-b}{12} = \frac{s-c}{13} = \frac{s}{36}$

If $\frac{s-a}{11} = \frac{s}{36}$

Then, $s - a = \frac{11s}{36}$

Similarly, $s - b = \frac{12s}{36}$

$s - c = \frac{13s}{36}$

Now, $\tan^2\frac{A}{2} = \frac{(s-b)(s-c)}{s(s-a)}$

$$= \frac{\frac{12s}{36} \times \frac{13s}{36}}{s \times \frac{11s}{36}}$$

$$= \frac{12}{36} \times \frac{13}{36} \times \frac{36}{11} = \frac{13}{33}$$

But $33\mu \cdot \tan^2\frac{A}{2} = 65$

$\Rightarrow \quad 33\mu \times \frac{13}{33} = 65$

$\Rightarrow \quad 33\mu = 65 \times \frac{33}{13}$

$\Rightarrow \quad 33\mu = 165$

$\therefore \quad \mu = 5$

24. (9) Let $a = 2 = QR$, $b = \frac{7}{2} = PR$ and $c = \frac{5}{2} = PQ$

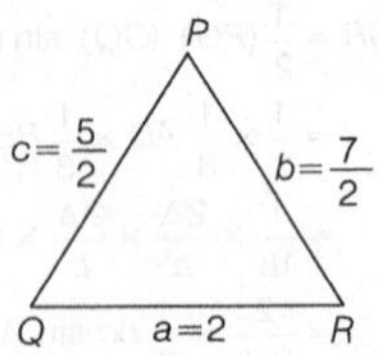

Then, $s = \frac{2 + \frac{7}{2} + \frac{5}{2}}{2} = 4$

We have, $\frac{\lambda}{16\Delta^2} = \frac{2\sin P - \sin 2P}{2\sin P + \sin 2P}$

$$= \frac{2\sin P - 2\sin P\cos P}{2\sin P + 2\sin P\cos P}$$

$$= \frac{2\sin P(1-\cos P)}{2\sin P(1+\cos P)}$$

$\Rightarrow \quad \frac{\lambda}{16\Delta^2} = \frac{1-\cos P}{1+\cos P} = \tan^2\frac{P}{2}$

$$= \frac{(s-b)(s-c)}{s(s-a)}$$

$$= \frac{(s-b)^2(s-c)^2}{[\sqrt{s(s-a)(s-b)(s-c)}]^2}$$

$$= \frac{(s-b)^2(s-c)^2}{\Delta^2} \quad \text{(using Hero's formula)}$$

$$= \frac{\left(4-\frac{7}{2}\right)^2\left(4-\frac{5}{2}\right)^2}{\Delta^2}$$

$\Rightarrow \quad \frac{\lambda}{16\Delta^2} = \frac{9}{16\Delta^2}$

$\therefore \quad \lambda = 9$

DPP-4 Half-Angle Formulae using 's', Heron's Formulae, Area of Triangle using Sides and Angles

1. (b) We know that,

$$\Delta = \frac{1}{2} \cdot \frac{a^2 \sin B \sin C}{\sin(B+C)}$$

$$= \frac{1}{2} \cdot \frac{(4)^2 \cdot \sin 30^\circ \sin 45^\circ}{\sin 75^\circ}$$

$$= \frac{8 \cdot \frac{1}{2} \cdot \frac{1}{\sqrt{2}}}{\frac{\sqrt{3}+1}{2\sqrt{2}}} = \frac{8}{\sqrt{3}+1}$$

$$= \frac{8(\sqrt{3}-1)}{3-1}$$

$$= 4(\sqrt{3}-1)\text{ cm}^2$$

2. (b) Let ΔABC is an isosceles triangle in which $b = c = 6$ cm, then

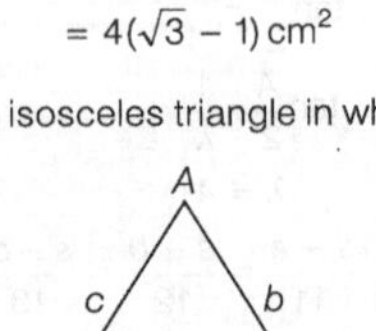

Area of $\Delta ABC = \frac{1}{2}bc\sin A$ (using SAS)

$$9 = \frac{1}{2} \times 6 \times 6 \cdot \sin A$$

$$\sin A = \frac{18}{36} = \frac{1}{2}$$

$$\sin A = \sin 30^\circ$$

$\therefore \quad A = 30^\circ$

3. (c) We have, $a = 1, b = 2, \angle C = 60^\circ$

$$\Delta = \frac{1}{2}ab\sin C = \frac{1}{2} \times 1 \times 2 \times \frac{\sqrt{3}}{2} = \frac{\sqrt{3}}{2} \text{ sq units}$$

4. (a) We know that,

$$\Delta = \frac{1}{2} \times \text{Base} \times \text{Corresponding altitude}$$

$$\Delta = \frac{1}{2} \times a \times p_1$$

$\therefore \quad p_1 = \frac{2\Delta}{a}$

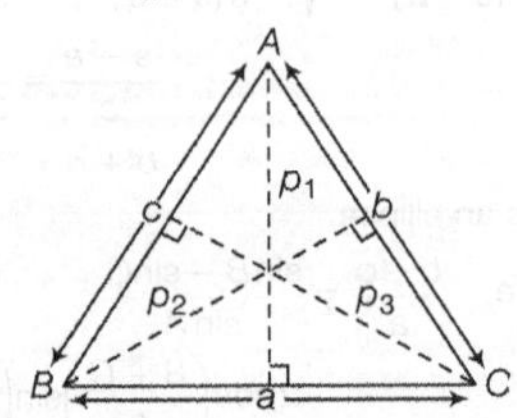

Similarly, $p_2 = \frac{2\Delta}{b}$

and $p_3 = \frac{2\Delta}{c}$

It is given that p_1, p_2, p_3 are in HP.

$\frac{2\Delta}{a}, \frac{2\Delta}{b}, \frac{2\Delta}{c}$ are in HP.

On dividing each by 2Δ, we get

$\frac{1}{a}, \frac{1}{b}, \frac{1}{c}$ are in HP.

Thus, a, b, c are in AP.

$\therefore \quad \sin A, \sin B, \sin C$ are also in AP. (using formula)

5. (c) We have

$$\text{Area of } \Delta ABC = \Delta$$

Now, $a^2 \sin 2B + b^2 \sin 2A$

$= 2k^2 \sin^2 A \sin B \cos B + 2k^2 \sin^2 B \sin A \cos A$ (using sine rule)

$= 2k^2 \sin A \sin B (\sin A \cos B + \cos A \sin B)$

$= 2(k \sin A)(k \sin B) \sin(A + B)$

$= 2ab \sin(\pi - C) = 2ab \sin C$ (using formula)

$= 4\left(\frac{1}{2} ab \sin C\right) = 4\Delta$

6. (a) Let a, b, c be the sides of a triangle, then

$$\Delta = 20\sqrt{3} \text{ cm}^2$$

We know that,

$$\Delta = \frac{1}{2} ab \sin C$$

$$\Rightarrow \quad 20\sqrt{3} = \frac{1}{2} ab \sin 60°$$

$$20\sqrt{3} = \frac{1}{2} ab \frac{\sqrt{3}}{2}$$

$$\Rightarrow \quad ab = 80 \quad \ldots(i)$$

We have,

$$a + b + c = 20$$

$$a + b = 20 - c$$

On squaring,

$$(a + b)^2 = (20 - c)^2 \quad \ldots(ii)$$

Now, $a^2 + b^2 = (a + b)^2 - 2ab$

$= (20 - c)^2 - 2 \times 80$ [from Eqs. (i) and (ii)]

$= 400 + c^2 - 40c - 160$

$$\Rightarrow \quad a^2 + b^2 = c^2 - 40c + 240 \quad \ldots(iii)$$

Using cosine rule, we get

$$\cos C = \frac{a^2 + b^2 - c^2}{2ab}$$

$$\Rightarrow \quad \cos 60° = \frac{c^2 - 40c + 240 - c^2}{2 \times 80}$$

[from Eqs. (i) and (iii)]

$$\Rightarrow \quad \frac{1}{2} = \frac{240 - 40c}{2 \times 80}$$

$$\Rightarrow \quad 240 - 40c = 80$$

$$\Rightarrow \quad 40c = 240 - 80$$

$$\therefore \quad c = \frac{160}{40} = 4 \text{ cm}$$

7. (b) We have,

$$\Delta = a^2 - (b - c)^2$$

and $2s = a + b + c$

Now, $\sqrt{s(s-a)(s-b)(s-c)} = (a + b - c)(a - b + c)$ (using Hero's formula)

$= (2s - 2c)(2s - 2b)$

$$\Rightarrow \quad \sqrt{s(s-a)(s-b)(s-c)} = 4(s - c)(s - b)$$

$$\Rightarrow \quad \frac{1}{4} = \sqrt{\frac{(s-b)(s-c)}{s(s-a)}}$$

$$\therefore \quad \tan\frac{A}{2} = \frac{1}{4}$$

We know that,

$$\tan A = \frac{2\tan\frac{A}{2}}{1 - \tan^2\frac{A}{2}} = \frac{2 \times \frac{1}{4}}{1 - \left(\frac{1}{4}\right)^2}$$

$$= \frac{\frac{1}{2}}{\frac{15}{16}} = \frac{1}{2} \times \frac{16}{15}$$

$$\therefore \quad \tan A = \frac{8}{15}$$

8. (c) We have,

$$AB = PQ$$

$$AC = QR$$

and $\angle A = 2\angle Q$

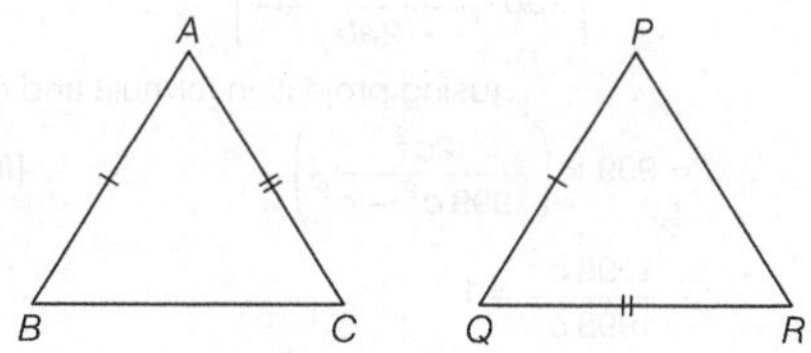

$\because \quad (\Delta ABC) = (\Delta PQR)$ (given)

$$\frac{1}{2} AB \cdot AC \sin A = \frac{1}{2} PQ \cdot QR \sin Q$$

$$\Rightarrow \quad \sin A = \sin Q$$

$$\Rightarrow \quad \sin 2Q = \sin Q$$

$$2 \sin Q \cos Q - \sin Q = 0$$

$$\sin Q (2\cos Q - 1) = 0$$

Either $\sin Q = 0$

$\Rightarrow \quad Q = 0°$ (not possible)

or $2\cos Q - 1 = 0$

$$\cos Q = \frac{1}{2}$$

$$\cos Q = \cos\frac{\pi}{3}$$

$$Q = \frac{\pi}{3}$$

$$\therefore \quad \angle A = 2\angle Q = 2 \times \frac{\pi}{3}$$

$$= \frac{2\pi}{3}$$

9. (a) Given, $a = 6$

$$b = 3$$

and $\cos(A + B) = \frac{4}{5}$

Now, $\sin(A + B) = \sqrt{1 - \left(\frac{4}{5}\right)^2} = \frac{3}{5}$

$$\Rightarrow \quad \sin(\pi - C) = \frac{3}{5} = \sin C$$

$$\therefore \quad \Delta = \frac{1}{2} ab \sin C$$

$$= \frac{1}{2} \times 6 \times 3 \times \frac{3}{5}$$

$$= \frac{54}{10} = 5.4 \text{ sq units}$$

10. (b) We have,

$$a^2 + b^2 = 1999c^2 \quad \dots(i)$$

Now, $999\left(\dfrac{\cot A + \cot B}{\cot C}\right) = 999\left[\dfrac{\dfrac{\cos A}{\sin A} + \dfrac{\cos B}{\sin B}}{\dfrac{\cos C}{\sin C}}\right]$

$$= 999\left[\dfrac{\dfrac{bc\cos A}{2\Delta} + \dfrac{ac\cos B}{2\Delta}}{\dfrac{ab\cos C}{2\Delta}}\right] \left(\text{using } \Delta = \frac{1}{2}bc\sin A\right)$$

$$= 999\left[\frac{c(b\cos A + a\cos B)}{ab\cos C}\right]$$

$$= 999\left[\dfrac{2\cdot c\cdot c}{2ab\cdot\dfrac{a^2+b^2-c^2}{2ab}}\right]$$

(using projection formula and cosine rule)

$$= 999 \times \left(\frac{2c^2}{1999c^2 - c^2}\right) \quad \text{[from Eq. (i)]}$$

$$= \frac{1998c^2}{1998c^2} = 1$$

11. (c) We know that, AM $\geq$ GM

Then, $\dfrac{(s-a)+(s-b)+(s-c)}{3} \geq [(s-a)(s-b)(s-c)]^{1/3}$

$$\Rightarrow \quad \frac{3s-(a+b+c)}{3} \geq \left[\frac{s(s-a)(s-b)(s-c)}{s}\right]^{1/3}$$

$$\Rightarrow \quad \left(\frac{s}{3}\right)^3 \geq \frac{\Delta^2}{s} \quad \left(\because a+b+c = 2s \text{ and Hero's formula}\right)$$

$$\Rightarrow \quad \Delta^2 \leq \frac{s^4}{(3)^3} = \frac{s^4}{27}$$

$$\Rightarrow \quad \Delta \leq \frac{s^2}{3\sqrt{3}}$$

But given that $\Delta \leq \dfrac{s^2}{\lambda\sqrt{3}}$, then comparing, we get

$$\lambda = 3$$

12. (a) We have,

$$\tan\frac{C}{2} = \frac{\cot\frac{A}{2}\cot\frac{B}{2} - 1}{\cot\frac{A}{2} + \cot\frac{B}{2}}$$

$$= \frac{\frac{6}{5}\cdot\frac{37}{20} - 1}{\frac{6}{5} + \frac{37}{20}} = \frac{222 - 100}{120 + 185}$$

$$= \frac{122}{305} = \frac{2}{5}$$

Also, $\tan\dfrac{A}{2}\tan\dfrac{C}{2} = \sqrt{\dfrac{(s-b)(s-c)}{s(s-a)}} \times \sqrt{\dfrac{(s-a)(s-b)}{s(s-c)}}$

$$\Rightarrow \quad \frac{5}{6}\cdot\frac{2}{5} = \frac{s-b}{s}$$

$$\Rightarrow \quad 3(s-b) = s$$

$$\Rightarrow \quad 2s = 3b$$

$$\Rightarrow \quad a + c = 2b$$

$\therefore$ a, b, c are in AP.

13. (a,b,c) In ΔABC, using sine rule

$$\frac{a}{\sin A} = \frac{b}{\sin B} = \frac{c}{\sin C}$$

$$\Rightarrow \quad \frac{\sqrt{3}+1}{\sin 105°} = \frac{b}{\sin 30°}$$

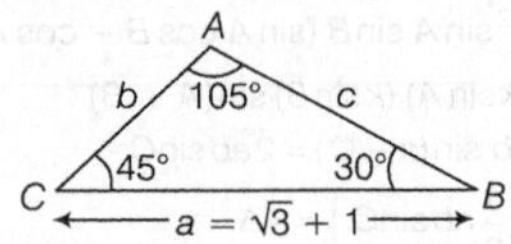

$$\Rightarrow \quad \frac{\sqrt{3}+1}{\sin 75°} = \frac{b}{\frac{1}{2}}$$

$$\Rightarrow \quad b = \frac{\sqrt{3}+1}{2}\cdot\frac{1}{\frac{\sqrt{3}+1}{2\sqrt{2}}} \quad \left[\because \sin 75° = \frac{\sqrt{3}+1}{2\sqrt{2}}\right]$$

$$= \frac{\sqrt{3}+1}{2}\cdot\frac{2\sqrt{2}}{\sqrt{3}+1} = \sqrt{2}$$

$$\therefore \quad b = \sqrt{2}$$

Now, area of $\Delta ABC = \dfrac{1}{2}ab\sin 45°$

$$= \frac{1}{2} \times (\sqrt{3}+1)\sqrt{2}\cdot\frac{1}{\sqrt{2}}$$

$$= \frac{\sqrt{3}+1}{2} \quad \text{[option (a)]}$$

$$= \sqrt{2}\cdot\frac{\sqrt{3}+1}{2\sqrt{2}}$$

$$= \sqrt{2}\cdot\sin 75° \quad \text{[option (c)]}$$

$$= \sqrt{2}\sin(90° - 15°)$$

$$= \sqrt{2}\cos 15° \quad \text{[option (b)]}$$

14. (a,b,c) $\because$ $\Delta = \dfrac{1}{2}ap_1$

$$\Rightarrow \quad \frac{1}{p_1} = \frac{a}{2\Delta}$$

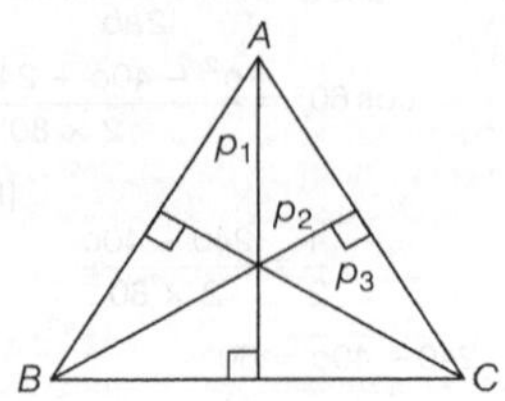

Similarly, $\dfrac{1}{p_2} = \dfrac{b}{2\Delta}$

and $\dfrac{1}{p_3} = \dfrac{c}{2\Delta}$

Now, $\dfrac{1}{p_1} + \dfrac{1}{p_2} - \dfrac{1}{p_3} = \dfrac{1}{2\Delta}(a+b-c)$ [option (b)]

$$= \frac{2s - 2c}{2\Delta} \quad (\because 2s = a+b+c)$$

$$= \frac{s-c}{\Delta} \quad \text{[option (a)]}$$

From option (c),

$$\frac{2ab}{(a+b+c)\Delta}\cdot\cos^2\frac{C}{2} = \frac{2ab}{2s\Delta}\cdot\frac{s(s-c)}{ab}$$

$$= \frac{s-c}{\Delta} = \frac{1}{p_1} + \frac{1}{p_2} - \frac{1}{p_3} \quad \text{[option (c)]}$$

15. (a,b,c,d) $ar(\Delta ABC) = ar(\Delta ABD) + ar(\Delta ACD)$

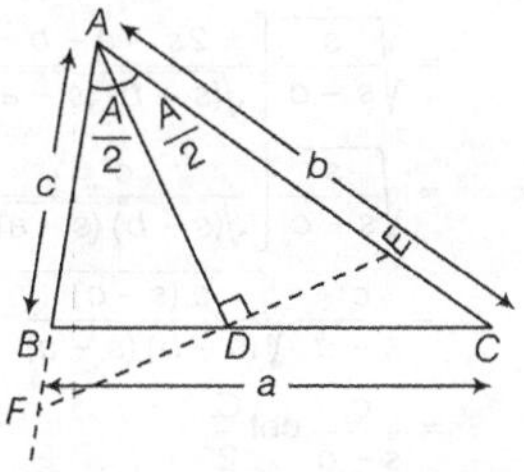

$$\Rightarrow \quad \frac{1}{2}bc\sin A = \frac{1}{2}c \cdot AD\sin\frac{A}{2} + \frac{1}{2}b \cdot AD\sin\frac{A}{2}$$

$$\Rightarrow \quad \frac{1}{2}bc \cdot 2\sin\frac{A}{2}\cos\frac{A}{2} = \frac{1}{2}(b+c)\,AD \cdot \sin\frac{A}{2}$$

$$\Rightarrow \quad AD = \frac{2bc}{b+c}\cos\frac{A}{2} \quad \ldots\text{(i)}$$

[option (b)]

Now, in ΔADE,

$$AE = AD\sec\frac{A}{2}$$

$$AE = \frac{2bc}{b+c} \quad \text{[from Eq. (i)]}$$

$\therefore$ AE is HM of b and c. [option (a)]

Now, $EF = ED + DF = 2DE$

$$= 2 \cdot AD\tan\frac{A}{2} \quad \left(\because \text{in } \Delta ADE,\ \tan\frac{A}{2} = \frac{DE}{AD}\right)$$

$$= 2 \cdot \frac{2bc\cos\frac{A}{2}}{b+c} \cdot \frac{\sin\frac{A}{2}}{\cos\frac{A}{2}}$$

$$EF = \frac{4bc}{b+c}\sin\frac{A}{2} \quad \text{[option (c)]}$$

$\therefore$ $AD \perp EF$, $DE = DF$ and AD is bisector.

$\therefore$ ΔAEF is isosceles.

16. (b) Let a, b, c be the sides of a triangle such that

$$b + c = p$$
$$bc = q$$
and
$$r = a$$

We know that, AM $\geq$ GM

$$\frac{b+c}{2} \geq \sqrt{bc}$$

$$p \geq 2\sqrt{q}$$

$\therefore$ $\angle A$ is the greatest angle.

It is given that, $(p + r)(p - r) = q$

$$\Rightarrow \quad (b + c + a)(b + c - a) = bc$$

$$\Rightarrow \quad 2s(2s - 2a) = bc$$

$$\Rightarrow \quad \frac{s(s-a)}{bc} = \frac{1}{4}$$

$$\Rightarrow \quad \sqrt{\frac{s(s-a)}{bc}} = \frac{1}{2}$$

$$\Rightarrow \quad \cos\frac{A}{2} = \cos 60°$$

$$\Rightarrow \quad \frac{A}{2} = 60°$$

$$\therefore \quad \angle A = 120°$$

17. (a) We know that, area of $\Delta = \frac{1}{2}bc\sin A$

$$= \frac{1}{2}q\sin 120° \quad (\because bc = q)$$

$$= \frac{1}{2}q \cdot \frac{\sqrt{3}}{2}$$

$$= \frac{\sqrt{3}}{4}q \text{ sq units}$$

18. (a) Let b and c be the roots, then equation is

$$x^2 - (b + c)x + bc = 0$$

$$x^2 - px + q = 0 \quad (\because b + c = p \text{ and } bc = q)$$

$$\therefore \quad x = \frac{p \pm \sqrt{p^2 - 4q}}{2}$$

Here, sides are $\frac{p \pm \sqrt{p^2 - 4q}}{2}, r$.

19. (5) Let a, $a + d$, $a + 2d$, $a + 3d$ be the sides of a quadrilateral in AP.

Then, according to the question,

$$a = 6$$

and $\quad a + 3d - a = 6 \quad$ (difference)

$$\Rightarrow \quad d = 2$$

$\therefore$ Sides are 6, 8, 10, 12.

Now, $\quad 2s = 6 + 8 + 10 + 12$

$$s = 18$$

Thus, $\quad A^2 = (18 - 6)(18 - 8)(18 - 10)(18 - 12)$

$$= 12 \times 10 \times 8 \times 6$$
$$= 5760$$

We have,

$$A^2 = 1152\lambda$$

$$\Rightarrow \quad 5760 = 1152\lambda$$

$$\Rightarrow \quad \lambda = \frac{5760}{1152}$$

$$\therefore \quad \lambda = 5$$

20. (2) $\because \cot\frac{A}{2} \cdot \cot\frac{B}{2} = c$

$$\sqrt{\frac{s(s-a)}{(s-b)(s-c)}} \cdot \sqrt{\frac{s(s-b)}{(s-a)(s-c)}} = c$$

$$\frac{s}{s-c} = c$$

$$\Rightarrow \quad \frac{1}{s-c} = \frac{c}{s} \quad \ldots\text{(i)}$$

Similarly, $\quad \cot\frac{B}{2}\cot\frac{C}{2} = a$

$$\Rightarrow \quad \frac{1}{s-a} = \frac{a}{s} \quad \ldots\text{(ii)}$$

and $\quad \cot\frac{C}{2}\cot\frac{A}{2} = b$

$$\Rightarrow \quad \frac{1}{s-b} = \frac{b}{s} \quad \ldots\text{(iii)}$$

Now, on adding Eqs. (i), (ii) and (iii), we get

$$\frac{1}{s-a} + \frac{1}{s-b} + \frac{1}{s-c} = \frac{a}{s} + \frac{b}{s} + \frac{c}{s}$$

$$= \frac{a+b+c}{s} = \frac{2s}{s} = 2$$

DPP-5 Properties of Triangles

1. LHS $= \dfrac{a\cos A + b\cos B + c\cos C}{a+b+c}$

$= \dfrac{k\sin A\cos A + k\sin B\cos B + k\sin C\cos C}{k\sin A + k\sin B + k\sin C}$

$= \dfrac{1}{2}\left[\dfrac{\sin 2A + \sin 2B + \sin 2C}{\sin A + \sin B + \sin C}\right]$ (using sine rule)

$= \dfrac{1}{2}\left[\dfrac{4\sin A\sin B\sin C}{4\cos\frac{A}{2}\cos\frac{B}{2}\cos\frac{C}{2}}\right]$

(using $\sin 2A + \sin 2B + \sin 2C = 4\sin A\sin B\sin C$)

$= \dfrac{1}{2}\cdot\dfrac{2\sin\frac{A}{2}\cos\frac{A}{2}\cdot 2\sin\frac{B}{2}\cos\frac{B}{2}\cdot 2\sin\frac{C}{2}\cos\frac{C}{2}}{\cos\frac{A}{2}\cos\frac{B}{2}\cos\frac{C}{2}}$

$= 4\sin\frac{A}{2}\sin\frac{B}{2}\sin\frac{C}{2} = \frac{r}{R} = \text{RHS}\left(\text{using } r = 4R\sin\frac{A}{2}\sin\frac{B}{2}\sin\frac{C}{2}\right)$

Hence proved.

2. LHS $= 2R^2(\sin 2A + \sin 2B + \sin 2C)$

$= 2R^2(4\sin A\sin B\sin C)$

$= 2[(2R\sin A)(2R\sin B)]\sin C \quad \left(\because R = \dfrac{a}{2\sin A} = \dfrac{b}{2\sin B} = \dfrac{c}{2\sin C}\right)$

$= 2ab\sin C = 2\times 2\left(\frac{1}{2}ab\sin C\right)$

$= 4\Delta = \text{RHS}$ **Hence proved.**

3. LHS $= r + r_3 + r_1 - r_2$

$= 4R\sin\frac{A}{2}\sin\frac{B}{2}\sin\frac{C}{2} + 4R\cos\frac{A}{2}\cos\frac{B}{2}\sin\frac{C}{2}$

$+ 4R\sin\frac{A}{2}\cos\frac{B}{2}\cos\frac{C}{2} - 4R\cos\frac{A}{2}\sin\frac{B}{2}\cos\frac{C}{2}$

$= 4R\left[\sin\frac{C}{2}\left(\cos\frac{A}{2}\cos\frac{B}{2} + \sin\frac{A}{2}\sin\frac{B}{2}\right)\right]$

$+ 4R\left[\cos\frac{C}{2}\left(\sin\frac{A}{2}\cos\frac{B}{2} - \cos\frac{A}{2}\sin\frac{B}{2}\right)\right]$

$= 4R\left[\sin\frac{C}{2}\cos\left(\frac{A-B}{2}\right) + \cos\frac{C}{2}\sin\left(\frac{A-B}{2}\right)\right]$

$= 4R\sin\left(\frac{C}{2} + \frac{A-B}{2}\right)$

$= 4R\sin\left(\frac{A+C-B}{2}\right) = 4R\sin\left(\frac{\pi - 2B}{2}\right)$

$= 4R\sin\left(\frac{\pi}{2} - B\right) = 4R\cos B$

But $r + r_3 + r_1 - r_2 = \lambda R\cos B$

$\therefore \quad \lambda = 4$

4. True, $\because \tan\frac{A}{2}\tan\frac{B}{2} = \sqrt{\dfrac{(s-b)(s-c)}{s(s-a)}}\cdot\sqrt{\dfrac{(s-a)(s-c)}{s(s-b)}}$

$= \dfrac{s-a}{s}$

5. True, $\because \cot\frac{A}{2} + \cot\frac{B}{2} = \sqrt{\dfrac{s(s-a)}{(s-b)(s-c)}} + \sqrt{\dfrac{s(s-b)}{(s-a)(s-c)}}$

$= \sqrt{\dfrac{s}{s-c}}\left[\sqrt{\dfrac{s-a}{s-b}} + \sqrt{\dfrac{s-b}{s-a}}\right]$

$= \sqrt{\dfrac{s}{s-c}}\left[\dfrac{s-a+s-b}{\sqrt{(s-b)(s-a)}}\right]$

$= \sqrt{\dfrac{s}{s-c}}\left[\dfrac{2s-a-b}{\sqrt{(s-b)(s-a)}}\right]$

$= \sqrt{\dfrac{s}{s-c}}\left[\dfrac{c}{\sqrt{(s-b)(s-a)}}\right]$

$= \dfrac{c}{s-c}\cdot\sqrt{\dfrac{s(s-c)}{(s-b)(s-a)}}$

$= \dfrac{c}{s-c}\cot\dfrac{C}{2}$

6. True, we know that,

$\Delta = \frac{1}{2}ab\sin C$

$\Rightarrow \quad \sin C = \dfrac{2\Delta}{ab}$

7. False, we know that,

$r = (s-c)\tan\frac{C}{2}$

$\therefore \quad r \neq (s-c)\tan\frac{A}{2}$

8. (4) Given, $r_1 = \lambda R\sin\frac{A}{2}\cos\frac{B}{2}\cos\frac{C}{2}$

We know that,

$r_1 = 4R\sin\frac{A}{2}\cos\frac{B}{2}\cos\frac{C}{2}$

But $\quad r_1 = \lambda R\sin\frac{A}{2}\cos\frac{B}{C}\cos\frac{C}{2}$

$\therefore \quad \lambda = 4$

9. (2) We know that,

$\Delta = \frac{1}{2}ab\sin C$

$\Rightarrow \quad \sqrt{s(s-a)(s-b)(s-c)} = \frac{1}{2}ab\sin C$

$\Rightarrow \quad \sin C = \dfrac{2\sqrt{s(s-a)(s-b)(s-c)}}{ab}$

and it is given that

$\sin C = \dfrac{\sqrt{s(s-a)(s-b)(s-c)}}{ab}\cdot\lambda$

$\therefore \quad \lambda = 2$

10. (a) In right angled ΔABC,

$\angle ABC = 90°,\ a = 6,\ R = 5$

Then, $\quad AC = OA + OC = 2OC \quad (\because OA = OC = R)$

$= 10$

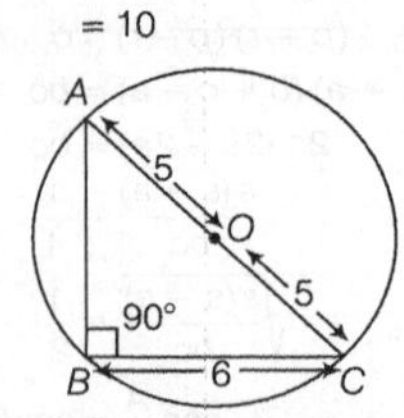

Now, using Pythagoras theorem,

$AB^2 = AC^2 - BC^2 = (10)^2 - (6)^2$

$= 100 - 36 = 64$

$AB = 8$

$\therefore \quad$ Area of $\Delta ABC = \frac{1}{2}\cdot AB\cdot BC$

$= \frac{1}{2}\times 8\times 6 = 24 \text{ cm}^2$

11. (d) In ΔABC, let $a = 3, b = 4, c = 5$, then

$$s = \frac{a+b+c}{2} = \frac{3+4+5}{2}$$
$$s = 6$$

and
$$\Delta = \sqrt{s(s-a)(s-b)(s-c)}$$
$$= \sqrt{6(6-3)(6-4)(6-5)}$$
$$\Delta = 6$$

We know that,
$$r = \frac{\Delta}{s} = \frac{6}{6} = 1$$

$\therefore$ Inradius $(r) = 1$

12. (d) We have, in ΔABC, $a = 5, b = 12, \angle C = 90°$, then using the cosine rule,

$$\cos C = \frac{a^2 + b^2 - c^2}{2ab}$$
$$\cos 90° = \frac{5^2 + 12^2 - c^2}{2 \times 5 \times 12}$$
$$0 = 169 - c^2 \qquad (\because \cos 90° = 0)$$
$$c = 13$$

Now, we know that,
$$\frac{a}{2\sin A} = \frac{b}{2\sin B} = \frac{c}{2\sin C} = R$$
$$\because \quad R = \frac{c}{2\sin C} = \frac{13}{2 \cdot \sin 90°}$$
$$\therefore \quad R = \frac{13}{2}$$

13. (b) Let r and R be the inradius and circumradius respectively, then we know that

$$r = 4R \sin\frac{A}{2} \sin\frac{B}{2} \sin\frac{C}{2}$$
$$= 4R \sin 30° \sin 30° \sin 30°$$
(ΔABC is an equilateral triangle, $\angle A = \angle B = \angle C = 60°$)
$$= 4R\left(\frac{1}{2}\right)^3 = \frac{R}{2}$$
$$\therefore \quad R = 2r$$

14. (c) We know that,
$$\frac{a}{2\sin A} = \frac{b}{2\sin B} = \frac{c}{2\sin C} = R$$

Then,
$$\sin A = \frac{a}{2R}, \sin B = \frac{b}{2R}, \sin C = \frac{c}{2R}$$

Now,
$$2R^2 \sin A \sin B \sin C = 2R^2 \cdot \frac{a}{2R} \cdot \frac{b}{2R} \cdot \frac{c}{2R}$$
$$= \frac{abc}{4R} = \Delta \qquad \left(\because R = \frac{abc}{4\Delta}\right)$$

15. (c) We know that,
$$\frac{a}{2\sin A} = \frac{b}{2\sin B} = \frac{c}{2\sin C} = R$$
$$\Rightarrow \quad \frac{a}{\sin A} = \frac{b}{\sin B} = \frac{c}{\sin C} = 2R \qquad \text{...(i)}$$

Now, $a\cot A + b\cot B + c\cot C$
$$= \frac{a}{\sin A} \cdot \cos A + \frac{b}{\sin B} \cdot \cos B + \frac{c}{\sin C} \cdot \cos C$$
$$= 2R(\cos A + \cos B + \cos C) \qquad \text{[from Eq. (i)]}$$
$$= 2R\left(1 + 4\sin\frac{A}{2}\sin\frac{B}{2}\sin\frac{C}{2}\right)$$
$$= 2R\left(1 + \frac{r}{R}\right) \qquad \left(\because r = 4R\sin\frac{A}{2}\sin\frac{B}{2}\sin\frac{C}{2}\right)$$
$$= 2(R + r)$$

16. (d) $\because$ In ΔABC, the radius of circumcircle $(R) = AB = AC$

We know that,
$$\frac{a}{2\sin A} = \frac{b}{2\sin B} = \frac{c}{2\sin C} = R (= b = c)$$

If
$$\frac{b}{2\sin B} = b$$
$$\Rightarrow \quad \sin B = \frac{1}{2} = \sin\frac{\pi}{6}$$
$$\therefore \quad B = \frac{\pi}{6} = C$$

Now,
$$A = \pi - (B + C)$$
$$= \pi - 2 \times \frac{\pi}{6} = \pi - \frac{\pi}{3} = \frac{2\pi}{3}$$

17. (a) We have, p, q, r be the sides of a ΔPQR, then
$$R = \frac{pqr}{4\Delta} \quad \text{and} \quad r = \frac{\Delta}{s}$$

Now,
$$\frac{R}{r} = \frac{pqr}{4\Delta} \cdot \frac{s}{\Delta} = \frac{pqr}{4\Delta^2} \cdot s \qquad \text{...(i)}$$

Let
$$p = 4\lambda, q = 5\lambda, r = 6\lambda$$

Then,
$$s = \frac{p+q+r}{2} = \frac{4\lambda + 5\lambda + 6\lambda}{2}$$
$$s = \frac{15\lambda}{2}$$

and using Hero's formula
$$\Delta^2 = s(s-p)(s-q)(s-r)$$
$$= \frac{15\lambda}{2}\left(\frac{15\lambda}{2} - 4\lambda\right)\left(\frac{15\lambda}{2} - 5\lambda\right)\left(\frac{15\lambda}{2} - 6\lambda\right)$$
$$= \frac{15\lambda}{2} \cdot \frac{7\lambda}{2} \cdot \frac{5\lambda}{2} \cdot \frac{3\lambda}{2}$$
$$\Delta^2 = \frac{225 \times 7\lambda^4}{16}$$

From Eq. (i),
$$\frac{R}{r} = \frac{4\lambda \cdot 5\lambda \cdot 6\lambda}{\frac{4 \times 225 \times 7\lambda^4}{16}} \cdot \frac{15\lambda}{2} = \frac{16}{7}$$
$$\therefore \quad R : r = 16 : 7$$

18. (b) In ΔABC, $\angle C = 90°$, then using Pythagoras theorem

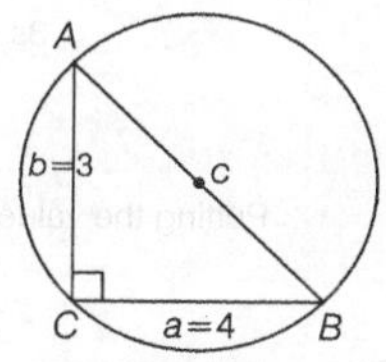

$$c = \sqrt{3^2 + 4^2}$$
$$c = 5$$

We know that,
$$R = \frac{c}{2\sin C} = \frac{5}{2 \times \sin 90°} = \frac{5}{2}$$
$$\because \quad s = \frac{a+b+c}{2} = \frac{3+4+5}{2} = 6$$

and
$$\Delta = \frac{1}{2} \times a \times b = \frac{1}{2} \times 3 \times 4 = 6$$
$$\because \quad r = \frac{\Delta}{s} = \frac{6}{6} = 1$$
$$\therefore \quad \frac{R}{r} = \frac{\frac{5}{2}}{1} = \frac{5}{2}$$

19. (a) In an n-sided regular polygon
$$\angle BOC = \frac{2\pi}{n}$$
$$BL = LC = \frac{1}{2}BC$$

and
$$\angle BOL = \angle COL = \frac{1}{2}\angle BOC$$
$$\therefore \quad \angle BOL = \angle COL = \frac{\pi}{n}$$

Let circumradius $(OB) = R$

and inradius $(OL) = r$

In ΔOLB,

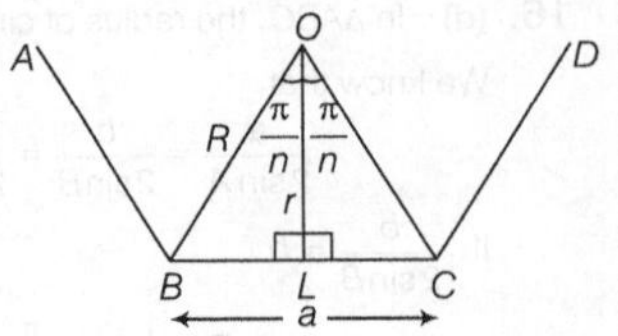

$$r = \frac{a}{2}\cot\frac{\pi}{n}$$

and $$R = \frac{a}{2}\operatorname{cosec}\frac{\pi}{n}$$

$\therefore$ $$r + R = \frac{a}{2}\cot\frac{\pi}{n} + \frac{a}{2}\operatorname{cosec}\frac{\pi}{n}$$

$$= \frac{a}{2}\left(\frac{\cos\frac{\pi}{n}}{\sin\frac{\pi}{n}} + \frac{1}{\sin\frac{\pi}{n}}\right)$$

$$= \frac{a}{2}\left(\frac{1+\cos\frac{\pi}{n}}{\sin\frac{\pi}{n}}\right) = \frac{a}{2}\cdot\frac{2\cos^2\frac{\pi}{2n}}{2\sin\frac{\pi}{2n}\cdot\cos\frac{\pi}{2n}}$$

$\therefore$ $$r + R = \frac{a}{2}\cot\frac{\pi}{2n}$$

20. (b,d) We know that,

$$r_1 = \frac{\Delta}{s-a},\ r_2 = \frac{\Delta}{s-b},\ r_3 = \frac{\Delta}{s-c}$$

$$2r_2 = \frac{2\Delta}{s-b},\ 3r_3 = \frac{3\Delta}{s-c}$$

We have,

$$r_1 = 2r_2 = 3r_3 = k \text{ (let)}$$

$\Rightarrow$ $$\frac{\Delta}{s-a} = \frac{2\Delta}{s-b} = \frac{3\Delta}{s-c} = k$$

$\Rightarrow$ $$s - a = \frac{\Delta}{k} \quad \text{...(i)}$$

$$s - b = \frac{2\Delta}{k} \quad \text{...(ii)}$$

$$s - c = \frac{3\Delta}{k} \quad \text{...(iii)}$$

Now, $$s - a + s - b + s - c = \frac{\Delta}{k} + \frac{2\Delta}{k} + \frac{3\Delta}{k}$$

$$3s - (a + b + c) = \frac{6\Delta}{k}$$

$$s = \frac{6\Delta}{k}$$

Putting the value of s in Eqs. (i), (ii) and (iii), we get

$$a = \frac{5\Delta}{k},\ b = \frac{4\Delta}{k},\ c = \frac{3\Delta}{k}$$

$\therefore$ $$a : b : c = 5 : 4 : 3$$

So, $$a : b = 5 : 4 \text{ and } a : c = 5 : 3$$

21. (b,d) $\because$ $\Delta = \frac{1}{2}ap_1 = \frac{1}{2}bp_2 = \frac{1}{2}cp_3$ (by figure)

$\Rightarrow$ $$p_1 = \frac{2\Delta}{a},\ p_2 = \frac{2\Delta}{b},\ p_3 = \frac{2\Delta}{c}$$

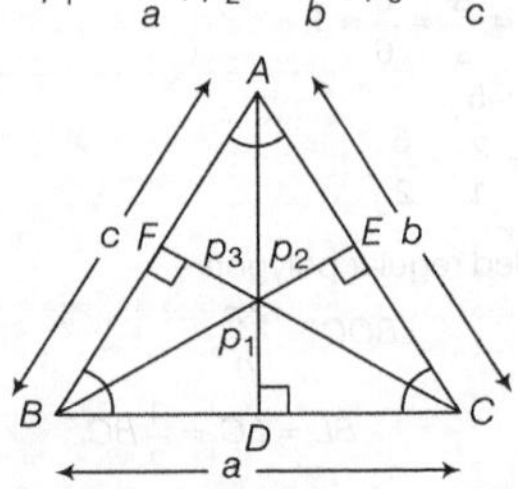

But given that a, b, c are in AP.

Then, $\frac{1}{a}, \frac{1}{b}, \frac{1}{c}$ are in HP.

$\Rightarrow$ $\frac{2\Delta}{a}, \frac{2\Delta}{b}, \frac{2\Delta}{c}$ are also in HP.

$\Rightarrow$ p_1, p_2, p_3 are in HP. [option (b)]

$\because$ $$p_1 = \frac{2\Delta}{a},\quad p_2 = \frac{2\Delta}{b},\quad p_3 = \frac{2\Delta}{c}$$

$\Rightarrow$ $$p_1 = \frac{2\Delta}{2R\sin A},\quad p_2 = \frac{2\Delta}{2R\sin B},\quad p_3 = \frac{2\Delta}{2R\sin C}$$

$\Rightarrow$ $$\frac{1}{p_1} = \frac{R\sin A}{\Delta},\quad \frac{1}{p_2} = \frac{R\sin B}{\Delta},\quad \frac{1}{p_3} = \frac{R\sin C}{\Delta}$$

$\because a, b, c$ are in AP.

Then $\sin A, \sin B, \sin C$ are also in AP.

$\Rightarrow$ $$2\sin B = \sin A + \sin C \quad \text{...(i)}$$

Now, $$\frac{1}{p_1} + \frac{1}{p_2} + \frac{1}{p_3} = \frac{R}{\Delta}(\sin A + \sin B + \sin C)$$

$$= \frac{R}{\Delta}3\sin B \quad \text{[from Eq. (i)]}$$

$$\le \frac{3R}{\Delta}$$

22. (a,c,d) $\because$ $$R = \frac{b}{2\sin\alpha} = \frac{b}{2}\operatorname{cosec}\alpha \quad \text{... (i)}$$

[option (a)]

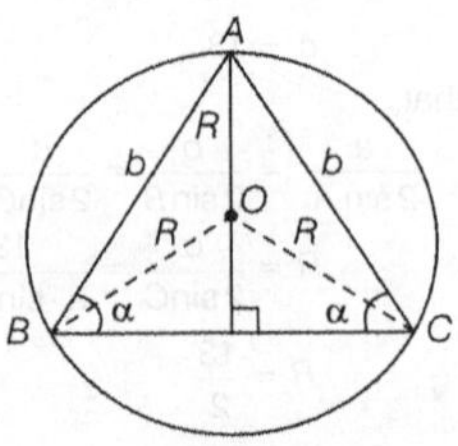

$\because$ $\angle B = \alpha$, $\angle C = \alpha$, then $\angle A = 180° - 2\alpha$

$\therefore$ $$\Delta = \frac{1}{2}b\cdot b\sin(180° - 2\alpha) = \frac{1}{2}b^2\sin 2\alpha$$

and $$r = 4R\sin\frac{\alpha}{2}\sin\frac{\alpha}{2}\sin(90° - \alpha)$$

$$= 4\cdot\frac{b}{2\sin\alpha}\cdot\sin^2\frac{\alpha}{2}\cdot\cos\alpha\cdot\frac{\sin\alpha}{\sin\alpha} \quad \text{[from Eq. (i)]}$$

$$= \frac{b\cdot 2\sin^2\frac{\alpha}{2}\cdot 2\sin\alpha\cdot\cos\alpha}{2\cdot\sin^2\alpha}$$

$$= \frac{b\,(1-\cos\alpha)\sin 2\alpha}{2\,(1-\cos^2\alpha)}$$

$$r = \frac{b\sin 2\alpha}{2(1+\cos\alpha)} \quad \text{[option (c)]}$$

$\because$ $$OI = \sqrt{R^2 - 2Rr} = R\sqrt{1 - \frac{2r}{R}}$$

$$= R\sqrt{1 - 4\cos\alpha + 4\cos^2\alpha}$$

$$= R\ (2\cos\alpha - 1)$$

$$= R\left[2\left(2\cos^2\frac{\alpha}{2} - 1\right) - 1\right]$$

$$= R\left[4\cos^2\frac{\alpha}{2} - 3\right]$$

$$= \frac{R\left[4\cos^3\frac{\alpha}{2} - 3\cos\frac{\alpha}{2}\right]}{\cos\frac{\alpha}{2}}$$

$$= \frac{b}{2\sin\alpha}\cdot\frac{\cos\frac{3\alpha}{2}}{\cos\frac{\alpha}{2}} \quad \text{[from Eq. (i)]}$$

$\therefore$ $$OI = \left|\frac{b\cos\frac{3\alpha}{2}}{2\sin\alpha\cos\frac{\alpha}{2}}\right| \quad \text{[option (d)]}$$

23. (c) We have, $x^2 + y^2 = 9$ ...(i)

and $y^2 = 8x$ ...(ii)

On solving Eqs. (i) and (ii), we get P and Q

i.e. $P \equiv (1, 2\sqrt{2})$ and $Q \equiv (1, -2\sqrt{2})$

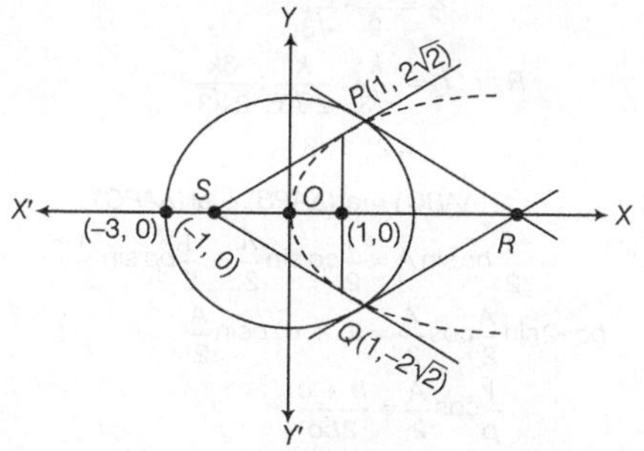

So, $PQ = \sqrt{(4\sqrt{2})^2 + 0} = 4\sqrt{2}$

$$\text{ar}(\Delta PQR) = \frac{1}{2} 4\sqrt{2} \cdot 8 = 16\sqrt{2}$$

$$\text{ar}(\Delta PQS) = \frac{1}{2} 4\sqrt{2} \cdot 2 = 4\sqrt{2}$$

$\therefore$ $\text{ar}(\Delta PQS) : \text{ar}(\Delta PQR) = 4\sqrt{2} : 16\sqrt{2}$

$= 1 : 4$

24. (b) Let $\angle PSR = \theta$, then $\sin\theta = \dfrac{2\sqrt{2}}{2\sqrt{3}}$

$\therefore$ $\sin\theta = \dfrac{PR}{2R}$

$\Rightarrow$ $PR = 6\sqrt{2} = 2R\sin\theta$

$\therefore$ $R = 3\sqrt{3}$

25. (d) $\because$ $\Delta = 16\sqrt{2}$

and $s = \dfrac{6\sqrt{2} + 6\sqrt{2} + 4\sqrt{2}}{2} = 8\sqrt{2}$

$\because$ $r = \dfrac{\Delta}{s} = \dfrac{16\sqrt{2}}{8\sqrt{2}}$

$\therefore$ $r = 2$

26. (b) By definition incentre of ΔABC is the orthocentre of the excentral $\Delta I_1I_2I_3$.

27. (a) Let $\angle I_3I_1I_2 = \theta$

Then, angle of pedal, $\Delta = \pi - 2\theta = A$

$\Rightarrow$ $\theta = \dfrac{\pi}{2} - \dfrac{A}{2}$

Similarly, other angles are $\dfrac{\pi}{2} - \dfrac{B}{2}$ and $\dfrac{\pi}{2} - \dfrac{C}{2}$.

28. (b) Sides of pedal Δ

$\Rightarrow$ $I_2I_3 \cos\theta = BC$

$$I_2I_3 = \frac{a}{\cos\left(\frac{\pi}{2} - \frac{A}{2}\right)}$$

$$I_2I_3 = 4R\cos\frac{A}{2}$$

Similarly, other sides are

$$4R\cos\frac{B}{2},\ 4R\cos\frac{C}{2}$$

29. (b) $II_1 = 4R\sin\dfrac{A}{2}$ and $I_2I_3 = 4R\cos\dfrac{A}{2}$

$\therefore$ $(II_1)^2 + (I_2I_3)^2 = 16R^2$

30. (3) It is given that $C_1L = C_2M = 1$

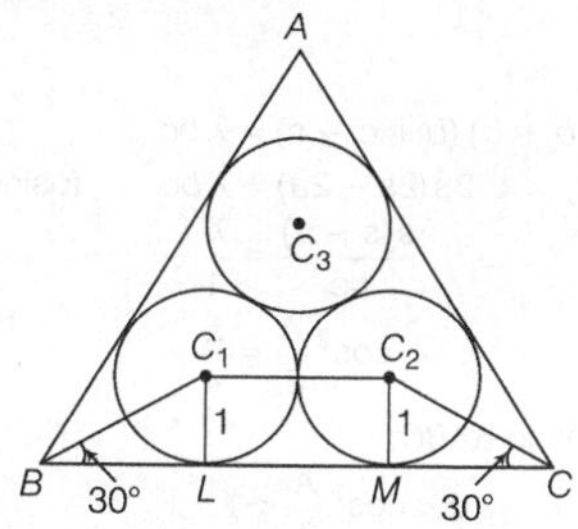

In ΔBLC_1 and ΔCMC_2

$$\cot 30° = \frac{BL}{C_1L} \quad \text{and} \quad \cot 30° = \frac{CM}{C_2M}$$

$$\sqrt{3} = \frac{BL}{1} \quad \text{and} \quad \sqrt{3} = \frac{CM}{1}$$

$\therefore$ $BL = CM = \sqrt{3}$

and $LM = C_1C_2 = 1 + 1 = 2$

$\therefore$ $BC = BL + LM + MC$

$= \sqrt{3} + 2 + \sqrt{3} = 2(1 + \sqrt{3})$

We know that area of equilateral Δ

$$= \frac{\sqrt{3}}{4}(\text{Side})^2$$

$$= \frac{\sqrt{3}}{4}[2(1 + \sqrt{3})]^2$$

$$= \frac{\sqrt{3}}{4} \cdot 4(1 + 3 + 2\sqrt{3})$$

$$= \sqrt{3}(4 + 2\sqrt{3})$$

$= 6 + 4\sqrt{3}$ sq units

But $\Delta = 6 + 4\sqrt{p}$ (given)

$\therefore$ $p = 3$

31. (3) We have, $r_1 = 2r_2 = 2r_3$

$$\frac{\Delta}{s-a} = \frac{2\Delta}{s-b} = \frac{3\Delta}{s-c} = \lambda \text{ (let)}$$

$\therefore$ $s - a = \dfrac{\Delta}{\lambda}, s - b = \dfrac{2\Delta}{\lambda}, s - c = \dfrac{3\Delta}{\lambda}$

Now, $s - a + s - b + s - c = \dfrac{\Delta}{\lambda} + \dfrac{2\Delta}{\lambda} + \dfrac{3\Delta}{\lambda}$

$$3s - (a + b + c) = \frac{6\Delta}{\lambda}$$

$s = \dfrac{6\Delta}{\lambda}$ $\quad (\because a + b + c = 2s)$

$\therefore$ $s - a = \dfrac{\Delta}{\lambda}$

$$a = s - \frac{\Delta}{\lambda} = \frac{6\Delta}{\lambda} - \frac{\Delta}{\lambda} = \frac{5\Delta}{\lambda}$$

$$b = s - \frac{2\Delta}{\lambda} = \frac{6\Delta}{\lambda} - \frac{2\Delta}{\lambda} = \frac{4\Delta}{\lambda}$$

$$c = s - \frac{3\Delta}{\lambda} = \frac{6\Delta}{\lambda} - \frac{3\Delta}{\lambda} = \frac{3\Delta}{\lambda}$$

$\therefore$ $a : b : c = 5 : 4 : 3$

Thus, $\dfrac{a}{b} + \dfrac{b}{c} + \dfrac{c}{a} = \dfrac{5}{4} + \dfrac{4}{3} + \dfrac{3}{5}$

$$= \frac{75 + 80 + 36}{60} = \frac{191}{60}$$

But $\dfrac{a}{b} + \dfrac{b}{c} + \dfrac{c}{a} = \dfrac{191}{20\lambda}$ (given)

$\therefore$ $\lambda = 3$

Revisal Problems (JEE Main)

1. (c) We have,

$$(a+b+c)(b+c-a) = \lambda bc$$

$$\Rightarrow \quad 2s(2s-2a) = \lambda bc \qquad (\text{using } 2s = a+b+c)$$

$$\Rightarrow \quad \frac{s(s-a)}{bc} = \frac{\lambda}{4}$$

$$\Rightarrow \quad \cos^2\frac{A}{2} = \frac{\lambda}{4}$$

We know that, in ΔABC,

$$0 < \cos^2\frac{A}{2} < 1$$

$$0 < \frac{\lambda}{4} < 1$$

$$\therefore \quad 0 < \lambda < 4$$

2. (a) It is given that, $r = r_1 + r_3 - r_2$

$$\frac{\Delta}{s} = \frac{\Delta}{s-a} + \frac{\Delta}{s-c} - \frac{\Delta}{s-b}$$

$$\frac{1}{s} + \frac{1}{s-b} = \frac{1}{s-a} + \frac{1}{s-c}$$

$$\frac{2s-b}{s(s-b)} = \frac{2s-a-c}{(s-a)(s-c)} \qquad (\because 2s = a+b+c)$$

$$\frac{2s-b}{b} = \frac{s(s-b)}{(s-a)(s-c)}$$

$$\frac{2s}{b} - 1 = \cot^2\frac{B}{2}$$

$$\Rightarrow \quad \frac{s}{b} = \frac{1}{2}\left(1+\cot^2\frac{B}{2}\right) \qquad \ldots\text{(i)}$$

We have, $\quad \dfrac{\pi}{3} < B < \pi$

$$\frac{\pi}{6} < \frac{B}{2} < \frac{\pi}{2}$$

$$\Rightarrow \quad 0 < \cot\frac{B}{2} < \sqrt{3}$$

$$\Rightarrow \quad 0 < \cot^2\frac{B}{2} < 3$$

$$\Rightarrow \quad \frac{1}{2} < \frac{1}{2}\left(1+\cot^2\frac{B}{2}\right) < 2$$

$$\therefore \quad \frac{1}{2} < \frac{s}{r} < 2, \text{ i.e. } \frac{s}{r} \in \left(\frac{1}{2}, 2\right)$$

3. (c) In equilateral ΔABC,

$$A = B = C = 60°$$

and $\quad AB = BC = CA = k$ (let)

$$\therefore \quad s = \frac{3k}{2}$$

Now, $\quad R = \dfrac{a}{2\sin A} = \dfrac{k}{2\sin 60°}$

$$\left(\text{using } \frac{a}{2\sin A} = \frac{b}{2\sin B} = \frac{c}{2\sin C} = R\right)$$

$$R = \frac{k}{2\cdot\frac{\sqrt{3}}{2}} = \frac{k}{\sqrt{3}}$$

$$R = \frac{k}{\sqrt{3}} \qquad \ldots\text{(i)}$$

We know that,

$$r = (s-a)\tan\frac{A}{2}$$

$$= \left(\frac{3k}{2} - k\right)\tan 30° = \frac{k}{2}\cdot\frac{1}{\sqrt{3}} = \frac{k}{2\sqrt{3}}$$

$$r = \frac{k}{2\sqrt{3}} \qquad \ldots\text{(ii)}$$

and $\quad r_2 = s\tan\dfrac{B}{2} = \dfrac{3k}{2}\tan 30°$

$$r_2 = \frac{3k}{2}\cdot\frac{1}{\sqrt{3}} \qquad \ldots\text{(iii)}$$

$$\therefore \quad R : r : r_2 = \frac{k}{\sqrt{3}} : \frac{k}{2\sqrt{3}} : \frac{3k}{2\sqrt{3}}$$

$$= 2 : 1 : 3$$

4. (c) $\because \quad \text{ar}(\Delta ABC) = \text{ar}(\Delta APB) + \text{ar}(\Delta APC)$

$$\Rightarrow \quad \frac{1}{2}bc\sin A = \frac{1}{2}cp\sin\frac{A}{2} + \frac{1}{2}bp\sin\frac{A}{2}$$

$$\Rightarrow \quad bc\cdot 2\sin\frac{A}{2}\cos\frac{A}{2} = (b+c)p\sin\frac{A}{2}$$

$$\Rightarrow \quad \frac{1}{p}\cos\frac{A}{2} = \frac{b+c}{2bc}$$

$$\Rightarrow \quad \frac{1}{p}\cos\frac{A}{2} = \frac{1}{2}\left(\frac{1}{b}+\frac{1}{c}\right)$$

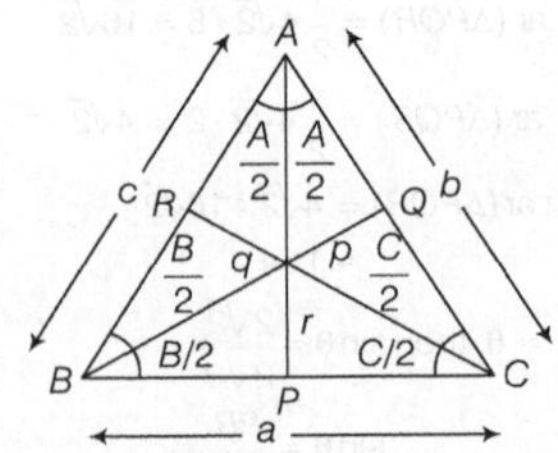

Similarly, $\quad \dfrac{1}{q}\cos\dfrac{B}{2} = \dfrac{1}{2}\left(\dfrac{1}{a}+\dfrac{1}{c}\right)$

and $\quad \dfrac{1}{r}\cos\dfrac{C}{2} = \dfrac{1}{2}\left(\dfrac{1}{a}+\dfrac{1}{b}\right)$

Now, $\quad \dfrac{1}{p}\cos\dfrac{A}{2} + \dfrac{1}{q}\cos\dfrac{B}{2} + \dfrac{1}{r}\cos\dfrac{C}{2} = \dfrac{1}{2}\left(\dfrac{2}{a}+\dfrac{2}{b}+\dfrac{2}{c}\right)$

$$= \frac{1}{a} + \frac{1}{b} + \frac{1}{c}$$

$$= a^{-1} + b^{-1} + c^{-1}$$

5. (a) We have, $\cos A\cos B + \sin A\sin B\sin C = 1$

$$\sin A\sin B\sin C = 1 - \cos A\cos B$$

$$\sin C = \frac{1-\cos A\cos B}{\sin A\sin B}$$

$$\therefore \quad \sin C \le 1$$

$$\Rightarrow \quad \frac{1-\cos A\cos B}{\sin A\sin B} \le 1 \qquad (\because \sin\theta \le 1)$$

$$1 \le \cos A\cos B + \sin A\sin B$$

$$\cos(A-B) \ge 1$$

But $\cos(A-B)$ cannot be greater than 1.

$$\therefore \quad \cos(A-B) = 1$$

$$A - B = 0$$

$$\Rightarrow \quad A = B$$

$$\therefore \quad \sin C = 1$$

$$\Rightarrow \quad \sin C = \sin 90°$$

$\Rightarrow C = 90°$ and $A = B$, then

$$A = B = 45°$$

Now, using sine rule

$$a : b : c = \sin A : \sin B : \sin C$$

$$= \sin 45° : \sin 45° : \sin 90°$$

$$= \frac{1}{\sqrt{2}} : \frac{1}{\sqrt{2}} : 1$$

$$\therefore \quad a : b : c = 1 : 1 : \sqrt{2}$$

6. (a) $\because$ $\sin A = x, \cos A = \pm\sqrt{1-x^2}$

$\cos B = y,\ \sin B = \pm\sqrt{1-y^2}$

$\because$ $\sin C = \sin(\pi - \overline{A+B})$

$= \sin(A+B)$

$= \sin A\cos B + \cos A \sin B$

$\sin C = xy \pm \sqrt{(1-x^2)(1-y^2)}$ ...(i)

and $\cos C = -\cos(A+B)$

$= -(\cos A\cos B - \sin A \sin B)$

$= \sin A \sin B - \cos A\cos B$

Now, $\cos C = \pm\sqrt{1-y^2}\cdot x \pm y\cdot\sqrt{1-x^2}$

$\Rightarrow$ $\cos^2 C = x^2(1-y^2) + y^2(1-x^2) \pm 2xy\sqrt{(1-x^2)(1-y^2)}$

$= x^2 + y^2 - 2x^2y^2 \pm 2xy\sqrt{(1-x^2)(1-y^2)}$

$x^2 + y^2 - 2xy\,[xy \pm \sqrt{(1-x^2)(1-y^2)}]$

$\cos^2 C = x^2 + y^2 - 2xy\sin C$ [from Eq. (i)]

7. (c) We have,

$2a^2 + 4b^2 + c^2 = 4ab + 2ac$

$2a^2 + 4b^2 + c^2 - 4ab - 2ac = 0$

$[a^2 + (2b)^2 - 4ab] + (a^2 + c^2 - 2ac) = 0$

$(a-2b)^2 + (a-c)^2 = 0$

It is possible only when $a - 2b = 0$ and $a - c = 0$

$a = 2b = c$

$\frac{a}{2} = \frac{b}{1} = \frac{c}{2} = \lambda$ (let)

$a = 2\lambda,\ b = \lambda,\ c = 2\lambda$

Now, using cosine rule, we get

$\cos B = \frac{a^2 + c^2 - b^2}{2ac}$

$= \frac{4\lambda^2 + 4\lambda^2 - \lambda^2}{2\times 2\lambda \times 2\lambda} = \frac{7}{8}$

8. (c) We have, $\lambda = \frac{(a+b+c)^2}{ab+bc+ca}$ ...(i)

We know that, in ΔABC,

$a + b > c,\ b + c > a,\ c + a > b$

$\therefore$ $a > |b - c|,\ b > |c - a|,\ c > |a - b|$

On squaring and adding, we get

$a^2 + b^2 + c^2 > 2(a^2 + b^2 + c^2) - 2(ab + bc + ca)$

$\Rightarrow$ $a^2 + b^2 + c^2 < 2(ab + bc + ca)$

$\Rightarrow$ $a^2 + b^2 + c^2 + 2(ab + bc + ca) < 4(ab + bc + ca)$

$\Rightarrow$ $\frac{(a+b+c)^2}{ab+bc+ca} < 4$

$\therefore$ $\lambda < 4$ [from Eq. (i)]

Now, we know that,

$(a-b)^2 + (b-c)^2 + (c-a)^2 \geq 0$

$2(a^2 + b^2 + c^2) - 2(ab + bc + ca) \geq 0$

$a^2 + b^2 + c^2 + 2(ab + bc + ca) \geq 3(ab + bc + ca)$

$\frac{(a+b+c)^2}{ab+bc+ca} \geq 3$

$\lambda \geq 3$ [from Eq. (i)]

$\therefore$ $\lambda \in [3, 4)$

9. (c) Circles with radii 3, 4, 5 touch each other externally and the point of intersection of tangents is P, then P is the incentre of $\Delta C_1C_2C_3$.

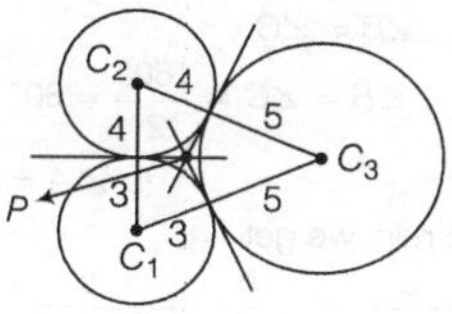

So, distance of point P from the points of contact = Inradius of $\Delta C_1C_2C_3$

and $C_1C_2 = 7,\ C_2C_3 = 9,\ C_3C_1 = 8$

$\Rightarrow$ $a = 7,\ b = 9,\ c = 8$

We know that, $r = \frac{\Delta}{s} = \frac{\sqrt{s(s-a)(s-b)(s-c)}}{s}$

$r = \sqrt{\frac{(s-a)(s-b)(s-c)}{s}}$

$= \sqrt{\frac{(12-7)(12-9)(12-8)}{12}}$ $\left(\because s = \frac{a+b+c}{2} = \frac{7+9+8}{2}\right.$ $\left.\therefore s = 12\right)$

$\therefore$ $r = \sqrt{\frac{5\cdot 3\cdot 4}{12}} = \sqrt{5}$

10. (c) We have, $a(b-c)x^2 + b(c-a)x + c(a-b) = 0$

If we put $x = 1$ and it satisfies the equation, then $x = 1$ is its root, so, other root is also 1.

Now, product of roots $= \frac{c(a-b)}{a(b-c)}$

$1 = \frac{ca - bc}{ab - ca}$

$\Rightarrow$ $ab - ca = ca - bc$

$\Rightarrow$ $2ca = ab + bc$

$\Rightarrow$ $b = \frac{2ca}{a+c}$

$\therefore$ a, b, c are in HP.

$\Rightarrow \frac{1}{a}, \frac{1}{b}, \frac{1}{c}$ are in AP.

$\Rightarrow \frac{s}{a}, \frac{s}{b}, \frac{s}{c}$ are also in AP.

$\Rightarrow \frac{s}{a} - 1, \frac{s}{b} - 1, \frac{s}{c} - 1$ are also in AP.

$\Rightarrow \frac{s-a}{a}, \frac{s-b}{b}, \frac{s-c}{c}$ are also in AP.

Now, multiplying in each term by $\frac{abc}{(s-a)(s-b)(s-c)}$, we get

$\Rightarrow$ $\frac{bc}{(s-b)(s-c)}, \frac{ca}{(s-a)(s-c)}, \frac{ab}{(s-a)(s-b)}$

are also in AP.

$\Rightarrow$ $\frac{(s-b)(s-c)}{bc}, \frac{(s-c)(s-a)}{ca}, \frac{(s-a)(s-b)}{ab}$

are also in HP.

$\therefore \sin^2\frac{A}{2}, \sin^2\frac{B}{2}, \sin^2\frac{C}{2}$ are also in HP.

Revisal Problems (JEE Advanced)

1. (c) We have, in ΔABC, $b = c$, then

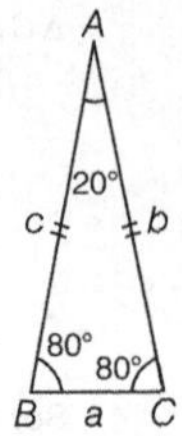

$$\angle B = \angle C$$

$$\therefore \quad \angle B = \angle C = \frac{160°}{2} = 80°$$

$$(\because \angle A = 20°, \text{ given})$$

Now, using sine rule, we get

$$\frac{a}{\sin 20°} = \frac{b}{\sin 80°} = \frac{c}{\sin 80°}$$

If $\quad \dfrac{a}{\sin 20°} = \dfrac{b}{\sin 80°} = \dfrac{b}{\sin(90° - 10°)}$

$$\Rightarrow \quad \frac{a}{\sin 20°} = \frac{b}{\cos 10°}$$

$$\Rightarrow \quad a = \frac{b \sin 20°}{\cos 10°} = \frac{2b \cdot \sin 10° \cos 10°}{\cos 10°}$$

$$\Rightarrow \quad a = 2b \sin 10° \quad \ldots \text{(i)}$$

Thus, $\quad a^3 + b^3 = (2b \sin 10°)^3 + b^3$

$$= 8b^3 \sin^3 10° + b^3$$

$$= 2b^3 (4\sin^3 10°) + b^3$$

$$= 2b^3 (3\sin 10° - \sin 30°) + b^3$$

$$(\because \sin 3A = 3\sin A - 4\sin^3 A)$$

$$= 3b^2 (2b \sin 10°) - b^3 + b^3$$

$$= 3b^2 \cdot a \quad \text{[from Eq. (i)]}$$

$$= 3c^2 a \quad (\because b = c)$$

2. (b) $\because a, b, c$ are in AP, then

$$2b = a + c$$

$$2 \times 5 = 3 + c \quad (\because a = 3,\ b = 5)$$

$$c = 10 - 3 = 7$$

$$\because \quad \cos C = \frac{a^2 + b^2 - c^2}{2ab}$$

$$= \frac{3^2 + 5^2 - 7^2}{2 \times 3 \times 5}$$

$$= \frac{-1}{2} = \cos 120°$$

$\therefore C = 120°$, then $A, B < 60°$, then $\tan A, \tan B > 0$

$$\Rightarrow \quad \tan A + \tan B > 0 \quad \ldots \text{(i)}$$

We know that,

$$\tan A + \tan B + \tan C = \tan A \tan B \tan C$$

$$\Rightarrow \quad \tan A + \tan B + \tan 120° = \tan A \tan B \tan 120°$$

$$\Rightarrow \quad \tan A + \tan B - \sqrt{3} = -\sqrt{3} \tan A \tan B$$

$$\Rightarrow \quad \tan A + \tan B = \sqrt{3}(1 - \tan A \tan B) \quad \ldots \text{(ii)}$$

$$\therefore \quad \sqrt{3}(1 - \tan A \tan B) > 0 \quad \text{[from Eq. (i)]}$$

$$\tan A \tan B < 1 \quad \ldots \text{(iii)}$$

We know that,

$$\text{AM} \ge \text{GM}$$

$$\frac{\tan A + \tan B}{2} \ge \sqrt{\tan A \tan B}$$

$$\sqrt{3}(1 - \tan A \tan B) \ge 2\sqrt{\tan A \tan B} \quad \text{[from Eq. (ii)]}$$

Let $\tan A \tan B = y$

$$\sqrt{3}(1 - y) \ge 2\sqrt{y}$$

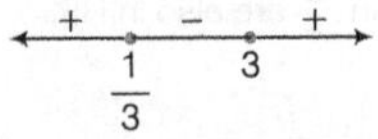

$$3(1 + y^2 - 2y) \ge 4y$$

$$3y^2 - 10y + 3 \ge 0$$

$$(3y - 1)(y - 3) \ge 0$$

$$y \le \frac{1}{3} \quad \text{and} \quad y \ge 3$$

$$\tan A \tan B \le \frac{1}{3}, \ \tan A \tan B \ge 3 \quad \text{(neglecting)}$$

$\therefore$ Maximum value of $\tan A \tan B = \dfrac{1}{3}$

3. (c) $\because D$ divides AB in 1 : 3.

Let $AB = a$, then $AD = \dfrac{a}{4}$, $BD = \dfrac{3}{4}a$

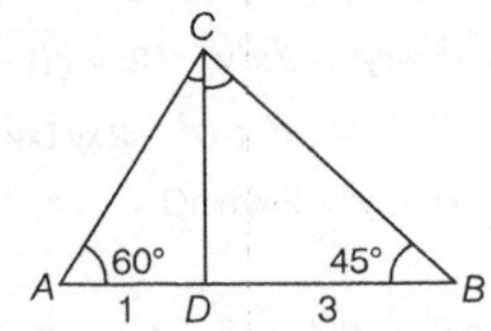

Using sine rule, in ΔACD and ΔBCD,

$$\frac{CD}{\sin 60°} = \frac{AD}{\sin(\angle ACD)}$$

and $\quad \dfrac{CD}{\sin 45°} = \dfrac{BD}{\sin(\angle BCD)}$

$$CD = \frac{\sqrt{3}}{2} \cdot \frac{a}{4\sin(\angle ACD)} \quad \ldots \text{(i)}$$

and $\quad CD = \dfrac{1}{\sqrt{2}} \cdot \dfrac{3a}{4\sin(\angle BCD)} \quad \ldots \text{(ii)}$

From Eqs. (i) and (ii), we get

$$\frac{1}{\sqrt{2}} \cdot \frac{3a}{4\sin(\angle BCD)} = \frac{\sqrt{3}a}{2 \times 4 \sin(\angle ACD)}$$

$$\frac{\sin(\angle ACD)}{\sin(\angle BCD)} = \frac{1}{\sqrt{6}}$$

$$\therefore \quad \frac{\sqrt{6}\sin(\angle ACD)}{\sin(\angle BCD)} = 1$$

4. (a) Let $BC = a$ and $CD = b$

$$\text{ar}(ABCD) = \text{ar}(\Delta ABD) + \text{ar}(\Delta BCD)$$

$$4\sqrt{3} = \frac{1}{2} \times 2 \times 5 \cdot \sin 60° + \frac{1}{2} ab \sin 120°$$

$$4\sqrt{3} = \frac{5\sqrt{3}}{2} + \frac{ab\sqrt{3}}{4}$$

$$ab = 6 \quad \ldots \text{(i)}$$

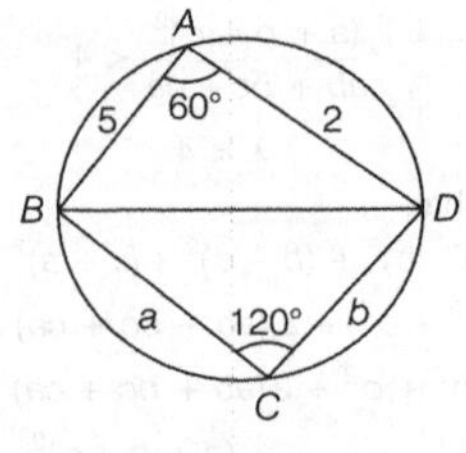

and we know that,

$$\cos A = \frac{2^2 + 5^2 - a^2 - b^2}{2(10 + ab)}$$

$$\Rightarrow \quad \cos 60° = \frac{29 - a^2 - b^2}{2(10 + ab)}$$

$$\frac{1}{2} = \frac{29 - a^2 - b^2}{2(10 + 6)} \quad \text{[from Eq. (i)]}$$

$$\Rightarrow \quad a^2 + b^2 = 13 \quad \ldots\text{(ii)}$$

$$\because \quad (a + b)^2 = a^2 + b^2 + 2ab$$

$$= 13 + 2 \times 6 = 25$$

Now, $a + b = 5$ and $ab = 6$

On solving, we get

$$a = 3, b = 2$$

or $$a = 2, b = 3$$

5. (c) We know that,

$$(a - b)^2 + (b - c)^2 + (c - a)^2 \geq 0$$

$$2(a^2 + b^2 + c^2) \geq 2(ab + bc + ca)$$

$$3(a^2 + b^2 + c^2) \geq a^2 + b^2 + c^2 + 2ab + 2bc + 2ca$$

$$3(a^2 + b^2 + c^2) \geq (a + b + c)^2 > d^2$$

$$\frac{a^2 + b^2 + c^2}{d^2} > \frac{1}{3}$$

6. (a) We have

$$p = \sin A \sin B \sin C$$

$$q = \cos A \cos B \cos C$$

$$\therefore \quad \tan A \tan B \tan C = \frac{p}{q} \quad \ldots\text{(i)}$$

and $\tan A \tan B + \tan B \tan C + \tan C \tan A$

$$= \frac{\sin A \sin B \cos C + \sin B \sin C \cos A + \sin A \cos B \sin C}{\cos A \cos B \cos C}$$

$$= \frac{\sin B(\sin A \cos C + \cos A \sin C) + \sin A \cos B \sin C}{\cos A \cos B \cos C}$$

$$= \frac{\sin B \sin(A + C) + \sin A \cos B \sin C}{\cos A \cos B \cos C}$$

$$= \frac{\sin^2 B + \sin A \cos B \sin C}{\cos A \cos B \cos C}$$

$$= \frac{1 - \cos^2 B + \sin A \cos B \sin C}{\cos A \cos B \cos C}$$

$$= \frac{1 - \cos B(\cos B - \sin A \sin C)}{\cos A \cos B \cos C}$$

$$= \frac{1 + \cos B\,[\cos(A + C) + \sin A \sin C]}{\cos A \cos B \cos C}$$

$$= \frac{1 + \cos B[\cos A \cos C - \sin A \sin C + \sin A \sin C]}{\cos A \cos B \cos C}$$

$$= \frac{1 + \cos B \cos A \cos C}{\cos A \cos B \cos C}$$

$$= \frac{1 + q}{q}$$

We know that, in ΔABC

$$\tan A + \tan B + \tan C = \tan A \tan B \tan C = \frac{p}{q}$$

Let $\tan A = \alpha$, $\tan B = \beta$, $\tan C = \gamma$, then the equation whose roots are α, β, γ is

$$x^3 - (\alpha + \beta + \gamma)x^2 + (\alpha\beta + \beta\gamma + \gamma\alpha)x - \alpha\beta\gamma = 0$$

$$x^3 - \frac{p}{q}x^2 + \left(\frac{1 + q}{q}\right)x - \frac{p}{q} = 0$$

$$\therefore \quad qx^3 - px^2 + (1 + q)x - p = 0$$

7. (a) We have, $F(x + y) = F(x) \cdot F(y), \forall x, y$ and $F(1) = 2$

$$F(2) = F(1 + 1) = F(1) \cdot F(1) = 2 \times 2 = 4$$

$$F(3) = F(1 + 2) = F(1) \cdot F(2) = 2 \times 4 = 8$$

$$\therefore \quad a = F(3) = 8, b = F(1) + F(3) = 2 + 8 = 10$$

$$c = F(2) + F(3) = 4 + 8 = 12$$

Using cosine rule,

$$\cos A = \frac{b^2 + c^2 - a^2}{2bc} = \frac{(10)^2 + (12)^2 - (8)^2}{2 \times 10 \times 12}$$

$$= \frac{100 + 144 - 64}{240} = \frac{180}{240}$$

$$\cos A = \frac{3}{4}$$

$$\because \quad \cos 2A = 2\cos^2 A - 1 = 2\left(\frac{3}{4}\right)^2 - 1$$

$$\cos 2A = \frac{1}{8} = \cos C$$

$$\left(\because \cos C = \frac{a^2 + b^2 - c^2}{2ab} = \frac{64 + 100 - 144}{2 \times 8 \times 10} = \frac{1}{8}\right)$$

$$\therefore \quad 2A = C$$

8. (a) $\because r_1 = \dfrac{\Delta}{s - a}, r_2 = \dfrac{\Delta}{s - b}, r_3 = \dfrac{\Delta}{s - c}$

Now, $$\lambda\left(s\Sigma\frac{1}{a} - 3\right) = \frac{bc}{r_1} + \frac{ca}{r_2} + \frac{ab}{r_3}$$

$$= \frac{bc(s - a)}{\Delta} + \frac{ca(s - b)}{\Delta} + \frac{ab(s - c)}{\Delta}$$

$$= \frac{abc}{\Delta}\left(\frac{s - a}{a} + \frac{s - b}{b} + \frac{s - c}{c}\right)$$

$$= \frac{abc}{\Delta}\left(\frac{s}{a} + \frac{s}{b} + \frac{s}{c} - 3\right)$$

$$= \frac{abc}{\Delta}\left[s\left(\frac{1}{a} + \frac{1}{b} + \frac{1}{c}\right) - 3\right]$$

$$\lambda\left(s\Sigma\frac{1}{a} - 3\right) = \frac{abc}{\Delta}\left(s\,\Sigma\frac{1}{a} - 3\right)$$

$$\therefore \quad \lambda = \frac{abc}{\Delta}$$

9. (b) $\because \cos A + \cos B + \cos C = \dfrac{7}{4}$

$$1 + 4\sin\frac{A}{2}\sin\frac{B}{2}\sin\frac{C}{2} = \frac{7}{4}$$

$$\left(\text{using } \cos A + \cos B + \cos C = 1 + 4\sin\frac{A}{2}\sin\frac{B}{2}\sin\frac{C}{2}\right)$$

$$\sin\frac{A}{2}\sin\frac{B}{2}\sin\frac{C}{2} = \frac{3}{16} \quad \ldots\text{(i)}$$

We know that

$$r = 4R\sin\frac{A}{2}\sin\frac{B}{2}\sin\frac{C}{2}$$

$$= 4R \times \frac{3}{16} = \frac{3R}{4} \quad \text{[from Eq. (i)]}$$

$$\Rightarrow \quad \frac{R}{r} = \frac{4}{3}$$

$$\therefore \quad \frac{27R}{r} = 27 \cdot \frac{4}{3} = 36$$

10. (c) Let $AB = BC = k$

$$\Rightarrow \quad AC = \sqrt{2}k$$

and let θ be the obtuse angle formed by medians.

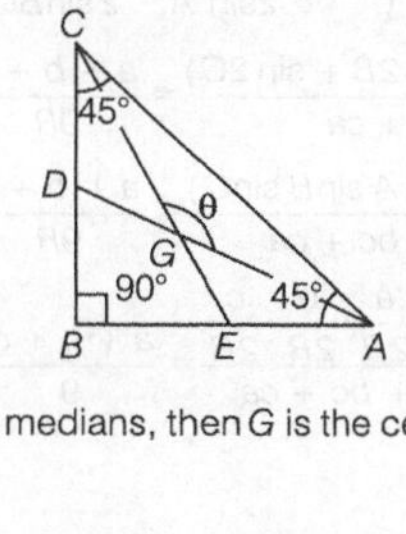

$\because AD$ and CE are medians, then G is the centroid and $BE = \dfrac{k}{2}$.

$\because \quad GA = GC = \frac{2}{3}CE = \frac{2}{3}\sqrt{(BC)^2 + (BE)^2}$

(using Pythagoras theorem in ΔBCE)

$$= \frac{2}{3}\sqrt{k^2 + \frac{k^2}{4}} = \frac{2}{3}\cdot\frac{\sqrt{5}}{2}k = \frac{\sqrt{5}}{3}k$$

Now, applying cosine rule, in ΔAGC,

$$\cos\theta = \frac{(AG)^2 + (GC)^2 - (AC)^2}{2(AG)(GC)}$$

$$= \frac{\frac{5}{9}k^2 + \frac{5}{9}k^2 - 2k^2}{2\cdot\frac{\sqrt{5}}{3}k\cdot\frac{\sqrt{5}}{3}k}$$

$$= \frac{-8k^2}{10k^2} = \frac{-4}{5}$$

11. (a) $\because$ Arc $(BC) = 3$, Arc $(CA) = 4$ and Arc $(AB) = 5$

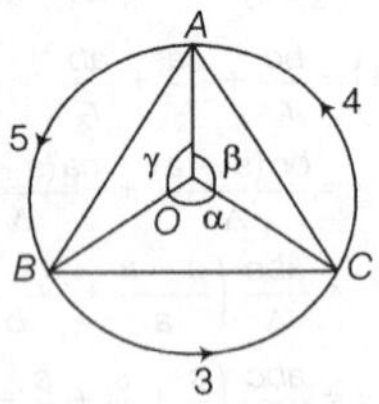

Let r be the radius of circle.

We know that, Arc = Radius × Angle

Then, $3 = r\alpha,\ 4 = r\beta,\ 5 = r\gamma$

$\Rightarrow \quad \alpha = \frac{3}{r}, \beta = \frac{4}{r}, \gamma = \frac{5}{r}$

and $\quad 2\pi r = 3 + 4 + 5$

$\Rightarrow \quad r = \frac{6}{\pi}$

$\because \quad$ ar(ΔABC) = ar(ΔOBC) + ar(ΔOAC) + ar(ΔOAB)

$$= \frac{1}{2}r^2\sin\alpha + \frac{1}{2}r^2\sin\beta + \frac{1}{2}r^2\sin\gamma$$

$$= \frac{r^2}{2}\left(\sin\frac{3}{r} + \sin\frac{4}{r} + \sin\frac{5}{r}\right)$$

$$= \frac{1}{2}\cdot\frac{36}{\pi^2}\left(\sin\frac{\pi}{2} + \sin\frac{2\pi}{3} + \sin\frac{5\pi}{6}\right)$$

$$= \frac{18}{\pi^2}\left(1 + \frac{\sqrt{3}}{2} + \frac{1}{2}\right)$$

$$= \frac{9\sqrt{3}(1+\sqrt{3})}{\pi^2} \text{ sq units}$$

12. (a,b,c) We have,

$$\frac{a\cos A + b\cos B + c\cos C}{a\sin B + b\sin C + c\sin A} = \frac{a+b+c}{9R}$$

$$\frac{2R\sin A\cos A + 2R\sin B\cos B + 2R\sin C\cos C}{a\cdot\frac{b}{2R} + b\cdot\frac{c}{2R} + c\cdot\frac{a}{2R}} = \frac{a+b+c}{9R}$$

$$\left(\text{using } \frac{a}{2\sin A} = \frac{b}{2\sin B} = \frac{c}{2\sin C} = R\right)$$

$$\frac{2R^2(\sin 2A + \sin 2B + \sin 2C)}{ab + bc + ca} = \frac{a+b+c}{9R}$$

$$\frac{2R^2(4\sin A\sin B\sin C)}{ab + bc + ca} = \frac{a+b+c}{9R}$$

$$\frac{8R^3\cdot\frac{a}{2R}\cdot\frac{b}{2R}\cdot\frac{c}{2R}}{ab + bc + ca} = \frac{a+b+c}{9}$$

$\Rightarrow \quad 9abc = (a+b+c)(ab+bc+ca)$

$\Rightarrow \quad a(b-c)^2 + b(c-a)^2 + c(a-b)^2 = 0$

which is possible only when $a - b = 0$,

$b - c = 0$,

and $\quad c - a = 0$

$\therefore \quad a = b = c$

$\therefore$ ΔABC is an equilateral triangle,

i.e. $\quad \angle A = \angle B = \angle C = 60°$

13. (a,b,d) $\because\ r = \frac{\Delta}{s},\ r_1 = \frac{\Delta}{s-a}$

Now, $\quad r_1\cdot r = \frac{\Delta^2}{s(s-a)} = \frac{s(s-a)(s-b)(s-c)}{s(s-a)}$

$$= (s-b)(s-c) = (s-b)^2$$

$$= \frac{(2s-2b)^2}{4} = \frac{(a+b+c-2b)^2}{4}$$

$$= \frac{a^4}{4} \quad \text{[option (d)]}$$

$$= \frac{4R^2\sin^2 A}{4} = R^2\sin^2 A \quad \text{[option (a)]}$$

Also, if $\quad \angle B = \theta$

$\Rightarrow \quad \angle A = \pi - 2\theta$

$\therefore \quad r_1\cdot r = R^2\sin^2(\pi - 2\theta)$

$$= R^2\sin^2 2\theta$$

$$= R^2\sin^2 2B \quad \text{[option (b)]}$$

14. (a,c,d) Given that, $s-a, s-b, s-c$ are in AP.

$\Rightarrow$ a, b, c are in AP. [option (c)]

$\Rightarrow \frac{\Delta}{s-a}, \frac{\Delta}{s-b}, \frac{\Delta}{s-c}$ are in HP.

$\Rightarrow r_1, r_2, r_3$ are in HP. [option (a)]

Also, $\quad \cos A = \frac{b^2 + c^2 - a^2}{2bc}$

$$= \frac{b^2 + c^2 - (2b-c)^2}{2bc}$$

$$= \frac{4c - 3b}{2c} \quad \text{[option (d)]}$$

15. (a,b,c) Let AD be the median, then $BD = CD = \frac{a}{2}$

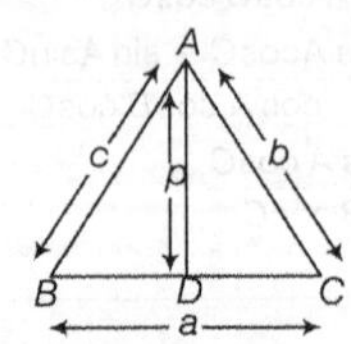

$\because \quad AB^2 + AC^2 = 2(AD^2 + BD^2)$ (using median theorem)

$$c^2 + b^2 = 2\left(p^2 + \frac{a^2}{4}\right)$$

$\Rightarrow \quad b^2 + c^2 = 2p^2 + \frac{a^2}{2}$

$\Rightarrow \quad 4p^2 = 2b^2 + 2c^2 - a^2$ [option (a)]

$$= b^2 + c^2 + (b^2 + c^2 - a^2)$$

$\Rightarrow \quad 4p^2 = b^2 + c^2 + 2bc\cos A$ [option (b)]

$$= (b^2 + c^2 - a^2) + 2bc\cos A + a^2$$

$$= 2bc\cos A + 2bc\cos A + a^2$$

$\therefore \quad 4p^2 = a^2 + 4bc\cos A$ [option (c)]

16. (b,d) Let a,b,c be the sides of ΔABC, then

$$\cos A = \frac{b^2 + c^2 - a^2}{2bc} \quad \text{(using cosine rule)}$$

$$\Rightarrow \quad c^2 - 2b\cos A \cdot c + b^2 - a^2 = 0 \quad \ldots\text{(i)}$$

(which is a quadratic equation in c)

Let c_1 and c_2 be two roots of Eq. (i), then

$$c_1 + c_2 = 2b\cos A \quad \ldots\text{(ii)}$$

$$c_1 \cdot c_2 = b^2 - a^2 \quad \ldots\text{(iii)}$$

We have, $c_1^2 + c_1 \cdot c_2 + c_2^2 = a^2$

$$c_1^2 + c_2^2 + c_1 \cdot c_2 = a^2$$

$$\Rightarrow \quad (c_1 + c_2)^2 - c_1 c_2 = a^2$$

$$\Rightarrow \quad 4b^2\cos^2 A - b^2 + a^2 = a^2 \quad \text{[from Eqs. (ii) and (iii)]}$$

$$\Rightarrow \quad \cos^2 A = \frac{1}{4}$$

$$\Rightarrow \quad \cos A = \pm\frac{1}{2}$$

$$\therefore \quad \angle A = 60^\circ \text{ and } 120^\circ$$

17. (b,c,d) We have, $r_1 = 8, r_2 = 12, r_3 = 24$, then

$$r_1 r_2 + r_2 r_3 + r_3 r_1 = 96 + 288 + 192 = 576$$

We know that,

$$a = \frac{r_1(r_2 + r_3)}{\sqrt{r_1 r_2 + r_2 r_3 + r_3 r_1}} = \frac{8(12 + 24)}{\sqrt{576}} = \frac{8 \times 36}{24} = 12 \quad \text{[option (b)]}$$

$$b = \frac{r_2(r_1 + r_3)}{\sqrt{r_1 r_2 + r_2 r_3 + r_3 r_1}} = \frac{12(8 + 24)}{24} = \frac{12 \times 32}{24} = 16 \quad \text{[option (c)]}$$

$$c = \frac{r_3(r_1 + r_2)}{\sqrt{r_1 r_2 + r_2 r_3 + r_3 r_1}} = \frac{24(8 + 12)}{24} = \frac{24 \times 20}{24} = 20 \quad \text{[option (d)]}$$

Now, $s = \frac{a + b + c}{2} = \frac{12 + 16 + 20}{2} = 24$

We know that, $r_1 r_2 r_3 = rs^2$

$$8 \times 12 \times 24 = r(24)^2$$

$$\therefore \quad r = \frac{8 \times 12 \times 24}{(24)^2} = 4 \text{ [option (a) is not correct]}$$

18. (a,b,c,d) Let r be the inradius and r_1, r_2, r_3 be the exradii, then

$$A = \pi r^2, A_1 = \pi r_1^2, A_2 = \pi r_2^2, A_3 = \pi r_3^2$$

Now, option (a), LHS $= \sqrt{A_1} + \sqrt{A_2} + \sqrt{A_3}$

$$= \sqrt{\pi} r_1 + \sqrt{\pi} r_2 + \sqrt{\pi} r_3$$

$$= \sqrt{\pi}(r_1 + r_2 + r_3) = \text{RHS}$$

Option (b), LHS $= \frac{1}{\sqrt{A_1}} + \frac{1}{\sqrt{A_2}} + \frac{1}{\sqrt{A_3}} = \frac{1}{\sqrt{\pi}}\left(\frac{1}{r_1} + \frac{1}{r_2} + \frac{1}{r_3}\right)$

$$= \frac{1}{\sqrt{\pi}} \cdot \frac{1}{r} = \frac{1}{\sqrt{\pi r^2}} \quad \left(\because \frac{1}{r} = \frac{1}{r_1} + \frac{1}{r_2} + \frac{1}{r_3}\right)$$

$$= \frac{1}{\sqrt{A}} = \text{RHS}$$

Option (c), LHS $= \frac{1}{\sqrt{A_1}} + \frac{1}{\sqrt{A_2}} + \frac{1}{\sqrt{A_3}} = \frac{1}{\sqrt{\pi} \cdot r}$ [using option (b)]

$$\text{RHS} = \frac{s^2}{\sqrt{\pi}\, r_1 r_2 r_3} = \frac{s^2}{\sqrt{\pi} \cdot \frac{\Delta}{s-a} \cdot \frac{\Delta}{s-b} \cdot \frac{\Delta}{s-c}}$$

$$= \frac{s^2 (s-a)(s-b)(s-c)}{\sqrt{\pi}\,\Delta^3} = \frac{s\Delta^2}{\sqrt{\pi}\,\Delta^3}$$

$$= \frac{1}{\sqrt{\pi}} \cdot \frac{s}{\Delta} = \frac{1}{\sqrt{\pi}} \cdot \frac{1}{r} = \text{LHS} \quad \left(\because r = \frac{\Delta}{s}\right)$$

Option (d), LHS $= \sqrt{A_1} + \sqrt{A_2} + \sqrt{A_3}$

$= \sqrt{\pi}(r_1 + r_2 + r_3)$ [using option (a)]

$= \sqrt{\pi}(4R + r)$ $(\because 4R = r_1 + r_2 + r_3 - r)$

$=$ RHS

19. (b) We know that

$$\frac{1}{r_1} + \frac{1}{r_2} + \frac{1}{r_3} = \frac{1}{r} \quad \ldots\text{(i)}$$

and AM $\geq$ HM

$$\frac{r_1 + r_2 + r_3}{3} \geq \frac{3}{\frac{1}{r_1} + \frac{1}{r_2} + \frac{1}{r_3}}$$

$$\frac{r_1 + r_2 + r_3}{3} \geq \frac{3}{\frac{1}{r}} \quad \text{[from Eq. (i)]}$$

$$\frac{r_1 + r_2 + r_3}{r} \geq 9$$

$\because$ Minimum value of $\frac{r_1 + r_2 + r_3}{r}$ is 9, Statement I is true.

Statement II is also true, but not the correct explanation of Statement I.

20. (d) **Statement II** In cyclic quadrilateral $ABCD$

$$A + C = 180^\circ \text{ and } B + D = 180^\circ$$

$$\therefore \quad \cos A = -\cos C \text{ and } \cos B = -\cos D$$

Now, $\Sigma\cos A = \cos A + \cos B + \cos C + \cos D$

$$= \cos A + \cos B - \cos A - \cos B = 0$$

Hence, Statement II is correct.

Statement I $\because A + C = 180^\circ$ and $B + D = 180^\circ$

$$\therefore \quad \sin A = \sin C \text{ and } \sin B = \sin D$$

Now, $\Sigma\sin A = \sin A + \sin B + \sin C + \sin D$

$$= 2(\sin A + \sin B) \neq 0$$

Hence, Statement I is incorrect.

21. (7) Let O and r be the centre and radius of circle passing through the vertices $A_1, A_2, A_3, \ldots, A_n$.

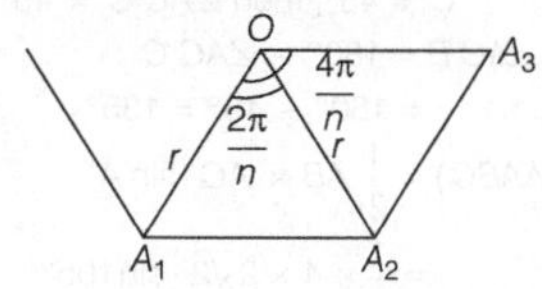

$$\because \quad OA_1 = OA_2 = r$$

and $$\cos\frac{2\pi}{n} = \frac{(OA_1)^2 + (OA_2)^2 - (A_1A_2)^2}{2OA_1 \cdot OA_2}$$

$$\cos\frac{2\pi}{n} = \frac{2r^2 - (A_1A_2)^2}{2r^2}$$

$$(A_1A_2)^2 = 2r^2 - 2r^2\cos\frac{2\pi}{n}$$

$$= 2r^2\left(1 - \cos\frac{2\pi}{n}\right) = 2r^2 \cdot 2\sin^2\frac{\pi}{n}$$

$$\therefore \quad A_1A_2 = 2r\sin\frac{\pi}{n} \quad \ldots\text{(i)}$$

Similarly, $$A_1A_3 = 2r\sin\frac{2\pi}{n} \quad \ldots\text{(ii)}$$

and $$A_1A_4 = 2r\sin\frac{3\pi}{n} \quad \ldots\text{(iii)}$$

Now, we have,

$$\frac{1}{A_1A_2} = \frac{1}{A_1A_3} + \frac{1}{A_1A_4}$$

$$\Rightarrow \quad \frac{1}{2r\sin\frac{\pi}{n}} = \frac{1}{2r\sin\frac{2\pi}{n}} + \frac{1}{2r\sin\frac{3\pi}{n}}$$

$$\Rightarrow \quad \frac{1}{\sin\frac{\pi}{n}} = \frac{\sin\frac{3\pi}{n} + \sin\frac{2\pi}{n}}{\sin\frac{3\pi}{n}\sin\frac{2\pi}{n}}$$

$$\Rightarrow \quad \sin\frac{3\pi}{n}\sin\frac{2\pi}{n} = \sin\frac{3\pi}{n}\sin\frac{\pi}{n} + \sin\frac{2\pi}{n}\sin\frac{\pi}{n}$$

$$\Rightarrow \quad \sin\frac{3\pi}{n}\sin\frac{2\pi}{n} - \sin\frac{2\pi}{n}\sin\frac{\pi}{n} = \sin\frac{3\pi}{n}\sin\frac{\pi}{n}$$

$$\Rightarrow \quad \sin\frac{2\pi}{n}\left(\sin\frac{3\pi}{n} - \sin\frac{\pi}{n}\right) = \sin\frac{3\pi}{n}\sin\frac{\pi}{n}$$

$$\Rightarrow \quad \left(2\sin\frac{2\pi}{n}\cos\frac{2\pi}{n}\right)\sin\frac{\pi}{n} = \sin\frac{3\pi}{n}\sin\frac{\pi}{n}$$

$$\Rightarrow \quad \sin\frac{4\pi}{n} = \sin\frac{3\pi}{n} \Rightarrow \sin\left(\frac{4\pi}{n}\right) = \sin\left(\pi - \frac{3\pi}{n}\right)$$

$$\Rightarrow \quad \frac{4\pi}{n} = \pi - \frac{3\pi}{n}$$

$$\Rightarrow \quad \frac{7}{n} = 1 \Rightarrow n = 7$$

22. (4) In ΔABC, using sine rule,

$$\frac{a}{\sin A} = \frac{b}{\sin B} = \frac{c}{\sin C}$$

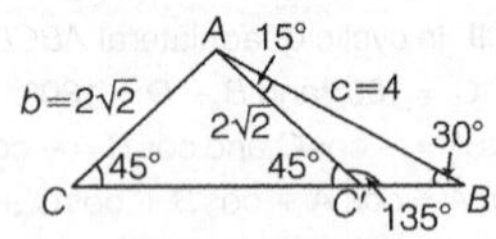

$$\frac{a}{\sin A} = \frac{2\sqrt{2}}{\sin 30°} = \frac{4}{\sin C}$$

If $\dfrac{2\sqrt{2}}{\sin 30°} = \dfrac{4}{\sin C}$

$$\Rightarrow \quad 4\sqrt{2} = \frac{4}{\sin C} \Rightarrow \sin C = \frac{1}{\sqrt{2}} = \sin 45°$$

$$\Rightarrow \quad C = 45°$$

$\therefore \quad A = 180° - (45° + 30°) = 105°$

$\because \quad C = 45°$, then $\angle ACC' = 45°$ $\quad (\because AC = AC')$

Now, $\quad \angle AC'B = 180° - \angle ACC'$

$= 180° - 45° = 135°$

Now, $\quad \text{ar}(\Delta ABC) = \dfrac{1}{2} AB \times AC \cdot \sin A$

$$= \frac{1}{2} \times 4 \times 2\sqrt{2} \cdot \sin 105°$$

$$= 4\sqrt{2} \cdot \frac{\sqrt{3}+1}{2\sqrt{2}} = 2(\sqrt{3}+1)$$

and $\quad \text{ar}(\Delta ABC') = \dfrac{1}{2} AB \times AC' \cdot \sin 15°$

$$= \frac{1}{2} \times 4 \times 2\sqrt{2}\left(\frac{\sqrt{3}-1}{2\sqrt{2}}\right) = 2(\sqrt{3}-1)$$

$\therefore$ Absolute difference of areas of triangle

$$= |2\sqrt{3} + 2 - 2\sqrt{3} + 2| = 4 \text{ sq units}$$

23. (5) Let $OL = r$ and $OA = R$, $\angle AOL = \dfrac{\pi}{n}$ and a be the length of side of a regular polygon, then $AL = BL = \dfrac{a}{2}$.

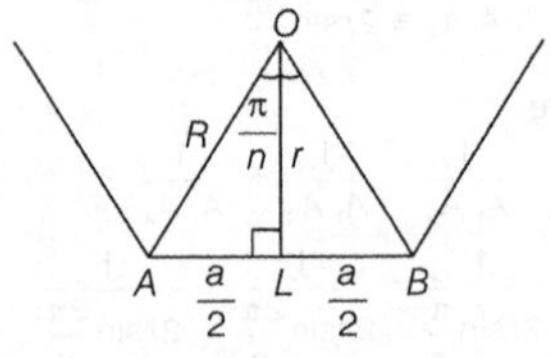

In ΔAOL,

$$\text{cosec}\frac{\pi}{n} = \frac{R}{\frac{a}{2}} \Rightarrow R = \frac{a}{2}\text{cosec}\frac{\pi}{2} \quad \ldots\text{(i)}$$

and $$\cot\frac{\pi}{2} = \frac{r}{\frac{a}{2}} \Rightarrow r = \frac{a}{2}\cot\frac{\pi}{n} \quad \ldots\text{(ii)}$$

Now, $\quad \dfrac{R}{r} = \sqrt{5} - 1$ $\quad$ (given)

$$\Rightarrow \quad \frac{\frac{a}{2}\cdot\text{cosec}\frac{\pi}{n}}{\frac{a}{2}\cdot\cot\frac{\pi}{n}} = \sqrt{5} - 1$$

$$\Rightarrow \quad \cos\frac{\pi}{n} = \frac{1}{\sqrt{5}-1} \times \frac{\sqrt{5}+1}{\sqrt{5}+1} = \frac{\sqrt{5}+1}{4}$$

$$\Rightarrow \quad \cos\frac{\pi}{n} = \cos\frac{\pi}{5} \quad \left(\because \cos\frac{\pi}{5} = \cos 36° = \frac{\sqrt{5}+1}{4}\right)$$

$\therefore \quad n = 5$

24. (3) We know that,

$$\text{ar}(\Delta ABC) = \frac{1}{2} ab\sin C$$

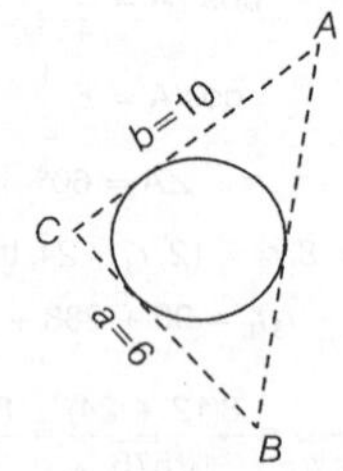

$$15\sqrt{3} = \frac{1}{2} \cdot 6 \cdot 10\sin C$$

$\sin C = \dfrac{\sqrt{3}}{2}$ and C is an obtuse angle.

$\therefore \quad C = 120°$

Now, $\quad c = \sqrt{a^2 + b^2 - 2ab\cos C}$

$$= \sqrt{6^2 + 10^2 - 2 \times 6 \times 10 \times \cos 120°}$$

$c = 14$

Then, $\quad s = \dfrac{a+b+c}{2} = \dfrac{6+10+14}{2} = 15$

We know that, $\quad r = \dfrac{\Delta}{s} = \dfrac{15\sqrt{3}}{15} = \sqrt{3}$

$\therefore \quad r^2 = 3$

25. (d) From Fig. (i),

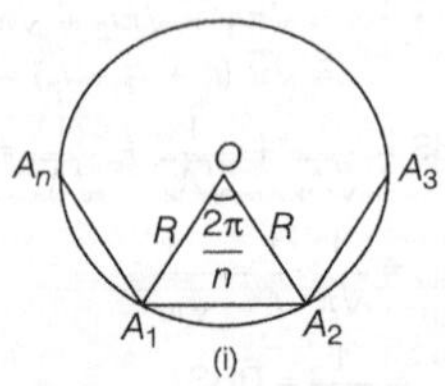

(i)

$$(A_1A_2)^2 = R^2 + R^2 - 2R \cdot R\cos\frac{2\pi}{n}$$

$$= 2R^2 - 2R^2\cos\frac{2\pi}{n}$$

$$= 2R^2\left(1 - \cos\frac{2\pi}{n}\right)$$

$$= 4R^2\sin^2\frac{\pi}{n}$$

$$A_1A_2 = 2R\sin\frac{\pi}{n}$$

$$\therefore \quad a = 2R\sin\frac{\pi}{n}$$

$$\Rightarrow \quad R = \frac{a}{2\sin\frac{\pi}{n}} \quad \ldots\text{(i)}$$

From Fig. (ii),

From in ΔA_1PI, $\tan\frac{\pi}{n} = \frac{A_1P}{PI}$

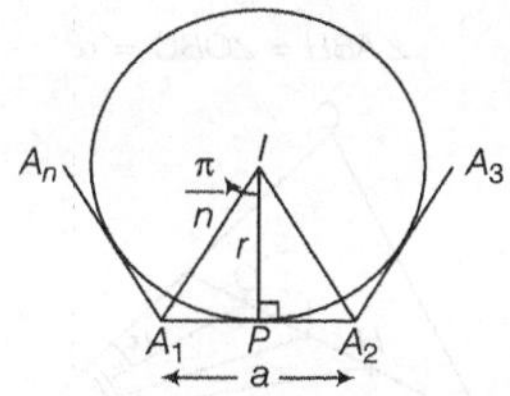

(ii)

$$\tan\frac{\pi}{n} = \frac{\frac{a}{2}}{r}$$

$$r = \frac{1}{2}a\cot\frac{\pi}{n} \quad \ldots\text{(ii)}$$

Now, $r + R = \frac{1}{2}a\cot\frac{\pi}{n} + \frac{a}{2\sin\frac{\pi}{n}}$

$$= \frac{a}{2}\cdot\frac{\cos\frac{\pi}{n}}{\sin\frac{\pi}{n}} + \frac{a}{2\sin\frac{\pi}{n}}$$

$$= \frac{a}{2\sin\frac{\pi}{n}}\left(1+\cos\frac{\pi}{n}\right)$$

$$= \frac{a}{2\cdot 2\sin\frac{\pi}{2n}\cos\frac{\pi}{2n}} \times 2\cos^2\frac{\pi}{2n}$$

$$= \frac{a}{2}\cot\frac{\pi}{2n}$$

26. (a) Area of the regular polygon

$$= n \times \text{ar}(\Delta A_1OA_2)$$

$$= n \times \frac{1}{2}OA_1\cdot OA_2\cdot \sin\angle A_1OA_2 \quad \text{[from Fig. (i)]}$$

$$= n \times \frac{1}{2}R\cdot R\cdot\sin\frac{2\pi}{n} = \frac{n}{2}R^2\sin\left(\frac{2\pi}{n}\right)$$

27. (d) We have,

$$\angle A + \angle C = 180°$$

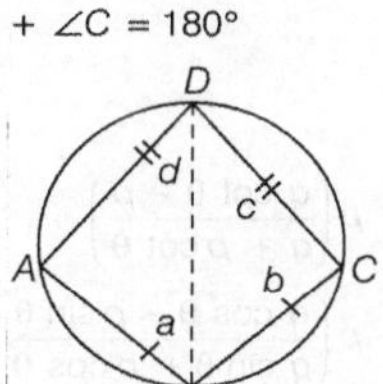

and $\angle B + \angle D = 180°$

Let $AB = BC = 6\text{ cm}$

and $CD = DA = 8\text{ cm}$

Then, $\Delta ABD \cong \Delta BDC$

$\therefore$ $\angle A = \angle C = \frac{180°}{2} = 90°$

$\therefore$ In ΔABD, using Pythagoras theorem,

$$BD = \sqrt{AB^2 + AD^2}$$

$$= \sqrt{6^2 + 8^2} = 10$$

$\therefore$ Radius of circumcircle $= \frac{BD}{2} = 5\text{ cm}$

28. (c) Here, $AB = BC = 5\text{ cm}, CD = DA = 12\text{ cm}$

$\therefore$ $\angle A = \angle C = 90°$, then

$$\text{ar}(ABCD) = \text{ar}(\Delta ABD) + \text{ar}(\Delta BCD)$$

$$= \frac{1}{2}\cdot AB\cdot AD + \frac{1}{2}\cdot BC\cdot CD$$

$$= \frac{1}{2}\times 5\times 12 + \frac{1}{2}\times 5 \times 12$$

$$= 60\text{ cm}^2$$

$\because$ $AB + AD = BC + CD$, then incircle can be drawn

$$s = \frac{a+b+c+d}{2} = \frac{5+5+12+12}{2} = 17\text{cm}$$

We know that,

$$\text{ar}(ABCD) = r\cdot s$$

$\Rightarrow$ $60 = r\cdot 17$

$\therefore$ $r = \frac{60}{17}\text{ cm}$

29. (a) We know that,

$$\cos A = \frac{a^2+d^2-b^2-c^2}{2(ad+bc)} \quad \ldots\text{(i)}$$

$\because$ A circle can be inscribed in a quadrilateral, then

$$a - d = b - c$$

$\Rightarrow$ $(a-d)^2 = (b-c)^2$

$\Rightarrow$ $a^2 + d^2 - 2ad = b^2 + c^2 - 2bc$

$\Rightarrow$ $a^2 + d^2 - b^2 - c^2 = 2(ad - bc)$

Put in Eq. (i), we get

$$\cos A = \frac{2(ad-bc)}{2(ad+bc)}$$

$$\frac{1}{\cos A} = \frac{ad+bc}{ad-bc}$$

$\therefore$ $\frac{1-\cos A}{1+\cos A} = \frac{ad+bc-ad+bc}{ad+bc+ad-bc}$

$$\tan^2\frac{A}{2} = \frac{bc}{ad}$$

30. (a) $\because$ A circle is inscribed in a quadrilateral $ABCD$, so it touches $ABCD$ at P, Q, R, S, then

$$AP = AS,\ BP = BQ,\ CQ = CR \text{ and } DR = DS$$

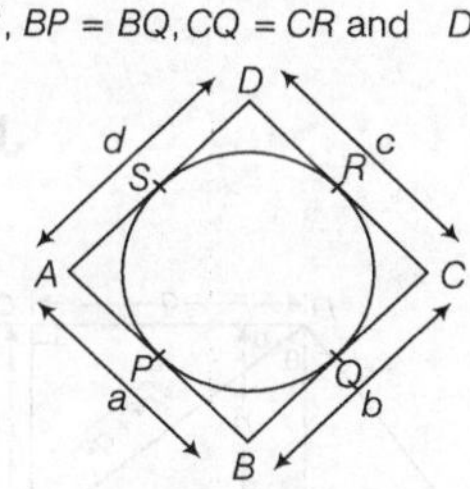

$\therefore$ $(AP + BP) + (CR + DR) = AS + BQ + CQ + DS$

$$AB + CD = BC + AD$$

$$a + c = b + d$$

$\therefore$ $s = \frac{(a+c)+(b+d)}{2} = a + c = b + d$

If $s = a + c$, then

$$s - a = c \text{ and } s - c = a$$

If $s = b + d$, then

$$s - b = d \text{ and } s - d = b$$

$\therefore$ $\text{ar}(ABCD) = \sqrt{(s-a)(s-b)(s-c)(s-d)} = \sqrt{abcd}$

31. (c) $\frac{bx}{c} + \frac{cy}{a} + \frac{az}{b} = b \sin B + c \sin C + a \sin A$

$$= \frac{b^2 + c^2 + a^2}{2R}$$

So, $k = 2R$

32. (c) $\because \cot A + \cot B + \cot C$

$$= \frac{R}{abc}(b^2 + c^2 - a^2 + c^2 + a^2 - b^2 + a^2 + b^2 - c^2)$$

$$= \frac{R}{abc}(a^2 + b^2 + c^2) = \frac{R}{abc}\left(\frac{4\Delta^2}{x^2} + \frac{4\Delta^2}{y^2} + \frac{4\Delta^2}{z^2}\right)$$

$$= \frac{4R\Delta^2}{abc}\left(\frac{1}{x^2} + \frac{1}{y^2} + \frac{1}{z^2}\right) = \frac{4R\Delta}{abc} \cdot \Delta\left(\frac{1}{x^2} + \frac{1}{y^2} + \frac{1}{z^2}\right)$$

$$= \Delta\left(\frac{1}{x^2} + \frac{1}{y^2} + \frac{1}{z^2}\right)$$

So, $k = \Delta$

33. (d) $\Sigma \frac{c \sin B + b \sin C}{x} = \Sigma \frac{x + x}{x} = 2 + 2 + 2 = 6$

34. (c) Since, A, H, I, O and B are concyclic, so

$$\angle AHB = \angle AIB = \angle AOB$$

(angles of chord AB subtended at same side of AB in corresponding circle)

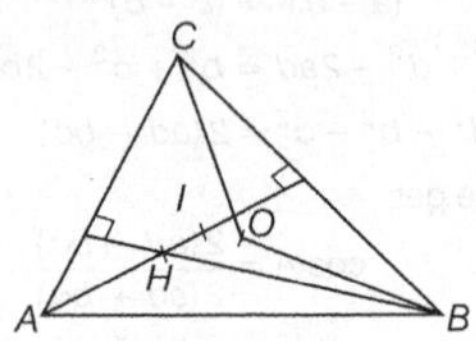

Here, $\angle AHB = \pi - \angle BAH - \angle HBA$

$$= \pi - \left(\frac{\pi}{2} - B\right) - \left(\frac{\pi}{2} - A\right)$$

$$= A + B = \pi - C$$

and $\angle AOB = 2C$

$\therefore \quad \pi - C = 2C \Rightarrow C = \frac{\pi}{3}$

35. (b) $\angle ABH = \frac{\pi}{2} - A$

and $\angle OBC = \frac{\pi}{2} - A$

$\therefore \quad \angle ABH = \angle OBC = \alpha$

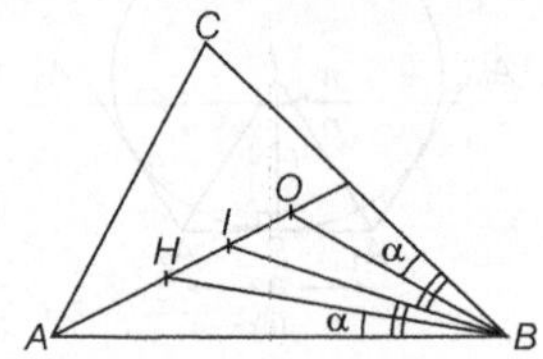

Also, $\angle IBA = \frac{B}{2} = \angle IBC$

$\therefore \quad ÐIBH = \angle IBO = \frac{B}{2} - \alpha$

$$= \frac{B}{2} - \left(\frac{\pi}{2} - A\right)$$

$\therefore$ HI and OI will be chords of equal lengths

Hence, $\frac{HI}{OI} = 1$

36. (c) Given, $AH = HI$, then

$$\alpha = \frac{A - C}{2}$$

$\Rightarrow \quad \frac{\pi}{2} - A = \frac{A - C}{2}$

$\Rightarrow \quad \pi - 2A = A - C$

$\Rightarrow \quad \pi + C = 3A$

$\Rightarrow \quad 3A = \pi + \frac{\pi}{3}$

$\Rightarrow \quad A = \frac{4\pi}{9}$

and $B = \pi - A - C = \frac{2\pi}{9}$

$\because$ A, C, B are in AP.

$\therefore$ Common difference $= C - B = \frac{\pi}{3} - \frac{2\pi}{9} = \frac{\pi}{9}$

JEE Main & AIEEE Archive

1. (a) Let $AB = x$

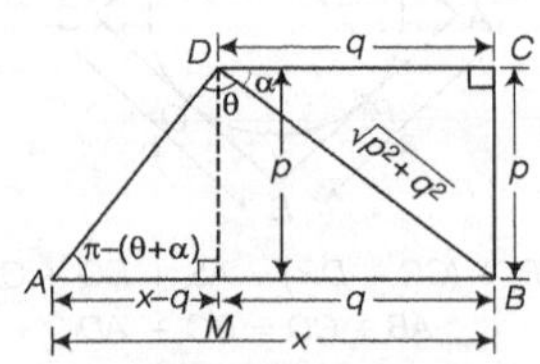

In ΔDAM, $\tan(\pi - \theta - \alpha) = \frac{p}{x - q}$

$\Rightarrow \quad \tan(\theta + \alpha) = \frac{p}{q - x}$

$\Rightarrow \quad q - x = p \cot(\theta + \alpha)$

$\Rightarrow \quad x = q - p \cot(\theta + \alpha)$

$$= q - p\left(\frac{\cot\theta \cot\alpha - 1}{\cot\alpha + \cot\theta}\right)$$

$$= q - p\left(\frac{\frac{q}{p}\cot\theta - 1}{\frac{q}{p} + \cot\theta}\right) \quad \left(\because \text{in } \Delta BDC, \cot\alpha = \frac{q}{p}\right)$$

$$= q - p\left(\frac{q\cot\theta - p}{q + p\cot\theta}\right)$$

$$= q - p\left(\frac{q\cos\theta - p\sin\theta}{q\sin\theta + p\cos\theta}\right)$$

$\Rightarrow \quad x = \frac{q^2\sin\theta + pq\cos\theta - pq\cos\theta + p^2\sin\theta}{p\cos\theta + q\sin\theta}$

$\therefore \quad AB = \frac{(p^2 + q^2)\sin\theta}{p\cos\theta + q\sin\theta}$

Alternate Method

Applying sine rule in ΔABD,

$$\frac{a}{\sin A} = \frac{b}{\sin B} = \frac{c}{\sin C}$$

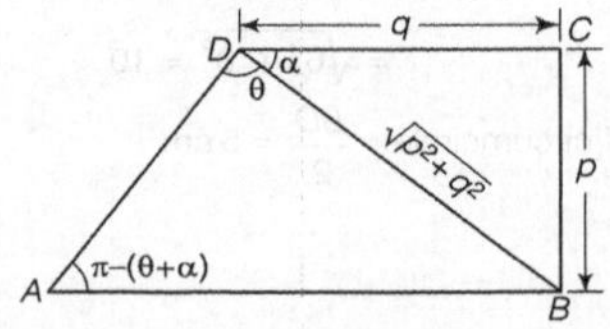

$$\frac{AB}{\sin\theta} = \frac{\sqrt{p^2+q^2}}{\sin\{\pi - (\theta+\alpha)\}}$$

$$\Rightarrow \frac{AB}{\sin\theta} = \frac{\sqrt{p^2+q^2}}{\sin(\theta+\alpha)}$$

$$\Rightarrow AB = \frac{\sqrt{p^2+q^2}\sin\theta}{\sin\theta\cos\alpha + \cos\theta\sin\alpha}$$

$$= \frac{(p^2+q^2)\sin\theta}{q\sin\theta + p\cos\theta}$$

$$\left(\because \cos\alpha = \frac{q}{\sqrt{p^2+q^2}} \text{ and } \sin\alpha = \frac{p}{\sqrt{p^2+q^2}}\right)$$

$$= \frac{(p^2+q^2)\sin\theta}{p\cos\theta + q\sin\theta}$$

2. (b) Given, ΔPQR such that

$$3\sin P + 4\cos Q = 6 \quad \ldots\text{(i)}$$

$$4\sin Q + 3\cos P = 1 \quad \ldots\text{(ii)}$$

On squaring and adding Eqs. (i) and (ii), we get

$$(3\sin P + 4\cos Q)^2 + (4\sin Q + 3\cos P)^2 = 36 + 1$$

$$\Rightarrow 9(\sin^2 P + \cos^2 P) + 16(\sin^2 Q + \cos^2 Q) + 2\times 3\times 4(\sin P\cos Q + \sin Q\cos P) = 37$$

$$\Rightarrow 24[\sin(P+Q)] = 37 - 25$$

$$\Rightarrow \sin(P+Q) = \frac{1}{2}$$

Since, P and Q are angles of ΔPQR, hence $0° < P, Q < 180°$.

$$\Rightarrow P + Q = 30° \text{ or } 150°$$

$$\Rightarrow R = 150° \text{ or } 30°$$

Hence, two cases arise here.

Case I $R = 150°$

$$\Rightarrow P + Q = 30°$$

$$\Rightarrow 0° < P, Q < 30°$$

$$\Rightarrow \sin P < \frac{1}{2}, \cos Q < 1$$

$$\Rightarrow 3\sin P + 4\cos Q < \frac{3}{2} + 4$$

$$\Rightarrow 3\sin P + 4\cos Q < \frac{11}{2} < 6$$

$\Rightarrow$ $3\sin P + 4\cos Q \Rightarrow 6$ is not possible.

Case II $R = 30°$

Hence, $R = 30°$, i.e. $\frac{\pi}{6}$ is the only possibility.

3. (c) Since, $\tan\frac{P}{2}$ and $\tan\frac{Q}{2}$ are the roots of equation

$$ax^2 + bx + c = 0$$

$$\therefore \tan\frac{P}{2} + \tan\frac{Q}{2} = -\frac{b}{a} \quad \ldots\text{(i)}$$

and

$$\tan\frac{P}{2}\tan\frac{Q}{2} = \frac{c}{a} \quad \ldots\text{(ii)}$$

Also,

$$\frac{P}{2} + \frac{Q}{2} + \frac{R}{2} = \frac{\pi}{2} \quad (\because P + Q + R = \pi)$$

$$\Rightarrow \frac{P+Q}{2} = \frac{\pi}{2} - \frac{R}{2}$$

$$\Rightarrow \frac{P+Q}{2} = \frac{\pi}{4} \quad \left(\because \angle R = \frac{\pi}{2}, \text{ given}\right)$$

$$\Rightarrow \tan\left(\frac{P}{2} + \frac{Q}{2}\right) = 1$$

$$\Rightarrow \frac{\tan\frac{P}{2} + \tan\frac{Q}{2}}{1 - \tan\frac{P}{2}\tan\frac{Q}{2}} = 1$$

$$\Rightarrow \frac{-\frac{b}{a}}{1 - \frac{c}{a}} = 1 \quad \text{[from Eqs. (i) and (ii)]}$$

$$\Rightarrow -\frac{b}{a} = 1 - \frac{c}{a}$$

$$\Rightarrow -b = a - c$$

$$\therefore c = a + b$$

Alternate Method

Since, $\angle R = \frac{\pi}{2}$

$$\Rightarrow \angle P + \angle Q = \frac{\pi}{2}$$

$$\Rightarrow \frac{\angle P}{2} = \frac{\pi}{4} - \frac{\angle Q}{2}$$

$$\therefore \tan\frac{P}{2} = \tan\left(\frac{\pi}{4} - \frac{Q}{2}\right) = \frac{\tan\frac{\pi}{4} - \tan\frac{Q}{2}}{1 + \tan\frac{\pi}{4}\tan\frac{Q}{2}}$$

$$\Rightarrow \tan\frac{P}{2} + \tan\frac{P}{2}\tan\frac{Q}{2} = 1 - \tan\frac{Q}{2}$$

$$\Rightarrow \tan\frac{P}{2} + \tan\frac{Q}{2} = 1 - \tan\frac{P}{2}\tan\frac{Q}{2}$$

$$\Rightarrow -\frac{b}{a} = 1 - \frac{c}{a} \Rightarrow -b = a - c \quad \text{(by given equation)}$$

$$\therefore c = a + b$$

4. (c) We know that,

$$\frac{c}{\sin C} = 2R$$

$$\Rightarrow c = 2R \quad (\because C = 90°) \ldots\text{(i)}$$

and

$$\tan\frac{C}{2} = \frac{r}{s-c}$$

$$\Rightarrow \tan\frac{\pi}{4} = \frac{r}{s-c}$$

$$\therefore r = s - c$$

$$\Rightarrow r = \frac{a+b+c}{2} - c$$

$$\Rightarrow a + b - c = 2r \quad \ldots\text{(ii)}$$

On adding Eqs. (i) and (ii), we get

$$2(r + R) = a + b$$

5. (b) Given that, $\cos^{-1} x - \cos^{-1}\frac{y}{2} = \alpha$

$$\Rightarrow \cos^{-1}\left(\frac{xy}{2} + \sqrt{1-x^2}\sqrt{1-\frac{y^2}{4}}\right) = \alpha$$

$$\Rightarrow \frac{xy}{2} + \sqrt{1-x^2}\sqrt{1-\frac{y^2}{4}} = \cos\alpha$$

$$\Rightarrow 2\sqrt{1-x^2}\sqrt{1-\frac{y^2}{4}} = 2\cos\alpha - xy$$

On squaring both sides, we get

$$\frac{4(1-x^2)(4-y^2)}{4} = 4\cos^2\alpha + x^2y^2 - 4xy\cos\alpha$$

$$\Rightarrow 4 - 4x^2 - y^2 + x^2y^2 = 4\cos^2\alpha + x^2y^2 - 4xy\cos\alpha$$

$$\therefore 4x^2 - 4xy\cos\alpha + y^2 = 4\sin^2\alpha$$

6. (c) In ΔBAD, $\cos(90° - B) = \dfrac{AD}{c}$

$\Rightarrow \quad AD = c \sin B$

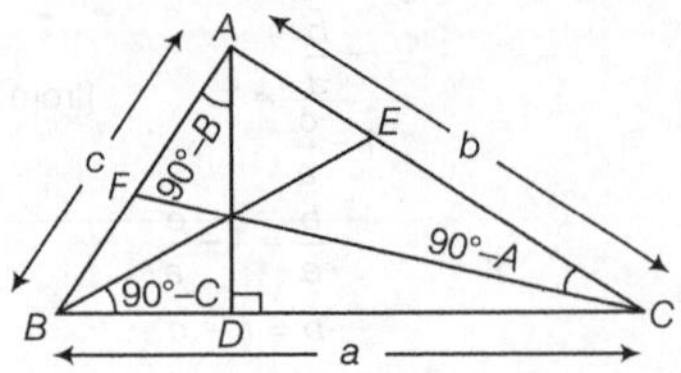

Similarly, $\quad BE = a \sin C$

and $\quad CF = b \sin A$

Since, AD, BE, CF are in HP.

So, $c \sin B, a \sin C, b \sin A$ are in HP.

$\Rightarrow \dfrac{1}{\sin C \sin B}, \dfrac{1}{\sin A \sin C}$ and $\dfrac{1}{\sin B \sin A}$ are in AP.

Hence, $\sin A, \sin B$ and $\sin C$ are in AP.

Alternate Method

$$\text{ar}(\Delta ABC) = \frac{1}{2} \times BC \times AD$$

$\Rightarrow \quad \Delta = \dfrac{1}{2} \times a \times AD$

$\Rightarrow \quad AD = \dfrac{2\Delta}{a}$

Similarly, $BE = \dfrac{2\Delta}{b}, CF = \dfrac{2\Delta}{c}$

Since, AD, BE and CF are in HP.

So, $\dfrac{1}{a}, \dfrac{1}{b}, \dfrac{1}{c}$ are in HP.

Hence, a, b, c are in AP.

$\therefore \sin A, \sin B, \sin C$ are in AP.

7. (c) Let $a = \sin \alpha, b = \cos \alpha$ and $c = \sqrt{1 + \sin \alpha \cos \alpha}$

Here, we see that the greatest side is c.

Now, $\quad \cos C = \dfrac{a^2 + b^2 - c^2}{2ab}$

$\Rightarrow \quad \cos C = \dfrac{\sin^2 \alpha + \cos^2 \alpha - 1 - \sin \alpha \cos \alpha}{2 \sin \alpha \cos \alpha}$

$\Rightarrow \quad \cos C = -\dfrac{\sin \alpha \cos \alpha}{2 \sin \alpha \cos \alpha}$

$\Rightarrow \quad \cos C = -\dfrac{1}{2} = \cos 120°$

$\therefore \quad \angle C = 120°$

8. (a) Given, $\quad a \cos^2 \dfrac{C}{2} + c \cos^2 \dfrac{A}{2} = \dfrac{3b}{2}$

$\Rightarrow \quad a\left[\dfrac{s(s-c)}{ab}\right] + c\left[\dfrac{s(s-a)}{bc}\right] = \dfrac{3b}{2}$

$\Rightarrow \quad \dfrac{s(s - c + s - a)}{b} = \dfrac{3b}{2}$

$\Rightarrow \quad 2s(2s - c - a) = 3b^2$

$\Rightarrow \quad 2s(a + b + c - c - a) = 3b^2$

$\Rightarrow \quad (a + b + c)b = 3b^2$

$\Rightarrow \quad a + b + c = 3b$

$\therefore \quad 2b = a + c$

Hence, a, b, c are in AP.

9. (c) Given, $AD = 4$ and $BD = DC$

Since, the centroid G divides the line AD in the ratio 2 : 1.

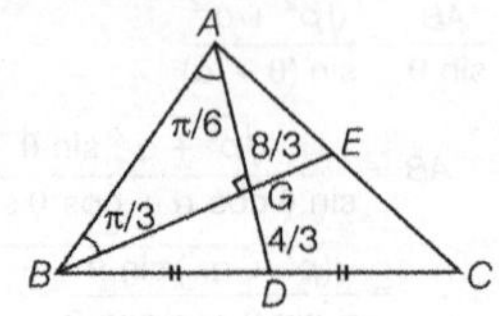

$\therefore \quad AG = \dfrac{8}{3}$ and $DG = \dfrac{4}{3}$

In ΔABG, $\tan \dfrac{\pi}{3} = \dfrac{AG}{BG}$

$\Rightarrow \quad BG = AG \cot \dfrac{\pi}{3}$

$\Rightarrow \quad BG = \dfrac{8}{3} \times \dfrac{1}{\sqrt{3}} = \dfrac{8}{3\sqrt{3}}$

$$\text{Area of } \Delta ADB = \frac{1}{2} \times AD \times BG$$

$$= \frac{1}{2} \times 4 \times \frac{8}{3\sqrt{3}} = \frac{16}{3\sqrt{3}}$$

Since, median divides a triangle into two triangles of equal area.

$\therefore \quad$ Area of $\Delta ABC = 2 \times$ Area of ΔADB

$$= 2 \times \frac{16}{3\sqrt{3}} = \frac{32}{3\sqrt{3}} \text{ sq units}$$

10. (b) Since, $\quad A + B + C = \pi$

$\Rightarrow \quad A + C = \pi - B$

$\quad \dfrac{A + C}{2} = \dfrac{\pi - B}{2}$

$\Rightarrow \quad \dfrac{A - B + C}{2} = \dfrac{\pi}{2} - B \left(\text{adding } -\dfrac{B}{2} \text{ both sides}\right)$

$\therefore \quad 2ca \sin\left(\dfrac{A - B + C}{2}\right) = 2ca \sin\left(\dfrac{\pi}{2} - B\right)$

$$= 2ac \cos B = 2ac\left(\frac{a^2 + c^2 - b^2}{2ac}\right)$$

$$= a^2 + c^2 - b^2$$

11. (a) Given that, $a = 4$, $b = 3$ and $\angle A = 60°$

Now, $\quad \cos 60° = \dfrac{c^2 + 9 - 16}{2 \times 3 \times c}$

$\Rightarrow \quad \dfrac{1}{2} = \dfrac{c^2 - 7}{2 \times 3c} \Rightarrow c^2 - 3c - 7 = 0$

12. (b) Given that,

$$\tan \frac{A}{2} = \frac{5}{6}$$

and $\quad \tan \dfrac{C}{2} = \dfrac{2}{5}$

Now, $\tan \dfrac{A}{2} \tan \dfrac{C}{2} = \dfrac{5}{6} \times \dfrac{2}{5}$

$\Rightarrow \quad \sqrt{\dfrac{(s-b)(s-c)}{s(s-a)}} \cdot \sqrt{\dfrac{(s-a)(s-b)}{s(s-c)}} = \dfrac{1}{3}$

$\Rightarrow \quad \dfrac{s-b}{s} = \dfrac{1}{3}$

$\Rightarrow \quad 2s = 3b$

$\therefore \quad a + c = 2b \quad (\because 2s = a + b + c)$

So, a, b, c are in AP.

JEE Advanced & IIT JEE Archive

1. (b) **Plan**

(i) $\cos C = \dfrac{a^2 + b^2 - c^2}{2ab}$

(ii) $R = \dfrac{abc}{4\Delta}, r = \dfrac{\Delta}{s}$

where, R, r, Δ denote the circumradius, inradius and area of triangle, respectively.

Let the sides of triangle be a, b and c.

Given, $x = a + b$, $y = ab$

Now,
$$x^2 - c^2 = y$$
$$\Rightarrow \quad (a+b)^2 - c^2 = ab$$
$$\Rightarrow \quad a^2 + b^2 + 2ab - c^2 = ab$$
$$\Rightarrow \quad a^2 + b^2 - c^2 = -ab$$
$$\Rightarrow \quad \frac{a^2 + b^2 - c^2}{2ab} = -\frac{1}{2} = \cos 120°$$
$$\Rightarrow \quad \angle C = \frac{2\pi}{3}$$
$$\because \quad R = \frac{abc}{4\Delta}, r = \frac{\Delta}{s}$$
$$\Rightarrow \quad \frac{r}{R} = \frac{4\Delta^2}{s(abc)} = \frac{4\left[\frac{1}{2} ab \sin\left(\frac{2\pi}{3}\right)\right]^2}{\frac{x+c}{2} \cdot y \cdot c}$$
$$\therefore \quad \frac{r}{R} = \frac{3y}{2c(x+c)}$$

2. (b,d) **Concept Involved** Whenever cosine of angle and sides are given or to find out, we should always use cosine law.

i.e.
$$\cos A = \frac{b^2 + c^2 - a^2}{2bc}$$
$$\cos B = \frac{a^2 + c^2 - b^2}{2ac}$$
and
$$\cos C = \frac{a^2 + b^2 - c^2}{2ab}$$

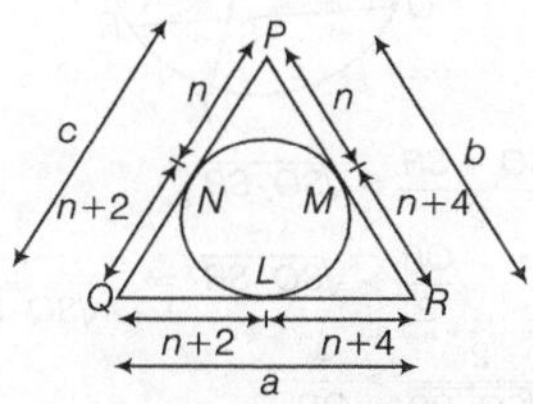

$$\therefore \quad \cos P = \frac{b^2 + c^2 - a^2}{2bc}$$
$$\Rightarrow \quad \frac{1}{3} = \frac{(2n+4)^2 + (2n+2)^2 - (2n+6)^2}{2(2n+4)(2n+2)} \quad \left(\because \cos P = \frac{1}{3}, \text{given}\right)$$
$$\Rightarrow \quad \frac{4n^2 - 16}{8(n+1)(n+2)} = \frac{1}{3}$$
$$\Rightarrow \quad \frac{n^2 - 4}{2(n+1)(n+2)} = \frac{1}{3}$$
$$\Rightarrow \quad \frac{(n-2)}{2(n+1)} = \frac{1}{3}$$
$$\Rightarrow \quad 3n - 6 = 2n + 2$$
$$\Rightarrow \quad n = 8$$

$\therefore$ Sides are $2n+2, 2n+4, 2n+6$, i.e. 18, 20, 22.

3. (c) **Concept Involved**

If ΔABC has sides a, b, c. Then,
$$\tan \frac{A}{2} = \sqrt{\frac{(s-b)(s-c)}{s(s-a)}}$$
where, $s = \dfrac{a+b+c}{2}$

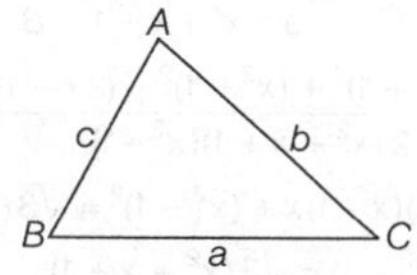

Given, $\quad a = 2, b = \dfrac{7}{2}, c = \dfrac{5}{2}$

Then, $\quad s = \dfrac{2 + \frac{7}{2} + \frac{5}{2}}{2} = 4$

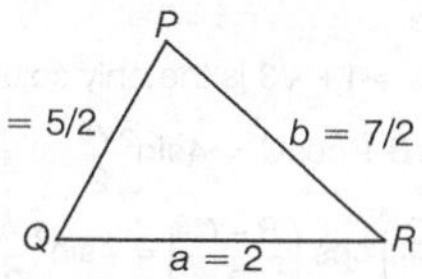

$$\therefore \quad \frac{2\sin P - \sin 2P}{2\sin P + \sin 2P} = \frac{2\sin P(1 - \cos P)}{2\sin P(1 + \cos P)}$$
$$= \frac{2\sin^2 \frac{P}{2}}{2\cos^2 \frac{P}{2}} = \tan^2 \frac{P}{2}$$
$$= \frac{(s-b)(s-c)}{s(s-a)} \times \frac{(s-b)(s-c)}{(s-b)(s-c)}$$
$$= \frac{(s-b)^2(s-c)^2}{\Delta^2} = \frac{\left(4 - \frac{7}{2}\right)^2\left(4 - \frac{5}{2}\right)^2}{\Delta^2}$$
$$= \left(\frac{3}{4\Delta}\right)^2$$

4. (d) Since, A, B, C are in AP.
$$\Rightarrow \quad 2B = A + C, \text{ i.e. } \angle B = 60°$$
$$\therefore \quad \frac{a}{c}(2\sin C\cos C) + \frac{c}{a}(2\sin A\cos A)$$
$$= 2k(a\cos C + c\cos A) \quad \left(\text{using, } \frac{a}{\sin A} = \frac{b}{\sin B} = \frac{c}{\sin C} = \frac{1}{k}\right)$$
$$= 2k(b) = 2\sin B \quad (\text{using, } b = a\cos C + c\cos A)$$
$$= \sqrt{3}$$

5. (3) Given, $\sin C = \dfrac{\sqrt{3}}{2}$ and C is to be obtuse.
$$\Rightarrow \quad C = \frac{2\pi}{3}$$
$$\because \quad c = \sqrt{a^2 + b^2 - 2ab\cos C}$$
$$= \sqrt{6^2 + 10^2 - 2 \times 6 \times 10 \times \cos\frac{2\pi}{3}} = 14$$
$$\therefore \quad r = \frac{\Delta}{s}$$
$$\Rightarrow \quad r^2 = \frac{225 \times 3}{\left(\frac{6 + 10 + 14}{2}\right)^2} = 3$$

6. (b) Using, $\cos C = \dfrac{a^2+b^2-c^2}{2ab}$

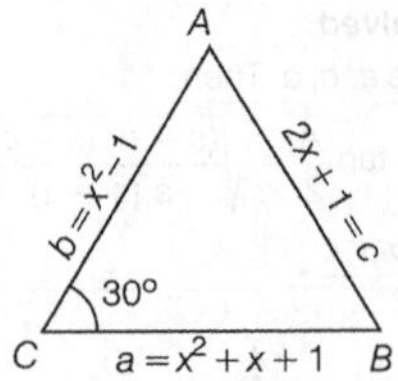

$$\Rightarrow \quad \frac{\sqrt{3}}{2} = \frac{(x^2+x+1)^2 + (x^2-1)^2 - (2x+1)^2}{2(x^2+x+1)(x^2-1)}$$

$$\Rightarrow \quad (x+2)(x+1)(x-1)x + (x^2-1)^2 = \sqrt{3}(x^2+x+1)(x^2-1)$$

$$\Rightarrow \quad x^2 + 2x + (x^2-1) = \sqrt{3}(x^2+x+1)$$

$$\Rightarrow \quad (2-\sqrt{3})x^2 + (2-\sqrt{3})x - (\sqrt{3}+1) = 0$$

$$\Rightarrow \quad x = -(2+\sqrt{3})$$

and $\quad x = 1 + \sqrt{3}$

But $\quad x = -(2+\sqrt{3})$

$\Rightarrow$ c is negative.

$\therefore \quad x = 1 + \sqrt{3}$ is the only solution.

7. (b,c) Given, $\cos B + \cos C = 4\sin^2 \dfrac{A}{2}$

$$\Rightarrow 2\cos\left(\frac{B+C}{2}\right)\cos\left(\frac{B-C}{2}\right) = 4\sin^2\frac{A}{2}$$

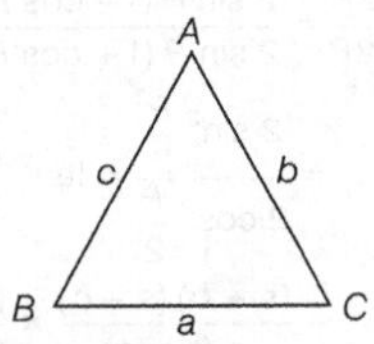

$$\Rightarrow \quad 2\sin\frac{A}{2}\left[\cos\left(\frac{B-C}{2}\right) - 2\sin\frac{A}{2}\right] = 0$$

$$\Rightarrow \quad \cos\left(\frac{B-C}{2}\right) - 2\cos\left(\frac{B+C}{2}\right) = 0$$

as $\quad \sin\dfrac{A}{2} \neq 0$

$$\Rightarrow \quad -\cos\frac{B}{2}\cos\frac{C}{2} + 3\sin\frac{B}{2}\sin\frac{C}{2} = 0$$

$$\Rightarrow \quad \tan\frac{B}{2}\tan\frac{C}{2} = \frac{1}{3}$$

$$\Rightarrow \quad \sqrt{\frac{(s-a)(s-c)}{s(s-b)}}\cdot\sqrt{\frac{(s-b)(s-a)}{s(s-c)}} = \frac{1}{3} \Rightarrow \frac{s-a}{s} = \frac{1}{3}$$

$$\Rightarrow \quad 2s = 3a$$

$\therefore \quad b + c = 2a$ [option (b)]

$\therefore$ Locus of A is an ellipse. [option (c)]

8. (4) In ΔABC, by sine rule,

$$\frac{a}{\sin A} = \frac{2\sqrt{2}}{\sin 30°} = \frac{4}{\sin C}$$

$$\Rightarrow \quad C = 45°, C' = 135°$$

when, $C = 45° \Rightarrow A = 180° - (45° + 30°) = 105°$

when, $C' = 135° \Rightarrow A = 180° - (135° + 30°) = 15°$

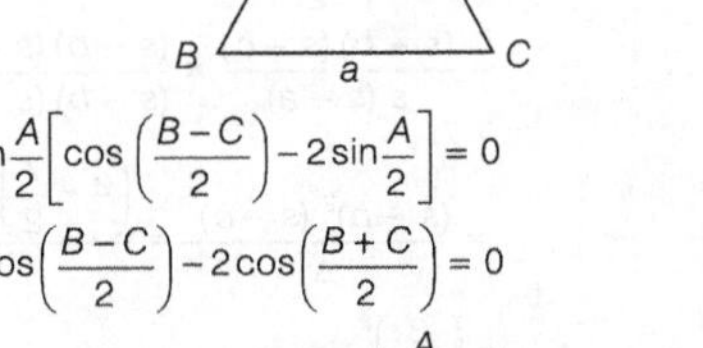

Area of $\Delta ABC = \dfrac{1}{2}AB \times AC \sin A = \dfrac{1}{2} \times 4 \times 2\sqrt{2}\sin(105°)$

$$= 4\sqrt{2} \times \frac{\sqrt{3}+1}{2\sqrt{2}} = 2(\sqrt{3}+1)$$

Area of $\Delta ABC' = \dfrac{1}{2}AB \times AC' \sin A$

$$= \frac{1}{2} \times 4 \times 2\sqrt{2}\sin(15°) = 2(\sqrt{3}-1)$$

$\therefore$ Difference of areas of triangle $= |2(\sqrt{3}+1) - 2(\sqrt{3}-1)| = 4$

Alternate Method

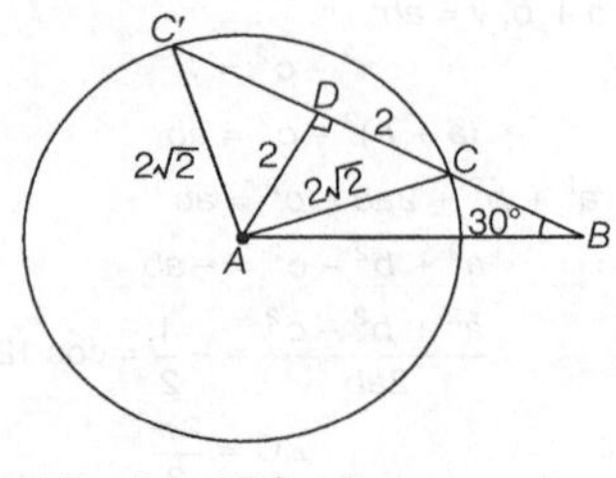

$$AD = 2 \text{ and } DC = 2$$

$\therefore$ Difference of areas of ΔABC and $\Delta ABC'$ = Area of $\Delta ACC'$

$$= \frac{1}{2} \times AD \times CC' = \frac{1}{2} \times 2 \times 4 = 4$$

9. (b,d) Let a straight line through the vertex P of a given ΔPQR intersects the side QR at the point S and the circumcircle of ΔPQR at the point T.

Points P, Q, R, T are concyclic, then $PS \cdot ST = QS \cdot SR$

We know that, AM > GM

$$\frac{PS + ST}{2} > \sqrt{PS \cdot ST}$$

and $\quad \dfrac{1}{PS} + \dfrac{1}{ST} > \dfrac{2}{\sqrt{PS \cdot ST}} = \dfrac{2}{\sqrt{QS \cdot SR}}$ [option (b)] ...(i)

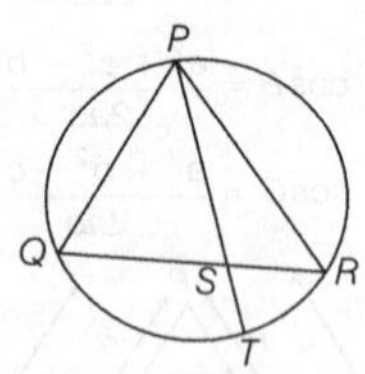

Also, $\quad \dfrac{SQ + SR}{2} > \sqrt{SQ \cdot SR}$

$$\Rightarrow \quad \frac{QR}{2} > \sqrt{SQ \cdot SR} \Rightarrow \frac{1}{\sqrt{SQ \cdot SR}} > \frac{2}{QR}$$

$\Rightarrow \quad \dfrac{2}{\sqrt{SQ \cdot SR}} > \dfrac{4}{QR}$ ...(ii)

From Eqs. (i) and (ii), we get

$$\frac{1}{PS} + \frac{1}{ST} > \frac{2}{\sqrt{QS \cdot SR}} > \frac{4}{QR}$$

10. (c) Coordinates of P and Q are $(1, 2\sqrt{2})$ and $(1, -2\sqrt{2})$, R and S are (9, 0) and (–1, 0), respectively.

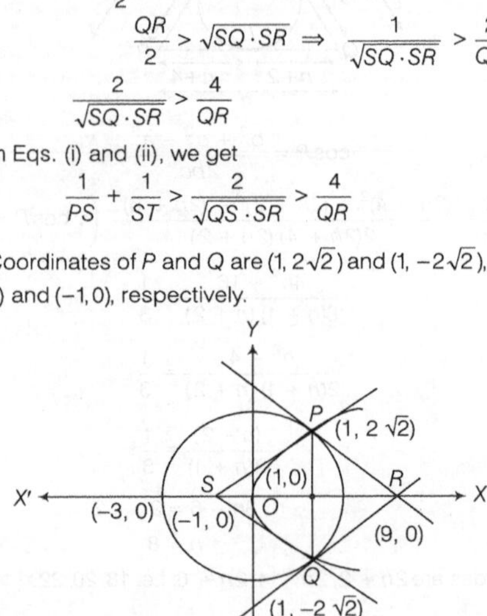

Now, $PQ = \sqrt{(4\sqrt{2})^2 + 0^2} = 4\sqrt{2}$

Area of $\Delta PQR = \frac{1}{2} \cdot 4\sqrt{2} \cdot 8 = 16\sqrt{2}$

Area of $\Delta PQS = \frac{1}{2} \cdot 4\sqrt{2} \cdot 2 = 4\sqrt{2}$

$\therefore$ Ratio of area of ΔPQS and ΔPQR is 1 : 4.

11. (b) Equation of circumcircle of ΔPRS is

$$(x+1)(x-9) + y^2 + \lambda y = 0$$

It will pass through $(1, 2\sqrt{2})$, then

$$-16 + 8 + \lambda \cdot 2\sqrt{2} = 0 \Rightarrow \lambda = \frac{8}{2\sqrt{2}} = 2\sqrt{2}$$

$\therefore$ Equation of circumcircle is

$$x^2 + y^2 - 8x + 2\sqrt{2}y - 9 = 0.$$

Hence, its radius is $3\sqrt{3}$.

Alternate Method

Let $\angle PSR = \theta \Rightarrow \sin\theta = \frac{2\sqrt{2}}{2\sqrt{3}}$

$\therefore$ $\sin\theta = \frac{PR}{SR}$

$\Rightarrow$ $PR = 6\sqrt{2} = 2R \cdot \sin\theta \Rightarrow R = 3\sqrt{3}$

12. (d) In ΔPQR, radius of incircle is $r = \frac{\Delta}{s}$

Since, $\Delta = 16\sqrt{2}$

Now, $s = \frac{6\sqrt{2} + 6\sqrt{2} + 4\sqrt{2}}{2} = 8\sqrt{2}$

$\therefore$ $r = \frac{16\sqrt{2}}{8\sqrt{2}} = 2$

13. (a,b,c,d) Since, $\Delta ABC = \Delta ABD + \Delta ACD$

$\Rightarrow$ $\frac{1}{2} bc \sin A = \frac{1}{2} c\, AD \sin\frac{A}{2} + \frac{1}{2} b\, AD \sin\frac{A}{2}$

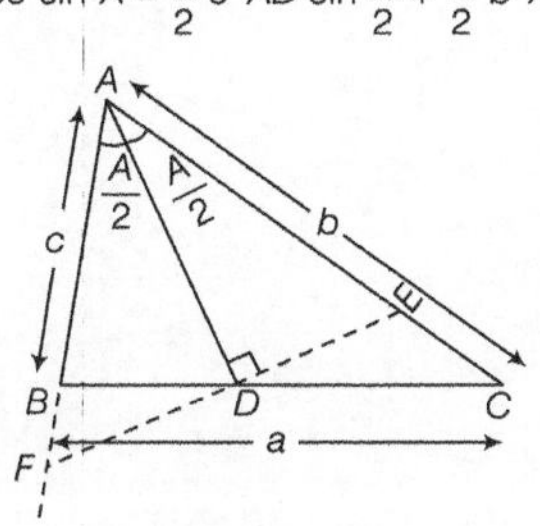

$\Rightarrow$ $AD = \frac{2bc}{b+c}\cos\frac{A}{2}$ [option (b)]

Again, $AE = AD\sec\frac{A}{2} = \frac{2bc}{b+c}$

$\Rightarrow AE$ is HM of b and c. [option (a)]

$$EF = ED + DF = 2DE$$

$$= 2AD\tan\frac{A}{2}$$

$$= 2 \cdot \frac{2bc}{b+c}\cos\frac{A}{2} \cdot \tan\frac{A}{2}$$

$\therefore$ $EF = \frac{4bc}{b+c}\sin\frac{A}{2}$ [option (c)]

Since, $AD \perp EF$ and $DE = DF$ and AD is bisector.

$\Rightarrow$ ΔAEF is an isosceles. [option (d)]

14. (c) Let $AB = AC = a$ and $\angle A = 120°$.

$\therefore$ Area of triangle $= \frac{1}{2}a^2 \sin 120°$

where, $a = AD + BD$

$$= \sqrt{3}\tan 30° + \sqrt{3}\cot 15°$$

$$= 1 + \frac{\sqrt{3}}{\tan(45° - 30°)}$$

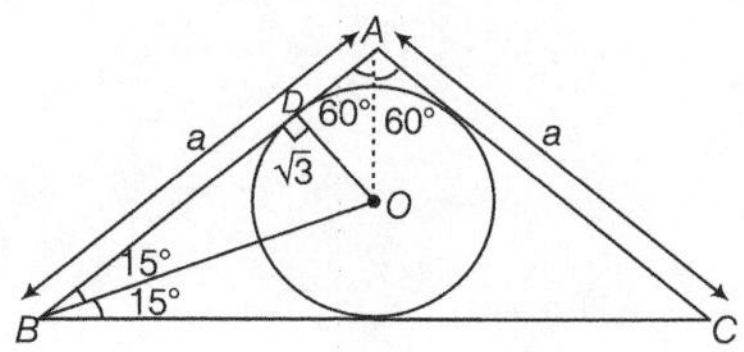

$$= 1 + \sqrt{3}\left(\frac{1 + \tan 45°\tan 30°}{\tan 45° - \tan 30°}\right)$$

$$= 1 + \sqrt{3}\left(\frac{\sqrt{3}+1}{\sqrt{3}-1}\right)$$

$\Rightarrow$ $a = 4 + 2\sqrt{3}$

$\therefore$ Area of triangle $= \frac{\sqrt{3}}{4}(4 + 2\sqrt{3})^2$

$$= (12 + 7\sqrt{3}) \text{ sq units}$$

15. (d) Let a, b, c be the sides of ΔABC.

Now, $\frac{b+c}{a} = \frac{k(\sin B + \sin C)}{k\sin A}$ (by sine rule)

$$= \frac{2\sin\left(\frac{B+C}{2}\right)\cos\left(\frac{B-C}{2}\right)}{2\sin\frac{A}{2}\cos\frac{A}{2}}$$

$\Rightarrow$ $\frac{b+c}{a} = \frac{\cos\left(\frac{B-C}{2}\right)}{\sin\frac{A}{2}}$

Also, $\frac{b-c}{a} = \frac{\sin\left(\frac{B-C}{2}\right)}{\cos\frac{A}{2}}$

$\Rightarrow$ $(b-c)\cos\frac{A}{2} = a\sin\left(\frac{B-C}{2}\right)$

Last 3 Years' Questions
JEE Main & Advanced
(2016-2018)

JEE Main

2016

1. If $0 \le x < 2\pi$, then the number of real values of x, which satisfy the equation $\cos x + \cos 2x + \cos 3x + \cos 4x = 0$, is **[Trigonometric Equations]**

(a) 3 (b) 5 (c) 7 (d) 9

2. A man is walking towards a vertical pillar in a straight path, at a uniform speed. At a certain point A on the path, he observes that the angle of elevation of the top of the pillar is 30°. After walking for 10 min from A in the same direction, at a point B, he observes that the angle of elevation of the top of the pillar is 60°. Then, the time taken (in minutes) by him, from B to reach the pillar, is **[Height and Distance]**

(a) 6 (b) 10 (c) 20 (d) 5

2017

3. If $5(\tan^2 x - \cos^2 x) = 2\cos 2x + 9$, then the value of $\cos 4x$ is **[Trigonometric Equations]**

(a) $-\frac{3}{5}$ (b) $\frac{1}{3}$

(c) $\frac{2}{9}$ (d) $-\frac{7}{9}$

4. Let a vertical tower AB have its end A on the level ground. Let C be the mid-point of AB and P be a point on the ground such that $AP = 2AB$. If $\angle BPC = \beta$, then $\tan \beta$ is equal to **[Height and Distance]**

(a) $\frac{6}{7}$ (b) $\frac{1}{4}$

(c) $\frac{2}{9}$ (d) $\frac{4}{9}$

2018

5. If sum of all the solutions of the equation

$$8\cos x \cdot \left(\cos\left(\frac{\pi}{6}+x\right)\cdot\cos\left(\frac{\pi}{6}-x\right)-\frac{1}{2}\right)=1 \text{ in } [0, \pi] \text{ is } k\pi, \text{ then } k \text{ is equal to}$$

[Trigonometric Equations]

(a) $\frac{2}{3}$ (b) $\frac{13}{9}$

(c) $\frac{8}{9}$ (d) $\frac{20}{9}$

6. PQR is a triangular park with $PQ = PR = 200$ m. A TV tower stands at the mid-point of QR. If the angles of elevation of the top of the tower at P, Q and R are respectively 45°, 30° and 30°, then the height of the tower (in m) is **[Height and Distance]**

(a) 100 (b) 50

(c) $100\sqrt{3}$ (d) $50\sqrt{2}$

JEE Advanced

2016

1. Let $S = \left\{x \in (-\pi, \pi): x \neq 0, \pm\frac{\pi}{2}\right\}$. The sum of all distinct solutions of the equation $\sqrt{3}\sec x + \operatorname{cosec} x + 2(\tan x - \cot x) = 0$ in the set S is equal to

[Trigonometric Equations, Type 1]

(a) $-\frac{7\pi}{9}$ (b) $-\frac{2\pi}{9}$ (c) 0 (d) $\frac{5\pi}{9}$

2. The value of $\sum_{k=1}^{13} \frac{1}{\sin\left(\frac{\pi}{4} + \frac{(k-1)\pi}{6}\right)\sin\left(\frac{\pi}{4} + \frac{k\pi}{6}\right)}$ is equal to

[Trigonometric Ratios and Identities, Type 1]

(a) $3 - \sqrt{3}$
(b) $2(3 - \sqrt{3})$
(c) $2(\sqrt{3} - 1)$
(d) $2(2 + \sqrt{3})$

2017

3. Let α and β be non zero real numbers such that $2(\cos\beta - \cos\alpha) + \cos\alpha\cos\beta = 1$. Then which of the following is/are true? **[Trigonometric Ratios and Identities, Type 2]**

(a) $\sqrt{3}\tan\left(\frac{\alpha}{2}\right) - \tan\left(\frac{\beta}{2}\right) = 0$ (b) $\tan\left(\frac{\alpha}{2}\right) - \sqrt{3}\tan\left(\frac{\beta}{2}\right) = 0$

(c) $\tan\left(\frac{\alpha}{2}\right) + \sqrt{3}\tan\left(\frac{\beta}{2}\right) = 0$ (d) $\sqrt{3}\tan\left(\frac{\alpha}{2}\right) + \tan\left(\frac{\beta}{2}\right) = 0$

2018

4. In a ΔPQR, let $\angle PQR = 30^\circ$ and the sides PQ and QR have lengths $10\sqrt{3}$ and 10, respectively. Then, which of the following statement(s) is (are) TRUE? **[Solutions of Triangles, Type 2]**

(a) $\angle QPR = 45^\circ$
(b) The area of the ΔPQR is $25\sqrt{3}$ and $\angle QRP = 120^\circ$
(c) The radius of the incircle of the ΔPQR is $10\sqrt{3} - 15$
(d) The area of the circumcircle of the ΔPQR is 100π

5. Let a, b, c be three non-zero real numbers such that the equation $\sqrt{3}\,a\cos x + 2b\sin x = c, x \in \left[-\frac{\pi}{2}, \frac{\pi}{2}\right]$, has two distinct real roots α and β with $\alpha + \beta = \frac{\pi}{3}$.

Then, the value of $\frac{b}{a}$ is ....

[Trigonometric Equations, Type 4]

Note *Type-1 Only one correct option, Type-2 More then one correct option, Type-3 Single digit Answer, Type-4 Numerical value based*

Answer *with* Explanations

JEE Main

1. *(c)* Given equation is

$$\cos x + \cos 2x + \cos 3x + \cos 4x = 0$$

$$\Rightarrow (\cos x + \cos 3x) + (\cos 2x + \cos 4x) = 0$$

$$\Rightarrow 2\cos 2x \cos x + 2\cos 3x \cos x = 0$$

$$\Rightarrow 2\cos x (\cos 2x + \cos 3x) = 0$$

$$\Rightarrow 2\cos x \left(2\cos \frac{5x}{2} \cos \frac{x}{2}\right) = 0$$

$$\Rightarrow \cos x \cdot \cos \frac{5x}{2} \cdot \cos \frac{x}{2} = 0$$

$$\Rightarrow \cos x = 0 \text{ or } \cos \frac{5x}{2} = 0$$

or $\cos \frac{x}{2} = 0$

Now, $\cos x = 0$

$$\Rightarrow x = \frac{\pi}{2}, \frac{3\pi}{2} \quad [\because 0 \le x < 2\pi]$$

$$\cos \frac{5x}{2} = 0$$

$$\Rightarrow \frac{5x}{2} = \frac{\pi}{2}, \frac{3\pi}{2}, \frac{5\pi}{2}, \frac{7\pi}{2} \frac{9\pi}{2}, \frac{11\pi}{2}, \ldots$$

$$\Rightarrow x = \frac{\pi}{5}, \frac{3\pi}{5}, \pi, \frac{7\pi}{5}, \frac{9\pi}{5} \quad [\because 0 \le x < 2\pi]$$

and $\cos \frac{x}{2} = 0$

$$\Rightarrow \frac{x}{2} = \frac{\pi}{2}, \frac{3\pi}{2}, \frac{5\pi}{2}, \ldots$$

$$\Rightarrow x = \pi \quad [\because 0 \le x < 2\pi]$$

Hence, $x = \frac{\pi}{2}, \frac{3\pi}{2}, \pi, \frac{\pi}{5}, \frac{3\pi}{5}, \frac{7\pi}{5}, \frac{9\pi}{5}$

2. *(d)* According to given information, we have the following figure

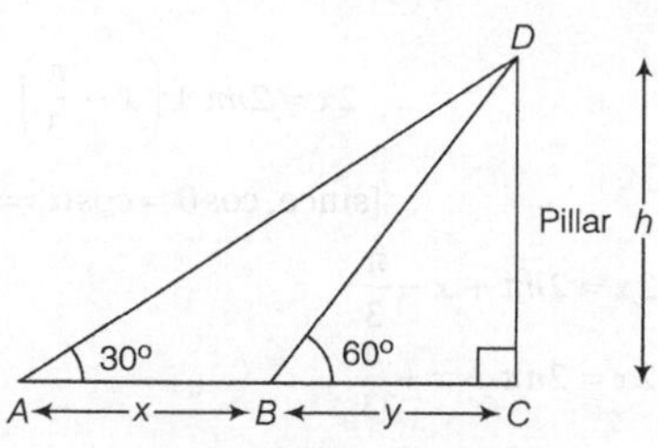

Now, from ΔACD and ΔBCD, we have

$$\tan 30° = \frac{h}{x+y}$$

and
$$\tan 60° = \frac{h}{y}$$

$$\Rightarrow h = \frac{x+y}{\sqrt{3}} \quad \ldots(i)$$

and
$$h = \sqrt{3}\, y \quad \ldots(ii)$$

From Eqs. (i) and (ii),

$$\frac{x+y}{\sqrt{3}} = \sqrt{3}\, y$$

$$\Rightarrow x + y = 3y$$

$$\Rightarrow x - 2y = 0 \Rightarrow y = \frac{x}{2}$$

$\because$ Speed is uniform.

$\therefore$ Distance y will be cover in 5 min.

$\because$ Distance x covered in 10 min.

$\therefore$ Distance $\frac{x}{2}$ will be cover in 5 min.

3. *(d)* Given, $5(\tan^2 x - \cos^2 x) = 2\cos 2x + 9$

$$\Rightarrow 5\left(\frac{1-\cos 2x}{1+\cos 2x} - \frac{1+\cos 2x}{2}\right) = 2\cos 2x + 9$$

Put $\cos 2x = y$, we have

$$5\left(\frac{1-y}{1+y} - \frac{1+y}{2}\right) = 2y + 9$$

$$\Rightarrow 5(2 - 2y - 1 - y^2 - 2y) = 2(1+y)(2y+9)$$

$$\Rightarrow 5(1 - 4y - y^2) = 2(2y + 9 + 2y^2 + 9y)$$

$$\Rightarrow 5 - 20y - 5y^2 = 22y + 18 + 4y^2$$

$$\Rightarrow 9y^2 + 42y + 13 = 0$$

$$\Rightarrow 9y^2 + 3y + 39y + 13 = 0$$

$$\Rightarrow 3y(3y+1) + 13(3y+1) = 0$$

$$\Rightarrow (3y+1)(3y+13) = 0$$

$$\Rightarrow y = -\frac{1}{3}, -\frac{13}{3}$$

$$\therefore \cos 2x = -\frac{1}{3}, -\frac{13}{3}$$

$$\Rightarrow \cos 2x = -\frac{1}{3} \quad \left[\because \cos 2x \ne -\frac{13}{3}\right]$$

Now, $\cos 4x = 2\cos^2 2x - 1$

$$= 2\left(-\frac{1}{3}\right)^2 - 1$$

$$= \frac{2}{9} - 1 = -\frac{7}{9}$$

4. *(c)* Let $AB = h$, then $AD = 2h$ and $AC = BC = \frac{h}{2}$

Again, let $\angle CPA = \alpha$

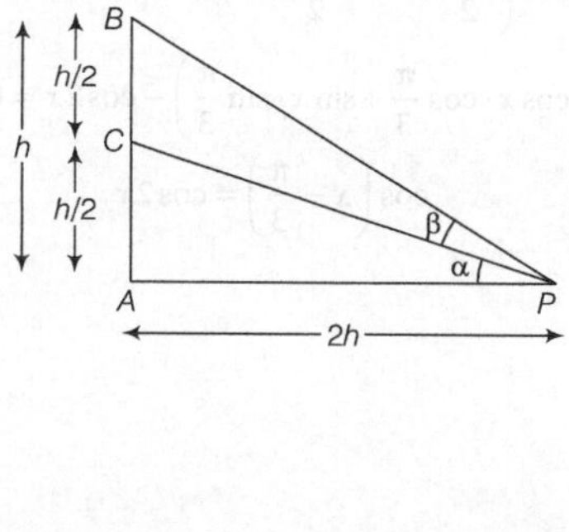

Now, in ΔABP, $\tan(\alpha+\beta)=\dfrac{AB}{AP}=\dfrac{h}{2h}=\dfrac{1}{2}$

Also, in ΔACP, $\tan\alpha=\dfrac{AC}{AP}=\dfrac{\frac{h}{2}}{2h}=\dfrac{1}{4}$

Now, $\quad \tan\beta=\tan[(\alpha+\beta)-\alpha]$

$$=\frac{\tan(\alpha+\beta)-\tan\alpha}{1+\tan(\alpha+\beta)\tan\alpha}$$

$$=\frac{\frac{1}{2}-\frac{1}{4}}{1+\frac{1}{2}\times\frac{1}{4}}=\frac{\frac{1}{4}}{\frac{9}{8}}=\frac{2}{9}$$

5. *(b)* **Key idea** *Apply the identity*

$$\cos(x+y)\cos(x-y)=\cos^2 x-\sin^2 y$$

and $\cos 3x=4\cos^3 x-3\cos x$

We have, $8\cos x\left(\cos\left(\frac{\pi}{6}+x\right)\cos\left(\frac{\pi}{6}-x\right)-\frac{1}{2}\right)=1$

$$\Rightarrow 8\cos x\left(\cos^2\frac{\pi}{6}-\sin^2 x-\frac{1}{2}\right)=1$$

$$\Rightarrow \quad 8\cos x\left(\frac{3}{4}-\sin^2 x-\frac{1}{2}\right)=1$$

$$\Rightarrow \quad 8\cos x\left(\frac{3}{4}-\frac{1}{2}-1+\cos^2 x\right)=1$$

$$\Rightarrow \quad 8\cos x\left(\frac{-3+4\cos^2 x}{4}\right)=1$$

$$\Rightarrow \quad 2(4\cos^3 x-3\cos x)=1$$

$$\Rightarrow \quad 2\cos 3x=1$$

$$\Rightarrow \quad \cos 3x=\frac{1}{2}$$

$$\Rightarrow \quad 3x=\frac{\pi}{3},\frac{5\pi}{3},\frac{7\pi}{3} \qquad [0\le 3x\le 3\pi]$$

$$\Rightarrow \quad x=\frac{\pi}{9},\frac{5\pi}{9},\frac{7\pi}{9}$$

Sum $\quad =\dfrac{\pi}{9}+\dfrac{5\pi}{9}+\dfrac{7\pi}{9}=\dfrac{13\pi}{9}$

$$\Rightarrow \quad k\pi=\frac{13\pi}{9}$$

Hence, $\quad k=\dfrac{13}{9}$

6. *(a)*

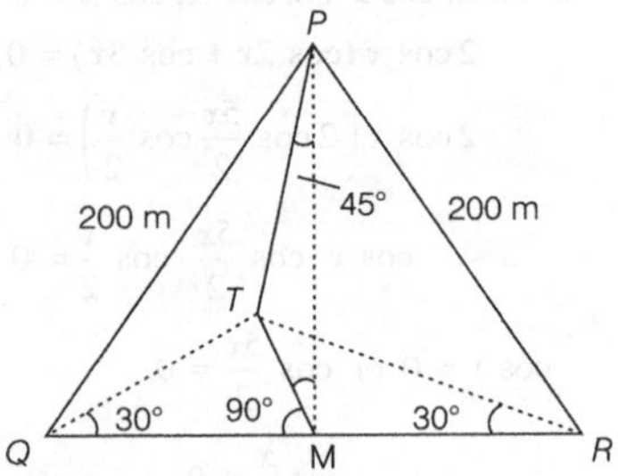

Let height of tower TM be h.

In $\quad \Delta PMT$, $\tan 45°=\dfrac{TM}{PM}$

$$\Rightarrow \quad 1=\frac{h}{PM}$$

$$\Rightarrow \quad PM=h$$

In ΔTQM, $\quad \tan 30°=\dfrac{h}{QM}$;

$$QM=\sqrt{3}h$$

In ΔPMQ, $PM^2+QM^2=PQ^2$

$$h^2+(\sqrt{3}h)^2=(200)^2$$

$$\Rightarrow \quad 4h^2=(200)^2$$

$$\Rightarrow \quad h=100\text{ m}$$

JEE Advanced

1. *(c)* Given, $\sqrt{3}\sec x+\operatorname{cosec} x+2(\tan x-\cot x)=0$,

$(-\pi<x<\pi)-\{0,\pm\pi/2\}$

$$\Rightarrow \sqrt{3}\sin x+\cos x+2(\sin^2 x-\cos^2 x)=0$$

$$\Rightarrow \quad \sqrt{3}\sin x+\cos x-2\cos 2x=0$$

Multiplying and dividing by $\sqrt{a^2+b^2}$, i.e. $\sqrt{3+1}=2$, we get

$$2\left(\frac{\sqrt{3}}{2}\sin x+\frac{1}{2}\cos x\right)-2\cos 2x=0$$

$$\Rightarrow \left(\cos x\cdot\cos\frac{\pi}{3}+\sin x\cdot\sin\frac{\pi}{3}\right)-\cos 2x=0$$

$$\Rightarrow \quad \cos\left(x-\frac{\pi}{3}\right)=\cos 2x$$

$$\therefore \quad 2x=2n\pi\pm\left(x-\frac{\pi}{3}\right)$$

[since, $\cos\theta=\cos\alpha\Rightarrow\theta=2n\pi\pm\alpha$]

$$\Rightarrow \quad 2x=2n\pi+x-\frac{\pi}{3}$$

or $\quad 2x=2n\pi-x+\dfrac{\pi}{3}$

$$\Rightarrow \quad x=2n\pi-\frac{\pi}{3}\text{ or }3x=2n\pi+\frac{\pi}{3}$$

$$\Rightarrow \quad x=2n\pi-\frac{\pi}{3}\text{ or }x=\frac{2n\pi}{3}+\frac{\pi}{9}$$

$$\therefore \quad x=\frac{-\pi}{3}\text{ or }x=\frac{\pi}{9},\frac{-5\pi}{9},\frac{7\pi}{9}$$

Now, sum of all distinct solutions

$$= \frac{-\pi}{3} + \frac{\pi}{9} - \frac{5\pi}{9} + \frac{7\pi}{9} = 0$$

2. *(c)* Here, $\displaystyle\sum_{k=1}^{13} \frac{1}{\sin\left\{\frac{\pi}{4} + \frac{(k-1)\pi}{6}\right\} \sin\left(\frac{\pi}{4} + \frac{k\pi}{6}\right)}$

Converting into differences, b y multiplying and dividing by

$$\sin\left[\left(\frac{\pi}{4} + \frac{k\pi}{6}\right) - \left\{\frac{\pi}{4} + \frac{(k-1)\pi}{6}\right\}\right], \text{ i.e. } \sin\left(\frac{\pi}{6}\right).$$

$$\therefore \sum_{k=1}^{13} \frac{\sin\left[\left(\frac{\pi}{4} + k\frac{\pi}{6}\right) - \left\{\frac{\pi}{4} + (k-1)\frac{\pi}{6}\right\}\right]}{\sin\frac{\pi}{6}\left\{\sin\left\{\frac{\pi}{4} + (k-1)\frac{\pi}{6}\right\} \sin\left(\frac{\pi}{4} + k\frac{\pi}{6}\right)\right\}}$$

$$= 2\sum_{k=1}^{13} \frac{\left[\sin\left(\frac{\pi}{4} + \frac{k\pi}{6}\right)\cos\left\{\frac{\pi}{4} + (k-1)\frac{\pi}{6}\right\} - \sin\left\{\frac{\pi}{4} + (k-1)\frac{\pi}{6}\right\}\cos\left(\frac{\pi}{4} + \frac{k\pi}{6}\right)\right]}{\sin\left\{\frac{\pi}{4} + (k-1)\frac{\pi}{6}\right\}\sin\left(\frac{\pi}{4} + k\frac{\pi}{6}\right)}$$

$$= 2\sum_{k=1}^{13}\left[\cot\left\{\frac{\pi}{4} + (k-1)\frac{\pi}{6}\right\} - \cot\left(\frac{\pi}{4} + k\frac{\pi}{6}\right)\right]$$

$$= 2\left[\left\{\cot\left(\frac{\pi}{4}\right) - \cot\left(\frac{\pi}{4} + \frac{\pi}{6}\right)\right\} + \left\{\cot\left(\frac{\pi}{4} + \frac{\pi}{6}\right) - \cot\left(\frac{\pi}{4} + \frac{2\pi}{6}\right)\right\} + \ldots + \left\{\cot\left(\frac{\pi}{4} + 12\frac{\pi}{6}\right) - \cot\left(\frac{\pi}{4} + 13\frac{\pi}{6}\right)\right\}\right]$$

$$= 2\left\{\cot\frac{\pi}{4} - \cot\left(\frac{\pi}{4} + 13\frac{\pi}{6}\right)\right\}$$

$$= 2\left[1 - \cot\left(\frac{29\pi}{12}\right)\right] = 2\left[1 - \cot\left(2\pi + \frac{5\pi}{12}\right)\right]$$

$$= 2\left[1 - \cot\frac{5\pi}{12}\right]\left[\because \cot\frac{5\pi}{12} = (2-\sqrt{3})\right]$$

$$= 2(1 - 2 + \sqrt{3}) = 2(\sqrt{3} - 1)$$

3. *(b,c)* We have,

$$2(\cos\beta - \cos\alpha) + \cos\alpha\cos\beta = 1$$

or $\quad 4(\cos\beta - \cos\alpha) + 2\cos\alpha\cos\beta = 2$

$\Rightarrow 1 - \cos\alpha + \cos\beta - \cos\alpha\cos\beta$

$$= 3 + 3\cos\alpha - 3\cos\beta - 3\cos\alpha\cos\beta$$

$$\Rightarrow (1-\cos\alpha)(1+\cos\beta) = 3(1+\cos\alpha)(1-\cos\beta)$$

$$\Rightarrow \frac{(1-\cos\alpha)}{(1+\cos\alpha)} = \frac{3(1-\cos\beta)}{1+\cos\beta}$$

$$\Rightarrow \tan^2\frac{\alpha}{2} = 3\tan^2\frac{\beta}{2}$$

$$\therefore \tan\frac{\alpha}{2} \pm \sqrt{3}\tan\frac{\beta}{2} = 0$$

4. *(b, c, d)* We have,

In ΔPQR

$$\angle PQR = 30°$$

$$PQ = 10\sqrt{3}$$

$$QR = 10$$

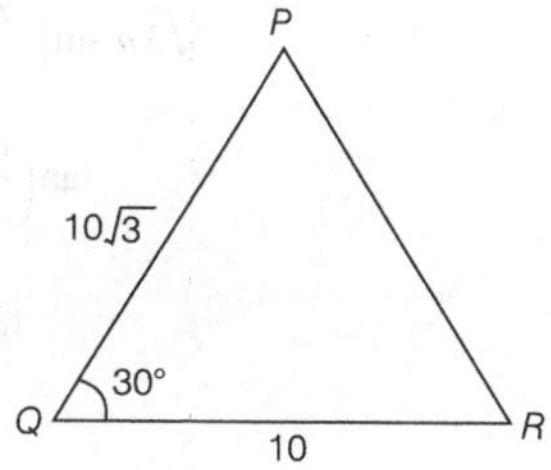

By cosine rule

$$\cos 30° = \frac{PQ^2 + QR^2 - PR^2}{2PQ \cdot QR}$$

$$\Rightarrow \frac{\sqrt{3}}{2} = \frac{300 + 100 - PR^2}{200\sqrt{3}}$$

$$\Rightarrow 300 = 300 + 100 - PR^2$$

$$\Rightarrow PR = 10$$

Since, $\quad PR = QR = 10$

$\therefore \quad \angle QPR = 30°$

and $\quad \angle QRP = 120°$

$$\text{Area of } \Delta PQR = \frac{1}{2}PQ \cdot QR \cdot \sin 30°$$

$$= \frac{1}{2} \times 10\sqrt{3} \times 10 \times \frac{1}{2} = 25\sqrt{3}$$

Radius of incircle of

$$\Delta PQR = \frac{\text{Area of } \Delta PQR}{\text{Semi-perimetre of } \Delta PQR}$$

i.e. $\quad r = \dfrac{\Delta}{s} = \dfrac{25\sqrt{3}}{\dfrac{10\sqrt{3} + 10 + 10}{2}}$

$$= \frac{25\sqrt{3}}{5(\sqrt{3}+2)}$$

$$\Rightarrow r = 5\sqrt{3}(2 - \sqrt{3})$$

$$= 10\sqrt{3} - 15$$

and radius of circumcircle

$$(R) = \frac{abc}{4\Delta} = \frac{10\sqrt{3} \times 10 \times 10}{4 \times 25\sqrt{3}} = 10$$

$\therefore$ Area of circumcircle of

$$\Delta PQR = \pi R^2 = 100\pi$$

Hence, option (b), (c) and (d) are correct answer.

5. *(0.5)* We have, α, β are the roots of

$$\sqrt{3}a\cos x + 2b\sin x = c$$

$\therefore$ $$\sqrt{3}a\cos\alpha + 2b\sin\alpha = c \quad \ldots(i)$$

and $$\sqrt{3}a\cos\beta + 2b\sin\beta = c \quad \ldots(ii)$$

On subtracting Eq. (ii) from Eq. (i), we get

$$\sqrt{3}a(\cos\alpha - \cos\beta) + 2b(\sin\alpha - \sin\beta) = 0$$

$$\Rightarrow \sqrt{3}a\left(-2\sin\left(\frac{\alpha+\beta}{2}\right)\right)\sin\left(\frac{\alpha-\beta}{2}\right) + 2b\left(2\cos\left(\frac{\alpha+\beta}{2}\right)\right)\sin\left(\frac{\alpha-\beta}{2}\right) = 0$$

$$\Rightarrow \sqrt{3}a\sin\left(\frac{\alpha+\beta}{2}\right) = 2b\cos\left(\frac{\alpha+\beta}{2}\right)$$

$$\Rightarrow \tan\left(\frac{\alpha+\beta}{2}\right) = \frac{2b}{\sqrt{3}a}$$

$$\Rightarrow \tan\left(\frac{\pi}{6}\right) = \frac{2b}{\sqrt{3}a} \qquad \left[\because \alpha+\beta = \frac{\pi}{3}, \text{ given}\right]$$

$$\Rightarrow \frac{1}{\sqrt{3}} = \frac{2b}{\sqrt{3}a}$$

$$\Rightarrow \frac{b}{a} = \frac{1}{2}$$

$$\Rightarrow \frac{b}{a} = 0.5$$